AF608465

Birkhäuser

# Operator Theory: Advances and Applications

**Volume 285**

**Founded in 1979 by Israel Gohberg**



More information about this series at http://www.springer.com/series/4850

Fritz Gesztesy • Andrei Martinez-Finkelshtein
Editors

# From Operator Theory to Orthogonal Polynomials, Combinatorics, and Number Theory

## A Volume in Honor of Lance Littlejohn's 70th Birthday

*Editors*
Fritz Gesztesy
Department of Mathematics
Baylor University
Waco, TX, USA

Andrei Martinez-Finkelshtein
Department of Mathematics
Baylor University
Waco, TX, USA

Department of Mathematics
University of Almería
Almería, Spain

ISSN 0255-0156 ISSN 2296-4878 (electronic)
Operator Theory: Advances and Applications
ISBN 978-3-030-75427-3 ISBN 978-3-030-75425-9 (eBook)
https://doi.org/10.1007/978-3-030-75425-9

Mathematics Subject Classification: 05A17, 11F11, 11F67, 11P81, 11P84, 26D10, 33C45, 33C47, 34A40, 34B05, 34L05, 34L10, 35J25, 41A50, 42C05, 42C15, 47A10, 47A20, 47D06, 47E05, 58J20, 60F05, 82C70



This book is published under the imprint Birkhäuser, www.birkhauser-science.com, by the registered company Springer Nature Switzerland AG.
The registered company address is: Gewerbestrasse 11, 6330 Cham, Switzerland

Lance L. Littlejohn

# Preface

Lance Littlejohn was born on November 14, 1951, to Disney and June Littlejohn in St. Thomas, Ontario. He grew up with three siblings, Ken (who passed away in 2009), Charles, and Karen (who passed away in 2018).

Growing up, Lance was an avid baseball player. His early dream, like many kids, was to play professional baseball. He was scouted by Reno Bertoia, a former Detroit Tigers player, who invited Lance to a professional tryout. Much to Lance's chagrin, Reno's recommendation to him after the tryout was to attend college and forget baseball! Still, Lance has very fond memories of his baseball-playing days and has maintained life-long friendships with several former teammates. He played on several championship teams in St. Thomas and London, including the London Majors baseball team that won the Intercounty Baseball League in 1975 and, subsequently, was inducted into the London Sports Hall of Fame in 2015. Seriously, how many mathematicians can boast of that?

After his mother passed away, Lance was in doubt about whether to attend college and he credits his older brother, Charles, for convincing him to start at The University of Western Ontario (now Western University) in 1970. Lance began as a physical education major but quickly realized that he cherished intellectual challenges and hence switched his major to studying mathematics at Western and, in his senior year, won the Annie R. Kingston Gold Medal in Mathematics. It was at Western that Lance first met Jon and Peter Borwein, sons of David Borwein who was head of the mathematics department at Western. Both Jon and Peter were life-long friends of Lance before their passings in 2016 and 2020, respectively. While an undergraduate student at Western, Lance first met Wolfgang Gawronski, a postdoc working with David Borwein. Thirty years later, Wolfgang and Lance began their on-going collaboration, later joined by George Andrews, of studying combinatorial properties of some newly discovered sequences of integers.

After graduating with his master's degree in mathematics at Western in 1976, Lance began his Ph.D. studies at Penn State under the supervision of the late Allan M. Krall. Allan's father, Harry Laverne Krall, was a well-known expert in orthogonal polynomials in the 1930s–1940s. Laverne had suggested to Allan that Lance work on studying a generalization of the Legendre polynomials, orthogonal on $[-1, 1]$ with respect to the Lebesgue measure on $(-1, 1)$, but with unequal jumps at the endpoints $\pm 1$, which he thought would be eigenfunctions of a minimal sixth-order differential equation. These polynomials [42] are known as the Krall polynomials, named after H. L. Krall; together, Allan and Lance published 13 joint papers before Allan's untimely death in 2008.

While at Penn State, Lance was introduced (by department chair George Andrews) to Wendy Diane Miller, and the two were married in 1983. They have two children: son Alexander who lives in the Phoenix area with his wife, Erin, and their daughters Melanie and Drew, and daughter Mary who lives in Waco. Wendy and Lance are immensely, and profoundly, proud of their children.

Lance graduated with his Ph.D. from Penn State in 1981 and took a position as assistant professor at The University of Texas at San Antonio. Lance loved UTSA and the city of San Antonio but, being a Canadian, had a difficult time dealing with the Texas heat (irony: in 2007, he returned to Texas!). In 1983, Lance and Wendy moved to Utah State University in Logan, Utah. He became an associate professor in 1987 and full professor in 1990 at USU.

In 1983, Lance was invited by F. V. (Derick) Atkinson to spend 2 weeks at the University of Toronto. While at Toronto, Lance met W. N. (Norrie) Everitt, at that time one of the world's leading authorities in the spectral theory of differential equations. This began a life-long mathematical collaboration with Norrie, and together they have published close to three dozen papers. Lance also spent a couple of sabbatical years in the UK working with Norrie (as well as with Des Evans and Malcolm Brown).

In 1986, Lance was a plenary speaker at the Second International Symposium on Orthogonal Polynomials, Special Functions and Applications in Segovia, Spain. This was the start of a long collaboration and friendship with several Spanish colleagues, including Paco Marcellán, Jorge Arvesú, Antonio García, Antonio Durán, Jesús Dehesa, Mirta Castro Smirnova, and Manuel Alfaro, leading to a sabbatical leave in 1999–2000 to Madrid. Lance was also a plenary speaker at the Fifth International Conference in Orthogonal Polynomials and Special Functions in Patras, Greece, in 1999.

Kil Kwon, of the Korea Advanced Institute of Science and Technology, met Lance in 1990 in Erice, Sicily, at the Third International Conference in Orthogonal Polynomials and Special Functions and thus began their joint work. They, together with several of Kil's graduate students, have published two dozen papers on various topics related to the classification of orthogonal polynomials satisfying differential equations. Part of their work was completed when Kil spent a sabbatical year at Utah State University in 1992–1993 and part was completed while Lance was visiting KAIST in the summers of 2002 and 2003.

To date, Lance has supervised or co-supervised a total of eight Ph.D. students at Utah State University and Baylor University; their success fills Lance with considerable pride. Lance maintains an active collaboration with his second student, Richard Wellman; together, they have written 15 papers, two of which, [45] and [47], are among Lance's favorites.

At the time of this writing, Lance has written more than 119 papers, with over 72 co-authors, and has made research contributions to:

- The classification of Lagrangian symmetrizable ordinary differential equations; see [43, 44].
- The study of orthogonal polynomials, classical and Sobolev, satisfying differential equations; see, for example [31–34, 36–40];
- The general theory of orthogonal polynomials and their properties; see [4, 10, 30].
- The spectral analysis of second-order and higher-order differential equations including those having orthogonal and exceptional orthogonal polynomial eigenfunctions; see for example [7, 12–14, 16–21, 29, 41, 48, 50].
- An abstract general left-definite theory of bounded below, self-adjoint operators; see for example [5, 15, 22, 23, 45, 46, 48, 49].
- The study of new combinatorial numbers, the Jacobi–Stirling numbers of the first and second kinds, discovered in the left-definite analysis of the classical Jacobi differential equation; see [1–3, 15, 22, 24, 25].
- The discovery of the orthogonality of the Laguerre polynomials $\{L_n^{\alpha}\}_{n=0}^{\infty}$, when $-\alpha \in \mathbb{N}$, with respect to a positive-definite Sobolev inner product and a subsequent spectral analysis; see [23, 35].
- A generalization of the Glazman–Krein–Naimark theory of self-adjoint extensions of Lagrangian symmetric differential operators with applications to orthogonal polynomials; see [8, 47].
- A new proof, extending an idea of the nineteenth century mathematician H. J. S. Smith, of the classic "Fermat two squares theorem" in number theory; see [11]. This was Lance's sojourn into the beautiful subject of number theory and he views this paper as one of his favorites.
- The study of integral inequalities, including various extensions of the classic Hardy integral inequality; see for example [6, 9, 26–28].

Lance became the chair of the Department of Mathematics at Baylor University in January 2007, a position he held until August 2019. Lance has always believed in the balance between pure and applied mathematics and proceeded to increase the visibility of applied mathematics within the department, pairing it with existing strengths in pure mathematics. Irrespective of the metrics applied, it is clear that the mathematics department made tremendous strides under Lance's leadership. As one of his lasting legacies, he not only put the Baylor Mathematics Department on the map, but also paved the way for continued success to come. The Mathematics Department at Baylor University is a very special place, and much credit naturally goes to Lance in this context.

Lance also served as associate dean for research in the Graduate School at Baylor University, under Larry Lyon, dean of the graduate school, starting in August 2012, a position he enjoyed very much but, due to health issues, was forced to step down from in August of 2016. During Lance's time in that position, he introduced an English speaking assistance program in the Graduate School aimed at helping international faculty and graduate students improve their English skills. He also took an active, but cautious, role in studying metrics of each graduate program on campus and annually discussed the results with department chairs and graduate program directors across campus.

As the title of this volume suggests, we were able to attract contributions from experts in a wide range of fields, reflecting Lance's varied interests in mathematics from operator theory, orthogonal polynomials, and combinatorics to number theory. Indeed, the contributions comprise compositions of integers and their close relations to Chebyshev polynomials; nonnegative extensions of Hamiltonian systems in the spirit of the Birman–Kreĭn–Višik theory; a proof of a version of Simon's Hausdorff dimension conjecture for orthogonal polynomials on the unit circle; a survey and new conjectures on hypergeometric functions of finite fields and modular forms; a proof of strong ballistic motion for $d$-dimensional periodic Jacobi operators; a survey on general left-definite theory in connection with self-adjoint extensions and singular perturbation theory; regular sampling theory in the range space of the analysis operator of a continuous frame; an extension of the notion of a coherent pair of measures of the second kind on the unit circle; a refinement of Hardy's inequality on finite intervals; a spectral theoretic view of exceptional Hermite polynomials; a discussion of occupation times for classical and quantum walks, orthogonal polynomials, and foci of ellipses inscribed in cyclic polygons; a differential analog of Favard's theorem motivated by numerical solutions of differential equations; the Burghelea–Friedlander–Kappeler-gluing formula for zeta determinants; new representations of the Laguerre–Sobolev and Jacobi–Sobolev orthogonal polynomials; the failure of compactness of the harmonic double layer potential on Lebesgue spaces in the presence of a single corner (or conical) singularity; weighted Chebyshev polynomials on compact subsets of the complex plane; and the Eichler integral of $E_2$ and $q$-brackets of $t$-hook functions.

We sincerely hope reading in this eclectic volume will cause much joy: *A Very Happy Birthday, Lance!*

Waco, TX, USA
March 2021

Fritz Gesztesy
Andrei Martinez-Finkelshtein

## References


1. G.E. Andrews, W. Gawronski, E. Egge, L.L. Littlejohn, The Jacobi–Stirling numbers. J. Combin. Theory Ser. A **120**, 288–303 (2013)
2. G.E. Andrews, W. Gawronski, L.L. Littlejohn, The Legendre–Stirling numbers. Discrete Math. **311**, 1255–1272 (2011)
3. G.E. Andrews, L.L. Littlejohn, A combinatorial interpretation of the Legendre–Stirling numbers. Proc. Am. Math. Soc. **137**, 2581–2590 (2009)

4. J. Arvesú, K. Driver, L.L. Littlejohn, Interlacing of zeros of Laguerre polynomials of equal and consecutive degree. Integral Transform. Spec. Funct. (2020). arXiv: arXiv:2009.10206
5. A. Bruder, L.L. Littlejohn, D. Tuncer, R. Wellman, Left-definite theory with applications to orthogonal polynomials. J. Comput. Appl. Math. **233**, 1380–1398 (2010)
6. R.C. Chisholm, W.N. Everitt, L.L. Littlejohn, An integral operator with applications. J. Inequal. Appl. **3**, 245–266 (1999)
7. J. Das, W.N. Everitt, D. Hinton, L.L. Littlejohn, C. Markett, The fourth-order Bessel-type differential equation. Appl. Anal. **83**, 325–362 (2004)
8. K. Elliott, L.L. Littlejohn, R. Wellman, Spectral analysis of the sixth-order Krall differential expression. Adv. Oper. Th. **5**, 1078–1099 (2020)
9. W.D. Evans, B.M. Brown, L.L. Littlejohn, Discrete inequalities, orthogonal polynomials and the spectral theory of difference operators. Proc. R. Soc. Lond. Series A **437**, 355–373 (1992)
10. W.D. Evans, L.L. Littlejohn, F. Marcellán, C. Markett, A. Ronveaux, On recurrence relations for Sobolev orthogonal polynomials. SIAM J. Math. Anal. **26**, 446–467 (1995)
11. W.N. Everitt, F.W. Clarke, L.L. Littlejohn, S.J.R. Vorster, H.J.S. Smith and the Fermat two squares theorem. Am. Math. Monthly **106**, 652–665 (1999)
12. W.N. Everitt, H. Kalf, L.L. Littlejohn, C. Markett, The fourth-order Bessel equation: eigenpackets and a generalized Hankel transform. Integral Transform. Spec. Funct. **17**, 845–862 (2006)
13. W.N. Everitt, K.H. Kwon, L.L. Littlejohn, J.K. Lee, S.C. Williams, Self-adjoint operators generated from non-Lagrangian symmetric differential equations having orthogonal polynomial eigenfunctions. Rocky Mountain J. Math. **31**, 899–937 (2001)
14. W.N. Everitt, K.H. Kwon, L.L. Littlejohn, R. Wellman, Orthogonal polynomial solutions of linear ordinary differential equations. J. Comput. Appl. Math. **133**, 85–109 (2001)
15. W.N. Everitt, K.H. Kwon, L.L. Littlejohn, R. Wellman, G.J. Yoon, Jacobi–Stirling numbers, Jacobi polynomials, and the left-definite analysis of the classical Jacobi differential expression. J. Comput. Appl. Math. **208**, 29–56 (2007)
16. W.N. Everitt, A.M. Krall, L.L. Littlejohn, On some properties of the Legendre type linear differential expression. Quaest. Math. **13**, 83–106 (1990)
17. W.N. Everitt, L.L. Littlejohn, Differential operators and the Legendre type polynomials. Differ. Integr. Equ. **1**, 97–116 (1988)
18. W.N. Everitt, L.L. Littlejohn, Orthogonal polynomials and spectral theory: a survey, in *Proceedings of Erice Conference on Orthogonal Polynomials and their Applications*, ed. by L. Gori, C. Brezinski, A. Ronveaux. Orthogonal Polynomials and Their Applications. IMACS Annals on Computing and Appl. Math. vol. 9 (J. C. Baltzer AG, Basel, Switzerland, 1991), pp. 21–55

19. W.N. Everitt, L.L. Littlejohn, S.M. Loveland, Some properties of the sixth-order Legendre-type differential expression. Rendiconti Mat. **13**, 773–799 (1993)
20. W.N. Everitt, L.L. Littlejohn, S.M. Loveland, Properties of the operator domains of the fourth-order Legendre-type differential expressions. Differ. Integr. Equ. **7**, 795–810 (1994)
21. W.N. Everitt, L.L. Littlejohn, V. Marić, On properties of the Legendre differential expression. Res. Math. **42**, 42–68 (2002)
22. W.N. Everitt, L.L. Littlejohn, R. Wellman, Legendre polynomials, Legendre–Stirling numbers, and the left-definite spectral analysis of the Legendre differential expression. J. Comput. Appl. **148**, 213–238 (2002)
23. W.N. Everitt, L.L. Littlejohn, R. Wellman, The Sobolev orthogonality and spectral analysis of the Laguerre polynomials $\{L_n^{-k}\}$ for positive integers $k$. J. Comput. Appl. Math. **171**, 199–234 (2004)
24. W. Gawronski, L.L. Littlejohn, T. Neuschel, Asymptotics of Stirling and Chebyshev–Stirling numbers of the second kind. Stud. Appl. Math **133**, 1–17 (2014)
25. W. Gawronski, L.L. Littlejohn, T. Neuschel, On the asymptotic normality of the Legendre–Stirling numbers of the second kind. Eur. J. Combin. **49** (2015), 218–231.
26. F. Gesztesy, L.L. Littlejohn, *Factorizations and Hardy–Rellich-type inequalities,* in Non-Linear Partial Differential Equations, Mathematical Physics, and Stochastic Analysis. The Helge Holden Anniversary Volume, ed. by F. Gesztesy, H. Hanche-Olsen, E.R. Jakobsen, Yu. Lyubarskii, N.H. Risebro, K. Seip. EMS Congress Reports, ETH-Zürich, 2018, pp. 207–226
27. F. Gesztesy, L.L. Littlejohn, I. Michael, M. Pang, Radial and logarithmic refinements of Hardy's inequality. St. Petersburg Math. J. **30**, 429–436 (2019)
28. F. Gesztesy, L.L. Littlejohn, I. Michael, R. Wellman, On Birman's sequence of Hardy–Rellich-type inequalities. J. Differ. Equ. **264**, 2761–2801 (2018)
29. F. Gesztesy, L.L. Littlejohn, R. Nichols, On self-adjoint boundary conditions for singular Sturm–Liouville operators bounded from below. J. Differ. Equ. **269**, 6448–6491 (2020)
30. S.S. Han, K.H. Kwon, L.L. Littlejohn, Zeros of orthogonal polynomials in certain discrete Sobolev spaces. J. Comput. Appl. Math. **67**, 309–325 (1996)
31. K.H. Kwon, I.H. Jung, L.L. Littlejohn, D.W. Lee, Sobolev orthogonal polynomials and spectral differential equations. Trans. Am. Math. Soc. **347**, 3629–3643 (1996)
32. K.H. Kwon, I.H. Jung, L.L. Littlejohn, D.W. Lee, Differential equations and Sobolev orthogonality. J. Comput. Appl. Math. **65**, 173–180 (1995)
33. K.H. Kwon, D.W. Lee, L.L. Littlejohn, Sobolev orthogonal polynomials and second order differential equations II. Bull. Korean Math. Soc. **33**, 135–170 (1996)
34. K.H. Kwon, J.K. Lee, L.L. Littlejohn, B.H. Yoo, Characterizations of classical type orthogonal polynomials. Proc. Am. Math. Soc. **120**, 485–493 (1994)

35. K.H. Kwon, L.L. Littlejohn, The orthogonality of the Laguerre polynomials $\{L_n^{-k}(x)\}$ for positive integers $k$. Ann. Num. Math. **2**, 289–303 (1995)
36. K.H. Kwon, L.L. Littlejohn, Classification of classical orthogonal polynomials. J. Korean Math. Soc. **34**, 973–1008 (1997)
37. K.H. Kwon, L.L. Littlejohn, Sobolev orthogonal polynomials and second-order differential equations. Rocky Mountain J. Math. **28**, 547–594 (1998)
38. K.H. Kwon, L.L. Littlejohn, B.H. Yoo, Characterizations of orthogonal polynomials satisfying differential equations. SIAM J. Math. Anal. **25**, 976–990 (1994)
39. K. H. Kwon, L. L. Littlejohn, and B. H. Yoo, New characterizations of classical orthogonal polynomials. Indag. Math. N.S. **7**, 199–213 (1996)
40. K.H. Kwon, L.L. Littlejohn, G.J. Yoon, Orthogonal polynomial solutions to spectral type differential equation: Magnus' conjecture. J. Approx. Th. **112**, 189–215 (2001)
41. C. Liaw, L.L. Littlejohn, R. Milson, J. Stewart, The spectral analysis of three families of exceptional Laguerre polynomials. J. Approx. Th. **202**, 5–41 (2016)
42. L.L. Littlejohn, The Krall polynomials: a new class of orthogonal polynomials. Quaest. Math. **5**, 255–265 (1982)
43. L.L. Littlejohn, Symmetry factors for differential equations. Am. Math. Monthly **90**, 462–464 (1983)
44. L.L. Littlejohn, D. Race, Symmetric and symmetrisable differential expressions. Proc. Lond. Math. Soc. (3) **60**, 344–364 (1990)
45. L.L. Littlejohn, R. Wellman, A general left-definite theory for certain self-adjoint operators with applications to differential equations. J. Differ. Equ. **181**, 280–339 (2002)
46. L.L. Littlejohn, R. Wellman, On the spectra of left-definite operators. Complex Anal. Oper. Th. **7**, 437–455 (2013)
47. L.L. Littlejohn, R. Wellman, Self-adjoint operators in extended Hilbert spaces $H \oplus W$: an application of the general GKN-EM theorem. Oper. Matrices **13**, 667–704 (2019)
48. L.L. Littlejohn, Q. Wicks, Glazman-Krein-Naimark theory, left-definite theory and the square of the Legendre polynomials differential operator. J. Math. Anal. Appl. **444**, 1–24 (2016)
49. L.L. Littlejohn, A. Zettl, Left-definite variations of the classical Fourier expansion theorem. Electron. Trans. Numer. Anal. **27**, 124–139 (2007)
50. L.L. Littlejohn, A. Zettl, The Legendre equation and its self-adjoint operators. Electron. J. Differ. Equ. **2011**(69), 1–33 (2011)

# Contents

# Compositions and Chebyshev Polynomials

**George E. Andrews**

*In honor of my good friend, Lance Littlejohn, on his 70th birthday.*

**Abstract** The Theory of Compositions of integers has mostly been relegated to the very basic aspects of combinatorics. The object of this paper is to reveal their close relation to the Chebyshev polynomials $T_n(x)$ and $U_n(x)$. As a result, interesting combinatorial questions arise for compositions that have not been examined previously.

**Keywords** Composition · Partition · Chebyshev polynomials

**AMS Classification** 11P81, 11P84, 05A17

## 1 Introduction

The theory of compositions of integers has always existed in the shadow of partitions of integers.

Compositions may be viewed as partitions in which order is taken into account. For example, there are five partitions of 4, namely 4, $2+2$, $2+1+1$, $1+1+1+1$, and there are eight compositions of 4, namely, 4, $3+1$, $1+3$, $2+2$, $2+1+1$, $1+2+1$, $1+1+2$, $1+1+1+1$.

The author "George E. Andrews" partially supported by the Simons Foundation.

G. E. Andrews (✉)
The Pennsylvania State University, University Park, PA, USA
e-mail: gea1@psu.edu



F. Gesztesy, A. Martinez-Finkelshtein (eds.), *From Operator Theory to Orthogonal Polynomials, Combinatorics, and Number Theory*, Operator Theory: Advances and Applications 285, https://doi.org/10.1007/978-3-030-75425-9_1

The subordinate role of composition has been well described in [10] by A. V. Sills where he first notes that if $c(n)$ denotes the number of compositions of $n$, then

$$c(n) = 2^{n-1}, \qquad \text{(cf. [3, p. 79])} \tag{1}$$

while $p(n)$, the number of partition of $n$ has been shown to be given by

$$p(n) = \frac{1}{\pi\sqrt{2}} \sum_{k=1}^{\infty} \sqrt{k} A_k(n) \frac{d}{dn} \left\{ \frac{\sinh\left(\frac{\pi}{k}\sqrt{\frac{2}{3}\left(n - \frac{1}{24}\right)}\right)}{\sqrt{n - \frac{1}{24}}} \right\}, \tag{2}$$

where $A_k(n)$ is a Kloosterman-type sum,

$$A_k(n) = \sum_{\substack{0 \le h < k \\ \gcd(h,k)=1}} \exp\left(\pi i s(h,k) - \frac{2\pi i n h}{k}\right),$$

where $s(h, k)$ is a Dedekind-sum,

$$s(h,k) = \sum_{j=1}^{k} \left(\left(\frac{j}{k}\right)\right)\left(\left(\frac{hj}{k}\right)\right),$$

with

$$((x)) := \begin{cases} x - \lfloor x \rfloor - \frac{1}{2}, & \text{if } x \neq \mathbb{Z} \\ 0, & \text{if } x \in \mathbb{Z} \end{cases}.$$

The contrast between (1) and (2) is also remarked upon by Sills [10]:

> The theory of partitions has a longer and more varied literature than that of compositions. Perhaps this is due to mathematicians feeling that the theory of partitions is deeper, and hence more interesting than that of compositions. After all, the generating function for $p(n)$ is a certain infinite product which (up to a trivial multiple) just so happens to be a modular form. On the other hand, the generating function for $c(n)$ can easily be seen, using (1) and then summing the geometric series to be
>
> $$\sum_{n=1}^{\infty} c(n)x^n = \sum_{n=1}^{\infty} 2^{n-1}x^n = \frac{x}{1-2x},$$
>
> a mere rational function.

Of course, some rational functions have a more storied history than others. To illustrate where we are headed, let us consider the following polynomial $P_n(x)$ defined as follows: for $n > 0$,

$$P_n(x) = \frac{1}{2} \sum_{c \in C(n)} (-2)^{\#e(c)} (2x)^{\#o(c)}, \tag{3}$$

where $C(n)$ is the set of compositions of $n$, $\#e(c)$ is the number of even parts in the composition $c$, and $\#o(c)$ is the number of odd parts.

$$C(1) = \{1\} \qquad P_1(x) = x$$

$$C(2) = \{1+1, 2\} \qquad P_2(x) = \frac{1}{2}\Big((-2)^0(2x)^2 + (-2)^1(2x)^0\Big) = 2x^2 - 1$$

$$C(3) = \{1+1+1, 1+2, 2+1, 3\}$$

$$P_3(x) = \frac{1}{2}\Big((-2)^0(2x)^3 + 2x(-2)^1(2x)^1 + (-2)^0(2x)^1\Big) = 4x^3 - 3x$$

$$C(4) = \{1+1+1+1, 1+1+2, 1+2+1, 2+1+1, 2+2, 1+3, 3+1, 4\}$$

$$P_4(x) = \frac{1}{2}\big((-2)^0(2x)^4 + 3x(-2)^1(2x)^2 + (-2)^2(2x)^0 + 2x(-2)^0(2x)^2 + (-2)^1(2x)^0\big) = 8x^4 - 8x^2 + 1$$

Clearly, at least for $n \leq 4$, $P_n(x)$ coincide with the $n$th Chebyshev polynomial of the first kind [8, p. 372], and the point of this paper is to reveal the full relationship between the Chebyshev polynomials of both the first and second kinds and compositions.

Indeed we consider two sequences of polynomials, $t_n(x, y)$ and $u_n(x, y)$ defined as follows

$$t_0(x, y) = u_0(x, y) = 1 \tag{4}$$

$$t_1(x, y) = u_1(x, y) = x, \tag{5}$$

and for $n > 1$,

$$u_n(x, y) = x u_{n-1}(x, y) + (1 + y) u_{n-2}(x, y), \tag{6}$$

$$t_n(x, y) = u_n(x, y) - u_{n-2}(x, y). \tag{7}$$

Given this recursive definition of $u_n$ and $t_n$, it is immediate (cf. [4, p. 117]) that

$$T_n(x) = \frac{1}{2} t_n(2x, -2) \tag{8}$$

$$U_n(x) = u_n(2x, -2), \tag{9}$$

where $T_n(x)$ and $U_n(x)$ are the Chebyshev polynomials of the first and second kind respectively [4, p. 117].

There have been a number of papers devoted to polynomial sequences equivalent via a change of variables to $u_n$ and $t_n$ (cf. [3, Ch. 3]). Indeed Hoggatt and Lind [6, 7] in two papers consider much more general sequences than these; however, their object is to link composition generating functions with Fibonacci and Lucas numbers. As a result, some of the more surprising Chebyshev-like identities (see below) do not turn up.

Here are the main results that we shall prove. First we require a further definition.

We define "0+composition" to extend the definitions of compositions to require the left-most entry to be non-negative and even. Hence the 0+compositions of 3 are $0+3, 0+2+1, 0+1+2, 0+1+1+1, 2+1$.

We now define

$$\tau_n(x, y) = \sum_{c \in C(n)} x^{\#o(c)} y^{\#e(c)} \tag{10}$$

$$\upsilon_n(x, y) = \sum_{c \in C^+(n)} x^{\#o(c)} y^{\#e(c)-1}, \tag{11}$$

where $C^+(n)$ is the set of all 0+compositions of $n$. Note that the exponent on $y$ is counting all the even parts of $c$ except for the first part (which is by definition even).

**Theorem 1** *For* $n \geq 0$,

$$t_n(x, y) = \tau_n(x, y) \tag{12}$$

$$u_n(x, y) = \upsilon_n(x, y). \tag{13}$$

**Note** Theorem 1 allows us to use $u_n(x, y)$ and $\upsilon_n(x, y)$ interchangably, and also $t_n(x, y)$ and $\tau_n(x, y)$.

**Theorem 2** *For* $n \geq 1$,

$$u_n(x, y) = \sum_{j,k \geq 0} \binom{n-j}{j} \binom{j}{k} x^{n-2j} y^k, \tag{14}$$

$$t_n(x, y) = \sum_{j,k \geq 0} \frac{(n-2j+k)}{(n-j)} \binom{n-j}{j} \binom{j}{k} x^{n-2j} y^k. \tag{15}$$

**Theorem 3** *For* $n \geq 2$, $m \geq 2$,

$$u_{n+m}(x, y) = u_n(x, y)u_m(x, y) + (1 + y)u_{n-1}(x, y)u_{m-1}(x, y), \tag{16}$$

$$t_{n+m}(x, y) = t_n(x, y)u_m(x, y) + (1 + y)t_{n-1}(x, y)u_{m-1}(x, y). \tag{17}$$

Each of Theorems 1–3 is provable using generating functions or recurrence arguments. The proofs in Sects. 2–4 will all be combinatorial enumerations of compositions. The object is to reveal the subtlety of compositions.

Subsequently we shall consider other results suggested by the original Chebyshev polynomials $T_n(x)$ and $U_n(x)$ such as the following results which will be directly deduced from Theorems 1–3

**Theorem 4** *For* $n \geq r - 1$, $m \geq n - r + 1$,

$$\begin{aligned} &u_n(x, y)u_m(x, y) - u_{n-r}(x, y)u_{m+r}(x, y) \\ &\quad =(-1)^{n-r+1}(1 + y)^{n-r+1}u_{r-1}(x, y)u_{m-n+r-1}(x, y), \end{aligned} \tag{18}$$

*and*

$$\begin{aligned} &t_n(x, y)t_m(x, y) - t_{n-r}(x, y)t_{m+r}(x, y) \\ &\quad =(-1)^{n-r-1}(1 + y)^{n-r-1}(y^2 - x^2)u_{r-1}(x, y)u_{m-n+r-1}(x, y). \end{aligned} \tag{19}$$

**Corollary 1** *For* $n \geq 1$,

$$u_n(x, y)^2 - u_{n-1}(x, y)u_{n+1}(x, y) = (-1)^n(1 + y)^n, \tag{20}$$

$$t_n(x, y)^2 - t_{n-1}(x, y)t_{n+1}(x, y) = (-1)^n(1 + y)^{n-2}(y^2 - x^2). \tag{21}$$

**Theorem 5** *For* $n \geq 2$,

$$\frac{\partial}{\partial x}u_n(x, y) = nu_{n-1}(x, y) - (y + 1)\frac{\partial}{\partial x}u_{n-2}(x, y), \tag{22}$$

$$\frac{\partial}{\partial x}t_n(x, y) = nu_{n-1}(x, y) - (y + 2)\frac{\partial}{\partial x}u_{n-2}(x, y). \tag{23}$$

**Theorem 6** *For* $n \geq 1$,

$$\frac{\partial}{\partial y}u_n(x, y) = \frac{\partial}{\partial x}u_{n-1}(x, y), \tag{24}$$

$$\frac{\partial}{\partial y}t_n(x, y) = \frac{\partial}{\partial x}t_{n-1}(x, y). \tag{25}$$

**Corollary 2** *The number of even parts in all the compositions of $n$ into $i$ odd parts and $j$ even parts equals the number of odd parts in all the compositions of $n-1$ into $i+1$ odd parts and $j-1$ even parts.*

For example, there are 10 compositions of 7 into 3 odd parts and 2 even parts: namely the ten rearrangements of 22111 making a total of 20 even parts. On the other hand, there are 5 compositions of 6 into 4 odds and 1 even; namely the five rearrangements of 21111, now making a total of 20 odd parts as predicted in Corollary 2.

**Corollary 3** *The number of even parts in the compositions of $n$ equals the number of odd parts in the compositions of $n-1$.*

It should be acknowldeged that all of these results can be directly deduced from the generating functions for these polynomials sequences. Indeed, a theory of generalized Lucas polynomials has been studied extensively in the literature (cf. [6]), and compositions have been studied in connection with these sequences (cf. [1, 5–8, 10]). Thus it is surprising that this deeper connection of compositions and Chebyshev polynomials has escaped notice. It would appear that tilings have been the combinatorial objects most associated with Chebyshev polynomials (cf. [4]).

We conclude with a look at the relevant generating functions and possible generalizations.

## 2 Proof of Theorem 1

We begin by noting that

$$u_0(x, y) = v_0(x, y) = 1$$

$$u_1(x, y) = v_(x, y) = x.$$

So to complete the proof of (13), we must show that $v_n(x, y)$ satisfies the recurrence (6).

To see this we note that $xv_{n-1}(x, y)$ may be viewed as the generating function (with $x$ marking odd parts and $y$ even parts after the first part) for compositions of $n$ with first parts zero and second part odd. Indeed, just add 1 to the first part of each composition counted by $v_{n-1}(x, y)$ and insert 0 as a new first part. Next if we add 2 to the first part of each composition enumerated by $v_{n-2}(x, y)$ we see we have all compositions enumerated by $v_{n-2}(x, y)$ in which the first part is positive. Lastly if we add 2 to the first part of each composition enumerated by $v_{n-2}(x, y)$ and then insert a zero first part (multiplying by $y$ to account for the fact that there is a new even part after the first part), we see that we have accounted for the compositions enumerated by $v_n(x, y)$ in which the first part is zero and the second part even. Thus taken together, the sum of $xv_{n-1}(x, y) + v_{n-2}(x, y) + yv_{n-2}(x, y)$ accounts for all the compositions enumerated by $\upsilon_n(x, y)$. I.e.,

$$\upsilon_n(x, y) = x\upsilon_{n-1}(x, y) + (1 + y)\upsilon_{n-2}(x, y),$$

and so (5) holds for $\upsilon_n(x, y)$. Thus (13) is proved.

As for (12), we begin by noting that

$$t_0(x, y) = \tau_0(x, y) = 1$$
$$t_1(x, y) = \tau_1(x, y) = x.$$

Thus it remains to verify (7). Now by adding 2 to the first part of the compositions enumerated by $\upsilon_{n-2}(x, y)$ we obtain the generating function for all compositions enumerated by $\upsilon_n(x, y)$ that have a positive first part. Consequently

$$\upsilon_n(x, y) - \upsilon_{n-2}(x, y)$$

is the generating function for compositions of $n$ of the type enumerated by $\upsilon_n(x, y)$ but with first part *equal* to zero. Delete the zero, and the result is precisely the generating function for all composition of $n$. I.e.,

$$\begin{aligned}\tau_n(x, y) &= \upsilon_n(x, y) - \upsilon_{n-2}(x, y)\\ &= u_n(x, y) - u_{n-2}(x, y),\end{aligned}$$

by (13). □

## 3 Proof of Theorem 2

To prove (14), we must show that there are

$$\binom{n-j}{j}\binom{j}{k}$$

compositions of $n$ of the type enumerated by $\upsilon_n(x, y)$ with $n - 2j$ odd parts and $k$ even parts (excluding the first part which is a non-negative even).

We start with MacMahon's observation [9, p 151], [2, p. 55] (proved by an elegant bjection) that the number of compositions of $N$ with $M$ parts is

$$\binom{N-1}{M-1}.$$

As we continue, we set

$$s = n - 2j,$$
$$T = s + k.$$

Thus $s$ is the number of odd parts, and $T + 1$ is the total number of parts (remember $k$ doesn't include the first part which is even).

Next, we claim that the number of compositions of $N$ into $M$ parts in which $s$ specified positions in the composition (e.g. the second place and ninth place) are allowed to be non-negative is given by

$$\binom{N+S-1}{M-1}.$$

To see this take the composition of $N + s$ into parts and subtract 1 from the parts in the specified positions.

Now we examine the compositions from $v_n(x, y)$ that have $s$ odd parts and $k$ even parts (excluding the first part). This leaves a composition of $n - s$ into $T + 1$ even parts with $s + 1$ positions (don't forget the first part) that are possibly non-negative. Dividing each part by 2 we see we have a composition of $(n - s)/2$ into $T+1$ parts with possible non-negative parts at $s+1$ position specified among $T+1$ possible positions. The total number of choices for the $s$ odd positions is

$$\binom{T}{s},$$

and the number of possible compositions with a specified $s + 1$ position is

$$\binom{((\frac{n-s}{2}) + s + 1) - 1}{(T+1) - 1}.$$

Hence the total is

$$\begin{aligned}\binom{T}{s}\binom{\frac{n+s}{2}}{T} &= \frac{\left(\frac{n+s}{2}\right)!}{s!(T-s)!\left(\frac{n+s}{2} - T\right)!}\\ &= \frac{(n-j)!}{(n-2j)!k!(j-k)!}\\ &= \binom{n-j}{j}\binom{j}{k},\end{aligned}$$

as desired. □

## 4 Proof of Theorem 3

We recall that the identity in question is (recalling Theorem 1):

$$\upsilon_{m+n}(x, y) = \upsilon_m(x, y)\upsilon_n(x, y) + (1 + y)\upsilon_{m-1}(x, y)\upsilon_{n-1}(x, y). \tag{26}$$

We now examine the compositions enumerated by $\upsilon_{m+n}(x, y)$, say

$$c_1\ c_2\ \dots\ c_h\ c_{h+1}\ \dots\ c_r,$$

and we have chosen $h$ so that

$$c_1 + c_2 + \cdots + c_h \leq m,$$

and

$$c_1 + c_2 + \cdots + c_{h+1} > m.$$

Let

$$t = m - c_1 - c_2 - \cdots - c_h.$$

**Case 1** $t = 0$
In this case, the corresponding pair is

$$(c_1, c_2, \dots, c_n)(0, c_{n+1}, \dots, c_2).$$

**Case 2** $t > 0$ and $c_{h+1}$ have the same parity.
In this case there is a perfect bijection between these compositions of $m + n$ and a pair made of a composition of $m$ and one of $n$:

$$(c_1, c_2, \dots, c_h, t)(c_{h+1} - t, c_{h+2}, \dots, c_r).$$

**Case 3** $t > 1$ and $c_{h+1}$ are of opposite parity.
Here we provide a bijection between these compositions and a pair made of a composition of $m - 1$ and one of $n - 1$:

$$(c_1, c_2, \dots, c_h, t - 1)(c_{h+1} - t - 1, c_{h+2}, \dots, c_r)$$

Note that $c_{h+1} - t - 1 \geq 0$ and even because $c_{+1} \geq t$ and they are of opposite parity.

**Case 4** $t = 1$ and $c_{h+1}$ is even.
The image pair for the bijection is now

$$(c_1, c_2, \dots, c_h - 1)(c_{h+1} - 2, c_{h+2}, \dots, c_r)$$

where the first is a composition of $m-1$ and the second of $n-1$. Note the fact that the count of evens in the two compositions is one less than in the original because $c_{h+1}$ was an interior even in the original, and now $c_{n+1}-2$ is an initial uncounted even.

Thus Cases 1 and 2 cover

$$\upsilon_m(x, y)\upsilon_n(x, y).$$

Case 3 covers

$$\upsilon_{m-1}(x, y)\upsilon_{n-1}(x, y),$$

and Case 4 covers

$$\upsilon_{m-1}(x, y)y\upsilon_{n-1}(x, y).$$

Adding all the cases together yields $\upsilon_{m+n}(x, y)$ proving (16).

Identity (17) is an immediate corollary of (16), namely

$$\begin{aligned}\tau_{n+m}(x, y) =&\upsilon_{n+m}(x, y) - \upsilon_{n+m-2}(x, y)\\ =&\upsilon_m(x, y)\upsilon_n(x, y) + (1+y)\upsilon_{m-1}(x, y)\upsilon_{n-1}(x, y)\\ &- \upsilon_m(x, y)\upsilon_{n-2}(x, y) - (1+y)\upsilon_{m-1}(x, y)\upsilon_{n-3}(x, y)\\ =&\tau_n(x, y)\upsilon_m(x, y) + (1+y)\tau_{n-1}(x, y)\upsilon_{n-1}(x, y)\end{aligned}$$

by (7) and Theorem 1. □

## 5 Proofs of Theorems 4 and Corollary 1

We start with Theorem 4 which is a direct Corollary of Theorem 3. First we must observe that the recurrence (6) is valid for $n \geq -1$ provided

$$u_{-1}(x, y) = 0. \tag{27}$$

By Theorem 3, for $n \geq r$

$$u_{m+n}(x, y) = u_n(x, y)u_m(x, y) + (1+y)u_{n-1}(x, y)u_{m-1}(x, y), \tag{28}$$

and

$$u_{n+m}(x, y) = u_{n-r}(x, y)u_{m+r}(x, y) + (1+y)u_{n-r-1}(x, y)u_{m-r-1}(x, y). \tag{29}$$

Subtracting (29) from (28), we find

$$u_n(x,y)u_m(x,y) - u_{n-r}(x,y)u_{m+r}(x,y) \tag{30}$$
$$= -(1+y)\,(u_{n-1}(x,y)u_{m-1}(x,y) - u_{n-r-1}(x,y)u_{rn+r-1}(x,y))\,.$$

Iterating (30) $n - r + 1$ times (which is valid because the boundary conditions continue to hold), we obtain

$$u_n(x,y)u_m(x,y) - u_{n-r}(x,y)u_{m+r}(x,y) \tag{31}$$
$$=(-1)^{n-r+1}(1+y)^{n-r+1}\big\{u_{r-1}(x,y)u_{m-n+r-1}(x,y)$$
$$- u_{-1}(x,y)u_{m+2r-1}(x,y)\big\},$$

and noting that $u_{-1}(x,y) = 0$, we obtain (8) as desired.

Now (9) is deduced from (8) as follows. We shall write $t_n$ for $t_n(x,y)$ and $u_n$ for $u_n(x,y)$ to simplify the resulting expressions

$$t_n t_m - t_{n-r}t_{m+r} \tag{32}$$
$$=(u_n - u_{n-2})(u_m - u_{m-2})$$
$$-(u_{n-r} - u_{n-r-2})(u_{m+r} - u_{m+r-2}) \qquad \text{(by (7))}$$
$$=(u_n u_m - u_{n-r}u_{m+r}) + (u_{n-2}u_{m-2} - u_{n-r-2}u_{m+r-2})$$
$$-(u_n u_{m-2} - u_{n-r}u_{m+r-2}) - (u_{n-2}u_m - u_{n-r-2}u_{m+r})$$
$$=(-1)^{n-r+1}(1+y)^{n-r+1}u_{r-1}u_{m-n+r-1}$$
$$+(-1)^{n-2-r+1}(1+y)^{n-r-1}u_{r-1}u_{m-n+r-1}$$
$$-(-1)^{n-r+1}(1+y)^{n-r+1}u_{r-1}u_{m-2-n+r-1}$$
$$-(-1)^{n-r+1}(1+y)^{n-2-r+1}u_{r-1}u_{m-n+2+r-1}$$
$$=(-1)^{n-r-1}(1+y)^{n-r-1}u_{r-1}$$
$$\times\left\{(1+y)^2 u_{m-n+r-1} + u_{m-n+r-1}\right.$$
$$\left.-(1+y)^2 u_{m-n+r-3} - u_{m-n+r+1}\right\}$$
$$=(-1)^{n-r-1}(1+y)^{n-r-1}u_{r-1}$$
$$\times\left\{(1+2y+y^2)u_{m-n+r-1} + u_{m-n+r-1}\right.$$
$$-(1+y)(u_{m-n+r-1} - xu_{m-n+r-2})$$
$$\left.-(xu_{m-n+r} + (1+y)u_{m-n+r-1})\right\} \qquad \text{(by (6))}$$
$$=(-1)^{n-r-1}(1+y)^{n-r-1}$$

$$\times \left\{y^2 u_{m-n+r-1} - x(u_{m-n+r} - (1+y)u_{m-n+r-2})\right\}$$
$$=(-1)^{n-r-1}(1+y)^{n-r-1}u_{r-1}(y^2-x^2)u_{m-n+r-1} \qquad \text{(by (6))}$$

Finally Corollary 1 is the case $m = n, r = 1$ of Theorem 4.

## 6 Proof of Theorem 6 and Corollaries

First we note that Theorem 6 and Corollary 2 are, in fact, the same result. Namely Theorem 6 is the analytic version. Corollary 2 is, effectively, the coefficient comparison version of Theorem 6.

In light of our emphasis on combinatorics we shall prove the assertion in Corollary2:
Namely, consider all the compositions of $n$ into $i$ odd and $j$ even parts. Take each such compositions of $n-1$ but subtracting 1 from each even part in turn (e.g. 2322 produces 1322, 2312 and 2321). This produces many copies of the same composition of $n-1$ into $i+1$ odd parts and $j-1$ evens. Indeed the first step produces as many compositions as there are even parts in the original compositions, and upon examining we see that we also have as many compositions of $n-1$ in $i+1$ odds and $j-1$ evens as there are odd parts in these compositions. Thus when $n=5, i=1, j=2$, the compositions are 221, 212, 122. Subtracting 1 in turn from each even yield 121, 211, 112, 211, 112,121, and now we have compositions of 4 with $i+1=2$ odds and $j-1=1$ even, and each composition is repeated as many times as there are odd parts.

Corollary 3 follows from Corollary 2 by summing overall $i$ and $j$.

## 7 Further Topics

As was remarked in the introduction, much as been written on the generalized Lucas sequence (cf. [3, 6–8]), and identities equivalent to most of our results appear in these works. The main new results are Theorem 1 and the light subsequently shone on the remaining theorems. As is quite obvious our $u_n(x, y)$ sequence is the sequence $u_{n+1}(x, 1+y)$ of [3, p. 45], consequently all of the identities listed there for $u_n$ are easily translated into results for our $u_m(x, y)$. It is worth noting that our Theorems 3–5 appear to have greater generality than those considered previously.

Generating function proofs are easily produced for each of Theorems 2–5. For example, Theorem 3 is easily deduced as follows:

From (6) and (7),

$$\sum_{n\geq 0} u_n(x, y)q^n = \frac{1}{1 - xq - (1+y)q^2}, \tag{33}$$

and

$$\sum_{n\geq 0} t_n(x,y)q^n = \frac{1-q^2}{1-xq-(1+y)q^2}. \tag{34}$$

Consequently

$$\begin{aligned}
&\sum_{m,n\geq 0} u_{m+n}(x,y)q_1^m q_2^n \qquad (35)\\
&=\sum_{N\geq 0} u_N(x,y)\sum_{n=0}^{N} q_1^{N-n}q_2^n\\
&=\sum_{n\geq 0} u_N(x,y)\frac{q_1^{N+1}-q_2^{N+1}}{(q_1-q_2)}\\
&=\frac{q_1}{(q_1-q_2)(1-xq_1-(1+y)q_1^2)}-\frac{q_2}{(q_1-q_2)(1-xq^2-(1+y)q_2^2)}\\
&=\frac{1+q_1q_2(1+y)}{(1-xq_1-(1+y)q_1^2)(1-xq_2-(1+y)q_2^2)}\\
&=\sum_{m,n\geq 0}(u_m(x,y)u_n(x,y)+(1+y)u_{m-1}(x,y)u_{n-1}(x,y))q_1^m q_2^n.
\end{aligned}$$

Comparing coefficients in the extremes of (35) yields Theorem 3. But, as was emphasized in the introduction, our object has been to illustrate the role played by the combinatorics of compositions.

It should be noted here that we discovered $\tau_n(x,y) = (t_n(x,y))$ as follows:

$$\begin{aligned}
&\sum_{n\geq 0}\tau_n(x,y)q^n \qquad (36)\\
&=\sum_{n=0}^{\infty}(xq+yq^2+xq^3+yq^4+\cdots)^n\\
&=\sum_{n\geq 0}\left(\frac{xq+yq^2}{1-q^2}\right)^n\\
&=\frac{1}{1-\frac{xq+yq^2}{1-q^2}}\\
&=\frac{1-q^2}{1-xq-(1+y)q^2},
\end{aligned}$$

which is (34). Examining what happens when the numerator $1 - q^2$ is replaced by 1 led directly to $\upsilon_n(x, y)$ $(= u_n(x, y))$.

Obviously one might keep track of the parts in compositions mod $k$ rather than just mod 2. Thus the relevant functions are

$$\sum_{n\geq 0} \tau_n(k; x_1, x_2, \ldots, x_k)q^n \tag{37}$$

$$= \frac{1 - q^k}{1 - x_1 q - x_2 q^2 - \cdots - x_{k-1}q^{k-1} - (1 + x_k)q^k},$$

and

$$\sum_{n\geq 0} \upsilon_n(k; x_1, x_2, \ldots, x_k)q^n \tag{38}$$

$$= \frac{1}{1 - x_1 q - x_2 q^2 - \cdots - x_{k-1}q^{k-1} - (1 + x_k)q^k},$$

where the $x_i$ marks parts congruent to $i$ mod $k$.

It appears that these generalizations do not possess the elegance of the case $k = 2$. For example, when $k = 3$, the analog of Theorem 3 is

$$u_{m+n} = u_m u_n + x_2 u_{m-1} u_{n-1} + (1 + x_3)(u_{m-1}u_{n-2} + u_{m-2}u_{n-1}).$$

## References


1. K. Alladi, V.E. Hoggatt, Compositions with ones and twos. Fibonacci Q. **13**, 233–238 (1975)
2. G.E. Andrews. *The Theory of Partitions* (Addison-Wesley, Reading, 1976); reissued Cambridge University Press, Cambridge, 1998
3. A. Benjamin, J. Quinn, *Proofs That Really Count* (Math. Assoc. America, Washington, 2003)
4. A. Benjamin, D. Walton, Counting on Chebyshev polynomials. Math. Mag. **82**, 117–126 (2009)
5. R.P. Grimaldi, Compositions with last summand odd. Ars Combin. **113A**, 299–319 (2014)
6. V.E. Hoggart, D.A. Lind, Fibonacci and binomial properties of weighted compositions. J. Comb. Th. **4**, 121–124 (1968)
7. V.E. Hoggart, D.A. Lind, Compositions and Fibonacci numbers. Fibonacci Q. **7**, 253–266 (1969)
8. J. Koshy, *Fibonacci and Lucas Numbers with Applications*, vol. 2 (Wiley, Hoboken, 2019)
9. P.A. MacMahon, *Combinatory Analysis*, vol. 1 (Cambridge University Press, Cambridge, 1916); reprinted by AMS-Chelsea, Providence, 1960
10. A.V. Sills, Compositions, partitions, and Fibonacci numbers. Fibonacci Q. **40**, 348–364 (2011)

# Non-negative Extensions of Hamiltonian Systems

**B. M. Brown, W. D. Evans, and I. G. Wood**

*The authors are delighted to dedicate this work to their long-time friend, Lance Littlejohn.*

**Abstract** This paper is focused on determining the non-negative self-adjoint extensions of a Hamiltonian system using the Kreĭn-Višik-Birman theory. As an example of this, the non-negative self-adjoint extensions of a fourth-order ODE are determined.

**Keywords** Hamiltonian system · KVB extension theory

**Subject Classifications** Primary 47A20; Secondary 34B05, 34L05, 47E05

## 1 Introduction

A densely defined, positive symmetric operator $T_0$ acting in a Hilbert space $H$ has 2 distinguished self-adjoint extensions, the Friedrichs extension $T_F$, and the Krein-von Neumann extension $T_K$. In [16], Krein established that $T_F$ and $T_K$ have an extremal role in that the set of non-negative self-adjoint extensions of $T_0$ is precisely the set of self-adjoint operators $S$ which satisfy $T_K \leq S \leq T_F$ in the quadratic form sense; in view of this $T_K$ and $T_F$ are referred to as the *soft* and *hard* extensions of

B. M. Brown (✉)
Cardiff School of Computer Science and Informatics, Queens Building, Cardiff University, Cardiff, Roath, UK
e-mail: BrownBM@cardiff.ac.uk

W. D. Evans
School of Mathematics, Cardiff University, Cardiff, UK

I. G. Wood
School of Mathematics, Statistics and Actuarial Science, University of Kent, Canterbury, UK



F. Gesztesy, A. Martinez-Finkelshtein (eds.), *From Operator Theory to Orthogonal Polynomials, Combinatorics, and Number Theory*, Operator Theory: Advances and Applications 285, https://doi.org/10.1007/978-3-030-75425-9_2

$T_0$, respectively. Recall that the form domain of a non-negative self-adjoint operator $A$ is the domain $Q(A) := D(A^{1/2})$ of its square root and $A \leq B$ in the quadratic form (or *form*) sense means that $Q(B) \subset Q(A)$ and $\|A^{1/2}u\| \leq \|B^{1/2}u\|$ for all $u \in Q(B)$; the quadratic form $t[u, v] := \langle A^{1/2}u, A^{1/2}v \rangle, u, v \in Q(A)$ will be referred to as the form of $A$ and $t[u, u]$ written as $t[u]$. Note that $t[u, v] = \langle Au, v \rangle$ for $u \in D(A), v \in Q(A)$.

The task of characterising all the non-negative self-adjoint extensions of a positive symmetric operator $T_0$ was initiated by Krein in [16] and further developed by Višik in [24] and then Birman in [6]. In this paper, we apply the Krein-Višik-Birman (KVB) theory to the problem of constructing the non-negative self-adjoint extensions in the case when $T_0$ is defined by a Hamiltonian system of equations on an interval $(a, b)$, $-\infty < a < b \leq \infty$, with $a$ a regular end-point and $b$ singular.

## 2 Preliminaries

The main focus of our work is the extension problem for linear Hamiltonian systems and (later in Sect. 5) the application to $2n$th order differential equations. This follows on from the work of [7] and [8] where all the non-negative extensions of certain Sturm–Liouville problems and elliptic partial differential operators were characterised in terms of boundary conditions. We first introduce Hamiltonian systems and, for more details, refer the interested reader to the book [3, Chapter 9], the paper [12] and the references contained therein. For $t \in (a, b)$, let

$$Ly(t) := Jy'(t) - Q(t)y(t),\ J = \begin{pmatrix} 0_n & -I_n \\ I_n & 0_n \end{pmatrix},\ Q(t) = \begin{pmatrix} -C(t) & A^*(t) \\ A(t) & B(t) \end{pmatrix}, \tag{1}$$

where $y = (x_y, u_y)^T$ and both $x_y$ and $u_y$ are $n \times 1$-column vectors, $0_n$, $I_n$ are the zero and identity $n \times n$-matrices, respectively, and $A(t)$, $B(t)$, $C(t)$ are real $n \times n$-matrices on $(a, b)$ satisfying

$$B(t) = B^*(t) \geq 0,\ C(t) = C^*(t). \tag{2}$$

Furthermore, let

$$W(t) = \begin{pmatrix} W_1(t) & 0_n \\ 0_n & 0_n \end{pmatrix},\ W_1(t) = W_1^*(t) \geq 0, \tag{3}$$

and introduce a semi-norm on the space of $(2n \times 1)$ vector-valued functions by

$$\|f\|_W^2 := \int_a^b f^*(s)W(s)f(s)ds. \tag{4}$$

We define the Hilbert space $L^2_W(a,b)$ to be the quotient space of $(2n \times 1)$ vector-valued, Lebesgue measurable functions $f$ on $(a,b)$ which satisfy $\|f\|_W < \infty$, with the equivalence relation $f \sim g$ taken to mean $\|f-g\|_W = 0$.

A linear relation $T$ in $L^2_W(a,b) \times L^2_W(a,b)$ is defined by

$$\{y, f\} \in T \Leftrightarrow y \in AC_{loc}(a,b) \text{ and } Ly = Wf, \tag{5}$$

which is single-valued if

$$f \in L^2_W, Ly = Wf \text{ and } Wy = 0 \Rightarrow y = 0. \tag{6}$$

Suppose hereafter that (6) is satisfied, so that in (5) the map $y \longmapsto f$ is an operator $T$ (the maximal operator) defined by

$$\begin{aligned} D(T) &= \{y \in AC_{loc}(a,b) \cap L^2_W(a,b) | \exists f \in L^2_W(a,b) \text{ such that } Ly = Wf\} \\ Ty &= f. \end{aligned} \tag{7}$$

From (7),

$$\langle y, Ty \rangle = \int_a^b y^*(t) W(t) f(t) dt$$

and

$$\|f\|^2_W = \int_a^b f^*(s) W(s) f(s) ds = \int_a^b x^*_f(s) W_1(s) x_f(s) ds.$$

Also,

$$Ly = \begin{pmatrix} L_1[x_y, u_y] \\ L_2[x_y, u_y] \end{pmatrix},$$

where

$$\begin{aligned} L_1[x_y, u_y] &:= -u'_y + Cx_y - A^* u_y, \\ L_2[x_y, u_y] &:= x'_y - Ax_y - Bu_y. \end{aligned}$$

It is known, see [27], that the restriction $T'_0$ of $T$ to $D(T) \cap C^\infty_0(a,b)$ is densely defined and closable in $L^2_W(a,b)$; its closure $T_0$ is the minimal operator.

**Lemma 1** *$T_0$ is symmetric with respect to $\langle \cdot, \cdot \rangle$ and $T^*_0 = T$.*

***Proof*** This lemma is a consequence of the following:

$$\langle y, T_0 z\rangle = \langle T_0 y, z\rangle \quad \text{for all } y, z \in D(T_0),$$
$$\langle y, T_0 z\rangle = \langle T y, z\rangle \quad \text{for all } y \in D(T),\ z \in D(T_0).$$

To verify these, let $y, z \in D(T_0)$ with $Ly = Wf,\ Lz = Wg$ for $f, g \in L^2_W(a, b)$. Then

$$L_1[x_z, u_z] = W_1 u_g,\ L_2[x_z, u_z] = 0,$$

and

$$\begin{aligned}\langle y, T_0 z\rangle &= \int_a^b x_y^* W_1 u_g ds = \int_a^b x_y^* \left(-u_z' + Cx_z - A^* u_z\right) ds \\ &= \int_a^b \{(x_y')^* u_z + x_y^* \left(Cx_z - A^* u_z\right)\} ds \\ &= \int_a^b \{\left(Ax_y + Bu_y\right)^* u_z + x_y^* \left(Cx_z - A^* u_z\right)\} ds \\ &= \int_a^b \left(u_y^* Bu_z + x_y^* Cx_z\right) ds =: D(y, z).\end{aligned} \tag{8}$$

Similarly

$$\langle z, T_0 y\rangle = \int_a^b \left(u_z^* Bu_y + x_z^* Cx_y\right) ds$$

and thus

$$\langle T_0 y, z\rangle = \langle y, T_0 z\rangle,$$

since $B$ and $C$ are symmetric. Hence $T_0$ is symmetric.

If $y, z \in D(T)$ with $Ly = Wf,\ Lz = Wg$ for $f, g \in L^2_W(a, b)$, then

$$\langle y, Tz\rangle = [-x_y^* u_z]_a^b + D(y, z). \tag{9}$$

Hence, for $y \in D(T),\ z \in D(T_0')$

$$\langle y, T_0 z\rangle = \langle Ty, z\rangle \tag{10}$$

whence $T_0^* = (T_0')^* = T$. □

## 3 The Friedrichs Extension $T_F$ of $T_0$

The Friedrichs extension $T_F$ of $T_0$ plays a major role in the KVB theory. For it to be defined, $T_0$ has to be bounded below, and this is guaranteed if the following hypothesis is satisfied:

***Hypothesis*** For some $\lambda_0 \in \mathbb{R}$, there exists $\mu > \lambda_0$ such that

$$Ly = \lambda W y \tag{11}$$

with $\lambda = \mu$ is disconjugate in a subinterval $(c, b)$ of $[a, b)$. □

Equation (11) is said to be disconjugate on a subinterval $(c, b)$ of $[a, b)$ if it does not contain an interval $(\alpha, \beta)$ which is such that there exists a non-trivial solution $y = (x, u)^T$ satisfying $x(\alpha) = x(\beta) = 0$ with $x(t) \neq 0$ on $(\alpha, \beta)$.

Lemma 2.4 in [28] implies that under these conditions the $2n \times n$ system

$$LY(t) = \lambda W(t)Y(t), \quad Y = \begin{pmatrix} X \\ U \end{pmatrix}$$

where $X$ and $U$ are $n \times n$ matrix-valued functions, has a principal solution in an interval $(c, b)$. A principal solution $\begin{pmatrix} \tilde{X} \\ \tilde{U} \end{pmatrix}$ is a solution of (11) such that $\tilde{X}(t)$ is invertible for $t$ near $b$ and

$$\lim_{t \to b} X^{-1}(t)\tilde{X}(t) = 0$$

for any other solution $Y = \begin{pmatrix} X \\ U \end{pmatrix}$, provided that $X^*(t)\tilde{U}(t) - U^*(t)\tilde{X}(t)$ is non-singular. Each matrix-valued solution $Y$ (including $\tilde{Y}$) in this definition is assumed to be a conjoined basis, which means that $X^*U$ is Hermitian and $Y$ has rank $n$.

In [28, Lemma 3.1], it is proved that if the disconjugacy hypothesis is assumed, then $T_0 - \lambda_0 I$ is positive, hence $T_0$ is positive if we can choose $\lambda_0 = 0$.

Since $T_0$ is symmetric, the assumption that it is non-negative implies that its deficiency indices are equal, this being the number $d$ of $L^2_W(a, b)$ solutions in $\mathbb{C} \setminus [0, \infty)$. The assumption that $a$ is a regular end-point of $(a, b)$ and $b$ singular (which is made in this paper) implies that $n \leq d \leq 2n$. This follows since the minimal operator is bounded below and that $2n$ is the maximum number of linearly independent solutions of (11), see [20]. We remark that a GKN type theory [27] has been developed for (1) which determines all the self-adjoint extensions of the minimal operator in terms of restrictions of functions in the maximal domain by boundary conditions formulated in terms of Lagrange brackets. (See also Gesztesy, Littlejohn and Nichols [10] for an up to date discussion of boundary conditions which achieve this for the Sturm–Liouville problem). We note that additional information on the spectral theory of Hamiltonian operators can be found in [4, 5, 13, 15, 22, 23], while further results on the Friedrichs extension of $2n$th order differential operators may be found in [14, 17, 19, 21, 26].

In the case of Hamiltonian systems regular at the left end-point and singular at the right, the Friedrichs extension is characterised by the following theorem, which we summarise from [28, Theorem 4.2].

**Theorem 2** *Suppose that the disconjugacy hypothesis holds for (11) with $\lambda_0 = 0$. Then*

$$D(T_F) = \left\{ y = (x, u)^T \in D(T) \;\middle|\; x(a) = 0,\ [y : \tilde{y}_j](b) = 0,\ j = 1, \cdots, d - n \right\}. \tag{12}$$

Here, $[\cdot : \cdot]$ is the Lagrangian bracket:

$$[y_1 : y_2] = y_2^* J y_1 = u_2^* x_1 - x_2^* u_1 \text{ for } y_1 = (x_1, u_1)^T,\ y_2 = (x_2, u_2)^T,$$

$d$ is the deficiency index of the minimal operator and the set of functions $\{\tilde{y}_j | j = 1, \ldots, d - n\}$ are those so-called LC-type solutions of (1) which form part of the principal solution. A full account of the construction of these boundary conditions is contained in [28, Sect. 4].

We remark that, by (8), the form associated with $T_F$ is

$$t_F[y, z] = \langle y, T_0 z\rangle = D(y, z), \tag{13}$$

and the form domain $Q(T_F) = Q(T_0)$ is $D(T_F^{1/2})$.

## 4 Characterisation of Non-negative Extensions $T_B$

We now use the KVB theory to determine the non-negative self-adjoint extensions of $T_0$ in the formulation of Alonso and Simon [1]. Since all such extensions are restrictions of the maximal operator $T$, this reduces to a question of characterising their domains.

We recall that the KVB theory establishes a one-one correspondence between the set of non-negative self-adjoint extensions $T_B$ of $T_0$ and the set of pairs $\{N_B, B\}$, where $N_B$ is a subspace of the kernel of $T$, and $B$ is a non-negative self-adjoint operator acting in $N_B$. The correspondence is determined by

$$t_B = t_F + b, \quad Q(T_B) = Q(T_F) \dotplus Q(B) \tag{14}$$

where $t_B, t_F, b$ are the forms of $T_B, T_F, B$ respectively and $Q(T_B), Q(T_F), Q(B)$ are their form domains.

**Theorem 3** *Suppose that the disconjugacy hypothesis holds with $\lambda_0 = 0$, and let a basis of the subspace $N_B$ be given by $\{\psi_1, \psi_2, \cdots, \psi_N\}$. For $z \in Q(T_B)$ we write $z = z_F + \sum_{k=1}^N z_k\psi_k$ with $z_F \in Q(T_F)$. Then*

$$D(T_B) = \left\{ z \in Q(T_B) \middle| [z : \psi_k]_a^b - \sum_{j=1}^N z_j b[\psi_k, \psi_j] = 0 \text{ for all } k = 1, \ldots, N \right\}. \tag{15}$$

***Proof*** Let $y, z \in Q(T_B)$ with $y = y_F + \sum_{k=1}^N y_k\psi_k$ and $z = z_F + \sum_{k=1}^N z_k\psi_k$ with $y_F, z_F \in Q(T_F)$. By (13),

$$\begin{aligned} t_F[y_F, z_F] &= t_F[y - \sum_{k=1}^N y_k\psi_k, z - \sum_{k=1}^N z_k\psi_k] \\ &= D(y, z) - \sum_{k=1}^N z_k D(y, \psi_k) - \sum_{k=1}^N \overline{y_k} D(\psi_k, z_F). \end{aligned} \tag{16}$$

On using the integration by parts formula (8), we have

$$D(y, \psi_k) = \langle y, T\psi_k \rangle + [x_y^* u_{\psi_k}]_a^b = [x_y^* u_{\psi_k}]_a^b$$

and similarly

$$D(\psi_k, z_F) = \overline{D(z_F, \psi_k)} = \overline{[x_{z_F}^* u_{\psi_k}]_a^b} = [u_{\psi_k}^* x_{z_F}]_a^b.$$

Thus

$$t_F[y_F, z_F] = D(y, z) - \sum_{k=1}^N z_k [x_y^* u_{\psi_k}]_a^b - \sum_{k=1}^N \overline{y_k} [u_{\psi_k}^* x_{z_F}]_a^b$$

and so

$$\begin{aligned} t_B[y, z] &= D(y, z) - \sum_{k=1}^N z_k [x_y^* u_{\psi_k}]_a^b \\ &\quad - \sum_{k=1}^N \overline{y_k} [u_{\psi_k}^* x_{z_F}]_a^b + \sum_{j,k=1}^N \overline{y_k} z_j b[\psi_k, \psi_j]. \end{aligned} \tag{17}$$

Now assume $z \in D(T_B)$. Then we also have

$$
\begin{aligned}
t_B[y,z] = \langle y, Tz\rangle &= -[x_y^* u_z]_a^b + D(y,z) \\
&= -[x_y^* u_{z_F}]_a^b - \sum_{k=1}^N z_k [x_y^* u_{\psi_k}]_a^b + D(y,z) \\
&= -[x_{y_F}^* u_{z_F}]_a^b - \sum_{k=1}^N \overline{y_k}[x_{\psi_k}^* u_{z_F}]_a^b - \sum_{k=1}^N z_k [x_y^* u_{\psi_k}]_a^b + D(y,z).
\end{aligned} \tag{18}
$$

A comparison of (17) and (18) gives

$$
-[x_{y_F}^* u_{z_F}]_a^b - \sum_{k=1}^N \overline{y_k}[x_{\psi_k}^* u_{z_F}]_a^b = -\sum_{k=1}^N \overline{y_k}[u_{\psi_k}^* x_{z_F}]_a^b + \sum_{j,k=1}^N \overline{y_k} z_j b[\psi_k, \psi_j]. \tag{19}
$$

Since $z \in D(T_B) \subset D(T)$, we have that

$$
z_F = z - \sum_{k=1}^N z_k \psi_k \in D(T) \cap Q(T_F) = D(T_F).
$$

Therefore,

$$
\begin{aligned}
t_F[y_F, z_F] &= \langle y_F, Tz_F\rangle \\
&= -[x_{y_F}^* u_{z_F}]_a^b + D(y_F, z_F) = -[x_{y_F}^* u_{z_F}]_a^b + t_F[y_F, z_F].
\end{aligned}
$$

This implies that $[x_{y_F}^* u_{z_F}]_a^b = 0$, and so (19) simplifies to

$$
-\sum_{k=1}^N \overline{y_k}[x_{\psi_k}^* u_{z_F}]_a^b = -\sum_{k=1}^N \overline{y_k}[u_{\psi_k}^* x_{z_F}]_a^b + \sum_{j,k=1}^N \overline{y_k} z_j b[\psi_k, \psi_j]. \tag{20}
$$

This must hold for every choice of $\{y_k\}$, and therefore, for $z$ to lie in $D(T_B)$, we require

$$
-[x_{\psi_k}^* u_{z_F}]_a^b = -[u_{\psi_k}^* x_{z_F}]_a^b + \sum_{j=1}^N z_j b[\psi_k, \psi_j] \quad \text{for all } k = 1, \ldots, N. \tag{21}
$$

Noting that

$$[x^*_{\psi_k} u_{z_F}]_a^b - [u^*_{\psi_k} x_{z_F}]_a^b = -[z_F : \psi_k]_a^b,$$

we can rewrite this as

$$[z_F : \psi_k]_a^b - \sum_{j=1}^{N} z_j b[\psi_k, \psi_j] = 0 \text{ for all } k = 1, \ldots, N. \tag{22}$$

Finally, for functions in $\ker T$, the Lagrange bracket $[\psi_k : \psi_j]$ is constant (see, [28, Corollary 2.1]), and so $[\psi_k : \psi_j]_a^b = 0$. This allows us to rewrite (22) as

$$[z : \psi_k]_a^b - \sum_{j=1}^{N} z_j b[\psi_k, \psi_j] = 0 \text{ for all } k = 1, \ldots, N, \tag{23}$$

which completes the proof. □

## 5 Example: A Fourth Order ODE

A motivation for this present work lies in characterising the extensions of the minimal operator generated by the $2n$th order formally symmetric expression

$$ly = \frac{1}{w(t)} \sum_{j=0}^{n} (-1)^j (r_j(t) y^{(j)})^{(j)} \tag{24}$$

where $r_n(t)^{-1}$, $r_j(t)$ for $1 \le j \le n-1$ and $w(t)$ are real-valued, locally integrable functions on $(a, b)$ such that $r_n(t)^{-1} > 0$, $w(t) > 0$, see [18] for further details. It has been well known for many years that such expressions can be rewritten in the form of Hamiltonian systems, e.g., [25]. For convenience, we follow the prescription in [28, Sect. 2] which transforms the scalar equation (24) into a first order system by defining

$$x = \begin{pmatrix} y \\ y' \\ \cdot \\ \cdot \\ \cdot \\ y^{(n-1)} \end{pmatrix} \qquad u = \begin{pmatrix} (-1)^{n-1}(r_n y^{(n)})^{(n-1)} + \ldots + r_1 y' \\ \cdot \\ \cdot \\ \cdot \\ -(r_n y^{(n)})' + r_{n-1} y^{(n-1)} \\ r_n y^{(n)} \end{pmatrix} \tag{25}$$

and

$$B = \operatorname{diag}\{0, \ldots, r_n^{-1}\}, \quad C = \operatorname{diag}\{r_0, \ldots, r_{n-1}\},$$

$$A = (\delta_{(i+1)j})_{n\times n}, \quad W = \operatorname{diag}\{w, \ldots, 0\}.$$

In an abuse of notation in this section, we also write $y = (x, u)^T$.

Next we examine the concrete example given by the operator

$$Ly := y^{(4)} + y \text{ on the interval } [0, \infty)$$

and look for its non-negative extensions which are charecterised by boundary values.

The minimal operator is in the minimal deficiency case (two $L^2(0,\infty)$-solutions, see [9] for further details). Two linearly independent $L^2(0,\infty)$-solutions, i.e. kernel elements, are

$$\phi_1 = e^{-\frac{x}{\sqrt{2}}} \cos\left(\frac{x}{\sqrt{2}}\right), \quad \phi_2 = e^{-\frac{x}{\sqrt{2}}} \sin\left(\frac{x}{\sqrt{2}}\right).$$

Our analysis is formulated in terms of the Hamiltonian system (1), which by (25) generates a vector $y = (z, z', -z''', z'')^T$ which we write symbolically as $(x, u)^T$ for 2-vectors $x = (z, z')^T$ and $u = (-z''', z'')^T$.

We first look at the case when the space $N_B$ is the whole kernel. In the context of this example we normalise $\phi_1$ and then apply the Gram-Schmidt method to $\phi_1, \phi_2$ to obtain the pair of orthonormal functions

$$\psi_1(x) = \frac{2^{5/4} e^{-\frac{x}{\sqrt{2}}} \cos\left(\frac{x}{\sqrt{2}}\right)}{\sqrt{3}} \tag{26}$$

and

$$\psi_2(x) = 2^{3/4}\sqrt{3}\left(e^{-\frac{x}{\sqrt{2}}} \sin\left(\frac{x}{\sqrt{2}}\right) - \frac{1}{3} e^{-\frac{x}{\sqrt{2}}} \cos\left(\frac{x}{\sqrt{2}}\right)\right). \tag{27}$$

These generate two Hamiltonian vectors

$$\boldsymbol{\psi}_1 = \begin{pmatrix} \dfrac{2^{5/4} e^{-\frac{x}{\sqrt{2}}} \cos\left(\frac{x}{\sqrt{2}}\right)}{\sqrt{3}} \\ -\dfrac{2^{3/4} e^{-\frac{x}{\sqrt{2}}} \sin\left(\frac{x}{\sqrt{2}}\right)}{\sqrt{3}} - \dfrac{2^{3/4} e^{-\frac{x}{\sqrt{2}}} \cos\left(\frac{x}{\sqrt{2}}\right)}{\sqrt{3}} \\ -\dfrac{2^{3/4}\left(e^{-\frac{x}{\sqrt{2}}} \cos\left(\frac{x}{\sqrt{2}}\right) - e^{-\frac{x}{\sqrt{2}}} \sin\left(\frac{x}{\sqrt{2}}\right)\right)}{\sqrt{3}} \\ \dfrac{2^{5/4} e^{-\frac{x}{\sqrt{2}}} \sin\left(\frac{x}{\sqrt{2}}\right)}{\sqrt{3}} \end{pmatrix}$$

and

$$\psi_2 = \begin{pmatrix} 2^{3/4}\sqrt{3}\left(e^{-\frac{x}{\sqrt{2}}}\sin\left(\frac{x}{\sqrt{2}}\right) - \frac{1}{3}e^{-\frac{x}{\sqrt{2}}}\cos\left(\frac{x}{\sqrt{2}}\right)\right) \\ 2^{3/4}\sqrt{3}\left(\frac{2}{3}\sqrt{2}e^{-\frac{x}{\sqrt{2}}}\cos\left(\frac{x}{\sqrt{2}}\right) - \frac{1}{3}\sqrt{2}e^{-\frac{x}{\sqrt{2}}}\sin\left(\frac{x}{\sqrt{2}}\right)\right) \\ -2^{3/4}\sqrt{3}\left(\frac{2}{3}\sqrt{2}e^{-\frac{x}{\sqrt{2}}}\sin\left(\frac{x}{\sqrt{2}}\right) + \frac{1}{3}\sqrt{2}e^{-\frac{x}{\sqrt{2}}}\cos\left(\frac{x}{\sqrt{2}}\right)\right) \\ 2^{3/4}\sqrt{3}\left(-\frac{1}{3}e^{-\frac{x}{\sqrt{2}}}\sin\left(\frac{x}{\sqrt{2}}\right) - e^{-\frac{x}{\sqrt{2}}}\cos\left(\frac{x}{\sqrt{2}}\right)\right) \end{pmatrix}$$

So in this Hamiltonian notation the components of the basis kernel functions become

$$x_1 = \begin{pmatrix} \frac{2\sqrt[4]{2}e^{-\frac{x}{\sqrt{2}}}\cos\left(\frac{x}{\sqrt{2}}\right)}{\sqrt{3}} \\ -\frac{2^{3/4}e^{-\frac{x}{\sqrt{2}}}\sin\left(\frac{x}{\sqrt{2}}\right)}{\sqrt{3}} - \frac{2^{3/4}e^{-\frac{x}{\sqrt{2}}}\cos\left(\frac{x}{\sqrt{2}}\right)}{\sqrt{3}} \end{pmatrix}$$

and

$$u_1 = \begin{pmatrix} -\frac{2^{3/4}\left(e^{-\frac{x}{\sqrt{2}}}\cos\left(\frac{x}{\sqrt{2}}\right) - e^{-\frac{x}{\sqrt{2}}}\sin\left(\frac{x}{\sqrt{2}}\right)\right)}{\sqrt{3}} \\ \frac{2\sqrt[4]{2}e^{-\frac{x}{\sqrt{2}}}\sin\left(\frac{x}{\sqrt{2}}\right)}{\sqrt{3}} \end{pmatrix}.$$

Similarly,

$$x_2 = \begin{pmatrix} 2^{3/4}\sqrt{3}\left(e^{-\frac{x}{\sqrt{2}}}\sin\left(\frac{x}{\sqrt{2}}\right) - \frac{1}{3}e^{-\frac{x}{\sqrt{2}}}\cos\left(\frac{x}{\sqrt{2}}\right)\right) \\ 2^{3/4}\sqrt{3}\left(\frac{2}{3}\sqrt{2}e^{-\frac{x}{\sqrt{2}}}\cos\left(\frac{x}{\sqrt{2}}\right) - \frac{1}{3}\sqrt{2}e^{-\frac{x}{\sqrt{2}}}\sin\left(\frac{x}{\sqrt{2}}\right)\right) \end{pmatrix}$$

and

$$u_2 = \begin{pmatrix} -2^{3/4}\sqrt{3}\left(\frac{2}{3}\sqrt{2}e^{-\frac{x}{\sqrt{2}}}\sin\left(\frac{x}{\sqrt{2}}\right) + \frac{1}{3}\sqrt{2}e^{-\frac{x}{\sqrt{2}}}\cos\left(\frac{x}{\sqrt{2}}\right)\right) \\ 2^{3/4}\sqrt{3}\left(-\frac{1}{3}e^{-\frac{x}{\sqrt{2}}}\sin\left(\frac{x}{\sqrt{2}}\right) - e^{-\frac{x}{\sqrt{2}}}\cos\left(\frac{x}{\sqrt{2}}\right)\right) \end{pmatrix}.$$

In this notation, with the condition on $z$ to lie in the domain of a non-negative extension (23), the entries in the Lagrangian bracket are zero at $x = \infty$ (as $z$ lies in the fourth order Sobolev space $H^4(0, \infty)$) and at zero are

$$\begin{pmatrix} [z : \psi_1](0) \\ [z : \psi_2](0) \end{pmatrix} = \begin{pmatrix} \frac{\sqrt[4]{2}\left(\sqrt{2}z(0) - 2z'''(0) - \sqrt{2}z''(0)\right)}{\sqrt{3}} \\ \frac{\sqrt[4]{2}\left(2z(0) + 3\sqrt{2}z'(0) + \sqrt{2}z'''(0) + 4z''(0)\right)}{\sqrt{3}} \end{pmatrix}.$$

Thus from (23) we get

$$\frac{\sqrt[4]{2}\left(\sqrt{2}z(0) - 2z'''(0) - \sqrt{2}z''(0)\right)}{\sqrt{3}} + \sum_{j=1}^{2} z_j b[\psi_1, \psi_j] = 0 \tag{28}$$

and

$$\frac{\sqrt[4]{2}\left(2z(0) + 3\sqrt{2}z'(0) + \sqrt{2}z'''(0) + 4z''(0)\right)}{\sqrt{3}} + \sum_{j=1}^{2} z_j b[\psi_2, \psi_j] = 0. \tag{29}$$

We next determine the coefficients $z_1$ and $z_2$ in terms of the boundary values of $z$. For this example the Friedrichs domain given by (12) merely requires $x(0) = 0$, i.e. the function $z$ and its derivative vanish at zero, as there are no LC-type principal solutions. This means that setting $z = z_F + z_1\psi_1 + z_2\psi_2$ gives

$$z(0) = z_1\psi_1(0) + z_2\psi_2(0) \quad \text{and} \quad z'(0) = z_1\psi_1'(0) + z_2\psi_2'(0).$$

Solving for $z_1$ and $z_2$ gives

$$\begin{pmatrix} z_1 \\ z_2 \end{pmatrix} = \frac{1}{\psi_1(0)\psi_2'(0) - \psi_1'(0)\psi_2(0)} \begin{pmatrix} \psi_2'(0) & -\psi_2(0) \\ -\psi_1'(0) & \psi_1(0) \end{pmatrix} \begin{pmatrix} z(0) \\ z'(0) \end{pmatrix}. \tag{30}$$

In the special case of the extension indexed by $B = 0$, that is the Kreĭn extension, we get

$$z(0) - \sqrt{2}z'''(0) - z''(0) = \sqrt{2}z(0) + 3z'(0) + z'''(0) + 2\sqrt{2}z''(0) = 0$$

as the two conditions which generate it. We remark that this extension, for even more general situations, is of some importance in itself. For further information see the survey article [2] and the seminal work of Grubb [11].

We next examine the extensions when the subspace $N_B$ of the kernel is of dimension 1. Let $N_B$ be spanned by

$$\psi = a_1 e^{-\frac{x}{\sqrt{2}}} \cos\left(\frac{x}{\sqrt{2}}\right) + a_2 e^{-\frac{x}{\sqrt{2}}} \sin\left(\frac{x}{\sqrt{2}}\right), \tag{31}$$

where $a_1, a_2$ are constants. Putting this into Hamiltonian form we get for $\psi = (x, u)^T$

$$x = \begin{pmatrix} e^{-\frac{x}{\sqrt{2}}}\left(a_1 \cos\left(\frac{x}{\sqrt{2}}\right) + a_2 \sin\left(\frac{x}{\sqrt{2}}\right)\right) \\ \frac{e^{-\frac{x}{\sqrt{2}}}\left((a_2 - a_1)\cos\left(\frac{x}{\sqrt{2}}\right) - (a_1 + a_2)\sin\left(\frac{x}{\sqrt{2}}\right)\right)}{\sqrt{2}} \end{pmatrix}$$

and

$$u = \begin{pmatrix} e^{-\frac{x}{\sqrt{2}}}\left(a_1\cos\left(\frac{x}{\sqrt{2}}\right) + a_2\sin\left(\frac{x}{\sqrt{2}}\right)\right) \\ \frac{e^{-\frac{x}{\sqrt{2}}}\left((a_2-a_1)\cos\left(\frac{x}{\sqrt{2}}\right)-(a_1+a_2)\sin\left(\frac{x}{\sqrt{2}}\right)\right)}{\sqrt{2}} \end{pmatrix}.$$

So, as in the two dimensional case, the Lagrangian bracket $[z : \psi]$ is zero at infinity and at 0 it is

$$\frac{1}{2}\left(a_1\left(-2z(0) + \sqrt{2}z'(0) - 2z'''(0) - \sqrt{2}z''(0)\right) + \sqrt{2}a_2(z''(0) - z'(0))\right).$$

This leads to the condition

$$\frac{\sqrt{2}}{2}\left(a_1\left(\sqrt{2}z(0) - z'(0) + \sqrt{2}z'''(0) + z''(0)\right) + a_2(z'(0) - z''(0))\right) = B\alpha, \tag{32}$$

where $z = z_F + \alpha\psi$ with $z_F \in D(T_F)$.

We next determine $\alpha$ in terms of the boundary values of $z$. We have $z(0) = \alpha\psi(0)$ and $z'(0) = \alpha\psi'(0)$ and distinguish two cases:

(A) $\psi(0) = 0$. In this case, $a_1 = 0$, $\alpha = \frac{z'(0)}{\psi'(0)}$ with $\psi'(0) = a_2/\sqrt{2}$ and the two boundary conditions for the extension $T_B$ become $z(0) = 0$ and

$$\psi'(0)(z'(0) - z''(0)) = B\frac{z'(0)}{\psi'(0)}. \tag{33}$$

(B) $\psi(0) \neq 0$. In this case, $\alpha = \frac{z(0)}{\psi(0)}$ with $\psi(0) = a_1$ and the two boundary conditions for the extension $T_B$ become $z'(0)\psi(0) - z(0)\psi'(0) = 0$ and

$$\frac{\sqrt{2}}{2}\left(a_1\left(\sqrt{2}z(0) - z'(0) + \sqrt{2}z'''(0) + z''(0)\right) + a_2(z'(0) - z''(0))\right) = B\frac{z(0)}{a_1}. \tag{34}$$

To conclude, we summarise the results for our example in a theorem.

**Theorem 4** *The non-negative extensions $T_B$ of the minimal operator associated with*

$$Ly := y^{(4)} + y \text{ on } [0, \infty)$$

*are given by one of the following two forms:*

1. *Let* $\psi_1$ *and* $\psi_2$ *be given by the equations* (26) *and* (27) *and let* $b$ *be a non-negative form on* $N$. *Then*

$$D(T_B) = \{z \in H^4(0,\infty) | z \text{ satisfies } (28) \text{ and } (29)\},$$

*where* $z_1, z_2$ *are given by* (30).

2. *Let* $\psi$ *be given by* (31) *and let* $B$ *be a non-negative number.*

   (a) *If* $a_1 = 0$, *then*

$$D(T_B) = \{z \in H^4(0,\infty) | z(0) = 0 \text{ and } z \text{ satisfies } (33)\}.$$

   (b) *If* $a_1 \neq 0$, *then*

$$D(T_B) = \{z \in H^4(0,\infty) | z'(0)\psi(0) - z(0)\psi'(0) = 0 \text{ and } z \text{ satisfies } (34)\}.$$

## References

1. A. Alonso, B. Simon, The Birman-Kreĭn-Vishik theory of selfadjoint extensions of semi-bounded operators. J. Operator Theory **4**(2), 251–270 (1980)
2. M.S. Ashbaugh, F. Gesztesy, M. Mitrea, R. Shterenberg, G. Teschl, A survey on the Krein–von Neumann extension, the corresponding abstract buckling problem, and Weyl-type spectral asymptotics for perturbed Krein Laplacians in nonsmooth domains, in *Mathematical Physics, Spectral Theory and Stochastic Analysis*, vol. 232. Oper. Theory Adv. Appl. (Birkhäuser/Springer Basel AG, Basel, 2013), pp. 1–106
3. F.V. Atkinson, *Discrete and Continuous Boundary Problems*. Mathematics in Science and Engineering, vol. 8 (Academic Press, New York-London, 1964)
4. H. Behncke, D. Hinton, Spectral theory of Hamiltonian systems with almost constant coefficients. J. Differ. Equ. **250**(3), 1408–1426 (2011)
5. H. Behncke, D.B. Hinton, Spectral theory of higher order differential operators by examples. J. Spectr. Theory **3**(3), 361–398 (2013)
6. M. Birman, On the theory of self-adjoint extensions of positive definite operators. Doklady Akad. Nauk SSSR (N.S.) **91**, 189–191 (1953)
7. B.M. Brown, W.D. Evans, Selfadjoint and *m* sectorial extensions of Sturm-Liouville operators. Integr. Equ. Oper. Theory **85**(2), 151–166 (2016)
8. B.M. Brown, W.D. Evans, I.G. Wood, Positive self-adjoint operator extensions with applications to differential operators. Integr. Equ. Oper. Theory **91**(5), 41 (2019)
9. W.N. Everitt, On the limit-point classification of fourth-order differential equations. J. Lond. Math. Soc. **44**, 273–281 (1969)
10. F. Gesztesy, L.L. Littlejohn, R. Nichols, On self-adjoint boundary conditions for singular Sturm-Liouville operators bounded from below. J. Differ. Equ. **269**(9), 6448–6491 (2020)
11. G. Grubb, Spectral asymptotics for the "soft" selfadjoint extension of a symmetric elliptic differential operator. J. Oper. Theory **10**(1), 9–20 (1983)
12. D.B. Hinton, J.K. Shaw, On the spectrum of a singular Hamiltonian system. Quaest. Math. **5**(1), 29–81 (1982/1983)
13. D. Hinton, A.M. Krall, K. Shaw, Boundary conditions for differential operators with intermediate deficiency index. Appl. Anal. **25**(1–2), 43–53 (1987)

14. H. Kalf, A characterization of the Friedrichs extension of Sturm-Liouville operators. J. Lond. Math. Soc. (2) **17**(3), 511–521 (1978)
15. A.M. Krall, A limit-point criterion for linear Hamiltonian systems. Appl. Anal. **61**(1–2), 115–119 (1996)
16. M.G. Kreĭn, The theory of self-adjoint extensions of semi-bounded Hermitian transformations and its applications. II. Mat. Sbornik N.S. **21**(63), 365–404 (1947)
17. M. Marletta, A. Zettl, The Friedrichs extension of singular differential operators. J. Differ. Equ. **160**(2), 404–421 (2000)
18. M.A. Naĭmark, *Linear differential operators. Part II: Linear differential operators in Hilbert space*. With additional material by the author, and a supplement by V. È. Ljance. Translated from the Russian by E. R. Dawson. English translation ed. by W.N. Everitt (Frederick Ungar Publishing, New York, 1968)
19. H.-D. Niessen, A. Zettl, Singular Sturm-Liouville problems: the Friedrichs extension and comparison of eigenvalues. Proc. Lond. Math. Soc. (3) **64**(3), 545–578 (1992)
20. Y. Shi, On the rank of the matrix radius of the limiting set for a singular linear Hamiltonian system. Linear Algebra Appl. **376**, 109–123 (2004)
21. H. Sun, Qingkai Kong, Yuming Shi, Essential spectrum of singular discrete linear Hamiltonian systems. Math. Nachr. **289**(2–3), 343–359 (2016)
22. H. Sun, Y. Shi, Self-adjoint extensions for singular linear Hamiltonian systems. Math. Nachr. **284**(5–6), 797–814 (2011)
23. H. Sun, Y. Shi, On essential spectra of singular linear Hamiltonian systems. Linear Algebra Appl. **469**, 204–229 (2015)
24. M. I. Višik, On general boundary problems for elliptic differential equations. Trudy Moskov. Mat. Obšč. **1**, 187–246 (1952)
25. P.W. Walker, A vector-matrix formulation for formally symmetric ordinary differential equations with applications to solutions of integrable square. J. Lond. Math. Soc. (2) **9**, 151–159 (1974/1975)
26. A. Wang, J. Sun, A. Zettl, Characterization of domains of self-adjoint ordinary differential operators. J. Differ. Equ. **246**(4), 1600–1622 (2009)
27. Z. Zheng, S. Chen, GKN theory for linear Hamiltonian systems. Appl. Math. Comput. **182**(2), 1514–1527 (2006)
28. Z. Zheng, Q. Kong, Friedrichs extensions for singular Hamiltonian operators with intermediate deficiency indices. J. Math. Anal. Appl. **461**(2), 1672–1685 (2018)

# On Simon's Hausdorff Dimension Conjecture

**David Damanik, Jake Fillman, Shuzheng Guo, and Darren C. Ong**

**Abstract** Barry Simon conjectured in 2005 that the Szegő matrices, associated with Verblunsky coefficients $\{\alpha_n\}_{n \in \mathbb{Z}_+}$ obeying $\sum_{n=0}^{\infty} n^{\gamma} |\alpha_n|^2 < \infty$ for some $\gamma \in (0, 1)$, are bounded for values $z \in \partial\mathbb{D}$ outside a set of Hausdorff dimension no more than $1 - \gamma$. Three of the authors recently proved this conjecture by employing a Prüfer variable approach that is analogous to work Christian Remling did on Schrödinger operators. This paper is a companion piece that presents a simple proof of a weak version of Simon's conjecture that is in the spirit of a proof of a different conjecture of Simon.

**Keywords** Orthogonal polynomials · Verblunsky coefficients · Szegő matrices

## 1 Introduction

This paper is concerned with Barry Simon's Hausdorff dimension conjecture for orthogonal polynomials on the unit circle (OPUC).

D. Damanik (✉)
Rice University, Houston, TX, USA
e-mail: damanik@rice.edu

J. Fillman
Texas State University, San Marcos, TX, USA
e-mail: fillman@txstate.edu

S. Guo
Rice University, Houston, TX, USA
Ocean University of China, Shandong, Qingdao, China

D. C. Ong
Xiamen University Malaysia, Sepang, Selangor Darul Ehsan, Malaysia
e-mail: darrenong@xmu.edu.my



F. Gesztesy, A. Martinez-Finkelshtein (eds.), *From Operator Theory to Orthogonal Polynomials, Combinatorics, and Number Theory*, Operator Theory: Advances and Applications 285, https://doi.org/10.1007/978-3-030-75425-9_3

There has been a large amount of activity studying OPUC in the past two decades, primarily due to the monographs [7, 8] by Simon, to which we refer the reader for general background material. In these monographs, Simon makes a number of conjectures, which are listed for the reader's convenience in [8, Appendix D]. Our main purpose here is to prove a result related to one of these conjectures, the Hausdorff dimension conjecture; see [8, Conjecture D.3.5, p. 982].

Before stating this conjecture, let us describe the general setting of OPUC. Suppose $\mu$ is a non-trivial (i.e., not finitely supported) probability measure on the unit circle $\partial\mathbb{D} = \{z \in \mathbb{C} : |z| = 1\}$. By the non-triviality assumption, the functions $1, z, z^2, \cdots$ are linearly independent in the Hilbert space $\mathscr{H} = L^2(\partial\mathbb{D}, \mu)$, and hence one can form, via the Gram-Schmidt procedure, the *monic orthogonal polynomials* $\Phi_n(z)$, whose Szegő dual is defined by

$$\Phi_n^* = z^n \overline{\Phi_n(1/\overline{z})}.$$

There are constants $\{\alpha_n\}_{n\in\mathbb{Z}_+}$ in $\mathbb{D} = \{z \in \mathbb{C} : |z| < 1\}$, called the *Verblunsky coefficients*, so that

$$\Phi_{n+1}(z) = z\Phi_n(z) - \overline{\alpha}_n \Phi_n^*(z), \qquad \text{for } n \in \mathbb{Z}_+, \tag{1}$$

which is the so-called *Szegő recurrence* (here and throughout the paper, $\mathbb{Z}_+$ denotes the set of all non-negative integers and $\mathbb{N}$ will denote the set of positive integers). Conversely, every sequence $\{\alpha_n\}_{n\in\mathbb{Z}_+}$ in $\mathbb{D}$ arises as the sequence of Verblunsky coefficients for a suitable nontrivial probability measure on $\partial\mathbb{D}$.

If we consider instead the orthonormal polynomials

$$\varphi_n(z) = \frac{\Phi_n(z)}{\|\Phi_n(z)\|_\mu},$$

where $\|\cdot\|_\mu$ is the norm of $\mathscr{H}$, one can verify that (1) becomes

$$\rho_n\varphi_{n+1}(z) = z\varphi_n(z) - \overline{\alpha}_n\varphi_n^*(z), \text{ for } n \in \mathbb{Z}_+, \tag{2}$$

where $\rho_n = (1 - |\alpha_n|^2)^{1/2}$.

The Szegő recurrence can be written in a matrix form as follows:

$$\begin{bmatrix}\varphi_{n+1}(z)\\ \varphi_{n+1}^*(z)\end{bmatrix} = \frac{1}{\rho_n}\begin{bmatrix} z & -\overline{\alpha}_n\\ -\alpha_n z & 1\end{bmatrix}\begin{bmatrix}\varphi_n(z)\\ \varphi_n^*(z)\end{bmatrix}, \text{ for } n \in \mathbb{Z}_+. \tag{3}$$

Alternatively, one can consider a different initial condition and derive the *orthogonal polynomials of the second kind*, by setting $\psi_0(z) = 1$ and then

$$\begin{bmatrix}\psi_{n+1}(z)\\ -\psi_{n+1}^*(z)\end{bmatrix} = \frac{1}{\rho_n}\begin{bmatrix} z & -\overline{\alpha}_n\\ -\alpha_n z & 1\end{bmatrix}\begin{bmatrix}\psi_n(z)\\ -\psi_n^*(z)\end{bmatrix}, \text{ for } n \in \mathbb{Z}_+. \tag{4}$$

In particular, the sequence $\{\psi_n(z)\}_{n\in\mathbb{Z}_+}$ is precisely the same as the first-kind polynomials for the measure $\widetilde{\mu}$ with Verblunsky coefficients $\widetilde{\alpha}_n = -\alpha_n$; compare the discussion in [7, Sect. 3.2, p. 222], especially Equations (3.2.2) and (3.2.3). Define

$$T_n(z) = \frac{1}{2}\begin{bmatrix} \varphi_n(z) + \psi_n(z) & \varphi_n(z) - \psi_n(z) \\ \varphi_n^*(z) - \psi_n^*(z) & \varphi_n^*(z) + \psi_n^*(z) \end{bmatrix}, \text{ for } n \in \mathbb{Z}_+. \tag{5}$$

Wc can now statc [8, Conjecture D.3.5, p. 982]:

**Hausdorff Dimension Conjecture** Assuming that the Verblunsky coefficients obey a decay estimate of the form

$$\sum_{n=0}^{\infty} n^\gamma |\alpha_n|^2 < \infty \tag{6}$$

for some $\gamma \in (0, 1)$, the associated Szegő matrices $T_n(z)$ are bounded in $n \in \mathbb{Z}_+$ for values $z \in \partial\mathbb{D}$ outside a set of Hausdorff dimension no more than $1 - \gamma$.

The primary interest in such a statement comes from the general principle that says that the singular part of $\mu$ is supported by the set of those $z \in \partial\mathbb{D}$ for which the associated Szegő matrices are unbounded, that is, such a result will imply that the singular part of the measure, $\mu_{\text{sing}}$, has a support of Hausdorff dimension at most $1-\gamma$. As discussed in [8], the upper bound on the Hausdorff dimension of the exceptional set of spectral parameters for which the Szegő matrices are unbounded is tight, that is, it cannot be improved.

Three of the authors recently proved Simon's Hausdorff dimension conjecture [3]. While the proof of the full conjecture (which is inspired by work of Remling [5, 6]) is not short and somewhat technical, we show here that if one is willing to give up a factor 2, then there is a rather short proof. In other words, we will show how to quickly establish the following result:

**Theorem 1.1** *Suppose that $\mu$ is such that the associated Verblunsky coefficients satisfy* (6). *Then there is a set $S \subset \partial\mathbb{D}$ of Hausdorff dimension at most* $2(1-\gamma)$ *so that for $z \in \partial\mathbb{D} \setminus S$,*

$$\sup_{n\geq 0} \|T_n(z)\| < \infty.$$

*In particular, $\mu_{\text{sing}}$ is supported by a set of dimension at most* $2(1-\gamma)$.

Of course, the theorem is only meaningful for $\gamma \in (1/2, 1)$, so by giving up a factor of 2 in the estimate, one not only loses the tightness of the estimate, one also reduces the size of the relevant range of $\gamma$'s by half. Nevertheless, we feel it is still worthwhile to point out that this weaker result can be obtained via a short argument, which avoids the lengthy and intricate arguments used in [3]. This short argument is inspired by the work [2], which incidentally proves another conjecture of Simon, namely [8, Conjecture D.3.4, p. 982]. Earlier related work can be found in [4]. We will prove Theorem 1.1 in Sect. 2.

Let us briefly remark that one can extend Theorem 1.1 to cover the case of logarithmic divergence.

**Corollary 1.2** *Suppose that $\mu$ is such that the associated Verblunsky coefficients satisfy*

$$\sum_{n=0}^{N} n^{\gamma} |\alpha_n|^2 \leq A(\log N)^B \tag{7}$$

*for all $N \geq 2$, where $A, B > 0$, $\gamma \in (0, 1)$ are constants. Then there is a set $S \subset \partial\mathbb{D}$ of Hausdorff dimension at most $2(1-\gamma)$ so that for $z \in \partial\mathbb{D} \setminus S$,*

$$\sup_{n \geq 0} \|T_n(z)\| < \infty.$$

*In particular, $\mu_{\text{sing}}$ is supported by a set of dimension at most $2(1-\gamma)$.*

All four of the authors of this manuscript have ties to Texas, so we are familiar with and grateful for Lance's contributions to advancing mathematics in this state in his role at Baylor University. Many happy returns, Lance!

## 2 A Weak Version of Simon's Hausdorff Dimension Conjecture

### 2.1 A Basic Estimate

We start by deducing a basic consequence of the assumption (6). Define $d := 1-\gamma$.

**Lemma 2.1** *Under the assumption* (6), $\{n^{-(d/2+\varepsilon/4)}\alpha_n\}_{n\in\mathbb{N}} \in \ell^1(\mathbb{N})$ *for all $\varepsilon > 0$.*

***Proof*** Applying the Cauchy-Schwarz inequality to dyadic blocks, for example, we see that

$$\begin{aligned}
\sum_{n=1}^{\infty} n^{-(d/2+\varepsilon/4)}|\alpha_n| &= \sum_{k=0}^{\infty} \sum_{n=2^k}^{2^{k+1}-1} n^{-(d/2+\varepsilon/4)}|\alpha_n| \\
&= \sum_{k=0}^{\infty} \left( \sum_{n=2^k}^{2^{k+1}-1} n^{-(1/2+\varepsilon/4)} n^{\gamma/2}|\alpha_n| \right)
\end{aligned}$$

$$\leq \sum_{k=0}^{\infty} \left( \sum_{n=2^k}^{2^{k+1}-1} n^{-(1+\varepsilon/2)} \right)^{1/2} \left( \sum_{n=2^k}^{2^{k+1}-1} n^{\gamma} |\alpha_n|^2 \right)^{1/2}$$

$$\lesssim \sum_{k=0}^{\infty} 2^{-k\varepsilon/4}$$

$$< \infty,$$

as desired. □

## 2.2 Prüfer Variables

Let $\{\alpha_n\}_{n\in\mathbb{Z}_+}$ be the Verblunsky coefficients of a nontrivial probability measure $\mu$ on $\partial\mathbb{D}$. As mentioned above, the $\alpha$'s give rise to a sequence $\{\Phi_n(z)\}_{n\in\mathbb{Z}_+}$ of monic polynomials (via the Szegő recurrence) that are orthogonal with respect to $\mu$. For $\beta \in [0, 2\pi)$, we also consider the monic polynomials $\{\Phi_n(z, \beta)\}_{n\in\mathbb{Z}_+}$ that are associated in the same way with the rotated Verblunsky coefficients $\{e^{i\beta}\alpha_n\}_{n\in\mathbb{Z}_+}$.

Let $\eta \in [0, 2\pi)$. Define the Prüfer variables by

$$\Phi_n(e^{i\eta}, \beta) = R_n(\eta, \beta) \exp\left[i(n\eta + \theta_n(\eta, \beta))\right],$$

where $R_n > 0$, $\theta_0 \in [0, 2\pi)$, and $|\theta_{n+1} - \theta_n| < \pi/2$ (compare [8, Corollary 10.12.2]). These variables obey the following pair of equations:

$$\frac{R_{n+1}^2(\eta,\beta)}{R_n^2(\eta,\beta)} = 1 + |\alpha_n|^2 - 2\mathrm{Re}\left(\alpha_n e^{i[(n+1)\eta+\beta+2\theta_n(\eta,\beta)]}\right),$$

$$e^{-i(\theta_{n+1}(\eta,\beta)-\theta_n(\eta,\beta))} = \frac{1 - \alpha_n e^{i[(n+1)\eta+\beta+2\theta_n(\eta,\beta)]}}{\left[1 + |\alpha_n|^2 - 2\mathrm{Re}\left(\alpha_n e^{i[(n+1)\eta+\beta+2\theta_n(\eta,\beta)]}\right)\right]^{1/2}}.$$

We also define $r_n(\eta, \beta) = |\varphi_n(\eta, \beta)|$.

When $\{\alpha_n\} \in \ell^2$,

$$r_n(\eta,\beta) \sim R_n(\eta,\beta) \sim \exp\left( -\sum_{j=0}^{n-1} \mathrm{Re}\left(\alpha_j e^{i[(j+1)\eta+\beta+2\theta_j(\eta,\beta)]}\right) \right). \tag{8}$$

(We write $f_n \sim g_n$ if there is $C > 1$ such that $C^{-1} g_n \leq f_n \leq C g_n$ for all $n$.) For the Prüfer equations and (8), see [8, Theorems 10.12.1 and 10.12.3].

## 2.3 Unboundedness and Infinite Energy

In this section, we will prove that the set of $\eta \in (0, 2\pi)$ for which the radius is unbounded has Hausdorff dimension no more than $2d$, which is stated as follows. The overall strategy in our proof of this statement will be inspired by Damanik-Killip [4].

**Proposition 2.2** *Assume* (6). *Then the set*

$$S = \{\eta \in [0, 2\pi) : R_n(\eta, \beta) \text{ is unbounded for some } \beta\}$$

*has Hausdorff dimension no more than* $2d = 2(1 - \gamma)$.

By (6), $\{\alpha_n\}_{n \in \mathbb{Z}_+} \in \ell^2$. Therefore, because of (8), it suffices for our purposes to show that

$$A(n, \eta, \beta) = \sum_{j=0}^{n-1} \alpha_j e^{i[(j+1)\eta + \beta + 2\theta_j(\eta, \beta)]}$$

is a bounded function of $n$ for all $\beta$, provided that $\eta$ is away from a set of Hausdorff dimension at most $2d$.

**Lemma 2.3** *If*

$$\widehat{\alpha}(\eta, n) = \lim_{N \to \infty} \sum_{j=n}^{N} \alpha_j e^{ij\eta}$$

*exists and obeys*

$$\sum_{j=1}^{\infty} |\widehat{\alpha}(\eta, j)\alpha_{j-1}| < \infty, \tag{9}$$

*then* $\eta \notin S$.

***Proof*** We will show that $A(n, \eta, \beta)$ is bounded (in $n$) for every $\beta \in [0, 2\pi)$ when (9) holds. The assertion then follows from (8).

Write $\gamma_j(\eta, \beta) = (j + 1)\eta + \beta + 2\theta_j(\eta, \beta)$. We have

$$\begin{aligned} A(n, \eta, \beta) &= \sum_{j=0}^{n-1} [\widehat{\alpha}(\eta, j) - \widehat{\alpha}(\eta, j + 1)]\, e^{i\gamma_j(\eta, \beta) - ij\eta} \\ &= \sum_{j=1}^{n-1} \widehat{\alpha}(\eta, j) \left[e^{i\gamma_j(\eta, \beta)} - e^{i(\gamma_{j-1}(\eta, \beta) + \eta)}\right] e^{-ij\eta} + O(1). \end{aligned}$$

Since

$$\begin{aligned}|e^{i\gamma_j(\eta,\beta)} - e^{i(\gamma_{j-1}(\eta,\beta)+\eta)}| &\le |\gamma_j(\eta,\beta) - \gamma_{j-1}(\eta,\beta) - \eta| \\ &= 2|\theta_j(\eta,\beta) - \theta_{j-1}(\eta,\beta)| \\ &\lesssim |\alpha_{j-1}|,\end{aligned}$$

where the first inequality follows from the mean value theorem and the last inequality follows from [8, Corollary 10.12.2] as well as the fact that $\alpha$'s are uniformly bounded away from 1, boundedness of $A(n,\eta,\beta)$ follows. □

**Lemma 2.4** *Let $\nu$ be a positive measure on $[0,2\pi)$. For each $s \in (0,1)$ and every measurable function $m : [0,2\pi) \to \mathbb{Z}_+$,*

$$\left\{\int \left|\sum_{n=0}^{m(\eta)} c_n e^{-in\eta}\right| d\nu(\eta)\right\}^2 \lesssim \mathscr{E}_s(\nu) \sum_{n=0}^{\infty} (n+1)^{1-s} \left|c_n\right|^2,$$

*where $\mathscr{E}_s(\nu)$ denotes the $s$-energy of $\nu$, which is defined by*

$$\mathscr{E}_s(\nu) = \iint (1 + |x-y|^{-s})\, d\nu(x)\, d\nu(y).$$

***Proof*** This follows by slightly adjusting the calculation from [9, §XIII.11, p. 196] (see also [1, §V.5]). □

***Proof of Proposition 2.2*** We will apply the criterion of Lemma 2.3. Let us first note that by the theorem of Salem-Zygmund [9, Theorem XIII.11.3(2)] and the connection between capacity and Hausdorff measure [1, §IV.1], the series defining $\widehat{\alpha}$ converges off a set of Hausdorff dimension $d$. Therefore, we may exclude from consideration those values of $\eta$ for which $\widehat{\alpha}$ is not defined.

By Lemma 2.1, $\{n^{-(d/2+\varepsilon/4)}\alpha_n\}_{n\in\mathbb{N}} \in \ell^1(\mathbb{N})$ for all $\varepsilon > 0$. Hence the proposition will follow from Lemma 2.3 once we prove that for all $\varepsilon > 0$, the set of $\eta$ for which $n^{(d/2+\varepsilon/4)}\widehat{\alpha}(\eta,n) = n^{(2d+\varepsilon)/4}\widehat{\alpha}(\eta,n)$ is unbounded is of Hausdorff dimension no more than $2d+\varepsilon$.

Recall that $d = 1-\gamma$. We consider the case where $2d < 1$, as otherwise there is nothing to prove. Choose $\varepsilon > 0$ small enough so that $2d+\varepsilon < 1$. Let $m(\eta)$ be a measurable $\mathbb{Z}_+$-valued function on $[0,2\pi)$. Because of (6), applying Lemma 2.4 with $s = 2d+\varepsilon$ yields

$$\begin{aligned}\int \left|\sum_{n=m_l(\eta)}^{2^{l+1}-1} \alpha_n e^{in\eta}\right| d\nu(\eta) &= \int \left|\sum_{n=0}^{\widetilde{m}_l(\eta)} \alpha_{2^{l+1}-1-n} e^{-in\eta}\right| d\nu(\eta) \\ &\lesssim \left\{\sum_{n=2^l}^{2^{l+1}-1} (n+1)^{1-(2d+\varepsilon)} \left|\alpha_n\right|^2\right\}^{1/2} \sqrt{\mathscr{E}_{2d+\varepsilon}(\nu)}\end{aligned}$$

$$= \left\{ \sum_{n=2^l}^{2^{l+1}-1} (n+1)^{-(d+\varepsilon)} (n+1)^{\gamma} \left|\alpha_n\right|^2 \right\}^{1/2} \sqrt{\mathscr{E}_{2d+\varepsilon}(\nu)}$$
$$\lesssim 2^{-(d+\varepsilon)l/2} \sqrt{\mathscr{E}_{2d+\varepsilon}(\nu)}$$

where $m_l(\eta) = \max\{m(\eta), 2^l\}$, $\widetilde{m}_l(\eta) = \min\{2^l - 1, 2^{l+1} - 1 - m(\eta)\}$, and sums with lower index greater than their upper index are to be treated as zero. Multiplying both sides by $2^{(2d+\varepsilon)l/4}$, summing this over $l$, and applying the triangle inequality on the left gives

$$\int \left| m(\eta)^{(2d+\varepsilon)/4} \sum_{n=m(\eta)}^{\infty} \alpha_n e^{in\eta} \right| d\nu(\eta) \lesssim \sqrt{\mathscr{E}_{2d+\varepsilon}(\nu)}.$$

That is, for any measurable integer-valued function $m(\eta)$, and any finite measure $\nu$ on $[0, 2\pi)$,

$$\int m(\eta)^{(2d+\varepsilon)/4} \left|\widehat{\alpha}(\eta, m(\eta))\right| d\nu \lesssim \sqrt{\mathscr{E}_{2d+\varepsilon}(\nu)}.$$

This implies that the set on which $n^{(2d+\varepsilon)/4}\widehat{\alpha}(\eta, n)$ is unbounded must be of zero $(2d+\varepsilon)$-capacity (i.e., it does not support a measure of finite $(2d+\varepsilon)$-energy).

As the Hausdorff dimension of sets of zero $(2d+\varepsilon)$-capacity is less than or equal to $2d+\varepsilon$ (see [1, §IV.1]), this completes the proof of the fact that $S$ has Hausdorff dimension no more than $2d$. □

## 2.4 *Proof of Theorem 1.1 and Corollary 1.2*

We are now in a position to prove Theorem 1.1.

***Proof of Theorem 1.1*** By Proposition 2.2, we obtain that the set

$$S = \{\eta \in [0, 2\pi) : R_n(\eta, \beta) \text{ is unbounded for some } \beta\}$$

has Hausdorff dimension no more than $2d = 2(1-\gamma)$. By (8), $R_n \sim r_n$. Since $r_n(\eta, 0) = |\phi_n(e^{i\eta})|$ and $r_n(\eta, \pi) = |\psi_n(e^{i\eta})|$, we see that $\phi_n$ and $\psi_n$ are bounded away from the set $S$. In view of (5), the first assertion follows. The second assertion follows since $\mu_{\text{sing}}$ is supported on the set $S$; compare [8, Corollary 10.8.4]. □

***Proof of Corollary 1.2*** If (7) holds, then, for any $\tau < \gamma$, writing $\delta = \gamma - \tau > 0$, we have the following after partitioning into dyadic blocks:

$$\begin{aligned}\sum_{n=1}^{\infty} n^{\tau} |\alpha_n|^2 &= \sum_{k=0}^{\infty} \sum_{n=2^k}^{2^{k+1}-1} n^{-(\gamma-\tau)} n^{\gamma} |\alpha_n|^2 \\ &\leq \sum_{k=0}^{\infty} 2^{-\delta k} \sum_{n=2^k}^{2^{k+1}-1} n^{\gamma} |\alpha_n|^2 \\ &\lesssim \sum_{k=0}^{\infty} 2^{-\delta k} \cdot (k+1)^B \\ &< \infty.\end{aligned}$$

Thus, (6) holds for all $\tau < \gamma$. Consequently, Theorem 1.1 implies that the set of $z \in \partial\mathbb{D}$ for which $T_n(z)$ is unbounded has Hausdorff dimension at most $2(1 - \tau)$ for all $\tau < \gamma$, so this set of $z$ has Hausdorff dimension bounded above by $2(1 - \gamma)$, as desired. □

**Acknowledgments** D.D. was supported in part by NSF grant DMS–1700131 and by an Alexander von Humboldt Foundation research award. J.F. was supported in part by Simons Foundation Collaboration Grant #711663. S.G. was supported by CSC (No. 201906330008) and NSFC (No. 11571327). D.O. was supported in part by two grants from the Fundamental Research Grant Scheme from the Malaysian Ministry of Education (Grant Numbers: FRGS/1/2018/STG06/XMU/02/1 and FRGS/1/2020/STG06/XMU/02/1) and a Xiamen University Malaysia Research Fund (Grant Number: XMUMRF/2020-C5/IMAT/0011).

## References

1. L. Carleson, *Selected Problems on Exceptional Sets* (D. Van Nostrand Co., Inc., Princeton, 1967)
2. D. Damanik, Verblunsky coefficients with Coulomb-type decay. J. Approx. Theory **139**, 257–268 (2006)
3. D. Damanik, S. Guo, D. Ong, Simon's OPUC Hausdorff dimension conjecture, preprint arXiv:2011.01411
4. D. Damanik, R. Killip, Half-line Schrödinger operators with no bound states. Acta Math. **193**, 31–72 (2004)
5. C. Remling, The absolutely continuous spectrum of one-dimensional Schrödinger operators with decaying potentials. Commun. Math. Phys. **193**, 151–170 (1998)
6. C. Remling, Bounds on embedded singular spectrum for one-dimensional Schrödinger operators. Proc. Am. Math. Soc. **128**, 161–171 (2000)
7. B. Simon, *Orthogonal Polynomials on the Unit Circle. Part 1. Classical Theory*. AMS Colloquium Publications, vol. 54, Part 1 (American Mathematical Society, Province, 2005)
8. B. Simon, *Orthogonal Polynomials on the Unit Circle. Part 2. Spectral Theory*. AMS Colloquium Publications, vol. 54, Part 2 (American Mathematical Society, Province, 2005)
9. A. Zygmund, *Trigonometric Series*, vols. I, II, 3rd edn. (Cambridge University Press, Cambridge, 2002)

# Hypergeometric Functions over Finite Fields and Modular Forms: A Survey and New Conjectures

**Madeline Locus Dawsey and Dermot McCarthy**

*In honor of Lance Littlejohn on his 70th birthday.*

**Abstract** Hypergeometric functions over finite fields were introduced by Greene in the 1980s as a finite field analogue of classical hypergeometric series. These functions, and their generalizations, naturally lend themselves to, and have been widely used in, character sum evaluations and counting points on algebraic varieties. More interestingly, perhaps, are their links to Fourier coefficients of modular forms. In this paper, we outline the main results in this area and also conjecture 13 new relations.

**Keywords** Hypergeometric functions · Modular forms

## 1 Introduction

Hypergeometric functions over finite fields were introduced by Greene [16] as a finite field analogue of classical hypergeometric series. Much of Greene's early work on these functions focused on developing transformation and summation formulas which mirror those of the classical series. These functions have a nice character sum representation and so the transformation and summation formulas can be interpreted as relations to simplify and evaluate complex character sums. Using character sums to count points on certain algebraic varieties over finite fields

M. L. Dawsey (✉)
Department of Mathematics, The University of Texas at Tyler, Tyler, TX, USA
e-mail: mdawsey@uttyler.edu

D. McCarthy
Department of Mathematics & Statistics, Texas Tech University, Lubbock, TX, USA
e-mail: dermot.mccarthy@ttu.edu

 
F. Gesztesy, A. Martinez-Finkelshtein (eds.), *From Operator Theory to Orthogonal Polynomials, Combinatorics, and Number Theory*, Operator Theory: Advances and Applications 285, https://doi.org/10.1007/978-3-030-75425-9_4

is well-established and, consequently, hypergeometric functions over finite fields naturally lend themselves to this endeavor.

Via modularity results, we then find relations between hypergeometric functions over finite fields and the Fourier coefficients of modular forms. As we will see, these relations are striking in their simplicity. While the modularity theorem and connections between hypergeometric functions over finite fields and elliptic curves yield infinitely many such relations with weight two newforms, relations in higher weights are rare. The main source of relations between hypergeometric functions over finite fields and the Fourier coefficients of modular forms of weight greater than two is the supercongruence conjectures of Rodriguez Villegas [36] which yield 14 such relations. In a recent paper studying generalized Paley graphs, the authors discovered evidence for two new such relations. Since then we have conducted a more extensive search where we have found a further 13 possible relations. The main purpose of this paper is to present the details of these conjectural relations. For context, we also outline the other main results in this area.

This paper is organized as follows. In Sect. 2, we define hypergeometric functions over finite fields and also a $p$-adic extension. We then outline the main results linking hypergeometric functions over finite fields and Fourier coefficients of modular forms, categorized by weight, with weight two in Sect. 3 and higher weights in Sect. 4. In Sect. 5, we discuss the Eichler–Selberg trace formula which is one of the main tools used to prove such results. Finally, in Sect. 6 we describe our 13 new conjectural relations.

## 2 Preliminaries

While hypergeometric functions over finite fields were originally defined by Greene [16], in this paper we will use a normalized version defined by the second author [27, 29], which allow many of the results we are interested in to be stated in a slightly more streamlined fashion. Throughout, let $p$ be a prime, and let $q$ be a prime power. Let $\mathbb{F}_q$ be the finite field with $q$ elements, and let $\widehat{\mathbb{F}_q^*}$ be the group of multiplicative characters of $\mathbb{F}_q^*$. We extend the domain of $\chi \in \widehat{\mathbb{F}_q^*}$ to $\mathbb{F}_q$ by defining $\chi(0) := 0$ (including for the trivial character $\varepsilon$) and denote $\overline{\chi}$ as the inverse of $\chi$. We denote by $\varphi$ the character of order two in $\widehat{\mathbb{F}_q^*}$ when $q$ is odd. More generally, for $k > 2$ a positive integer, we let $\chi_k \in \widehat{\mathbb{F}_q^*}$ denote a character of order $k$ when $q \equiv 1 \pmod{k}$. Let $\theta$ be a fixed non-trivial additive character of $\mathbb{F}_q$, and for $\chi \in \widehat{\mathbb{F}_q^*}$ define the Gauss sum $g(\chi) := \sum_{x \in \mathbb{F}_q} \chi(x)\theta(x)$. We define the finite field hypergeometric function as follows.

**Definition 1 ([27], Def. 1.4; [29], Def 2.4)** For $A_1, A_2, \ldots, A_m, B_1, B_2 \ldots, B_m \in \widehat{\mathbb{F}_q^*}$ and $x \in \mathbb{F}_q$,

$$
{}_mF_m\left(\begin{matrix} A_1, & A_2, & \ldots, & A_m \\ B_1, & B_2, & \ldots, & B_m \end{matrix} \,\middle|\, x\right)_q
:= \frac{-1}{q-1} \sum_{\chi \in \widehat{\mathbb{F}_q^*}} \prod_{i=1}^{m} \frac{g(A_i\chi)}{g(A_i)} \frac{g(\overline{B_i\chi})}{g(\overline{B_i})} \chi(-1)^m \chi(x).
$$

If $B_1 = \varepsilon$, as is often the case, then it is usually omitted from the list of parameters in ${}_mF_m$ and the notation is written as ${}_mF_{m-1}$. See [27, Prop. 2.5] and surrounding discussion for a precise description of the relationship between ${}_mF_m$ and Greene's function. In most of the relations connecting ${}_mF_m$ to Fourier coefficients of modular forms, all the $B_i$'s are trivial. In this case, ${}_mF_{m-1}$ equals $(-q)^{m-1}$ times Greene's function, with the same parameters. Many of the results concerning hypergeometric functions over finite fields that we quote in this paper, from other articles, were originally stated using Greene's function. If this is the case, note then that we have reformulated them in terms of ${}_mF_m$, as defined above.

All the results we will see relating the $p$-th Fourier coefficients of modular forms to ${}_mF_m(\cdots)_p$ will require characters of certain orders. Consequently, this restricts these results to $p$ in certain congruence classes. In some cases, these results can be extended to all odd primes using a function which extends ${}_mF_m(\cdots)$ to the $p$-adic setting. Let $\mathbb{Z}_p$ denote the ring of $p$-adic integers, $\Gamma_p(\cdot)$ denote Morita's $p$-adic gamma function, and $\omega$ denote the Teichmüller character of $\mathbb{F}_p$, with $\overline{\omega}$ denoting its character inverse. For $x \in \mathbb{Q}$ we let $\lfloor x \rfloor$ denote the greatest integer less than or equal to $x$ and $\langle x \rangle$ denote the fractional part of $x$, i.e. $x - \lfloor x \rfloor$.

**Definition 2 ([25], Def. 2.1; [28], Def. 1.1)** Let $p$ be an odd prime. For $a_1, a_2, \ldots, a_m, b_1, b_2 \ldots, b_m \in \mathbb{Q} \cap \mathbb{Z}_p$ and $x \in \mathbb{F}_p$,

$$
{}_mG_m\left[\begin{matrix} a_1, & a_2, & \ldots, & a_m \\ b_1, & b_2, & \ldots, & b_m \end{matrix} \,\middle|\, x\right]_p := \frac{-1}{p-1} \sum_{j=0}^{p-2} (-1)^{jm}\, \overline{\omega}^j(x)
$$

$$
\times \prod_{i=1}^{m} \frac{\Gamma_p(\langle a_i - \frac{j}{p-1} \rangle)}{\Gamma_p(\langle a_i \rangle)} \frac{\Gamma_p(\langle -b_i + \frac{j}{p-1} \rangle)}{\Gamma_p(\langle -b_i \rangle)} (-p)^{-\lfloor \langle a_i \rangle - \frac{j}{p-1} \rfloor - \lfloor \langle -b_i \rangle + \frac{j}{p-1} \rfloor}.
$$

A "$q$ version" of ${}_mG_m[\cdots]$ also exists [28, Def. 5.1] but is not needed here. There is a simple relationship between ${}_mF_m(\cdots)_p$ and ${}_mG_m[\cdots]_p$.

**Lemma 1 ([28], Lemma 3.3; [29], Lemma 2.5)** *For a fixed odd prime p, let $A_i, B_j \in \widehat{\mathbb{F}_p^*}$ be given by $\overline{\omega}^{a_i(p-1)}$ and $\overline{\omega}^{b_j(p-1)}$ respectively, where $\omega$ is the Teichmüller character. Then*

$$ {}_mF_m\left(\begin{matrix} A_1, A_2, \ldots, A_m \\ B_1, B_2, \ldots, B_m \end{matrix} \middle| \, t\right)_p = {}_mG_m\left[\begin{matrix} a_1, a_2, \ldots, a_m \\ b_1, b_2, \ldots, b_m \end{matrix} \middle| \, t^{-1}\right]_p. $$

We recall Dedekind's eta function, which will be used to describe some of the modular forms in this paper: $\eta(z) := q^{\frac{1}{24}} \prod_{n\geq 1}(1-q^n)$, where $q := e^{2\pi i z}$.

## 3 Weight Two Newforms

For each elliptic curve $E/\mathbb{Q}$, with conductor $N_E$, the modularity theorem guarantees the existence of a weight two newform of level $N_E$ whose Fourier coefficients are given by the coefficients of the Hasse–Weil $L$-function of $E$, $L(E,s) = \sum_{n\geq 1} a(E,n)\, n^{-s}$. This function is completely determined by its coefficients at the primes, $a(E,p)$, which are related to the number of rational points $N(E,p)$ on the reduction of $E$ modulo $p$ via the formula $N(E,p) = p + 1 - a(E,p)$.

As mentioned in the introduction, finite field hypergeometric functions naturally lend themselves to counting points on algebraic varieties over finite fields. In particular, $N(E,p)$, and consequently $a(E,p)$, as we will see below, can be evaluated by ${}_2F_1$ finite field hypergeometric functions. Passing through the modularity theorem then results in formulas for the $p$-th Fourier coefficients of weight two newforms in terms of these ${}_2F_1$ evaluations, and we get infinitely many such connections.

The first results relating $a(E,p)$ to ${}_2F_1$ finite field hypergeometric functions were due to Koike [20]. He examined various families of curves, including the Legendre family, which yields the following result.

**Theorem 1 (Koike [20])** *Let $\lambda \in \mathbb{Q} \setminus \{0,1\}$. Consider the elliptic curve $E_\lambda : y^2 = x(x-1)(x-\lambda)$ over $\mathbb{Q}$. If $p \geq 3$ is a prime with $\mathrm{ord}_p(\lambda(\lambda-1)) = 0$, then*

$$ a(E_\lambda, p) = \varphi(-1) \cdot {}_2F_1\left(\begin{matrix} \varphi, \varphi \\ \varepsilon \end{matrix} \middle| \, \lambda\right)_p. $$

*Example 1* We take $\lambda = -1$ in the above result. The curve $E_{-1} : y^2 = x^3 - x$ [21, 32.a3] is related via the modularity theorem to the modular form $\eta(4z)^2\eta(8z)^2 = \sum_{n\geq 1} a_1(n)\, q^n \in S_2^{\text{new}}(\Gamma_0(32))$ [21, 32.2.a.a]. Combining with Theorem 1 we get that for all odd primes

$$ a_1(p) = \varphi(-1) \cdot {}_2F_1\left(\begin{matrix} \varphi, \varphi \\ \varepsilon \end{matrix} \middle| \, -1\right)_p. $$

Fuselier [13] examined the family $E_t : y^2 = 4x^3 - \frac{27}{1-t}x - \frac{27}{1-t}$ and expressed $a(E_t, p)$ in terms of a ${}_2F_1$, whose parameters include characters of order 12, when $p \equiv 1 \pmod{12}$. Lennon [23] generalized Fuselier's result to evaluate $a(E, p)$ for any elliptic curve $E$, when the reduction of $E$ modulo $p$ is an elliptic curve over $\mathbb{F}_p$ with $j$-invariant not equal to 0 or 1728.

**Theorem 2 (Lennon [23] § 2.2)** *Let $E/\mathbb{Q}$ be an elliptic curve. Let $p \equiv 1 \pmod{12}$ be a prime such that $E_p : y^2 = x^3 + ax + b$ is an elliptic curve over $\mathbb{F}_p$ and $j(E_p) \neq 0, 1728$. Then*

$$a(E, p) = \chi_4\left(-\frac{a^3}{27}\right) {}_2F_1\left(\begin{matrix}\chi_{12}, \chi_{12}^5 \\ \varepsilon\end{matrix} \;\middle|\; \frac{1728}{j(E_p)}\right)_p .$$

Theorem 2 is independent of the model for $E_p$. The results in [23] are in fact over $\mathbb{F}_q$, for $q \equiv 1 \pmod{12}$ a prime power, and hence allow calculation of $a(E, p)$ up to sign when $p \not\equiv 1 \pmod{12}$ via the relation $a(E, p)^2 = a(E, p^2) + 2p$. Theorem 1.2 of [28] extends Theorem 2 to the $p$-adic setting, giving a direct evaluation of $a(E, p)$ for all primes $p > 3$ and resolves this sign issue.

**Theorem 3 (McCarthy [28] Thm 1.2)** *Let $E/\mathbb{Q}$ be an elliptic curve. Let $p > 3$ be a prime such that $E_p : y^2 = x^3 + ax + b$ is an elliptic curve over $\mathbb{F}_p$ and $j(E_p) \neq 0, 1728$. Then*

$$a(E, p) = \varphi(b) \cdot p \cdot {}_2G_2\left[\begin{matrix}\frac{1}{4}, \frac{3}{4} \\ \frac{1}{3}, \frac{2}{3}\end{matrix} \;\middle|\; 1 - \frac{1728}{j(E_p)}\right]_p .$$

Again, Theorem 3 is independent of the model for $E_p$.

*Example 2* Consider the elliptic curve $E : y^2 = x^3 + 27x - 27$ [21, 540.a2]. This is related via the modularity theorem to the modular form $q - q^5 - 4q^7 + 6q^{11} - 4q^{13} - 3q^{17} - 7q^{19} - 9q^{23} + q^{25} + \cdots = \sum_{n\geq 1} a_2(n)\, q^n \in S_2^{\text{new}}(\Gamma_0(540))$, [21, 540.2.a.a]. Then, by Theorem 3, we get that for all primes $p > 3$

$$a_2(p) = \left(\tfrac{-3}{p}\right) \cdot p \cdot {}_2G_2\left[\begin{matrix}\frac{1}{4}, \frac{3}{4} \\ \frac{1}{3}, \frac{2}{3}\end{matrix} \;\middle|\; -\tfrac{1}{4}\right]_p .$$

The theorem covers $p > 5$. We manually check that the relation also holds for $p = 5$.

Significant other contributions in this area, where connections between various families of elliptic curves and finite field hypergeometric functions are established, can be found in [5, 6, 10, 18, 22, 33]. Between all these results it should be possible to evaluate the $p$-th Fourier coefficients of all weight two newforms, with integer coefficients, using finite field hypergeometric functions (or their $p$-adic extensions). For newforms with non-integral coefficients, things are less straightforward and little is known. However, we have the following conjectural relations due to Evans [11].

*Conjecture 1 (Evans [11])* Consider the newforms $\sum_{n\geq 1} a_3(n)q^n \in S_2(\Gamma_0(972))$, [21, 972.2.a.e], and $\sum_{n\geq 1} a_4(n)q^n \in S_2(\Gamma_0(768))$, [21, 768.2.a.j], with coefficient fields $\mathbb{Q}\left(\sqrt{2}\right)$ and $\mathbb{Q}\left(\sqrt{3}\right)$ respectively.

1. If $q \equiv 1 \pmod 6$, then

$$\overline{\chi_6}(12) J(\chi_6, \chi_6) - \overline{\chi_6}(3)\, J\left(\chi_6^2, \chi_6^2\right) \cdot {}_3F_2\left(\begin{matrix} \overline{\chi_6}, & \varphi, & \chi_6 \\ & \varphi\chi_6, & \varphi\chi_6 \end{matrix} \middle| \frac{1}{4}\right)_q$$

$$= \begin{cases} a_3(p), & \text{if } q = p,\ p \equiv 1 \pmod 6 \\ a_3(p)^2 + 2p, & \text{if } q = p^2,\ p \equiv 5 \pmod 6. \end{cases}$$

2. If $q \equiv 1 \pmod 8$, then

$$\chi_8(-4) J(\chi_8, \chi_8) - \chi_8(-4) J\left(\chi_8^2, \chi_8^3\right) \cdot {}_3F_2\left(\begin{matrix} \overline{\chi_8}, & \chi_8^3, & \chi_8 \\ & \overline{\chi_8}^2, & \varphi\chi_8 \end{matrix} \middle| \frac{1}{4}\right)_q$$

$$= \begin{cases} a_4(p), & \text{if } q = p,\ p \equiv 1 \pmod 8 \\ a_4(p)^2 + 2p^2, & \text{if } q = p^2,\ p \not\equiv 1 \pmod 8. \end{cases}$$

# 4 Higher Weight Newforms

The lack of universal modularity results for algebraic varieties of dimension greater than one means that connections to modular forms of weight greater than two are somewhat ad hoc. In this section we outline the main results linking finite field hypergeometric functions and Fourier coefficients of modular forms of weight greater than two. We start with the connections coming from Rodriguez Villegas's supercongruence conjectures.

## *4.1 The Conjectures of Rodriguez Villegas*

In [36], Rodriguez Villegas examined the relationship between the number of points over $\mathbb{F}_p$ on certain Calabi–Yau manifolds and truncated classical hypergeometric series which correspond to a particular period of the manifold. In doing so, he identified numerically 22 possible supercongruences, 18 of which relate truncated classical hypergeometric series to Fourier coefficients of modular forms of weights three and four. One of these relations had previously been conjectured in [37]. The 14 cases involving weight four modular forms relate to Calabi–Yau threefolds, 13 of which were studied in [7]. The book of Meyer [31] contains a nice description of these threefolds.

While the supercongruence relations of Rodriguez Villegas are congruences involving classical hypergeometric series, it became obvious from the work of Mortenson [32] and Kilbourn [19] on proving the first few of these conjectures, and also from known connections between finite field hypergeometric functions and some of the modular forms in question, due to Ono [33] and Ahlgren and Ono [3], that corresponding to each of Rodriguez Villegas's conjectures was a linear relation between finite field hypergeometric functions and the Fourier coefficients of these modular forms. In fact, following the work of the second author in [25], which precisely describes the relationship between the truncated classical hypergeometric series appearing in Rodriguez Villegas's conjectures and the $p$-adic function defined in Definition 2 above, proving relationships between the relevant ${}_mG_m[\cdots]$ and the Fourier coefficients of the modular forms in question would suffice to prove the conjectures of Rodriguez Villegas. A list of all these relations is shown in Table 1. The Rodriguez Villegas conjectures corresponding to cases 1-5, 7 and 15 were proved individually [1, 15, 17, 19, 26, 32, 38]. A proof of all 14 weight four conjectures (cases 5–18) is offered in [24] (as yet unpublished) and, as a consequence of the results therein, all the relations in Table 1 should now be known.

**Table 1** Relations arising from the conjectures of Rodriguez Villegas

| | Hyp Series | Newform $f(z)=\sum a(n)q^n$ | | Connection | |
|---|---|---|---|---|---|
| | Parameters | Space | LMFDB | Relationship | When |
| 1. | $\left[\frac{1}{2},\frac{1}{2},\frac{1}{2};1,1,1\mid 1\right]$ | $S_3(\Gamma_0(16),(\frac{-4}{\cdot}))$ | 16.3.c.a | $a(p)=G[\cdots]_p$ | $p>2$ |
| 2. | $\left[\frac{1}{2},\frac{1}{3},\frac{2}{3};1,1,1\mid 1\right]$ | $S_3(\Gamma_0(12),(\frac{-3}{\cdot}))$ | 12.3.c.a | $a(p)=G[\cdots]_p$ | $p>3$ |
| 3. | $\left[\frac{1}{2},\frac{1}{4},\frac{3}{4};1,1,1\mid 1\right]$ | $S_3(\Gamma_0(8),(\frac{-2}{\cdot}))$ | 8.3.d.a | $a(p)=G[\cdots]_p$ | $p>2$ |
| 4. | $\left[\frac{1}{2},\frac{1}{6},\frac{5}{6};1,1,1\mid 1\right]$ | $S_3(\Gamma_0(144),(\frac{-4}{\cdot}))$ | 144.3.g.a | $a(p)=G[\cdots]_p$ | $p>3$ |
| 5. | $\left[\frac{1}{2},\frac{1}{2},\frac{1}{2},\frac{1}{2};1,1,1,1\mid 1\right]$ | $S_4(\Gamma_0(8))$ | 8.4.a.a | $a(p)=G[\cdots]_p-p$ | $p>2$ |
| 6. | $\left[\frac{1}{2},\frac{1}{2},\frac{1}{3},\frac{2}{3};1,1,1,1\mid 1\right]$ | $S_4(\Gamma_0(36))$ | 36.4.a.a | $a(p)=G[\cdots]_p-(\frac{12}{p})p$ | $p>3$ |
| 7. | $\left[\frac{1}{2},\frac{1}{2},\frac{1}{4},\frac{3}{4};1,1,1,1\mid 1\right]$ | $S_4(\Gamma_0(16))$ | 16.4.a.a | $a(p)=G[\cdots]_p-(\frac{8}{p})p$ | $p>2$ |
| 8. | $\left[\frac{1}{2},\frac{1}{2},\frac{1}{6},\frac{5}{6};1,1,1,1\mid 1\right]$ | $S_4(\Gamma_0(72))$ | 72.4.a.b | $a(p)=G[\cdots]_p-p$ | $p>3$ |
| 9. | $\left[\frac{1}{3},\frac{2}{3},\frac{1}{3},\frac{2}{3};1,1,1,1\mid 1\right]$ | $S_4(\Gamma_0(27))$ | 27.4.a.a | $a(p)=G[\cdots]_p-p$ | $p\neq 3$ |
| 10. | $\left[\frac{1}{3},\frac{2}{3},\frac{1}{4},\frac{3}{4};1,1,1,1\mid 1\right]$ | $S_4(\Gamma_0(9))$ | 9.4.a.a | $a(p)=G[\cdots]_p-(\frac{24}{p})p$ | $p>3$ |
| 11. | $\left[\frac{1}{3},\frac{2}{3},\frac{1}{6},\frac{5}{6};1,1,1,1\mid 1\right]$ | $S_4(\Gamma_0(108))$ | 108.4.a.a | $a(p)=G[\cdots]_p-(\frac{12}{p})p$ | $p>3$ |
| 12. | $\left[\frac{1}{4},\frac{3}{4},\frac{1}{4},\frac{3}{4};1,1,1,1\mid 1\right]$ | $S_4(\Gamma_0(32))$ | 32.4.a.a | $a(p)=G[\cdots]_p-p$ | $p>2$ |
| 13. | $\left[\frac{1}{4},\frac{3}{4},\frac{1}{6},\frac{5}{6};1,1,1,1\mid 1\right]$ | $S_4(\Gamma_0(144))$ | 144.4.a.f | $a(p)=G[\cdots]_{p}.-(\frac{8}{p})p$ | $p>3$ |
| 14. | $\left[\frac{1}{6},\frac{5}{6},\frac{1}{6},\frac{5}{6};1,1,1,1\mid 1\right]$ | $S_4(\Gamma_0(216))$ | 216.4.a.c | $a(p)=G[\cdots]_p-p$ | $p>3$ |
| 15. | $\left[\frac{1}{5},\frac{2}{5},\frac{3}{5},\frac{4}{5};1,1,1,1\mid 1\right]$ | $S_4(\Gamma_0(25))$ | 25.4.a.b | $a(p)=G[\cdots]_p-(\frac{5}{p})p$ | $p\neq 5$ |
| 16. | $\left[\frac{1}{8},\frac{3}{8},\frac{5}{8},\frac{7}{8};1,1,1,1\mid 1\right]$ | $S_4(\Gamma_0(128))$ | 128.4.a.b | $a(p)=G[\cdots]_p-(\frac{8}{p})p$ | $p>2$ |
| 17. | $\left[\frac{1}{10},\frac{3}{10},\frac{7}{10},\frac{9}{10};1,1,1,1\mid 1\right]$ | $S_4(\Gamma_0(200))$ | 200.4.a.f | $a(p)=G[\cdots]_p-p$ | $p\neq 2,5$ |
| 18. | $\left[\frac{1}{12},\frac{5}{12},\frac{7}{12},\frac{11}{12};1,1,1,1\mid 1\right]$ | $S_4(\Gamma_0(864))$ | 864.4.a.a | $a(p)=G[\cdots]_p-p$ | $p>3$ |

As noted above, a couple of the relationships described in Table 1 were known independently of the conjectures of Rodriguez Villegas [3, 33]. For example, case 5 was first proved by Ahlgren and Ono [3].

**Theorem 4 (Ahlgren and Ono [3], Thm. 6)** *Consider the weight four newform* $\eta^4(2z)\,\eta^4(4z) = \sum_{n\geq 1} a_5(n)q^n$ *in* $S_4\,(\Gamma_0(8))$. *If* $p$ *is an odd prime, then*

$$a_5(p) = {}_4F_3\left(\begin{matrix}\varphi,\ \varphi,\ \varphi,\ \varphi\\ \varepsilon,\ \varepsilon,\ \varepsilon\end{matrix}\;\middle|\;1\right)_p - p$$

Theorem 4 is equivalent to case 5 in Table 1 via Lemma 1.

## 4.2 Conjectures of Evans

In addition to Conjecture 1, Evans also provides three conjectural relations to weight three newforms. The following is one example.

*Conjecture 2 (Evans [11])* Consider the newform $\sum_{n\geq 1} a_6(n)q^n \in S_3\left(\Gamma_0(12), \left(\frac{-1}{\cdot}\right)\right)$, [21, 12.3.d.a], with coefficient field $\mathbb{Q}\left(\sqrt{-3}\right)$. If $q \equiv 1 \pmod 4$, then

$$-q - J\left(\overline{\chi_4}, \overline{\chi_4}\right)\cdot {}_3F_2\left(\begin{matrix}\overline{\chi_4},\ \overline{\chi_4},\ \overline{\chi_4}\\ \varepsilon,\ \chi_4\end{matrix}\;\middle|\;\frac{1}{4}\right)_q = \begin{cases} a_6(p), & \text{if } q = p,\ p \equiv 1 \pmod 4,\\ a_6(p)^2 + 2p^2, & \text{if } q = p^2,\ p \equiv 3 \pmod 4. \end{cases}$$

The other two conjectures are similar and relate to the newforms in $S_3\left(\Gamma_0(243), \left(\frac{-3}{\cdot}\right)\right)$, [21, 243.3.b.d], and $S_3\left(\Gamma_0(972), \left(\frac{\cdot}{3}\right)\right)$ [21, 972.3.c.f], both with coefficient field $\mathbb{Q}\left(\sqrt{-1}\right)$.

## 4.3 Relations with Ramanujan's τ-Function

Ramanujan's $\tau$-function, $\tau(n)$, can be defined as the coefficients of the unique normalized cusp form of weight 12 on the full modular group. i.e., $\eta(z)^{24} =: \sum_{n\geq 1} \tau(n)\, q^n$. The first result linking the $\tau$-function to finite field hypergeometric functions was given by Papanikolas.

**Theorem 5 (Papanikolas [35], Theorem 1.1)** *Let* $p$ *be an odd prime. Choose* $a, b \geq 0$ *satisfying* $p = a^2 + b^2$, *if* $p \equiv 1 \pmod 4$, *or* $a = b = 0$, *if* $p \equiv 3$

(mod 4). *Then*

$$\tau(p) = -1 - \left(1 + \tfrac{3}{2}\varphi(-1)\right) p^5 + 40p^3a^2b^2 - 128pa^4b^4$$
$$- \tfrac{1}{2}\sum_{\lambda=2}^{p-1} R\left(p, \varphi(1-\lambda)\cdot {}_3F_2\left(\begin{matrix}\varphi, \varphi, \varphi \\ \varepsilon, \varepsilon\end{matrix}\,\Big|\,\lambda\right)_p\right),$$

*where* $R(p, x) = x^5 - 4px^4 + 2p^2x^3 + 5p^3x^2 - 2p^4x - p^5$.

A similar result by Fuselier [13], involving powers of ${}_2F_1(\cdots)_p$ with characters of order 12, followed. Both of these results were established using the Eichler–Selberg trace formula, which we will discuss in Sect. 5.

## 4.4 Other Relations

In this section we mention some other noteworthy relations. Frechette, Ono and Papanikolas [12] provide the following relation between the Fourier coefficients of a weight 6 newform and a linear combination of a ${}_6F_5$ and a ${}_4F_3$.

**Theorem 6 (Frechette et al. [12], Corollary 1.2)** *Let* $\eta(z)^8\eta(4z)^4 + 8\eta(4z)^{12} = \sum_{n\geq 1} b(n)q^n$ *be the unique newform in* $S_6(\Gamma_0(8))$ *[21, 6.8.a.a]. If* $p$ *is an odd prime, then*

$$b(p) = {}_6F_5\left(\begin{matrix}\varphi, \varphi, \varphi, \varphi, \varphi, \varphi \\ \varepsilon, \varepsilon, \varepsilon, \varepsilon, \varepsilon\end{matrix}\,\Big|\,1\right)_p - p\cdot {}_4F_3\left(\begin{matrix}\varphi, \varphi, \varphi, \varphi \\ \varepsilon, \varepsilon, \varepsilon\end{matrix}\,\Big|\,1\right)_p + (1-\varphi(-1))\,p^2.$$

This is the only result, that we're aware of, involving a modular form of weight greater than four which can expressed via a simple linear relation of finite field hypergeometric functions.

In [30], Papanikolas and the second author provide evidence that the eigenvalues, of index $p$, of a certain Siegel eigenform can be evaluated by the function ${}_4F_3\left(\phi, \phi, \phi, \phi; \varepsilon, \varepsilon, \varepsilon \mid -1\right)_p$. In the course of their work they prove the following.

**Theorem 7 (McCarthy and Papanikolas [30], Theorem 1.8)** *Consider the newform* $\sum_{n\geq 1} c(n)\, q^n = q + 4iq^3 + 2q^5 - 8iq^7 + \cdots$ *in* $S_3\left(\Gamma_0(32), \left(\frac{-4}{\cdot}\right)\right)$ *[21, 32.3.c.a]. If* $p \equiv 1 \pmod 4$ *is prime, then*

$$c(p) = {}_3F_2\left(\begin{matrix}\chi_4, \varphi, \varphi \\ \varepsilon, \varepsilon\end{matrix}\,\Big|\,1\right)_p.$$

Both Theorems 6 and 7 were proved using the Eichler–Selberg trace formula, which we will discuss in Sect. 5.

In [34], Ono provides relations for the Fourier coefficients of the only four weight three newforms which can be expressed as eta-products, all of which have complex multiplication. One such relation is as follows.

**Theorem 8 (Ono [34], Corollary 11.20)** *Consider the newform* $\eta(z)^3\eta(7z)^3 = \sum_{n\geq 1} d(n)\, q^n \in S_3\left(\Gamma_0(7), \left(\frac{-7}{\cdot}\right)\right)$, *[21, 7.3.b.a]. If* $p \notin \{2, 3, 7\}$ *is prime, then*

$$d(p) = \varphi_p(-7) \cdot {}_3F_2\left(\begin{matrix} \varphi, \varphi, \varphi \\ \varepsilon, \varepsilon \end{matrix} \middle| \, 64\right)_p - \varphi_p(-7)\, p.$$

The results for the other three newforms are similar. The eta-products for these newforms are $\eta(4z)^6$, $\eta(2z)^3\eta(6z)^3$ and $\eta(z)^2\eta(2z)\eta(4z)\eta(8z)^2$ and these are the newforms in cases 1–3 of Table 1 respectively.

## 5 Trace Formulas for Hecke Operators

There have been two main ways in which relations between finite field hypergeometric functions and Fourier coefficients of modular forms have been established. The first is via the (known or independently established) modularity of some variety, as we saw in the case of elliptic curves in Sect. 3. The second is via the Eichler–Selberg trace formula for Hecke operators. These traces are connected to hypergeometric values by counting isomorphism classes of members of certain families of elliptic curves with prescribed torsion. This is a long and tedious process and works best when the dimension of the space in question is small, allowing the Fourier coefficients of specific forms to be isolated. The trace formula has also been used to establish modularity of certain varieties [2, 4], with the connection to hypergeometric functions following later, as was the case in Theorem 4.

For a positive integer $n$, let $\mathrm{Tr}_k\left(\Gamma_0(N), n\right)$ denote the trace of the $n$-th Hecke operator acting on $S_k\left(\Gamma_0(N)\right)$. A typical result relating $\mathrm{Tr}_k\left(\Gamma_0(N), n\right)$ to finite field hypergeometric functions is as follows.

**Theorem 9 (Papanikolas [35], Theorem 3.2)** *Let* $p$ *be an odd prime. Choose* $a, b \geq 0$ *satisfying* $p = a^2 + b^2$, *if* $p \equiv 1 \pmod 4$, *or* $a = b = 0$, *if* $p \equiv 3 \pmod 4$. *Define the polynomial*

$$G_k(s, p) = \sum_{j=0}^{\frac{k}{2}-1} (-1)^j \binom{k-2-j}{j} p^j s^{k-2j-2}.$$

*Let*

$$\delta_k(p) := \begin{cases} \frac{1}{2} G_k(p, 2a) + \frac{1}{2} G_k(p, 2b), & \text{if } p \equiv 1 \pmod 4, \\ (-p)^{\frac{k}{2}-1}, & \text{if } p \equiv 3 \pmod 4) \end{cases}$$

*and*

$$R_k(p,x) := \sum_{k=0}^{\frac{k}{2}-1} c_d\left(\tfrac{k}{2}-1\right) p^{\frac{k}{2}-1-d} x^d,$$

*where $c_d(r)$ is defined by the generating function $\frac{x+1}{(x^2+x+1)^{d+1}} = \sum_{j=-d}^{\infty} c_d(d+j)x^j$. For $k \geq 4$ even,*

$$\mathrm{Tr}_k\left(\Gamma_0(2), p\right) = -2 - \delta_k(p) - \sum_{\lambda=2}^{p-1} R_k\left(p, \varphi(1-\lambda)\cdot {}_3F_2\left(\begin{matrix}\varphi,\ \varphi,\ \varphi\\ \varepsilon,\ \varepsilon\end{matrix}\,\middle|\,\lambda\right)_p\right).$$

Taking $k = 12$ in Theorem 9, and using the fact that $\eta(z)^{24}$ and $\eta(2z)^{24}$ form a basis for $S_{12}(\Gamma_0(2))$, yields Theorem 5. By taking $k = 8$ and $k = 10$ in Theorem 9, Papanikolas also provides formulas, similar to that in Theorem 5, for the coefficients of the unique newforms in $S_8(\Gamma_0(2))$ and $S_{10}(\Gamma_0(2))$ respectively.

Similar evaluations of the traces of the $p$-th Hecke operators acting on the following spaces have also been produced.

- $S_k(\Gamma_0(4))$, for $k \geq 4$ even [2, 12];
- $S_k(\Gamma_0(8))$, for $k \geq 4$ even [4, 12];
- $S_k(\Gamma)$, for $k \geq 4$ even [13, 14];
- $S_k(\Gamma_0(3))$, $S_k(\Gamma_0(9))$, for $k \geq 4$ even [22]; and
- $S_3\left(\Gamma_0(16), \left(\frac{-4}{\cdot}\right)\right)$, $S_3\left(\Gamma_0(32), \left(\frac{-4}{\cdot}\right)\right)$ [30].

Lennon [22] uses the evaluation for $\mathrm{Tr}_k(\Gamma_0(9), p)$, when $k = 4$, to give another formula for the $p$-th Fourier coefficients of the newform in case 10 of Table 1.

**Theorem 10 (Lennon [22], Corollary 1.8)** *Let $\eta(3z)^8 = \sum_{n\geq 1} h(n)\, q^n \in S_4^{new}(\Gamma_0(9))$, [21, 9.4.a.a]. For $p \equiv 1 \pmod 3$,*

$$h(p) = {}_2F_1\left(\begin{matrix}\chi_3,\ \overline{\chi_3}\\ \varepsilon\end{matrix}\,\middle|\, 9\cdot 8^{-1}\right)_{p^3}.$$

## 6 New Relations

In recent work [9], we examined the number of complete subgraphs of order four contained in generalized Paley graphs. Let $k \geq 2$ be an integer. Let $q$ be a prime power such that $q \equiv 1 \pmod k$ if $q$ is even, or, $q \equiv 1 \pmod{2k}$ if $q$ is odd. The generalized Paley graph of order $q$, $G_k(q)$, is the graph with vertex set $\mathbb{F}_q$ where $ab$ is an edge if and only if $a - b$ is a $k$-th power residue. We provided a formula, in terms of ${}_3F_2$ finite field hypergeometric functions, for the number of complete subgraphs of order four contained in $G_k(q)$, which holds for all $k$. This formula includes all

$$ {}_3F_2\left(\begin{matrix} \chi_k^{t_1}, \chi_k^{t_2}, \chi_k^{t_3} \\ \chi_k^{t_4}, \chi_k^{t_5} \end{matrix} \,\middle|\, 1\right)_q, $$

as $(t_1, t_2, t_3, t_4, t_5)$ ranges over all tuples in $(\mathbb{Z}/k\mathbb{Z})^5$. We also showed that many of these terms can be simplified and many are equal to each other. We gave explicit determinations for $k \leq 4$ and noticed that many of the ${}_3F_2$'s that remained were known to be related to Fourier coefficients of weight three modular forms. We also found numerically two new possible relations. Specifically, consider the newform $g_1(z) = q + 3iq^2 - 5q^4 - 3iq^5 + 5q^7 - 3iq^8 + \cdots = \sum_{n=1}^{\infty} \beta_1(n)q^n \in S_3(\Gamma_0(27), (\frac{-3}{\cdot}))$ [21, 27.3.b.b]. Then numerical evidence suggests that, for $p \equiv 1 \pmod 6$,

$$ {}_3F_2\left(\begin{matrix} \chi_3, \chi_3, \overline{\chi_3} \\ \varepsilon, \varepsilon \end{matrix} \,\middle|\, 1\right)_p = \beta_1(p). \tag{1} $$

Also, consider the newform $g_2(z) = q + (2\zeta_8 - 2\zeta_8^3)q^3 + 4\zeta_8^2 q^5 + (8\zeta_8 + 8\zeta_8^3)q^7 - q^9 + \cdots = \sum_{n=1}^{\infty} \beta_2(n)q^n \in S_3(\Gamma_0(128), (\frac{-8}{\cdot}))$ [21, 128.3.d.c] , for a primitive eighth root of unity $\zeta_8$. Then, for $p \equiv 1 \pmod 4$, we observed

$$ {}_3F_2\left(\begin{matrix} \chi_4, \chi_4\ \overline{\chi_4} \\ \varepsilon, \varepsilon \end{matrix} \,\middle|\, 1\right)_p = \pm\beta_2(p). \tag{2} $$

Since then we have carried out a more extensive search. We examined the ${}_3F_2$'s coming from the results in [9] for all $k \leq 12$. We focused our search based on what appeared to be desirable characteristics that we observed in the small $k$ cases. The ${}_3F_2$'s can be sorted into orbits (see [9] for precise details) and all the new relations we found were where the ${}_3F_2$ and its conjugate were in the same orbit. All the new conjectural relations we have found are summarized in Table 2. To simplify the table we have listed the parameters of the ${}_3F_2$'s using rational numbers according to the convention that the fraction $\frac{t}{k}$ represents the character $\chi_k^t$. Interestingly, our search yielded the relations in cases 1–4 of Table 1 and the relation in Theorem 7, but as they are already known, we have not included them in Table 2. However, for completeness, we have included the relations from (1) and (2). They appear as cases 4 and 8 respectively. Similar to the relation in (2), many of the new relations involve a sign which doesn't seem to be resolvable by a simple twist by a Dirichlet character. So we first define some functions to explain these signs.

For $p \equiv 1 \pmod 4$, write $p = x^2 + y^2$ for integers $x$ and $y$, such that $x$ is odd and $y$ is even. For $p \equiv 1 \pmod{12}$, note that either $3 \mid x$ or $3 \mid y$ and define

$$ S_x(p) = \begin{cases} +1 & \text{if } 3 \mid y; \\ -1 & \text{if } 3 \mid x. \end{cases} $$

Note that $S_x(p)$ equals $c_{12}^2$, where $c_{12}$ is the quantity described in [8, Ch. 3.5]. If $p \equiv 1 \pmod{20}$, then either $5 \mid x$ or $5 \mid y$ and so we define

$$S_{20}(p) = \begin{cases} +1 & \text{if } 5 \mid y, \\ -1 & \text{if } 5 \mid x. \end{cases}$$

Now, for $p \equiv 1 \pmod 6$, define

$$S_6(p) = \begin{cases} S_x(p) & \text{if } p \equiv 1 \pmod{12}, \\ \pm 1 & \text{if } p \equiv 7 \pmod{12}. \end{cases}$$

For $p \equiv 1 \pmod 8$, write $p = u^2 + 2v^2$ for integers $u$ and $v$, such that $u \equiv 3 \pmod 4$ and $v$ is even. For $p \equiv 1 \pmod 4$, define

$$S_4(p) = \begin{cases} +1 & \text{if } p \equiv 1 \pmod 8 \text{ and } v \equiv 0 \pmod 4, \text{ or, } p \equiv 13 \pmod{16}, \\ -1 & \text{if } p \equiv 1 \pmod 8 \text{ and } v \equiv 2 \pmod 4, \text{ or, } p \equiv 5 \pmod{16}. \end{cases}$$

When $p \equiv 1 \pmod{12}$, define

$$S_u(p) = \begin{cases} +1 & \text{if } u \equiv 2 \pmod 3, \\ -1 & \text{if } u \equiv 1 \pmod 3, \end{cases} \quad \text{and,} \quad S_{12}(p) = \begin{cases} S_u(p) & \text{if } p \equiv 1 \pmod{24}, \\ \pm 1 & \text{if } p \equiv 13 \pmod{24}. \end{cases}$$

For $p \equiv 1 \pmod{10}$, write $p = a^2 + 5b^2 + 5c^2 + 5d^2$ for integers $a, b, c, d$ such that $a \equiv 4 \pmod 5$ and $ab = d^2 - c^2 - cd$. We note that $a$ is unique up to sign [8, Thm. 3.7.2]. Define

$$S_{10}(p) = \begin{cases} +1 & \text{if } 4 \nmid a; \\ -1 & \text{if } 4 \mid a. \end{cases}$$

$S_{10}(p)$ relates to case 14 in Table 2. The modular form in that case has CM by $\mathbb{Q}\left(\sqrt{-5}\right)$ and so its Fourier coefficients $a(p)$ vanish when $p \equiv 11 \pmod{20}$. Thus we only need the sign at $p \equiv 1 \pmod{20}$. When $p \equiv 1 \pmod{20}$ it appears $S_{10}(p) = S_{20}(p)$.

As we have seen, the functions $S_6$ and $S_{12}$, which affect cases 1, 2, 7, 9, 10 and 12 in Table 2, are not fully described when $p \equiv 7 \pmod{12}$ and $p \equiv 13 \pmod{24}$ respectively. Unfortunately, we were unable to ascribe a simple formula to the sign in those classes for those cases. Also, the choice of character is important in those cases. This is best explained using case 1 as an example. Combining [16, (4.25)]

**Table 2** New conjectural relations

| | Hyp Series | Newform $f(z)=\sum a(n)q^n$ | | Connection | |
|---|---|---|---|---|---|
| | Parameters | Space | LMFDB | Relationship | Conditions |
| 1. | $\left[\frac{1}{3}, \frac{1}{2}, \frac{1}{2}; 1, 1, 1 \mid 1\right]$ | $S_3(\Gamma_0(48), (\frac{-4}{\cdot}))$ | 48.3.g.a | $a(p) = S_6(p) \cdot F(\cdots)_p$ | $p \equiv 1 \pmod 6$ |
| 2. | $\left[\frac{1}{6}, \frac{1}{2}, \frac{1}{2}; 1, 1, 1 \mid 1\right]$ | $S_3(\Gamma_0(12), (\frac{-4}{\cdot}))$ | 12.3.d.a | $a(p) = S_6(p) \cdot F(\cdots)_p$ | $p \equiv 1 \pmod 6$ |
| 3. | $\left[\frac{1}{8}, \frac{1}{2}, \frac{1}{2}; 1, 1, 1 \mid 1\right]$ | $S_3(\Gamma_0(64), (\frac{-8}{\cdot}))$ | 64.3.d.a | $a(p) = F(\cdots)_p$ | $p \equiv 1 \pmod 8$ |
| 4. | $\left[\frac{1}{3}, \frac{1}{3}, \frac{2}{3}; 1, 1, 1 \mid 1\right]$ | $S_3(\Gamma_0(27), (\frac{-3}{\cdot}))$ | 27.3.b.b | $a(p) = F(\cdots)_p$ | $p \equiv 1 \pmod 6$ |
| 5. | $\left[\frac{1}{4}, \frac{1}{3}, \frac{2}{3}; 1, 1, 1 \mid 1\right]$ | $S_3(\Gamma_0(36), (\frac{-4}{\cdot}))$ | 36.3.d.a | $a(p) = F(\cdots)_p$ | $p \equiv 1 \pmod{12}$ |
| 6. | $\left[\frac{1}{6}, \frac{1}{3}, \frac{2}{3}; 1, 1, 1 \mid 1\right]$ | $S_3(\Gamma_0(108), (\frac{-3}{\cdot}))$ | 108.3.c.b | $a(p) = F(\cdots)_p$ | $p \equiv 1 \pmod 6$ |
| 7. | $\left[\frac{1}{3}, \frac{1}{4}, \frac{3}{4}; 1, 1, 1 \mid 1\right]$ | $S_3(\Gamma_0(576), (\frac{-24}{\cdot}))$ | 576.3.h.b | $a(p) = S_{12}(p) \cdot F(\cdots)_p$ | $p \equiv 1 \pmod{12}$ |
| 8. | $\left[\frac{1}{4}, \frac{1}{4}, \frac{3}{4}; 1, 1, 1 \mid 1\right]$ | $S_3(\Gamma_0(128), (\frac{-8}{\cdot}))$ | 128.3.d.c | $a(p) = S_4(p) \cdot F(\cdots)_p$ | $p \equiv 1 \pmod 4$ |
| 9. | $\left[\frac{1}{6}, \frac{1}{4}, \frac{3}{4}; 1, 1, 1 \mid 1\right]$ | $S_3(\Gamma_0(576), (\frac{-24}{\cdot}))$ | 576.3.h.a | $a(p) = S_{12}(p) \cdot F(\cdots)_p$ | $p \equiv 1 \pmod{12}$ |
| 10. | $\left[\frac{1}{3}, \frac{1}{6}, \frac{5}{6}; 1, 1, 1 \mid 1\right]$ | $S_3(\Gamma_0(432), (\frac{-4}{\cdot}))$ | 432.3.g.a | $a(p) = S_6(p) \cdot F(\cdots)_p$ | $p \equiv 1 \pmod 6$ |
| 11. | $\left[\frac{1}{4}, \frac{1}{6}, \frac{5}{6}; 1, 1, 1 \mid 1\right]$ | $S_3(\Gamma_0(288), (\frac{-4}{\cdot}))$ | 288.3.g.a | $a(p) = F(\cdots)_p$ | $p \equiv 1 \pmod{12}$ |
| 12. | $\left[\frac{1}{6}, \frac{1}{6}, \frac{5}{6}; 1, 1, 1 \mid 1\right]$ | $S_3(\Gamma_0(108), (\frac{-4}{\cdot}))$ | 108.3.d.a | $a(p) = S_6(p) \cdot F(\cdots)_p$ | $p \equiv 1 \pmod 6$ |
| 13. | $\left[\frac{1}{5}, \frac{1}{5}, \frac{4}{5}; 1, 1, 1 \mid 1\right]$ | $S_3(\Gamma_0(25), \chi)$ | 25.3.c.a | $a(p) = F(\cdots)_p$ | $p \equiv 1 \pmod 5$ |
| 14. | $\left[\frac{1}{2}, \frac{1}{10}, \frac{9}{10}; 1, 1, 1 \mid 1\right]$ | $S_3(\Gamma_0(20), (\frac{-20}{\cdot}))$ | 20.3.d.a | $a(p) = S_{10}(p) \cdot F(\cdots)_p$ | $p \equiv 1 \pmod{10}$ |
| 15. | $\left[\frac{1}{2}, \frac{1}{12}, \frac{11}{12}; 1, 1, 1 \mid 1\right]$ | $S_3(\Gamma_0(24), (\frac{-24}{\cdot}))$ | 24.3.h.a | $a(p) = F(\cdots)_p$ | $p \equiv 1 \pmod{12}$ |

*Note:* $\chi$ on row 13 is the Dirichlet character of conductor 5, with $2 \mapsto i$

with [27, Prop. 2.5] we see that, for $p \equiv 1 \pmod 6$,

$$
{}_3F_2\left(\begin{matrix} \chi_3, & \varphi, & \varphi \\ & \varepsilon, & \varepsilon \end{matrix} \middle| \, 1\right)_p = {}_3F_2\left(\begin{matrix} \overline{\chi_3}, & \varphi, & \varphi \\ & \varepsilon, & \varepsilon \end{matrix} \middle| \, 1\right)_p \times \begin{cases} +1 & \text{if } p \equiv 1 \pmod 4, \\ -1 & \text{if } p \equiv 3 \pmod 4. \end{cases}
$$

There are two characters of order three when $p \equiv 1 \pmod 6$ and they are conjugates of each other. So, when $p \equiv 1 \pmod{12}$, the ${}_3F_2$ in case 1 is independent of the choice of $\chi_3$. However, when $p \equiv 7 \pmod{12}$, the choice of character will determine the sign. Similar behavior is observed in cases 2, 7, 9, 10 and 12.

It doesn't appear that the relations in Table 2 can be extended to all primes in a simple way using ${}_mG_m[\cdots]_p$. It may be possible, however, with the introduction of extra factors which equal $\pm 1$ when $p$ is in the equivalence class outlined in the table.

**Acknowledgments** The first author is supported by an AMS-Simons travel grant from the American Mathematical Society and the Simons Foundation. The second author is supported by a grant from the Simons Foundation (#353329, Dermot McCarthy).

## References


1. S. Ahlgren, Gaussian hypergeometric series and combinatorial congruences, in *Symbolic Computation, Number Theory, Special Functions, Physics and Combinatorics* (Gainesville, FL, 1999), 1–12, Dev. Math., vol. 4 (Kluwer, Dordrecht, 2001)
2. S. Ahlgren, The points of a certain fivefold over finite fields and the twelfth power of the eta function. Finite Fields Appl. **8**(1), 18–33 (2002)
3. S. Ahlgren, K. Ono, A Gaussian hypergeometric series evaluation and Apéry number congruences. J. Reine Angew. Math. **518**, 187–212 (2000)
4. S. Ahlgren, K. Ono, Modularity of a certain Calabi–Yau threefold. Monatsh. Math. **129**(3), 177–190 (2000)
5. R. Barman, G. Kalita, Hypergeometric functions over $\mathbb{F}_q$ and traces of Frobenius for elliptic curves. Proc. Am. Math. Soc. **141**(10), 3403–3410 (2013)
6. R. Barman, G. Kalita, Elliptic curves and special values of Gaussian hypergeometric series. J. Number Theory **133**(9), 3099–3111 (2013)
7. V. Batyrev, D. van Straten, Generalized hypergeometric functions and rational curves on Calabi–Yau complete intersections in toric varieties. Commun. Math. Phys. **168**(3), 493–533 (1985)
8. B. Berndt, R. Evans, K. Williams, Gauss and Jacobi sums. Canadian Mathematical Society Series of Monographs and Advanced Texts (A Wiley-Interscience Publication, John Wiley & Sons, New York, 1998)
9. M.L. Dawsey, D. McCarthy, Generalized Paley graphs and their complete subgraphs of orders three and four. Res. Math. Sci. **8**, 18 (2021)
10. A. El-Guindy, K. Ono, Hasse invariants for the Clausen elliptic curves. Ramanujan J. **31**, 3–13 (2013)
11. R. Evans, Hypergeometric ${}_3F_2(1/4)$ evaluations over finite fields and Hecke eigenforms. Proc. Am. Math. Soc. **138**(2), 517–531 (2010)
12. S. Frechette, K. Ono, M. Papanikolas, Gaussian hypergeometric functions and traces of Hecke operators. Int. Math. Res. Not. **60**, 3233–3262 (2004)
13. J. Fuselier, Hypergeometric functions over $\mathbb{F}_p$ and relations to elliptic curves and modular forms. Proc. Am. Math. Soc. **138**(1), 109–123 (2010)
14. J. Fuselier, Traces of Hecke operators in level 1 and Gaussian hypergeometric functions. Proc. Am. Math. Soc. **141**(6), 1871–1881 (2013)
15. J.G. Fuselier, D. McCarthy, Hypergeometric type identities in the $p$-adic setting and modular forms. Proc. Am. Math. Soc. **144**, 1493–1508 (2016)
16. J. Greene, Hypergeometric series over finite fields. Trans. Am. Math. Soc. **301**, 77–101 (1987)
17. T. Ishikawa, Super congruence for the Apéry numbers. Nagoya Math. J. **118**, 195–202 (1990)
18. P. Kewat, R. Kumar, Hypergeometric functions and algebraic curves $y^e = x^d + ax + b$. J. Ramanujan Math. Soc. **34**(3), 325–342 (2019)
19. T. Kilbourn, An extension of the Apéry number supercongruence. Acta Arith. **123**(4), 335–348 (2006)
20. M. Koike, Orthogonal matrices obtained from hypergeometric series over finite fields and elliptic curves over finite fields. Hiroshima Math. J. **25**, 43–52 (1995)
21. LMFDB—The L-functions and Modular Forms Database. www.lmfdb.org
22. C. Lennon, Trace formulas for Hecke operators, Gaussian hypergeometric functions, and the modularity of a threefold. J. Number Theory **131**(12), 2320–2351 (2011)
23. C. Lennon, Gaussian hypergeometric evaluations of traces of Frobenius for elliptic curves. Proc. Am. Math. Soc. **139**(6), 1931–1938 (2011)
24. L. Long, F. Tu, N. Yui, W. Zudilin, Supercongruences for rigid hypergeometric Calabi–Yau threefolds. arXiv:1705.01663
25. D. McCarthy, Extending Gaussian hypergeometric series to the $p$-adic setting. Int. J. Number Theory **8**(7), 1581–1612 (2012)

26. D. McCarthy, On a supercongruence conjecture of Rodriguez-Villegas. Proc. Am. Math. Soc. **140**(7), 2241–2254 (2012)
27. D. McCarthy, Transformations of well-poised hypergeometric functions over finite fields. Finite Fields Appl. **18**(6), 1133–1147 (2012)
28. D. McCarthy, The trace of Frobenius of elliptic curves and the $p$-adic gamma function. Pac. J. Math. **261**(1), 219–236 (2013)
29. D. McCarthy, The number of $\mathbb{F}_p$-points on Dwork hypersurfaces and hypergeometric functions. Res. Math. Sci. **4**, 4 (2017)
30. D. McCarthy, M. Papanikolas, A finite field hypergeometric function associated to eigenvalues of a Siegel eigenform. Int. J. Number Theory **11**(8), 2431–2450 (2015)
31. C. Meyer, Modular Calabi–Yau Threefolds. Fields Institute Monographs, vol. 22 (American Mathematical Society, Providence, 2005)
32. E. Mortenson, Supercongruences for truncated ${}_{n+1}F_n$ hypergeometric series with applications to certain weight three newforms. Proc. Am. Math. Soc. **133**(2), 321–330 (2005)
33. K. Ono, Values of Gaussian hypergeometric series. Trans. Am. Math. Soc. **350**(3), 1205–1223 (1998)
34. K. Ono, The web of modularity: Arithmetic of the coefficients of modular forms and $q$-series, in *CBMS Regional Conference Series in Mathematics*, vol. 102. Published for the Conference Board of the Mathematical Sciences, Washington, DC (American Mathematical Society, Providence, 2004)
35. M. Papanikolas, A formula and a congruence for Ramanujan's $\tau$-function. Proc. Am. Math. Soc. **134**(2), 333–341 (2005)
36. F. Rodriguez Villegas, Hypergeometric families of Calabi–Yau manifolds, in *Calabi–Yau Varieties and Mirror Symmetry (Toronto, Ontario, 2001), Fields Inst. Commun.*, vol. 38 (Amer. Math. Soc., Providence, 2003), pp. 223–231
37. J. Stienstra, F. Beukers, On the Picard-Fuchs equation and the formal Brauer group of certain elliptic $K3$ surfaces. Math. Ann. **271**, 269–304 (1985)
38. L. Van Hamme, Proof of a conjecture of Beukers on Apéry numbers, in *Proceedings of the Conference on p-Adic Analysis* (Houthalen, 1987) (Vrije Univ. Brussel, Brussels, 1986), pp. 189–195

# Ballistic Transport for Periodic Jacobi Operators on $\mathbb{Z}^d$

**Jake Fillman**

**Abstract** In this expository work, we collect some background results and give a short proof of the following theorem: periodic Jacobi matrices on $\mathbb{Z}^d$ exhibit strong ballistic motion.

**Keywords** Jacobi operators · Ballistic motion · Quantum dynamics

## 1 Introduction

This expository note is concerned with the properties of Jacobi operators on $\mathbb{Z}^d$. Concretely, we fix the *dimension* $d \in \mathbb{N}$ and consider linear operators $J = J_{a,b} : \ell^2(\mathbb{Z}^d) \to \ell^2(\mathbb{Z}^d)$ given by

$$[Ju]_{\boldsymbol{x}} = \sum_{\substack{\boldsymbol{y} \in \mathbb{Z}^d \\ \|\boldsymbol{x}-\boldsymbol{y}\|_1 = 1}} a_{\boldsymbol{x},\boldsymbol{y}} u_{\boldsymbol{y}} + b_{\boldsymbol{x}} u_{\boldsymbol{x}}, \quad u \in \ell^2(\mathbb{Z}^d),\ \boldsymbol{x} \in \mathbb{Z}^d, \tag{1}$$

where $a$ and $b$ are bounded, $b$ is real-valued, and $a_{\boldsymbol{y},\boldsymbol{x}} = a^*_{\boldsymbol{x},\boldsymbol{y}}$ for all $\boldsymbol{x}$ and $\boldsymbol{y}$ in $\mathbb{Z}^d$ for which $\|\boldsymbol{x} - \boldsymbol{y}\|_1 = 1$. For convenience, we will write $\boldsymbol{x} \sim \boldsymbol{y}$ for $\boldsymbol{x}, \boldsymbol{y} \in \mathbb{Z}^d$ to mean $\|\boldsymbol{x} - \boldsymbol{y}\|_1 = 1$. We will also insist that

$$a_{\boldsymbol{x},\boldsymbol{y}} \neq 0 \text{ for all } \boldsymbol{x}, \boldsymbol{y} \text{ such that } \boldsymbol{x} \sim \boldsymbol{y}. \tag{2}$$

J. Fillman (✉)
Texas State University, San Marcos, TX, USA
e-mail: fillman@txstate.edu



F. Gesztesy, A. Martinez-Finkelshtein (eds.), *From Operator Theory to Orthogonal Polynomials, Combinatorics, and Number Theory*, Operator Theory: Advances and Applications 285, https://doi.org/10.1007/978-3-030-75425-9_5

In the case $d = 1$, we simply write $a_x := a_{x,x+1}$ and one obtains the familiar *Jacobi matrix* on $\ell^2(\mathbb{Z})$:

$$J = \begin{bmatrix} \ddots & \ddots & \ddots & & & \\ & a_0^* & b_1 & a_1 & & \\ & & a_1^* & b_2 & a_2 & \\ & & & \ddots & \ddots & \ddots \end{bmatrix}.$$

If one takes $a \equiv 1$, the operator (1) becomes a *discrete Schrödinger operator*. Jacobi matrices have inspired intense study over the years owing to their close connections with orthogonal polynomials, integrable systems, and mathematical physics; see, e.g., [1, 3, 4, 6, 14, 15, 17, 19] and references therein.

We will be interested in the case in which $J$ is *periodic*, i.e., there exists a full-rank subgroup $\mathbb{L} \subseteq \mathbb{Z}^d$ such that $U^{\boldsymbol{\ell}} J U^{-\boldsymbol{\ell}} = J$ for all $\boldsymbol{\ell} \in \mathbb{L}$, where $U^{\boldsymbol{\ell}}$ denotes the shift $\delta_{\boldsymbol{x}} \mapsto \delta_{\boldsymbol{x}+\boldsymbol{\ell}}$. Equivalently,

$$a_{\boldsymbol{x}+\boldsymbol{\ell},\boldsymbol{y}+\boldsymbol{\ell}} = a_{\boldsymbol{x},\boldsymbol{y}} \text{ and } b_{\boldsymbol{x}+\boldsymbol{\ell}} = b_{\boldsymbol{x}} \text{ for all } \boldsymbol{x} \sim \boldsymbol{y} \in \mathbb{Z}^d \text{ and all } \boldsymbol{\ell} \in \mathbb{L}. \tag{3}$$

We will prove what we need and will not prove what we do not need. For a fuller exposition of periodic operators, we recommend Kuchment's beautiful survey [13].

Of course, if $\{\boldsymbol{\ell}_1, \ldots, \boldsymbol{\ell}_d\}$ are linearly independent elements of $\mathbb{Z}^d$ generating a lattice $\mathbb{L}$ for which (3) holds, then it is a straightforward calculation to show that (3) also holds for the lattice $\mathbb{L}' = r\mathbb{Z}^d = \{r\boldsymbol{n} : \boldsymbol{n} \in \mathbb{Z}^d\}$, where $r = |\det(\boldsymbol{\ell}_1|\cdots|\boldsymbol{\ell}_d)|$ is the volume of $\mathbb{R}^d/\mathbb{L}$; consequently, no real generality is lost in considering lattices generated by multiples of the standard basis vectors, so we shall consider precisely this scenario in the present note. To that end, given $\boldsymbol{q} \in \mathbb{N}^d$, say $J$ is $\boldsymbol{q}$-periodic if

$$U^{\boldsymbol{mq}} J U^{-\boldsymbol{mq}} = J \text{ for all } \boldsymbol{m} \in \mathbb{Z}^d,$$

where $\boldsymbol{mq} = (m_1q_1, m_2q_2, \ldots, m_dq_d)$. Equivalently, $J$ is $\boldsymbol{q}$ periodic if (3) holds for the lattice $\mathbb{L} = \bigoplus_{j=1}^d q_j\mathbb{Z}$.

The goal of the note is to discuss quantum dynamics associated with such periodic operators. In particular, we focus on the growth of the position observables. For a linear operator $O$, we denote by $O(t) = e^{itJ}Oe^{-itJ}$ the corresponding time evolution with respect to $J$. For $1 \le j \le d$, the $j$th position operator $X_j$ is given by $X_j\delta_{\boldsymbol{x}} = x_j\delta_{\boldsymbol{x}}$, where

$$D(X_j) = \left\{\psi \in \ell^2(\mathbb{Z}^d) : X_j\psi \in \ell^2(\mathbb{Z}^d)\right\}.$$

The *vector* position operator $\boldsymbol{X} : D(\boldsymbol{X}) = \bigcap_{j=1}^d D(X_j) \to \ell^2(\mathbb{Z}^d) \otimes \mathbb{C}^d$ given by

$$\boldsymbol{X}\psi = (X_1\psi, \ldots, X_d\psi).$$

The primary phenomenon that we will discuss is that of *ballistic motion*, i.e., linear growth of the position observable(s). This was established for continuum Schrödinger operators by Asch–Knauf [2] and was later extended to Jacobi matrices in $d = 1$ by Damanik–Lukić–Yessen [5]. The result we want to discuss is the generalization of [5] to the case of general $d \geq 1$.

**Theorem 1.1** *If $J$ is periodic, then it exhibits ballistic motion in the following sense. There are bounded, self-adjoint operators $Q_k$, $1 \leq k \leq d$, such that*

$$\lim_{t\to\infty} \frac{X_k(t)}{t} = Q_k$$

*in the strong sense, and $\bigcap_{k=1}^{d} \ker(Q_k) = \{0\}$. In particular,*

$$Q := \lim_{t\to\infty} \frac{X(t)}{t}$$

*in the strong sense and $\ker(Q) = \{0\}$.*

Naturally, since Asch–Knauf already worked in higher dimension, the result is not surprising and could indeed be considered a folklore result, since it is simply a convex combination of [2] and [5]. Nevertheless, we felt it would be worthwhile to have an essentially self-contained exposition of the proof somewhere in the literature.

In recent years, there has also been substantial interest in studying the phenomenon of ballistic motion in specific aperiodic models. For instance, this has been established for limit-periodic and quasi-periodic models [7–11, 20].

In Sect. 2 we discuss a direct integral decomposition of $J$, and then explain in Sect. 3 how to use this to prove Theorem 1.1. Since the paper is expository in nature, we aim to supply proofs so that the article is self-contained, modulo background facts from functional analysis and analytic perturbation theory.

Let me end the introduction on a personal note. While I was an undergraduate at Baylor in the late '00s, I got to see first-hand the positive impact of Lance Littlejohn's leadership of the department. I have nothing but fond memories of my time in the Baylor mathematics department, and that starts with the collegial culture that Lance worked hard to cultivate while also working to enhance the deparment's academic reputation. Congratulations Lance on all you have achieved, and many happy returns!

## 2 Decomposition of $J$

We first explain how to decompose $J$ as a direct integral of operators on the fundamental domain with suitable self-adjoint boundary conditions. The reader is referred to [16] for additional background about direct integrals. More precisely, let

$\mathbb{T}^d = \mathbb{R}^d/\mathbb{Z}^d$, put

$$\Gamma = \mathbb{Z}^d \cap \prod_{j=1}^{d} [0, q_j) = \{0, 1, \ldots, q_1 - 1\} \times \cdots \times \{0, 1, \ldots, q_d - 1\}, \tag{4}$$

and consider $\mathscr{H}(\boldsymbol{\theta}) = \mathscr{H}(\boldsymbol{\theta}, \boldsymbol{q}) \subset \ell^\infty(\mathbb{Z}^d)$ comprising all those $\psi : \mathbb{Z}^d \to \mathbb{C}$ such that

$$\psi_{\boldsymbol{x}+\boldsymbol{n}\boldsymbol{q}} = e^{2\pi i \langle \boldsymbol{\theta}, \boldsymbol{n} \rangle} \psi_{\boldsymbol{x}}. \tag{5}$$

With the inner product

$$\langle \psi, \varphi \rangle_{\mathscr{H}(\boldsymbol{\theta})} = \sum_{\boldsymbol{x} \in \Gamma} \overline{\psi_{\boldsymbol{x}}} \varphi_{\boldsymbol{x}},$$

$\mathscr{H}(\boldsymbol{\theta})$ becomes a Hilbert space of dimension $\bar{q} = \#\Gamma = \prod_{j=1}^d q_j$. We will use $d\boldsymbol{\theta}$ to denote the Lebesgue measure on $\mathbb{T}^d$.

**Lemma 2.1** *If $J$ is $\boldsymbol{q}$-periodic, then $J$ maps $\mathscr{H}(\boldsymbol{\theta}, \boldsymbol{q})$ into itself for all $\boldsymbol{\theta}$.*

***Proof*** This is a short calculation. If $\psi \in \mathscr{H}(\boldsymbol{\theta})$, then

$$\begin{aligned} [J\psi]_{x+nq} &= \sum_{x \sim y} a_{x+nq, y+nq} \psi_{y+nq} + b_{x+nq} \psi_{x+nq} \\ &= e^{2\pi i \langle \boldsymbol{\theta}, \boldsymbol{n} \rangle} \sum_{x \sim y} a_{x,y} \psi_y + e^{2\pi i \langle \boldsymbol{\theta}, \boldsymbol{n} \rangle} b_x \psi_x \\ &= e^{2\pi i \langle \boldsymbol{\theta}, \boldsymbol{n} \rangle} [J\psi]_x, \end{aligned}$$

whence $J\psi \in \mathscr{H}(\boldsymbol{\theta})$. □

In view of the lemma, we may define $J(\boldsymbol{\theta}) = J|_{\mathscr{H}(\boldsymbol{\theta})}$ for each $\boldsymbol{\theta} \in \mathbb{T}^d$. Writing $\mathbb{C}^\Gamma$ for the space of functions $\Gamma \to \mathbb{C}$, one can view $\mathscr{H}(\boldsymbol{\theta}) \cong \mathbb{C}^\Gamma$ via the identification

$$\mathbb{C}^\Gamma \ni \delta_{\boldsymbol{x}} \mapsto \sum_{\boldsymbol{n} \in \mathbb{Z}^d} e^{2\pi i \langle \boldsymbol{\theta}, \boldsymbol{n} \rangle} \delta_{\boldsymbol{x}+\boldsymbol{n}\boldsymbol{q}} \in \mathscr{H}(\boldsymbol{\theta}), \tag{6}$$

so we also freely consider $J(\boldsymbol{\theta})$ as a linear operator on $\mathbb{C}^\Gamma$.

To describe the decomposition of $J$, define

$$\mathscr{H}_1 = \int_{\mathbb{T}^d}^{\oplus} \mathscr{H}(\boldsymbol{\theta})\, d\boldsymbol{\theta},$$

which consists of measurable functions $f$ mapping $\mathbb{T}^d$ into $\bigcup_{\theta \in \mathbb{T}^d} \mathscr{H}(\theta)$ such that $f(\theta) \in \mathscr{H}(\theta)$ for all $\theta$ and

$$\|f\|_{\mathscr{H}_1}^2 := \int_{\mathbb{T}^d} \|f(\theta)\|_{\mathscr{H}(\theta)}^2 \, d\theta < \infty.$$

Equipped with the inner product

$$\langle f, g \rangle_{\mathscr{H}_1} = \int_{\mathbb{T}^d} \langle f(\theta), g(\theta) \rangle_{\mathscr{H}(\theta)} \, d\theta,$$

$\mathscr{H}_1$ is a Hilbert space; see [16] for details. Write $f(\theta, x)$ for the $x$th coordinate of $f(\theta)$. Identifying the fibers of $\mathscr{H}_1$ with $\mathbb{C}^\Gamma$ as in (6), we can also view $\mathscr{H}_1$ simply as the collection of square-integrable maps $\mathbb{T}^d \to \mathbb{C}^\Gamma$, which we shall do freely when it is convenient to do so.

For $\psi \in \ell^1(\mathbb{Z}^d)$, define

$$[\mathscr{F}\psi](\theta, x) = \sum_{m \in \mathbb{Z}^d} \psi_{x+mq} e^{2\pi i \langle \theta, m \rangle}.$$

**Lemma 2.2** *For every $\psi \in \ell^1(\mathbb{Z}^d)$, $\mathscr{F}\psi \in \mathscr{H}_1$, $\|\mathscr{F}\psi\|_{\mathscr{H}_1} = \|\psi\|_{\ell^2(\mathbb{Z}^d)}$, and the image of $\ell^1(\mathbb{Z}^d)$ is dense in $\mathscr{H}_1$. In particular, $\mathscr{F}$ extends to a unitary operator $\mathscr{F} : \ell^2(\mathbb{Z}^d) \to \mathscr{H}_1$.*

***Proof*** For $x \in \Gamma$ and $n \in \mathbb{Z}^d$, denote $\mathscr{F}\delta_{x+nq} = \varphi_{x,n}$ and note that

$$\varphi_{x,n}(\theta) = \sum_{m \in \mathbb{Z}^d} e^{2\pi i \langle \theta, n-m \rangle} \delta_{x+mq}. \tag{7}$$

Since $\left\{ e^{2\pi i \langle \cdot, n \rangle} : n \in \mathbb{Z}^d \right\}$ is an orthonormal basis of $L^2(\mathbb{T}^d)$, one can check that $\left\{ \varphi_{x,n} : x \in \Gamma, \, n \in \mathbb{Z}^d \right\}$ is an orthonormal basis of $\mathscr{H}_1$, so the lemma follows immediately. □

The unitary operator $\mathscr{F}$ "diagonalizes" $J$ in the sense that it transforms $J$ to a (matrix) multiplication operator given by pointwise multiplication by $J(\theta)$ on $\mathscr{H}_1$. Concretely, define a linear operator $\widehat{J} : \mathscr{H}_1 \to \mathscr{H}_1$ by

$$[\widehat{J}g](\theta) = J(\theta)g(\theta). \tag{8}$$

It is convenient to use the direct integral notation for operators enjoying a decomposition as in (8); for instance, we will write

$$\widehat{J} = \int_{\mathbb{T}^d}^{\oplus} J(\theta) \, d\theta.$$

**Theorem 2.3** $\widehat{J} = \mathscr{F} J \mathscr{F}^*$.

***Proof*** This follows from a direct calculation. Recall $\varphi_{x,n} = \mathscr{F}\delta_{x+nq}$ for $x \in \Gamma$ and $n \in \mathbb{Z}^d$. Since $\varphi_{x,n}(\theta) \in \mathscr{H}(\theta)$ for each $\theta \in \mathbb{T}^d$, (7) yields

$$\begin{aligned}
J(\theta)\varphi_{x,n}(\theta) &= J\varphi_{x,n}(\theta) \\
&= \sum_{m\in\mathbb{Z}^d} e^{2\pi i\langle\theta,n-m\rangle} J\delta_{x+mq} \\
&= \sum_{m\in\mathbb{Z}^d} e^{2\pi i\langle\theta,n-m\rangle} \left( \sum_{z\sim x+mq} a_{x+mq,z}\delta_z + b_{x+mq}\delta_{x+mq} \right) \\
&= \sum_{m\in\mathbb{Z}^d} e^{2\pi i\langle\theta,n-m\rangle} \left( \sum_{y\sim x} a_{x,y}\delta_{y+mq} + b_x\delta_{x+mq} \right) \\
&= \sum_{y\sim x} a_{x,y}\varphi_{y,n}(\theta) + b_x\varphi_{x,n}(\theta).
\end{aligned}$$

On the other hand, a direct calculation from the definitions yields

$$\begin{aligned}
[\mathscr{F} J \mathscr{F}^* \varphi_{x,n}](\theta) &= [\mathscr{F} J\delta_{x+nq}](\theta) \\
&= \left[ \mathscr{F} \left( \sum_{y\sim x} a_{x,y}\delta_{y+nq} + b_x\delta_{x+nq} \right) \right](\theta) \\
&= \sum_{y\sim x} a_{x,y}\varphi_{y,n}(\theta) + b_x\varphi_{x,n}(\theta).
\end{aligned}$$

Thus $\widehat{J}\varphi_{x,n} = \mathscr{F} J \mathscr{F}^* \varphi_{x,n}$ for all $x$ and $n$. Since $\{\varphi_{x,n} : x \in \Gamma,\ n \in \mathbb{Z}^d\}$ is a basis of $\mathscr{H}_1$, the theorem is proved. □

**Corollary 2.4** *Let* $\bar{q} = \prod_{j=1}^d q_j$, *let* $E_1(\theta) \le \cdots \le E_{\bar{q}}(\theta)$ *denote the eigenvalues of* $J(\theta)$, *and define*

$$I_k = \{E_k(\theta) : \theta \in \mathbb{T}^d\}.$$

*Then*

$$\sigma(H) = \bigcup_{k=1}^{\bar{q}} I_k = \bigcup_{\theta\in\mathbb{T}^d} \sigma(J(\theta)).$$

***Proof*** Since $E_k(\theta)$ depends continuously on $\theta$ for every $1 \le k \le \bar{q}$, this is an immediate consequence of Theorem 2.3. □

We end the section by collecting a few spectral properties of $J$. We first show that $J$ has no square-summable eigenvectors. This is likely well-known, but I am unaware of a precise reference in the current setting.

**Lemma 2.5** *Suppose $J$ is periodic. Then $J$ has no eigenvalues.*

***Proof*** Suppose on the contrary that $J\psi = \lambda\psi$ for some $\lambda \in \mathbb{R}$ and $\psi \in \ell^2(\mathbb{Z}^d)$ with $\|\psi\| = 1$. The short version of the argument is this: the presence of a nontrivial eigenvector coupled with translation-invariance leads to a discontinuity of the density of states (DOS), but, since the volume of the boundary of a box grows more slowly than the volume of the box, the DOS is necessarily continuous, a contradiction.

Since we have not defined the density of states, let us spell all of this out in a little more detail. Denote by $\mathbb{L} = \bigoplus_{j=1}^{d} q_j\mathbb{Z}$ the lattice of periods of $J$. By periodicity of $J$, $U^{\boldsymbol{\ell}}\psi$ is an eigenfunction of $J$ for every $\boldsymbol{\ell} \in \mathbb{L}$. Let $P_\lambda$ denote the projection onto the eigenspace of $J$ corresponding to $\lambda$ and $\chi_N$ the projection onto the box

$$\Gamma_N = \mathbb{Z}^d \cap \prod_{k=1}^{d}[0, Nq_k).$$

Note that $\Gamma_1 = \Gamma$, the fundamental domain from (4). By replacing $\psi$ with $U^{\boldsymbol{\ell}_0}\psi$ with a suitable $\boldsymbol{\ell}_0$, we may assume $\psi$ does not vanish on $\Gamma$, which implies $c := \|\chi_1\psi\|^2 > 0$. Produce an orthonormal basis $\{\psi_j\}_{0\le j<M}$ of $\mathrm{Ran}(P_\lambda)$ such that $\psi_0 = \psi$ (in principle, one may have $M$ finite or $M = \infty$). Exploiting cyclicity of the trace, one has

$$\mathrm{tr}(\chi_1 P_\lambda) = \sum_{0\le j<M} \langle\psi_j, \chi_1\psi_j\rangle = \sum_{0\le j<M} \|\chi_1\psi_j\|^2 \ge c > 0.$$

By translation invariance $\mathrm{tr}(\chi_N P_\lambda) = N^d \mathrm{tr}(\chi_1 P_\lambda)$ and in particular

$$\mathrm{tr}(\chi_N P_\lambda) \ge cN^d \tag{9}$$

with $c > 0$. However, the values of an eigenfunction on $\Gamma_N$ are entirely determined by its values on

$$\widetilde{\partial}\Gamma_N = \{\boldsymbol{x} \in \Gamma_N : \mathrm{dist}(\boldsymbol{x}, \mathbb{Z}^d \setminus \Gamma_N) \le 2\}.$$

As such, $\mathrm{tr}(\chi_N P_\lambda) \lesssim N^{d-1}$, contradicting (9). □

**Lemma 2.6** *For each $k$ and each fixed choice of $\{\theta_j : j \neq k\}$ $E_j(\boldsymbol{\theta})$ is a piecewise real-analytic function of $\theta_k$. For a.e. $\boldsymbol{\theta} \in \mathbb{T}^d$ and each $j$, $\nabla E_j(\boldsymbol{\theta})$ exists. For each $j$, $E_j$ is a non-constant function.*

***Proof*** This follows from analytic eigenvalue perturbation theory [12]. We describe the broad strokes and leave the details to the reader.

Let $1 \le k \le d$, choose and fix $\theta_j \in \mathbb{T}$ for $j \neq k$. For $s \in \mathbb{T}$, define $\boldsymbol{\theta}^{(k)}(s) \in \mathbb{T}^d$ by $\theta_j^{(k)}(s) = \theta_j$ for $j \neq k$ and $\theta_k^{(k)}(s) = s$. Extending into the complex plane, we have that

$$A(s) := J(\boldsymbol{\theta}^{(k)}(s)), \quad s \in \mathbb{C}$$

is an analytic family of matrices. Consequently, one can choose branches[1] of the eigenvalues $\lambda_1(s), \ldots, \lambda_t(s)$ and the associated eigenprojections which are real-analytic functions of $s$. In fact, these will be holomorphic functions of $s \in \mathbb{C}$ away from a discrete set of points. We note that the multiplicity $m_j$ of $\lambda_j$ is constant, again away from a discrete set. We refer the reader to Kato [12] for details about analytic perturbation theory. See especially [12, Theorem II.6.1]; see also [18, Theorem 1.4.1 and Corollary 1.4.5]. In particular, piecewise analyticity and existence of $\nabla E_j(\boldsymbol{\theta})$ for a.e. $\boldsymbol{\theta}$ follows.

Let us make a quick observation. For each fixed $\lambda \in \mathbb{R}$, the function

$$g_\lambda(\boldsymbol{\theta}) := \det(\lambda - J(\boldsymbol{\theta}))$$

is real-analytic as a function of $\boldsymbol{\theta} \in \mathbb{T}^d$. Thus, by general principles, there are two possibilities: either $g_\lambda$ is identically zero or $g_\lambda \neq 0$ a.e.[2]

To conclude the argument, let us show that the former cannot occur, i.e., that $g_\lambda$ does not vanish identically. Let us suppose $g_\lambda \equiv 0$, i.e., $\lambda$ is an eigenvalue of $J(\boldsymbol{\theta})$ for every $\boldsymbol{\theta}$. In particular, for each $\boldsymbol{\theta}$, measureably choosing a normalized eigenvector $u(\boldsymbol{\theta})$ of $J(\boldsymbol{\theta})$ having eigenvalue $\lambda$, we see that $\mathscr{F}^* u$ is a normalized eigenvector of $J$ itself corresponding to the eigenvalue $\lambda$. However, since $a_{\mathbf{x},\mathbf{y}} \neq 0$ and $J$ is periodic, $J$ has empty point spectrum by Lemma 2.5, so it follows that $g_\lambda \neq 0$ for a.e. $\lambda$, and thus $E_j$ must be nonconstant for each $j$. □

## 3 Ballistic Motion

Let us make a few observations. Formally, for each $1 \le k \le d$,

$$\frac{d}{dt} X_k(t) = iJe^{itJ} X_k e^{-itJ} - ie^{itJ} X_k J e^{-itJ} = P_k(t),$$

[1] Note that we use the different notation $\lambda$ to emphasize that in general, the analytic enumeration $\lambda_j(\boldsymbol{\theta})$ need not necessarily coincide with the ordered enumeration $E_j(\boldsymbol{\theta})$.

[2] This follows from the general theory of analytic functions of a single variable if $d = 1$. The statement in higher dimensions follows by induction. Suppose $d \ge 2$ and $f : \mathbb{T}^d \to \mathbb{R}$ is real-analytic and vanishes on a set of positive $d$-dimensional measure. By Fubini's theorem, there is a positive-measure set of $t$ such that the restriction of $f$ to the set $\theta_d = t$ vanishes on a set of positive $d-1$-dimensional measure and hence vanishes everywhere on the fiber $\theta_d = t$. Thus, for each choice of $(\theta_1, \ldots, \theta_{d-1})$, $f(\theta_1, \ldots, \theta_{d-1}, t)$ vanishes for a positive-measure set of $t$ and hence vanishes everywhere.

where $P_k = i[J, X_k] = i(JX_k - X_kJ)$. Thus,

$$\frac{X_k(t)}{t} = \frac{X_k}{t} + \frac{1}{t}\int_0^s P_k(s)\,ds, \tag{10}$$

so one wants to understand the time averages of $P_k$. Since the integral equation (10) involves the unbounded operator $X_k$, some care is needed, so let us make this precise. For each $N \in \mathbb{N}$, let $X_{k,N}$ denote the bounded operator given by

$$X_{k,N}\delta_{\boldsymbol{x}} = \begin{cases} x_k\delta_{\boldsymbol{x}} & |x_k| \le N \\ N\delta_{\boldsymbol{x}} & \text{otherwise,} \end{cases}$$

and define $P_{k,N} = i[J, X_{k,N}]$.

**Proposition 3.1** *Let $J$ be a bounded Jacobi matrix on $\ell^2(\mathbb{Z}^d)$. For all $k$ and $t$, $D(X_k(t)) = D(X_k)$ and one has*

$$X_k(t)\psi = X_k\psi + \int_0^t P_k(s)\psi\,ds. \tag{11}$$

The following representation of $P_k$ in the standard basis will be helpful.

**Proposition 3.2** *For $\boldsymbol{x} \in \mathbb{Z}^d$,*

$$P_k\delta_{\boldsymbol{x}} = ia_{\boldsymbol{x},\boldsymbol{x}-\boldsymbol{e}_k}\delta_{\boldsymbol{x}-\boldsymbol{e}_k} - ia_{\boldsymbol{x},\boldsymbol{x}+\boldsymbol{e}_k}\delta_{\boldsymbol{x}+\boldsymbol{e}_k}.$$

*In particular, $\|P_k\| \le 2\|a\|_\infty$.*

***Proof*** For $\boldsymbol{x} \in \mathbb{Z}^d$, observe that

$$J\delta_{\boldsymbol{x}} = \sum_{j=1}^d \left(a_{\boldsymbol{x},\boldsymbol{x}+\boldsymbol{e}_j}\delta_{\boldsymbol{x}+\boldsymbol{e}_j} + a_{\boldsymbol{x},\boldsymbol{x}-\boldsymbol{e}_j}\delta_{\boldsymbol{x}-\boldsymbol{e}_j}\right) + b_{\boldsymbol{x}}\delta_{\boldsymbol{x}},$$

so

$$X_kJ\delta_{\boldsymbol{x}} = \sum_{j=1}^d \left((x_k + \delta_{j,k})a_{\boldsymbol{x},\boldsymbol{x}+\boldsymbol{e}_j}\delta_{\boldsymbol{x}+\boldsymbol{e}_j} + (x_k - \delta_{j,k})a_{\boldsymbol{x},\boldsymbol{x}-\boldsymbol{e}_j}\delta_{\boldsymbol{x}-\boldsymbol{e}_j}\right) + x_kb_{\boldsymbol{x}}\delta_{\boldsymbol{x}}.$$

Subtracting this from $JX_k\delta_{\boldsymbol{x}} = x_kJ\delta_{\boldsymbol{x}}$, we get

$$P_k\delta_{\boldsymbol{x}} = ia_{\boldsymbol{x},\boldsymbol{x}-\boldsymbol{e}_k}\delta_{\boldsymbol{x}-\boldsymbol{e}_k} - ia_{\boldsymbol{x},\boldsymbol{x}+\boldsymbol{e}_k}\delta_{\boldsymbol{x}+\boldsymbol{e}_k},$$

as desired. □

***Proof of Proposition 3.1*** This follows from the same argument as [5, Theorem 2.1]. Since $X_{k,N}$ is bounded, a direct calculation shows that (11) holds with $X_k$ replaced by $X_{k,N}$, that is,

$$X_{k,N}(t)\psi = X_{k,N}\psi + \int_0^t P_{k,N}(s)\psi \, ds. \tag{12}$$

Since $X_{k,N} \to X_k$ and $P_{k,N} \to P_k$ strongly, (11) follows immediately from (12), the uniform bound

$$\|P_k\| \le 2\|a\|_\infty,$$

and dominated convergence. □

From the representation of $P_k$ in Proposition 3.2, we can see that $P_k$ is $\boldsymbol{q}$-periodic whenever $J$ is $\boldsymbol{q}$-periodic, so it too defines operators $P_k(\boldsymbol{\theta}) := P_k|_{\mathscr{H}(\boldsymbol{\theta})}$ for each $\boldsymbol{\theta} \in \mathbb{T}^d$.

**Theorem 3.4** *We have* $\mathscr{F} P_k \mathscr{F}^* = \widehat{P}_k$, *where* $P_k = \int_{\mathbb{T}^d}^{\oplus} P_k(\boldsymbol{\theta})\, d\boldsymbol{\theta}$, *that is,*

$$[\widehat{P}_k g](\boldsymbol{\theta}) = P_k(\boldsymbol{\theta}) g(\boldsymbol{\theta}).$$

***Proof*** This is essentially the same calculation as in the proof of Theorem 2.3. □

We now have all the necessary pieces in place in order to prove the main result.

***Proof of Theorem 1.1*** For $\psi \in \mathbb{C}^\Gamma$, $\boldsymbol{x} \in \Gamma$, and $\boldsymbol{\theta} \in \mathbb{T}^d$, define

$$\boldsymbol{x}/\boldsymbol{q} = (x_1/q_1, \dots, x_d/q_d)$$

and define the multiplication operator $M : \mathscr{H}_1 \to \mathscr{H}_1$ by $[Mg](\boldsymbol{\theta}) = M(\boldsymbol{\theta})g(\boldsymbol{\theta})$, where

$$[M(\boldsymbol{\theta})\psi]_{\boldsymbol{x}} = e^{2\pi i \langle \boldsymbol{\theta}, \boldsymbol{x}/\boldsymbol{q}\rangle} \psi_{\boldsymbol{x}}$$

Write $\widetilde{J}(\boldsymbol{\theta}) = M(\boldsymbol{\theta})^{-1} J(\boldsymbol{\theta}) M(\boldsymbol{\theta})$ and likewise for $\widetilde{P}_k(\boldsymbol{\theta})$. A direct calculation shows

$$\widetilde{P}_k(\boldsymbol{\theta}) = \frac{q_k}{2\pi} \frac{\partial}{\partial \theta_k} \widetilde{J}(\boldsymbol{\theta}),$$

and thus, denoting the projection onto the eigenspace of $E_j(\boldsymbol{\theta})$ by $\Pi_j(\boldsymbol{\theta})$, we have

$$\Pi_j(\boldsymbol{\theta}) P_k(\boldsymbol{\theta}) \Pi_j(\boldsymbol{\theta}) = \frac{q_k}{2\pi} \frac{\partial E_j}{\partial \theta_k}(\boldsymbol{\theta}) \Pi_j(\boldsymbol{\theta})$$

for a.e. $\boldsymbol{\theta}$ (where we have used Lemma 2.6 to assert the existence of $\partial E_j / \partial \theta_k$ for a.e. $\boldsymbol{\theta}$).

Thus, we have

$$\begin{aligned}\frac{X_k(t)}{t} &= \frac{X_k}{t} + \frac{1}{t}\int_0^t P_k(s)\,ds \\ &= \frac{X_k}{t} + \frac{1}{t}\int_0^t \int_{\mathbb{T}^d}^{\oplus} e^{isJ(\boldsymbol{\theta})} P_k(\boldsymbol{\theta}) e^{-isJ(\boldsymbol{\theta})}\,d\boldsymbol{\theta}\,ds \\ &\to \frac{q_k}{2\pi}\int_{\mathbb{T}^d}^{\oplus} \sum_{j=1}^{\bar{q}} \frac{\partial E_j}{\partial \theta_k}(\boldsymbol{\theta})\Pi_j(\boldsymbol{\theta})\,d\boldsymbol{\theta},\end{aligned}$$

where we applied dominated convergence to deduce the final line. By Lemma 2.6,

$$\bigcap_{k=1}^{d} \ker(Q_k) = \{0\},$$

as desired. □

**Acknowledgments** J.F. is grateful to Ilya Kachkovskiy and Milivoje Lukić for helpful conversations. Work supported in part by Simons Foundation Collaboration Grant #711663.

# References

1. M. Aizenman, S. Warzel, *Random Operators*. Graduate Studies in Mathematics, vol. 168 (American Mathematical Society, Providence, 2015). Disorder effects on quantum spectra and dynamics. MR 3364516
2. J. Asch, A. Knauf, Motion in periodic potentials. Nonlinearity **11**(1), 175–200 (1998) MR 1492956
3. R. Carmona, J. Lacroix, *Spectral Theory of Random Schrödinger Operators*. Probability and Its Applications (Birkhäuser, Boston, 1990). MR 1102675
4. H.L. Cycon, R.G. Froese, W. Kirsch, B. Simon, *Schrödinger Operators with Application to Quantum Mechanics and Global Geometry Study*. Texts and Monographs in Physics (Springer, Berlin, 1987). MR 883643
5. D. Damanik, M. Lukic, W. Yessen, Quantum dynamics of periodic and limit-periodic Jacobi and block Jacobi matrices with applications to some quantum many body problems. Commun. Math. Phys. **337**(3), 1535–1561 (2015). MR 3339185
6. D. Damanik, Schrödinger operators with dynamically defined potentials. Ergodic Theory Dyn. Syst. **37**(6), 1681–1764 (2017). MR 3681983
7. J. Fillman, Ballistic transport for limit-periodic Jacobi matrices with applications to quantum many-body problems. Commun. Math. Phys. **350**(3), 1275–1297 (2017). MR 3607475
8. L. Ge, I. Kachkovskiy, Ballistic transport for one-dimensional quasiperiodic Schrödinger operators (2020). arxiv:2009.02896
9. I. Kachkovskiy, On transport properties of isotropic quasiperiodic $XY$ spin chains. Commun. Math. Phys. **345**(2), 659–673 (2016). MR 3514955
10. I. Kachkovskiy, On the relation between strong ballistic transport and exponential dynamical localization (2020). arxiv:2001.01314

11. Y. Karpeshina, Y.-R. Lee, R. Shterenberg, G. Stolz, Ballistic transport for the Schrödinger operator with limit-periodic or quasi-periodic potential in dimension two. Commun. Math. Phys. **354**(1), 85–113 (2017). MR 3656513
12. T. Kato, *Perturbation Theory for Linear Operators*. Classics in Mathematics (Springer, Berlin, 1995), Reprint of the 1980 edition. MR 1335452
13. P. Kuchment, An overview of periodic elliptic operators. Bull. Am. Math. Soc. (N.S.) **53**(3), 343–414 (2016). MR 3501794
14. C.A. Marx, S. Jitomirskaya, Dynamics and spectral theory of quasi-periodic Schrödinger-type operators. Ergodic Theory Dyn. Syst. **37**(8), 2353–2393 (2017). MR 3719264
15. L. Pastur, A. Figotin, *Spectra of Random and Almost-Periodic Operators*. Grundlehren der Mathematischen Wissenschaften [Fundamental Principles of Mathematical Sciences], vol. 297 (Springer, Berlin, 1992). MR 1223779
16. M. Reed, B. Simon, *Methods of Modern Mathematical Physics*, vol. IV. Analysis of Operators (Academic Press [Harcourt Brace Jovanovich, Publishers], New York-London, 1979). MR 0493421
17. B. Simon, *Szegő's Theorem and Its Descendants*. M. B. Porter Lectures (Princeton University Press, Princeton, 2011). Spectral theory for $L^2$ perturbations of orthogonal polynomials. MR 2743058
18. B. Simon, *Operator Theory*. A Comprehensive Course in Analysis, Part 4 (American Mathematical Society, Providence, 2015). MR 3364494
19. G. Teschl, *Jacobi Operators and Completely Integrable Nonlinear Lattices*. Mathematical Surveys and Monographs, vol. 72 (American Mathematical Society, Providence, 2000). MR 1711536
20. Z. Zhang, Z. Zhao, Ballistic transport and absolute continuity of one-frequency Schrödinger operators. Commun. Math. Phys. **351**(3), 877–921 (2017). MR 3623240

# Perspectives on General Left-Definite Theory

**Dale Frymark and Constanze Liaw**

*The authors would like to congratulate Lance Littlejohn on the occasion of his 70th birthday, and thank him for all of the help over the years. Lance provides the inspiration and encouragement to dream big.*

**Abstract** In 2002, Littlejohn and Wellman developed a celebrated general left-definite theory for semi-bounded self-adjoint operators with many applications to differential operators. The theory starts with a semi-bounded self-adjoint operator and constructs a continuum of related Hilbert spaces and self-adjoint operators that are intimately related with powers of the initial operator. The development spurred a flurry of activity in the field that is still ongoing today.

The main goal of this expository (with the exception of Proposition 1) manuscript is to compare and contrast the complementary theories of general left-definite theory, the Birman–Krein–Vishik (BKV) theory of self-adjoint extensions and singular perturbation theory. In this way, we hope to encourage interest in left-definite theory as well as point out directions of potential growth where the fields are interconnected. We include several related open questions to further these goals.

**Keywords** Sturm-Liouville operators · Birman-Krein-Vishik theory · Left-definite theory · Self-adjoint extensions · Finite rank perturbations · Spectral theory

---

D. Frymark (✉)
Nuclear Physics Institute, Department of Theoretical Physics, Czech Academy of Sciences, Řež, Czech Republic
e-mail: frymark@ujf.cas.cz

C. Liaw
Department of Mathematical Sciences, University of Delaware, Newark, DE, USA

CASPER, Baylor University, Waco, TX, USA
e-mail: liaw@udel.edu



F. Gesztesy, A. Martinez-Finkelshtein (eds.), *From Operator Theory to Orthogonal Polynomials, Combinatorics, and Number Theory*, Operator Theory: Advances and Applications 285, https://doi.org/10.1007/978-3-030-75425-9_6

# 1 Introduction

The development of general left-definite theory by Littlejohn and Wellman had vast repercussions for differential operators. It extended the important concept of a left-definite Hilbert space to a continuum of associated Hilbert spaces.[1] We appeal to differential operators to further explain the motivation for this extension from the classical left-definite space by following [27] below. Note that some knowledge of self-adjoint extension theory is assumed throughout the manuscript; the reader is encouraged to consult the Appendix for some basic definitions and notions. The references [1, 31] can be used for more details. For some locally integrable positive weight function $w$, let

$$\ell[y](x) = \lambda w(x)y(x), \tag{1}$$

where $\ell$ is Lagrangian symmetric differential expression of order $2n$ given by

$$\ell^n[f](x) = \sum_{j=1}^{n}(-1)^j(a_j(x)f^{(j)}(x))^{(j)}, \ x \in (a,b), \tag{2}$$

with $-\infty \leq a < b \leq \infty$. For simplicity, we assume each coefficient $a_j(x)$ is smooth and positive on $(a,b)$. Due to the presence of $w(x)$ on the right hand side of Eq. (1), $\mathcal{H} = L^2[(a,b), w(x)]$ is referred to as the *right-definite* spectral setting for $w^{-1}\ell$. Two well-known formulas play a role in defining the left-definite space.

For $f, g \in \mathcal{D}_{\max}$, the maximal domain of $\ell$, Green's formula says that

$$\int_a^b \ell[f](x)\overline{g}(x)dx = \int_a^b f(x)\overline{\ell[g]}(x)dx + [f,g](x)\Big|_a^b, \tag{3}$$

where $[\cdot,\cdot]$ is the sesquilinear form for $\ell$. Dirichlet's formula is

$$\int_a^b \ell[f](x)\overline{g}(x)dx = \sum_{j=0}^{n}\int_a^b a_j(x)f^{(j)}(x)\overline{g}^{(j)}(x)dx + [\![f,g]\!](x)\Big|_a^b, \tag{4}$$

where $[\![\cdot,\cdot]\!]$ is a bilinear form closely related to the sesquilinear form $[\cdot,\cdot]$. Let $\mathbf{A}$ be a self-adjoint operator acting via $\ell$ on a domain $\mathcal{D}(\mathbf{A})$ such that for all

[1] All Hilbert spaces considered are assumed to be separable.

$f, g \in \mathcal{D}(\mathbf{A})$ the bilinear form $[\![f, g]\!](x)\big|_a^b = 0$. In this case, Eq. (4) simplifies to

$$\langle \mathbf{A}f, g\rangle_{\mathcal{H}} = \int_a^b \ell[f](x)\overline{g(x)}dx = \sum_{j=0}^{n} \int_a^b a_j(x) f^{(j)}(x)\overline{g}^{(j)}(x)dx. \tag{5}$$

Such an operator $\mathbf{A}$ is then semi-bounded (see Eq. (8)) thanks to the positivity of the coefficient functions $a_j(x)$. The expression $\ell$ thus generates an inner product defined by the right hand side of Eq. (5). Explicitly, for $f, g \in \mathcal{D}(\mathbf{A})$,

$$\langle \mathbf{A}f, g\rangle_{\mathcal{H}} = \langle f, g\rangle_1. \tag{6}$$

The closure of $\mathcal{D}(\mathbf{A})$ in the topology generated by the norm $\|\cdot\|_1 = \langle\cdot, \cdot\rangle_1^{1/2}$ is then denoted by $H_1$ and referred to as the left-definite setting for $w^{-1}\ell$, with $H_1$ the (first) left-definite space. This is due to the inner product $\langle\cdot, \cdot\rangle_1$ originating from the left hand side of Eq. (1). The terminology itself can be traced back to Weyl in 1910 [32].

This first left-definite space presented above is identified by Littlejohn and Wellman as $\mathcal{D}(\mathbf{A}^{1/2})$, along with a continuum of other associated left-definite spaces in Definition 1. Their new characterization makes connections to other areas of self-adjoint operator theory more clear. The core definitions and results of left-definite theory (we omit the term 'general' from now on) are included in Sect. 3. A key aspect of the theory is that it includes stability of spectral type, see Theorem 1, which can be viewed as an important advantage over the other theories discussed. Since left-definite theory was described in 2002, it has been the focus of many works. Here, we focus on techniques and results recently developed by the authors in order to present several related conjectures and open questions. It is the our hope that researchers may take inspiration from these problems to make contributions to the field.

The Birman–Krein–Vishik (BKV) theory of semi-bounded forms, which puts closed semi-bounded forms into a one-to-one correspondence with self-adjoint operators, agrees with general left-definite theory. In many ways, this theory could also be thought of as a generalization of Eq. (5): identifying $\langle \mathbf{A}f, g\rangle_{\mathcal{H}}$ as an inner product in another Hilbert space for general self-adjoint operators $\mathbf{A}$. BKV theory and its connections to left-definite theory are the subject of Sect. 4. Proposition 1 proves that a continuum of semi-bounded closed forms are actually associated with each semi-bounded self-adjoint operator via left-definite theory. Such freedom in choosing a form to work with may provide valuable flexibility in applications.

There is also a scale of Hilbert spaces that is used in singular self-adjoint perturbation theory, when the perturbation does not lie in the Hilbert space but can be shown to belong to an associated Hilbert space. This scale of Hilbert spaces is generated from a self-adjoint operator, and if this operator is semi-bounded they are equivalent to the continuum of left-definite spaces of the operators. However, the scale of Hilbert spaces is focused on linear bounded functionals which act on these left-definite spaces, thereby corresponding to left-definite spaces with negative indices that are defined through duality. Although independently defined, there are instances in which the definitions of the spaces occurring in left-definite theory are identical to those in the scale of Hilbert spaces. We find this intriguing, especially due to the fact that they are used for different, complementary purposes. The connection between this scale of spaces and left-definite theory is described in Sect. 5.

One of the main obstacles in both BKV theory and singular perturbation theory is that the associated spaces are difficult to describe with explicit boundary conditions. This is another area where connections to left-definite theory can be of benefit, as there are now clear descriptions even in some hard to handle examples, i.e. Sturm–Liouville operators with limit-circle endpoints.

Self-adjoint extensions of symmetric Sturm–Liouville operators with limit-circle endpoints were recently characterized in terms of a singular perturbation by the authors in [10]. The perturbation heavily exploited the connection between the scale of Hilbert spaces and BKV theory, along with boundary triples and boundary pairs (see e.g. [6]). In particular, we mention that boundary triples are partially defined by the Green's formula in Eq. (3) and boundary pairs essentially operate on two key spaces from the scale of Hilbert spaces.

The basics of this perturbation setup are described in Sect. 6 in order to formulate a conjecture about how left-definite theory may assist with extending the construction to powers of such operators.

While the applications in this manuscript are focused on ordinary differential operators, it should be pointed out that left-definite theory does apply to self-adjoint elliptic partial differential operators too. However, to the best knowledge of the authors this fact has not been thoroughly explored.

Overall, the three theories: left-definite, BKV and singular perturbation, are found to be complementary to each other. We hope this manuscript helps readers find new and interesting connections between fields they may or may not have encountered before and sparks their interest in building upon these connections. For those familiar with the theories already, their focus may be honed on the end of each section, where the perspective of each individual theory (and their nuanced setups) is used for comparison with the others and conjectures are formulated.

## 1.1 Notation

The notation $L_{\max} = \{\ell, \mathcal{D}_{\max}\}$ is used to say that the operator $L_{\max}$ acts via $\ell$ on the domain $\mathcal{D}_{\max}$. Self-adjoint operators are written in bold face, e.g. $\mathbf{A}$, for emphasis while all other operators are typeset normally.

Due to the three different theories in the manuscript, we clarify the notation for each here to help avoid any confusion. In left-definite theory, the inner product is denoted with a single subscript $r$, i.e. $\langle \cdot\,, \cdot \rangle_r$, and the Hilbert space associated with this inner product is written as $\mathcal{H}_r$ (see Definition 1). In the BKV theory of semi-bounded forms, the inner product and associated Hilbert space are both denoted with the form as a subscript, i.e. $\langle \cdot, \cdot \rangle_{\mathfrak{ts}_{-\gamma}}$ and $\mathcal{H}_{\mathfrak{ts}_{-\gamma}}$ (see Eq. (9)). Finally, when discussing the scale of Hilbert spaces associated with a self-adjoint operator $\mathbf{A}$, we use double subscript $\langle \cdot, \cdot \rangle_{s,-s}$ to denote the duality pairing and, e.g. $\mathcal{H}_{-s}(\mathbf{A})$ to denote the desired space (see Definition 3).

## 2 Sturm–Liouville Operators

Many of the theories and conjectures in this manuscript, especially those in Sect. 3 and 6, are illustrated by applications to Sturm–Liouville operators because they are so well studied, see e.g. [1, 5, 15, 33]. We briefly introduce the central concepts of these operators here.

Consider the classical Sturm–Liouville differential equation

$$\frac{d}{dx}\left[p(x)\frac{df}{dx}(x)\right] + q(x)f(x) = -\lambda w(x)f(x),$$

where $p(x), w(x) > 0$ a.e. on $(a, b)$ and $q(x)$ real-valued a.e. on $(a, b)$, with $a < b$ and $a, b \in \mathbb{R} \cup \{\pm\infty\}$. Furthermore, let $1/p(x), q(x), w(x) \in L^1_{\text{loc}}(a, b)$. The differential expression can be viewed as a linear operator, mapping a function $f$ to the function

$$\ell[f](x) := -\frac{1}{w(x)}\left(\frac{d}{dx}\left[p(x)\frac{df}{dx}(x)\right] + q(x)f(x)\right). \tag{7}$$

This unbounded operator acts on the Hilbert space $L^2[(a, b), w]$, endowed with the inner product $\langle f, g\rangle := \int_a^b f(x)\overline{g(x)}w(x)dx$. In this setting, the eigenvalue problem $\ell[f](x) = \lambda f(x)$ can be considered. However, the operator acting via $\ell[\,\cdot\,]$ on $L^2[(a, b), w]$ is not self-adjoint a priori; additional boundary conditions may be required to ensure this property. The definition here is slightly different than that of the differential operator in Eq. (2), where to explain the different left and right-definite settings the weight function was not included. Hence, the operator of interest ended up arising from $w^{-1}\ell$, which agrees with the expression in Eq. (7).

Endpoints are either in the limit-point or limit-circle case depending on whether one or two solutions are in $L^2[(a, b), w]$, respectively. When in the limit-circle case, we assume that endpoints are non-oscillatory, see e.g. [5, 33] for more.

Furthermore, the operator $\ell^n[\,\cdot\,]$ is defined as the order two operator $\ell[\,\cdot\,]$ composed with itself $n$ times, creating a differential operator of order $2n$. Every formally symmetric differential expression $\ell^n[\,\cdot\,]$ of order $2n$ with coefficients $a_k : (a, b) \to \mathbb{R}$ and $a_k \in C^k(a, b)$, for $k = 0, 1, \ldots, n$ and $n \in \mathbb{N}$, has the Lagrangian symmetric form given in Eq. (2). Further details can be found in [14, 17].

The classical differential expressions of Jacobi, Laguerre and Hermite are all semi-bounded and admit such a representation. Semi-boundedness is defined as the existence of a constant $k \in \mathbb{R}$ such that for all $x$ in the domain of the operator $A$ the following inequality holds:

$$\langle Ax, x\rangle \geq k\langle x, x\rangle. \tag{8}$$

This additional property, combined with self-adjointness, allows for a continuum of nested Hilbert spaces to be defined within $L^2[(a, b), w]$ via non-negative real powers of the expression $\ell$. Indeed, this continuum will provide a Hilbert scale, and many facts about the spectrum and the operators (see e.g. [12, 28]) can be deduced using this point of view. More details about Hilbert scales can be found in [2, 23]. This particular Hilbert scale with self-adjoint operators that are semi-bounded is the topic of left-definite theory [27].

## 3 Left-Definite Theory

In general left-definite theory powers of a semi-bounded self-adjoint differential operator are used to create a continuum of operators whereupon spectral properties can be studied. This spectral information is invariant in the sense that knowledge about the spectrum of one of the operators in this continuum allows us to obtain insight into all the other operators. We begin by reviewing some of the main definitions and results from the landmark paper of Littlejohn and Wellman [27] in order to introduce the theory.

Let $\mathcal{V}$ be a vector space over $\mathbb{C}$ with inner product $\langle\cdot, \cdot\rangle$ and norm $\|\cdot\|$. The resulting inner product space is denoted $(\mathcal{V}, \langle\cdot, \cdot\rangle)$.

**Definition 1 ([27, Theorem 3.1])** Suppose $\mathbf{A}$ on the Hilbert space $\mathcal{H} = (\mathcal{V}, \langle\cdot, \cdot\rangle)$ is a self-adjoint operator that is bounded below by $kI$, where $k > 0$. Let $0 < r \in \mathbb{R}$. Define $\mathcal{H}_r = (\mathcal{V}_r, \langle\cdot, \cdot\rangle_r)$ with

$$\mathcal{V}_r = \mathcal{D}(\mathbf{A}^{r/2})$$

and

$$\langle x, y\rangle_r = \langle\mathbf{A}^{r/2}x, \mathbf{A}^{r/2}y\rangle \quad \text{for} \quad x, y \in \mathcal{V}_r.$$

Then $\mathcal{H}_r$ is said to be the $r$-th *left-definite space* associated with the pair $(\mathcal{H}, \mathbf{A})$.

It was proved in [27, Theorem 3.1] that $\mathcal{H}_r = (\mathcal{V}_r, \langle\cdot, \cdot\rangle_r)$ is also described as the left-definite space associated with the pair $(\mathcal{H}, \mathbf{A}^r)$. Specifically, we have:

1. $\mathcal{H}_r$ is a Hilbert space,
2. $\mathcal{D}(\mathbf{A}^r)$ is a subspace of $\mathcal{V}_r$,
3. $\mathcal{D}(\mathbf{A}^r)$ is dense in $\mathcal{H}_r$,

4. $\langle x, x \rangle_r \geq k^r \langle x, x \rangle$ for $x \in \mathcal{V}_r$, and
5. $\langle x, y \rangle_r = \langle \mathbf{A}^r x, y \rangle$ for $x \in \mathcal{D}(\mathbf{A}^r)$, $y \in \mathcal{V}_r$.

The left-definite domains are defined as the domains of compositions of the self-adjoint operator **A**, but the operator acting on this domain is slightly more difficult to define.

**Definition 2 ([27, Definition 2.2/2.3])** Let $\mathcal{H} = (\mathcal{V}, \langle \cdot, \cdot \rangle)$ be a Hilbert space. Suppose $\mathbf{A} : \mathcal{D}(\mathbf{A}) \subset \mathcal{H} \to \mathcal{H}$ is a self-adjoint operator that is bounded below by $k > 0$. Let $1 \leq r \in \mathbb{R}$. If there exists a self-adjoint operator $\mathbf{A}_r : \mathcal{H}_r \to \mathcal{H}_r$ that is a restriction of **A** from the domain $\mathcal{D}(\mathbf{A})$ to $\mathcal{D}(\mathbf{A}_r)$, we call such an operator an $r$-th *left-definite operator associated with* $(\mathcal{H}, \mathbf{A})$. For $0 < r < 1$ we obtain an $r$-th *left-definite operator associated with* $(\mathcal{H}, \mathbf{A})$ analogously, but by taking the closure of the domain $\mathcal{D}(\mathbf{A})$ with respect to the norm induced by the inner product $\langle \cdot, \cdot \rangle_r$.

The connection between the $r$-th left-definite operator and the $r$-th composition of the self-adjoint operator **A** can be made explicit:

**Corollary 1 ([27, Corollary 3.3])** *Suppose* **A** *is a self-adjoint operator in the Hilbert space* $\mathcal{H}$ *that is bounded below by* $k > 0$*. For each* $r > 0$*, let* $\mathcal{H}_r = (\mathcal{V}_r, \langle \cdot, \cdot \rangle_r)$ *and* $\mathbf{A}_r$ *denote, respectively, the* $r$*-th left-definite space and the* $r$*-th left-definite operator associated with* $(\mathcal{H}, \mathbf{A})$*. Then*

*1.* $\mathcal{D}(\mathbf{A}^r) = \mathcal{V}_{2r}$*, in particular,* $\mathcal{D}(\mathbf{A}^{1/2}) = \mathcal{V}_1$ *and* $\mathcal{D}(\mathbf{A}) = \mathcal{V}_2$*;*
*2.* $\mathcal{D}(\mathbf{A}_r) = \mathcal{D}(\mathbf{A}^{(r+2)/2})$*, in particular,* $\mathcal{D}(\mathbf{A}_1) = \mathcal{D}(\mathbf{A}^{3/2})$ *and* $\mathcal{D}(\mathbf{A}_2) = \mathcal{D}(\mathbf{A}^2)$*.*

The left-definite theory is particularly important for self-adjoint differential operators that are bounded below, as they are generally unbounded. The theory is trivial for bounded operators, as shown in [27, Theorem 3.4]. (Indeed, for bounded operators, there is only one Hilbert space, since $\mathcal{H} = \mathcal{H}_r$ for all $r > 0$.)

Our applications of left-definite theory will be focused on differential operators which possess a complete orthogonal set of eigenfunctions in $\mathcal{H}$. In [27, Theorem 3.6] it was proved that the point spectrum of **A** coincides with that of $\mathbf{A}_r$, and similarly for the continuous spectrum and for the resolvent set. It is possible to say more, a complete set of orthogonal eigenfunctions will persist throughout each space in the Hilbert scale.

**Theorem 1 ([27, Theorem 3.7])** *If* $\{\varphi_n\}_{n=0}^{\infty}$ *is a complete orthogonal set of eigenfunctions of* **A** *in* $\mathcal{H}$*, then for each* $r > 0$*,* $\{\varphi_n\}_{n=0}^{\infty}$ *is a complete set of orthogonal eigenfunctions of the* $r$*-th left-definite operator* $\mathbf{A}_r$ *in the* $r$*-th left-definite space* $\mathcal{H}_r$*.*

Another perspective on the last theorem is that it gives us a valuable indicator for when a space is a left-definite space for a specific operator. Also, we note that left-definite theory can be extended to bounded below operators by applying shifts. Uniqueness is then given up to the chosen shift.

One of the main applications of the theory is to Sturm–Liouville operators. In particular, if the operator has a complete system of orthogonal eigenfunctions then proving boundedness from below is easier and it is possible to invoke Theorem 1. An example of how left-definite theory can be applied to such an operator is now briefly described, full details can be found in [27, Sect. 12].

*Example 1* For $\alpha > -1$, consider the classical self-adjoint Laguerre differential operator $\mathbf{A}$ acting on $\mathcal{H} = L^2\left[(0,\infty), x^\alpha e^{-x}\right]$ via

$$\ell[f](x) = \frac{1}{x^\alpha e^{-x}}\left[-x^{\alpha+1}e^{-x}f'(x)\right]',$$

such that dom($\mathbf{A}$) possesses the Laguerre polynomials as a complete set of orthogonal eigenfunctions. The $n$th left-definite Hilbert space associated with the pair $(\mathcal{H}, \mathbf{A})$, also possessing this complete set of eigenfunctions, is defined as $\mathcal{H}_n = (\mathcal{V}_n, \langle\cdot,\cdot\rangle_n)$, where

$$\mathcal{V}_n := \left\{ f : (0,\infty) \to \mathbb{C} \;\middle|\; f \in \mathrm{AC}^{(n-1)}_{\mathrm{loc}}(0,\infty);\ f^{(n)} \in L^2\left[(0,\infty), t^{\alpha+n}e^{-t}\right] \right\}$$

and

$$\langle p, q\rangle_n := \sum_{j=0}^{n} b_j(n,k) \int_0^\infty p^{(j)}(t)\overline{q^{(j)}(t)}t^{\alpha+j}e^{-t}dt \quad \text{for } (p, q \in \mathcal{P}),$$

where $\mathcal{P}$ is the space of all (possibly complex-valued) polynomials. The constants $b_j(n,k)$ are defined as

$$b_j(n,k) := \sum_{i=0}^{j} \frac{(-1)^{i+j}}{j!}\binom{j}{i}(k+i)^n.$$

For several years after the discovery of the general left-definite theory, descriptions of left-definite spaces were similar to those of Example 1, i.e. the boundary conditions were not classically expressed by GKN theory. Similar results for specific (mostly classical) operators can be found in [9, 16, 17] and their references. Some progress towards expressing left-definite spaces in terms of standard boundary conditions was made much later in [29] and then expanded upon in [18].

In order to present the main result from [18], we let $\mathbf{L}^n$ be a self-adjoint operator defined by left-definite theory on $L^2[(a,b), w]$ with domain $\mathcal{D}^n_{\mathbf{L}}$ that includes a complete system of orthogonal eigenfunctions. Enumerate the orthogonal eigenfunctions as $\{P_k\}_{k=0}^\infty$. Let $\mathbf{L}^n$ operate on its domain via $\ell^n[\,\cdot\,]$, a differential operator of order $2n$, $n \in \mathbb{N}$, generated by composing a Sturm–Liouville differential expression with itself $n$ times. Furthermore, let $\mathbf{L}^n$ be an extension of the minimal operator $L^n_{\min}$ that has deficiency indices $(n,n)$, and the associated maximal domain be denoted by $\mathcal{D}^n_{\max}$. See the Appendix for the general definitions of these domains.

This allows us to compare several different potential descriptions of the left-definite domain. Consider

$$\mathcal{A}_n := \Big\{ f \in \mathcal{D}^n_{\max} \;:\; f, f', \ldots, f^{(2n-1)} \in AC_{\text{loc}}(a,b); \\ (p(x))^n f^{(2n)} \in L^2[(a,b), w] \Big\},$$

$$\mathcal{B}_n := \Big\{ f \in \mathcal{D}^n_{\max} \;:\; [f, P_j]_n \Big|_a^b = 0 \text{ for } j = 0, 1, \ldots, n-1 \Big\},$$

$$\mathcal{C}_n := \Big\{ f \in \mathcal{D}^n_{\max} \;:\; [f, P_j]_n \Big|_a^b = 0 \text{ for any } n \text{ distinct } j \in \mathbb{N} \Big\}, \text{ and}$$

$$\mathcal{F}_n := \Big\{ f \in \mathcal{D}^n_{\max} \;:\; \Big[ a_j(x) f^{(j)}(x) \Big]^{(j-1)} \Big|_a^b = 0 \text{ for } j = 1, 2, \ldots, n \Big\}.$$

The function $p(x)$ above is from the standard definition of a Sturm–Liouville differential operator, given in Eq. (7), and the $a_j(x)$'s are from the Lagrangian symmetric form of the operator in Eq. (2). The following conjecture about the equality of these domains is found in [18], which was in turn adapted from [29].

*Conjecture 1* Let $\mathbf{L}^n$ be a self-adjoint operator defined by left-definite theory on $L^2[(a,b), w]$ with domain $\mathcal{D}^n_{\mathbf{L}}$ that includes a complete system of orthogonal polynomial eigenfunctions, that is, we use $\mathcal{D}^n_{\mathbf{L}} = \mathcal{A}_n$. Let $\mathbf{L}^n$ operate on its domain via the expression $\ell^n[\cdot]$, a differential operator of order $2n$, where $n \in \mathbb{N}$, generated by composing a Sturm–Liouville differential operator with itself $n$ times. Furthermore, let $\mathbf{L}^n$ be an extension of the minimal operator $L^n_{\min}$, which has deficiency indices $(n, n)$. Then $\mathcal{A}_n = \mathcal{B}_n = \mathcal{C}_n = \mathcal{F}_n = \mathcal{D}^n_{\mathbf{L}}$, $\forall n \in \mathbb{N}$.

The conjecture was partially answered in [18, Theorem 6.5] under several extra assumptions: that $\mathcal{A}_n = \mathcal{B}_n$ and that $f \in \mathcal{F}_n$ implies that $f'', \ldots, f^{(2n-2)} \in L^2(a,b)$. The two primary ideas of the proof were a careful analysis of the sesquilinear form for the operator $\mathbf{L}^n$ and the introduction of a matrix of boundary values to help determine when GKN conditions were satisfied or not. The conjecture was answered in the affirmative for the Jacobi differential operator in [20].

**Theorem 2** *Let $\mathbf{L}^n$, $n \in \mathbb{N}$, be a self-adjoint operator defined by left-definite theory on $L^2[(a,b), w]$ with the left-definite domain $\mathcal{D}^n_{\mathbf{L}}$. Let $\mathbf{L}^n$ operate on its domain via $\ell^n[\cdot]$, a classical Jacobi differential expression of order $2n$, with parameters $\alpha, \beta > 0$, generated by composing the Sturm–Liouville operator with itself $n$ times. Furthermore, let $\mathbf{L}^n$ be an extension of the minimal operator $L^n_{min}$, which has deficiency indices $(2n, 2n)$. Then $\mathcal{D}^n_{\mathbf{L}} = \mathcal{A}_n = \mathcal{B}_n = \mathcal{C}_n = \mathcal{F}_n$, $\forall n \in \mathbb{N}$.*

Note that the apparent discrepancy between deficiency indices here is merely due to the use of separated boundary conditions instead of connected ones. It is also worth pointing out that the methods of [20] differ greatly from those of [18]. In particular, solutions to the Jacobi differential equation are written as infinite sums and only certain terms are shown to belong to the maximal domain modulo the minimal domain. This finite, explicit decomposition of the deficiency spaces results in easier manipulations of the sesquilinear form to prove the equality of the domains. Conjecture 1 remains open for other operators; the methods of [20] are almost certainly applicable to wider classes of operators and may be helpful. In particular, operators which do not possess a complete system of orthogonal polynomial eigenfunctions have not been considered. A possible weakening of this hypothesis would be to replace the orthogonal polynomials with principal solutions. Principal solutions are intimately related to the Friedrichs extension of a symmetric operator, which is the subject of our last conjecture.

The Friedrichs extension is usually defined through the closed semi-bounded form associated with a self-adjoint operator, see Sect. 4 (specifically Eq. (10)) for more about these forms or [6, 10, 30] for complete details. However, it suffices to think of the extension as the "smallest" self-adjoint extension among all other self-adjoint extensions (in the sense that it has the smallest form domain); it is often called the "soft" extension for this reason.

*Conjecture 2* Let $A$ be a closed semi-bounded symmetric operator and $\mathbf{A}_F$ be its Friedrichs self-adjoint extension. Then the $2r$-th left-definite space of $\mathbf{A}_F$ coincides with the domain of the Friedrichs extension of the $r$-th power of $\mathbf{A}$. Explicitly,

$$\mathrm{dom}((\mathbf{A}_F)^r) = \mathrm{dom}((\mathbf{A}^r)_F).$$

In other words, the conjecture is suggesting that the action of taking powers of an operator commutes with the action of taking the Friedrichs extension. In every computed case the conjecture seems to hold but in the stated generality the status is unclear. Verification in the case where $A$ is the Jacobi differential operator can be found in [19, Cor. 5.1]. Left-definite theory need not be mentioned in the statement of the conjecture, clearly, but it is the authors' opinion that left-definite theory can nonetheless be of great help in proving the statement.

A related conjecture (that would build upon the spectral stability results of Theorem 1 if confirmed) posits that multiplicity of eigenvalues are invariant under left-definite theory.

*Conjecture 3* Let $\mathbf{A}$ be a semi-bounded self-adjoint operator in a Hilbert space $\mathcal{H}$. Let $\lambda$ be an eigenvalue of $\mathbf{A}$ with multiplicity $m$. Then $\lambda$ is also an eigenvalue for the $r$-th left-definite operator $\mathbf{A}_r$ in the $r$-th left-definite space $\mathcal{H}_r$ of multiplicity $m$.

The difficulty of determining multiplicity of eigenvalues restricts the amount of evidence available to support the conjecture. In the case when $\mathbf{A}$ is the Jacobi differential operator, left-definite operators and Weyl $m$-functions for their extensions can be found in [19] and the spectral analysis of [10] could be applied to potentially verify the conjecture for the example.

Conjecture 3 also has implications for the intertwining of eigenvalues between distinct extensions. Usually this intertwining takes place between the Friedrichs extension and a transversal extension, which can be seen e.g. in a comparison of Neumann and Dirichlet eigenvalues for Sturm–Liouville operators with two regular endpoints. Essentially, it would be interesting to see if eigenvalue intertwining between two self-adjoint extensions implies that the $r$-th left-definite operators of the two extensions also have eigenvalue intertwining.

# 4 Comparison with BKV Semi-Bounded Form Theory

We first introduce the basics of the so-called Birman–Krein–Vishik (BKV) theory of semi-bounded forms. A collection of results from the theory along with original references can be found in, e.g. [3]. The brief presentation here mostly follows that of [6, Chapter 5] and [22, Chapter 6], which can also be consulted for more details.

Let $\mathbf{A}$ be a semi-bounded self-adjoint operator with lower bound $m(\mathbf{A}) < \infty$. There is a natural way to identify $\mathbf{A}$ with a closed semi-bounded form $\mathfrak{t}$ in $\mathcal{H}$ with the same lower bound $m(\mathfrak{t}) = m(\mathbf{A})$ via the First and Second Representation Theorems, see e.g. [6, Theorem 5.1.18] and [6, Theorem 5.1.23]. Namely, let $\varphi \in \mathrm{dom}(\mathbf{A})$, $\psi \in \mathrm{dom}(\mathfrak{t})$, $\gamma < m(\mathbf{A})$ and define

$$\mathrm{dom}(\mathfrak{t}_{\mathbf{A}}) = \mathrm{dom}(\mathbf{A} - \gamma)^{1/2},$$
$$\mathfrak{t}_{\mathbf{A}}[\varphi, \psi] = \langle (\mathbf{A} - \gamma)\varphi, \psi \rangle + \gamma \langle \varphi, \psi \rangle.$$

Equivalently, if $\varphi, \psi \in \mathrm{dom}(\mathfrak{t}_{\mathbf{A}})$, then

$$\mathfrak{t}_{\mathbf{A}}[\varphi, \psi] = \langle (\mathbf{A} - \gamma)^{1/2}\varphi, (\mathbf{A} - \gamma)^{1/2}\psi \rangle + \gamma \langle \varphi, \psi \rangle.$$

The space $\mathrm{dom}(\mathfrak{t}_{\mathbf{A}})$ endowed with the inner product

$$\langle \varphi, \psi \rangle_{\mathfrak{t}_{\mathbf{A}-\gamma}} := \mathfrak{t}_{\mathbf{A}}[\varphi, \psi] - \gamma \langle \varphi, \psi \rangle, \text{ for } \varphi, \psi \in \mathrm{dom}(\mathfrak{t}), \tag{9}$$

is a Hilbert space, denoted $\mathcal{H}_{\mathfrak{t}_{\mathbf{A}-\gamma}}$.

More general semi-bounded symmetric operators $S$ (i.e. the minimal operator of a Sturm–Liouville expression) also have a form associated with them via

$$\mathfrak{t}_S[f, g] = \langle Sf, g \rangle, \text{ for } f, g \in \mathrm{dom}(S). \tag{10}$$

The semi-bounded self-adjoint operator $\mathbf{S}_F$ associated with the closure of the form $\mathfrak{t}_S$ in Eq. (10) is called the *Friedrichs extension* of $S$, see [6, Definition 5.3.2].

This theory is often used to distinguish or construct specific self-adjoint extensions from a symmetric operator, as it is often more convenient to define a closed semi-bounded form than a self-adjoint operator. An ordering of closed semi-bounded forms then corresponds to an ordering of self-adjoint extensions, is also useful for this purpose. Details can be found in e.g. [6, Sect. 5.2] and [10, Remark 3.5].

It is also possible to put densely defined, closed, sectorial forms $\mathfrak{t}_T$ into one-to-one correspondence with $m$-sectorial operators $T$, but this falls outside the scope of the current manuscript (see e.g. [4, 8, 11]). However, it demonstrates that the correspondence between forms and operators is usable in a wide variety of contexts.

In comparison with Sect. 3, it should be clear that given a semi-bounded self-adjoint operator $\mathbf{A}$ the first left-definite space automatically coincides with the domain of the associated closed semi-bounded form after an appropriate shift to make the operator positive, i.e. $\mathcal{V}_1 = \mathrm{dom}(\mathfrak{t}_{\mathbf{A}})$. Indeed, for $f, g \in \mathcal{V}_1$ the action of $\mathfrak{t}_{\mathbf{A}}[f, g]$ is equal to $\langle f, g \rangle_1$, the inner product in the first left-definite space.

If we restrict our attention to closed semi-bounded forms, then left-definite theory is a natural extension of BKV theory; instead of associating a single closed semi-bounded form with a self-adjoint operator, it is possible to associate a continuum of closed semi-bounded forms.

**Proposition 1** *Let* $\mathbf{A}$ *be a semi-bounded self-adjoint operator with lower bound* $m(\mathbf{A})$*. For each* $r \in \mathbb{N}$*, the form (we suppress the dependence on* $\mathbf{S}$ *here), with* $f, g \in \mathrm{dom}(\mathfrak{t}_r)$ *and* $\gamma < m(\mathbf{A})$*, given by*

$$\begin{aligned}\mathrm{dom}(\mathfrak{t}_r) &= \mathrm{dom}(\mathbf{A} - \gamma)^{r/2},\\ \mathfrak{t}_r[f, g] &= \langle ((\mathbf{A} - \gamma)^{r/2} f, (\mathbf{A} - \gamma)^{r/2} g\rangle + \gamma\langle f, g\rangle,\end{aligned}$$

*is closed and semi-bounded.*

***Proof*** Here, we consider the left-definite spaces associated with the shifted operator $\mathbf{A} - \gamma$. Semi-boundedness of $\mathfrak{t}_r$ is assured by item (4) after Definition 1. Lemma 5.1.9 of [6] says that the semi-bounded form $\mathfrak{t}_r$ is closed if and only if the space $\mathcal{H}_{(\mathbf{A}-\gamma)^r}$ is in fact a Hilbert space. Endow the space $\mathrm{dom}(\mathfrak{t}_r)$ with the inner product $\langle\cdot, \cdot\rangle_r$ from Definition 1 and notice that this coincides with the definition of the space $\mathcal{H}_{(\mathbf{A}-\gamma)^r}$ above. Since $\mathcal{H}_r$ is a Hilbert space, this immediately implies that $\mathfrak{t}_r$ is closed. □

While the BKV theory of semi-bounded forms gives valuable knowledge about self-adjoint extensions, in practice the domain of the operator $\mathbf{A}^{1/2}$ is often difficult to determine, even for elementary choices of $\mathbf{A}$. Explicit domains can, for example, be determined when $\mathbf{S}$ is a Sturm–Liouville operator though, see [6, Sect. 6.9]. Left-definite domains have some advantages and disadvantages in this area;

As far as advantages go, for even $r$, left-definite domains do not involve fractional powers of operators and therefore are somewhat natural to consider. The square of a self-adjoint operator is sometimes useful in applications and falls into this category. There is also some spectral stability inherent in left-definite operators, see Theorem 1, that is, to the best knowledge of the authors, not available in BKV form theory. Left-definite theory thus provides alternatives to the classical form used in BKV theory and these form domains can be expressed with classical boundary conditions, see e.g. Theorem 2.

The main disadvantage of the theory is that it is unclear whether the ordering of closed semi-bounded forms is preserved under left-definite theory. This is made more difficult by the fact that explicit boundary conditions are somewhat elusive, see Theorem 2 for an example where they were determined. A proof of Conjecture 2 would go a long way towards solving this discrepancy. BKV theory, as previously mentioned, is also applicable to a wider range of operators.

It is also important to note that, in contrast to Sect. 3, the discussion in this section was not concerned with differential operators but holds in the wider context of semi-bounded self-adjoint operators.

BKV theory was exploited in [10] to build a boundary pair for Sturm–Liouville operators with limit-circle endpoints that was compatible with a boundary triple. This allows the set up of a perturbation problem that describes all possible self-adjoint extensions of the minimal operator by using a scale of spaces that is introduced in Sect. 5.

# 5 Scale of Spaces from Singular Perturbation Theory

Let $\mathbf{A}$ and $\mathbf{T}$ be operators on a Hilbert space $\mathcal{H}$. In perturbation theory, we are interested in the following question: If properties of the operator $\mathbf{A}$ are known, what can we say about the formal operator $\mathbf{A}+\mathbf{T}$?

When the Hilbert space $\mathcal{H}$ is infinite, an immediate question is how the sum of two operators $\mathbf{A}+\mathbf{T}$ can be rigorously defined. To briefly illustrate the severity of potential problems, recall that, by the closed graph theorem, we know that the unbounded operators $\mathbf{A}$ and $\mathbf{T}$ are only defined on a dense subset of $\mathcal{H}$. Noticing this, we realize that it can easily happen that the intersection of their domains is empty. In this case, $\mathbf{A}+\mathbf{T}$ would not be interesting as its domain equals the empty set. Of course, less severe scenarios can also lead to serious issues with the meaning of $\mathbf{A}+\mathbf{T}$.

Throughout perturbation theory, this problem is dealt with by making situation or application dependent assumptions on $\mathbf{A}$ and/or $\mathbf{T}$. Most frequently, some smallness hypothesis on the perturbation $\mathbf{T}$ is imposed. In our application to Sturm–Liouville operators, both $\mathbf{A}$ and $\mathbf{T}$ will be unbounded self-adjoint operators (making their sum self-adjoint as well), though $\mathbf{A}$ will be bounded from below and $\mathbf{T}$ will be of finite rank, i.e. has finite dimensional range. While these assumptions do not imply that $\mathbf{A}+\mathbf{T}$ is well-defined, the fact that $\mathbf{T}$ is of finite rank will allow us to formulate this perturbation problem rather concretely. With this setup, the operator $\mathbf{A}$ will be one self-adjoint extensions of a Sturm–Liouville operator and the perturbed operators $\mathbf{A}+\mathbf{T}$ (if this makes sense) will stand in bijection to all possible self-adjoint extensions of the minimal operator with $\mathbf{T}$ encoding those boundary conditions.

The operator $\mathbf{T}$ can be chosen more generally to be only self-adjoint, but this would no longer be representative of self-adjoint extensions of ordinary differential operators. An infinite-dimensional self-adjoint $\mathbf{T}$ could be representative of self-adjoint extensions of partial differential operators, where deficiency indices are usually infinite, but this falls outside the scope of the current work, see e.g. [7]

for potential connections in this direction. The restrictions on the operator $\mathbf{A}$ are necessary to apply left-definite and BKV semi-bounded form theory.

Further, in our application the following key property holds: the range of $\mathbf{T}$ is contained in a Hilbert space generated by operator $\mathbf{A}$. It will turn out that $\mathbf{T}$ is relatively bounded with respect to $\mathbf{A}$, so that the domain of $\mathbf{A}+\mathbf{T}$ equals that of $\mathbf{A}$. And now operator $\mathbf{A}+\mathbf{T}$ makes sense rigorously. To carry out this plan, we now define this scale of Hilbert spaces and then discuss what $\mathbf{A}+\mathbf{T}$ looks like in our situation.

The following definition of these finite rank singular form bounded perturbations roughly follows that of [2]. Let $\mathbf{A}$ be a self-adjoint operator on $\mathcal{H}$. Consider the non-negative operator $|\mathbf{A}| = (\mathbf{A}^*\mathbf{A})^{1/2}$, whose domain coincides with the domain of $\mathbf{A}$. We introduce a scale of Hilbert spaces.

**Definition 3 (See, e.g. [2, Sect. 1.2.2])** For $s \geq 0$, define the space $\mathcal{H}_s(A)$ to consist of $\varphi \in \mathcal{H}$ for which the $s$-norm

$$\|\varphi\|_s := \|(|\mathbf{A}| + I)^{s/2}\varphi\|_{\mathcal{H}}, \tag{11}$$

is bounded. The space $\mathcal{H}_s(\mathbf{A})$ equipped with the norm $\|\cdot\|_s$ is complete. The adjoint spaces, formed by taking the linear bounded functionals on $\mathcal{H}_s(\mathbf{A})$, are used to define these spaces for negative indices, i.e. $\mathcal{H}_{-s}(\mathbf{A}) := \mathcal{H}_s^*(\mathbf{A})$. The corresponding norm in the space $\mathcal{H}_{-s}(\mathbf{A})$ is thus defined by (11) as well. The collection of these $\mathcal{H}_s(\mathbf{A})$ spaces will be called the *scale of Hilbert spaces associated with the self-adjoint operator* $\mathbf{A}$.

Alternatively, if $\mathbf{A}$ is semi-bounded with lower bound $m(\mathbf{A})$, then we can choose to consider $\mathbf{A} - \gamma$ for $\gamma < m(\mathbf{A})$ instead of $|\mathbf{A}| + I$. In particular, both options generate the same spaces with equivalent norms.

It is not difficult to see that the spaces satisfy the nesting properties

$$\ldots \subset \mathcal{H}_2(\mathbf{A}) \subset \mathcal{H}_1(\mathbf{A}) \subset \mathcal{H} = \mathcal{H}_0(\mathbf{A}) \subset \mathcal{H}_{-1}(\mathbf{A}) \subset \mathcal{H}_{-2}(\mathbf{A}) \subset \ldots,$$

and that for every two $s, t$ with $s < t$, the space $\mathcal{H}_t(\mathbf{A})$ is dense in $\mathcal{H}_s(\mathbf{A})$ with respect to the norm $\|\cdot\|_s$. Indeed, the operator $(\mathbf{A}+I)^{t/2}$ defines an isometry from $\mathcal{H}_s(\mathbf{A})$ to $\mathcal{H}_{s-t}(\mathbf{A})$. For $\varphi \in \mathcal{H}_{-s}(\mathbf{A})$, $\psi \in \mathcal{H}_s(\mathbf{A})$, we define the duality pairing

$$\langle \varphi, \psi \rangle_{s,-s} := \big\langle (|\mathbf{A}| + I)^{-s/2}\varphi, (|\mathbf{A}| + I)^{s/2}\psi \big\rangle.$$

The main perspective we are choosing here is the relation between these Hilbert scales with those spaces generated by left-definite theory, see Sect. 3. Outside of this perspective we mention on the side that throughout the literature of other fields similar constructions occur under different names. For instance, the pairing of $\mathcal{H}_1(\mathbf{A})$, $\mathcal{H}$, and $\mathcal{H}_{-1}(\mathbf{A})$ is sometimes referred to as a *Gelfand triple* or *rigged Hilbert space*. Also, when $\mathbf{A}$ is the derivative operator, these scales are simply Hilbert–Sobolev spaces. When $\mathbf{A}$ is a general differential operator, they can be closely related to Hilbert–Sobolev spaces if the coefficients are nice. A refinement of this vague notion of 'niceness' would be an interesting project. More details about Hilbert scales can be found in [23].

Aside, we also mention that finite-rank perturbations of a given operator $\mathbf{A}$ arise most commonly when the vectors $\varphi$ are bounded linear functionals on the domain of the operator $\mathbf{A}$; so, many applications are focused on $\mathcal{H}_{-2}(\mathbf{A})$. Here, we discuss the case $\varphi \in \mathcal{H}_{-1}(\mathbf{A})$ for the sake of simplicity, the so-called form bounded singular case. However, [2] contains information on extensions to $\varphi \in \mathcal{H}_{-2}(\mathbf{A})$, and the case when $\varphi \notin \mathcal{H}_{-2}(\mathbf{A})$ can be found in [13, 24].

To define rank-$d$ form bounded perturbations of a self-adjoint operator $\mathbf{A}$ on a Hilbert space $\mathcal{H}$, consider a coordinate mapping $\mathbf{B} : \mathbb{C}^d \to \operatorname{Ran}(\mathbf{B}) \subset \mathcal{H}_{-1}(\mathbf{A})$ that acts via multiplication by the row vector

$$\begin{pmatrix} f_1, & \dots, f_d \end{pmatrix} \quad \text{with } f_1, \dots, f_d \in \mathcal{H}_{-1}(\mathbf{A}).$$

Formally, the mapping $\mathbf{B}^* : \operatorname{Ran}(\mathbf{B}) \to \mathbb{C}^d$ acts by

$$\mathbf{B}^* \cdot = \begin{pmatrix} \langle \cdot , f_1 \rangle_{s,-s} \\ \vdots \\ \langle \cdot , f_d \rangle_{s,-s} \end{pmatrix}.$$

We say *formally*, because the inner products occurring in $\mathbf{B}^*$ are not defined on all of $\operatorname{Ran}(\mathbf{B})$. However, in accordance with the definition of $\mathcal{H}_{-1}(\mathbf{A})$, they do make sense as a duality pairing on the quadratic form space of the unperturbed operator $\mathbf{A}$. And that is all we need. Abusing notation slightly, we use the same notation $\mathbf{B}^*$ for the operator restricted to this form domain.

The quadratic form sense now gives rigorous meaning to the finite rank form bounded singular perturbation

$$\mathbf{A}_\Theta := \mathbf{A} + \mathbf{B}\Theta\mathbf{B}^*, \tag{12}$$

where $\Theta : \mathbb{C}^d \to \mathbb{C}^d$ is an $d \times d$ matrix (not a linear relation).

For interpretation and application it is easiest to fix a coordinate map $\mathbf{B}$. The definition in Eq. (12) can be extended to linear relations $\Theta$. It is well-known that self-adjoint boundary conditions then stand in bijection to the self-adjoint linear relation $\Theta$. One way to access the explicit translation between boundary conditions and $\Theta$ is via boundary triplets. We decided not to include this information due to accessibility, see e.g. [10]. More information on finite-rank perturbations can be found in e.g. [21, 25, 26].

Given this setup, the main problem usually becomes determining which space a desired perturbation vector comes from. In practice, it can be very difficult to actually compute the norm in Definition 3 so it does not appear there are many tools available for this purpose. However, connections with left-definite theory and BKV semi-bounded form theory can be exploited.

For a semi-bounded self-adjoint operator $\mathbf{A}$ with lower bound $m(\mathbf{A})$, choose $\gamma < m(\mathbf{A})$. Given the note after Definition 3, it is then clear that $\mathcal{H}_1(\mathbf{A} - \gamma)$ coincides with $\mathcal{H}_{\mathbf{A}-\gamma}$ (inner product given in Eq. (9)) so that the underlying spaces are the same. Hence, showing that a perturbation vector belongs to the class $\mathcal{H}_{-1}(\mathbf{A}-\gamma)$ is the same as showing it is a linear bounded functional on the first left-definite space $\mathcal{V}_1$ or the form domain $\mathfrak{t}_{\mathbf{A}-\gamma}$.

Here we find another limitation of BKV semi-bounded form theory, as it is somewhat natural to consider what spaces are analogously related for more singular perturbations. Fortunately, the connection with left-definite theory remains valid in these cases and provides good indicators for what spaces perturbation vectors might be in. For instance, a perturbation vector belongs to the class $\mathcal{H}_{-2}(\mathbf{A}-\gamma)$ if and only if it is a linear bounded functional on $\mathcal{H}_2(\mathbf{A}-\gamma)$, whose underlying space is simply the second left-space $\mathcal{V}_2$, and so on and so forth for higher values of $s$ in accordance with Corollary 1. Left-definite theory is thus a fundamental subject in the study of singular perturbation theory.

And singular perturbation theory is thus complementary to left-definite theory. It provides a rigorous analysis of linear bounded functionals on left-definite spaces, thereby expanding the framework to consider $-r \in \mathbb{N}$, and many useful applications. However, because singular perturbation theory is for general self-adjoint operators, it is not immediately clear how to obtain the key spectral stability results of Theorem 1 and explicit descriptions of the domains involved. If the self-adjoint operator is strictly positive, both left-definite and singular perturbation theory can be applied directly.

## 6 Perturbation Setup

The connection between BKV semi-bounded form theory and singular perturbation theory was recently used by the authors in [10] to obtain a characterization of all possible self-adjoint extensions of Sturm–Liouville differential operators with one or two limit-circle endpoints. As explained in the previous two sections, the connection these theories have with left-definite theory mean that the left-definite space $\mathcal{V}_1$ is also inherently involved.

The perturbation setup itself is a bit technical, so we refer the reader to the full manuscript [10] for details and cover only the broad strokes here.

The operators $\mathbf{B}$ and $\mathbf{B}^*$ are as in Sect. 5 with the perturbation vectors $f_1$ and $f_2$ chosen to have their duality pairing act like the sesquilinear form for the Sturm–Liouville operator with one input being the principal solution. Namely, $f_1$ and $f_2$ are elements from $\mathcal{H}_{-1}(\mathbf{A})$ defined so that

$$\langle \cdot, f_1\rangle_{1,-1} := [\,\cdot\,, u_a](x)\Big|_{x=a} \quad \text{and} \quad \langle \cdot, f_2\rangle_{1,-1} := [\,\cdot\,, u_b](x)\Big|_{x=b}, \tag{13}$$

where $u_a$ and $u_b$ are principal solutions to the eigenvalue problem for some $\lambda \in \mathbb{R}$ near the endpoints $x = a$ and $x = b$ of the differential equation, respectively. The self-adjoint extension $\mathbf{A}_0$ of the minimal operator is chosen to be transversal to the domain of the Friedrichs extension and defined via a boundary triple. Namely, if $\mathbf{A}_F$

is the Friedrichs extension, then the span of $\mathrm{dom}(\mathbf{A}_F) \cup \mathrm{dom}(\mathbf{A}_0)$ is the maximal domain and $\mathrm{dom}(\mathbf{A}_F) \cap \mathrm{dom}(\mathbf{A}_0)$ is the minimal domain.

**Theorem 3 ([10, Theorem 3.6])** *Let $\Theta$ be a self-adjoint linear relation in $\mathbb{C}^2$. Define $\mathbf{A}_\Theta$ as the singular rank-two perturbation:*

$$\mathbf{A}_\Theta := \mathbf{A}_0 + \mathbf{B}\Theta\mathbf{B}^*. \tag{14}$$

*Then every self-adjoint extension of the minimal operator $L_{min}$ can be written as $\mathbf{A}_\Theta$ for some $\Theta$.*

In an application, the operator $\mathbf{A}_\Theta$ being well-defined reduces to showing that the somewhat abstractly defined $f_1, f_2$ belong to $\mathcal{H}_{-1}(\mathbf{A}_0)$. It turns out that this is essentially a consequence of constructing a boundary pair. In this context, a boundary pair consists of

- a bounded operator from $\mathcal{H}_1(\mathbf{A}_0)$ to $\mathbb{C}^2$ that is surjective and whose kernel is equal to the domain of the Friedrichs extension,
- and the space $\mathbb{C}^2$.

Such a map from $\mathcal{H}_1(\mathbf{A}_0)$ to $\mathbb{C}^2$ is then a linear bounded functional on $\mathcal{H}_1(\mathbf{A}_0)$ and hence generated by a function from $\mathcal{H}_{-1}(\mathbf{A}_0)$ via the duality pairing. Indeed, the proof uses this fact, as $f_1$ and $f_2$ are chosen to naturally fit this requirement. The other part of the theorem—that Eq. (14) establishes a one-to-one relationship between self-adjoint extensions and self-adjoint linear relations $\Theta$—relies on a parameterization stemming from the theory of boundary triples. Note that the use of linear relations here is in contrast to the matrices used in Sect. 6.

We also note that the perturbation from Theorem 3 can only be rigorously interpreted by appealing to BKV semi-bounded form theory, see [10, Remark 3.7] for details.

Ideally, the perturbation setup could be adapted to work for powers of Sturm–Liouville operators. This might provide a nice class of examples of finite rank perturbations. The following conjecture states what a formulation would look like.

*Conjecture 4* Let $\mathbf{L}_0^n$, $n \in \mathbb{N}$, be a self-adjoint extension of the minimal operator $L_{\min}^n$ on $L^2[(a,b), w]$, which has deficiency indices $(m,m)$ with $m \in \{n, 2n\}$. Let $\mathbf{L}^n$ operate on its domain via $\ell^n[\,\cdot\,]$ (defined in Eq. (2)), a semi-bounded Sturm–Liouville differential expression of order $2n$ generated by composing the Sturm–Liouville operator with itself $n$ times. Let $\Theta$ be a self-adjoint relation in $\mathbb{C}^m$. Then there exist a choice of $\mathbf{L}_0^n$, $\mathbf{B}$ and $\mathbf{B}^*$ such that the singular rank-$m$ perturbation

$$\mathbf{L}_\Theta := \mathbf{L}_0^n + \mathbf{B}\Theta\mathbf{B}^*$$

is well-defined and every self-adjoint extension of the minimal operator $L_{\min}^n$ can be written as $\mathbf{L}_\Theta$ for some $\Theta$.

The parametrization of self-adjoint linear relations again relates to the boundary conditions imposed by an underlying boundary triple here, so obtaining all self-adjoint extensions via the perturbation should not be an overly difficult problem. The choice of the operator $\mathbf{L}_0^n$ allows for some freedom; it was the result of a boundary triple construction in Theorem 3 but these are not unique and the only requirement needed here is that the chosen extension is transversal to the Friedrichs domain.

The real issue is finding perturbation vectors $f_1, \ldots, f_m$ from $\mathcal{H}_{-1}(\mathbf{L}_0^n)$ so that $\mathbf{B}$ and $\mathbf{B}^*$ are well-defined. In Theorem 3 these roles were essentially played by principal solutions via Eq. (13), but an obvious analog does not immediately present itself in this case. For any $\lambda \in \mathbb{R}$ there exist general solutions to the eigenvalue problem for the uncomposed Sturm–Liouville differential operator near an endpoint, one principal and one non-principal if it is limit-circle. Hence, there are $n$ principal solutions to use in Eq. (13), but it does not appear that these functions can create a boundary triple like in the base case, see e.g. [6, Chapter 6]. This problem was solved in [19] for powers of the Jacobi differential operator by applying a modified Gram–Schmidt process to these principal solutions, but a generalization is outstanding.

A proof of Conjecture 2 would also be helpful here, as an analogous statement may be true for the transversal self-adjoint extension $\mathbf{L}_0$. A possible intermediary goal would then be to state the problem in terms of the uncomposed operator $\mathbf{L}_0$: the operator $\mathbf{L}_0^n$ would be the $2n$-th left-definite operator associated with $\mathbf{L}_0$. Hence, the main problem in Conjecture 4 would be to show that the perturbation vectors in $\mathbf{B}^*$ are in $\mathcal{H}_{-n}(\mathbf{L}_0)$. This seemingly small change could have a large impact.

## Appendix: Extension Theory

For readers not familiar with classical self-adjoint extension theory for symmetric operators we include some basic definitions and notions as applied to ordinary differential operators. The classical references [1, 31] can also be consulted for further details.

Let $\ell$ be a Sturm–Liouville differential expression. It is important to reiterate that the analysis of self-adjoint extensions does not at all involve changing the differential expression associated with the operator, merely the domain of definition by applying boundary conditions.

**Definition 4 (See, e.g. [31, Sect. 17.4])** The *maximal domain* of $\ell[\,\cdot\,]$ is given by

$$\mathcal{D}_{\max} = \mathcal{D}_{\max}(\ell) := \big\{ f : (a,b) \to \mathbb{C} \;:\; f, pf' \in \mathrm{AC}_{\mathrm{loc}}(a,b); \\ f, \ell[f] \in L^2[(a,b), w] \big\}.$$

The designation of "maximal" is appropriate in this case because $\mathcal{D}_{\max}(\ell)$ is the largest possible subset of $L^2[(a,b), w]$ that $\ell$ maps back into $L^2[(a,b), w]$. For

$f, g \in \mathcal{D}_{\max}(\ell)$ and $a < a_0 \leq b_0 < b$ the *sesquilinear form* associated with $\ell$ by

$$[f, g]\Big|_{a_0}^{b_0} := \int_{a_0}^{b_0} \left\{ \ell[f(x)]\overline{g(x)} - \ell[\overline{g(x)}]f(x) \right\} w(x)dx. \qquad \text{(A.1)}$$

The notation for the sesquilinear form does not involve $\ell$ explicitly, rather the differential expression will be clear from context.

**Theorem 4 (See, e.g. [31, Sect. 17.4])** *The limits* $[f, g](b) := \lim_{x\to b^-}[f, g](x)$ *and* $[f, g](a) := \lim_{x\to a^+}[f, g](x)$ *exist and are finite for* $f, g \in \mathcal{D}_{max}(\ell)$.

Equation (A.1) is *Green's formula* for $\ell[\,\cdot\,]$, and in the case of Sturm–Liouville operators (7) it can be explicitly computed using integration by parts to be the modified Wronskian

$$[f, g]\Big|_a^b := p(x)[f'(x)g(x) - f(x)g'(x)]\Big|_a^b.$$

**Definition 5 (See, e.g. [31, Sect. 17.4])**
The *minimal domain* of $\ell[\,\cdot\,]$ is given by

$$\mathcal{D}_{\min} = \mathcal{D}_{\min}(\ell) := \left\{ f \in \mathcal{D}_{\max}(\ell) \ : \ [f, g]\big|_a^b = 0 \ \forall g \in \mathcal{D}_{\max}(\ell) \right\}.$$

The maximal and minimal operators associated with the expression $\ell[\,\cdot\,]$ are then defined as $L_{\min} = \{\ell, \mathcal{D}_{\min}\}$ and $L_{\max} = \{\ell, \mathcal{D}_{\max}\}$, respectively. By [31, Sect. 17.2], these operators are adjoints of one another, i.e. $(L_{\min})^* = L_{\max}$ and $(L_{\max})^* = L_{\min}$. The operator $L_{\min}$ is thus symmetric.

Note that the self-adjoint extensions of a symmetric linear operator coincide with those of the closure of the symmetric linear operator [14, Theorem XII.4.8], so without loss of generality we assume that $L_{\min}$ is closed.

**Definition 6 (Variation of [31, Sect. 14.2])** Define the *positive defect space* and the *negative defect space*, respectively, by

$$\mathcal{D}_+ := \{f \in \mathcal{D}_{\max} \ : \ L_{\max} f = if\} \quad \text{and} \quad \mathcal{D}_- := \{f \in \mathcal{D}_{\max} \ : \ L_{\max} f = -if\}.$$

The dimensions $\dim(\mathcal{D}_+) = m_+$ and $\dim(\mathcal{D}_-) = m_-$, called the *positive* and *negative deficiency indices of* $L_{\min}$ respectively, play an important role and are usually conveyed as the pair $(m_+, m_-)$. The symmetric operator $L_{\min}$ has self-adjoint extensions if and only if its deficiency indices are equal [31, Sect. 14.8.8].

The classical *von Neumann formula* then says that

$$\mathcal{D}_{\max} = \mathcal{D}_{\min} \dotplus \mathcal{D}_+ \dotplus \mathcal{D}_-.$$

The decomposition can be made into an orthogonal direct sum by using the graph norm, see [18]. If the operator $L_{\min}$ has any self-adjoint extensions, then the deficiency indices of $L_{\min}$ have the form $(m, m)$, where $0 \le m \le 2$ [31, Sect. 14.8.8]. Hence, Sturm–Liouville expressions that generate self-adjoint operators must have deficiency indices $(0, 0)$, $(1, 1)$ or $(2, 2)$. If a differential expression is either in the limit-circle case or regular at the endpoint $a$, it requires a boundary condition at $a$. If it is in the limit-point case at the endpoint $a$, it does not require a boundary condition. The analogous statements are true at the endpoint $b$.

**Acknowledgments** Since August 2020, Liaw, C. has been serving as a Program Director in the Division of Mathematical Sciences at the National Science Foundation (NSF), USA, and as a component of this position, she received support from NSF for research, which included work on this paper. Any opinions, findings, and conclusions or recommendations expressed in this material are those of the authors and do not necessarily reflect the views of the NSF.

## References

1. N. Akhiezer, I. Glazman, *Theory of Linear Operators in Hilbert Space* (Dover Publications, New York, 1993)
2. S. Albeverio, P. Kurasov, *Singular Perturbations of Differential Operators*. London Mathematical Society Lecture Note Series, vol. 271 (Cambridge University Press, Cambridge, 2000)
3. A. Alonso, B. Simon, The Birman–Krein–Vishik theory of self-adjoint extensions of semibounded operators. J. Oper. Theory **4**, 251–270 (1980)
4. Y. Arlinskii, Maximal sectorial extensions and closed forms associated with them. Ukr. Math J. **48**, 723–739 (1996)
5. P. Bailey, W. Everitt, A. Zettl, Algorithm 810: the SLEIGN2 Sturm–Liouville code. ACM Trans. Math. Softw. **27**, 143–192 (2001)
6. J. Behrndt, S. Hassi, H. de Snoo, *Boundary Value Problems, Weyl Functions, and Differential Operators*. Monographs in Mathematics, vol. 108 (Birkhäuser, Basel, 2020)
7. J. Behrndt, M. Langer, V. Lotoreichik, J. Rohleder, Quasi boundary triples and semi-bounded self-adjoint extensions. Proc. R. Soc. Edinb. Sect. A **147**, 895–916 (2017)
8. S. Di Bella, C. Trapani, Some representation theorems for sesquilinear forms. J. Math. Anal. Appl. **451**, 64–83 (2017)
9. A. Bruder, L. Littlejohn, D. Tuncer, R. Wellman, Left-definite theory with applications to orthogonal polynomials. J. Comput. Appl. Math. **233**, 1380–1398 (2010)
10. M. Bush, D. Frymark, C. Liaw, Singular boundary conditions for Sturm–Liouville operators via perturbation theory (2020). Preprint, arXiv:2011.03388
11. R. Corso, A Kato's second type representation theorem for solvable sesquilinear forms. J. Math. Anal. Appl. **462**, 982–998 (2018)
12. V. Domínguez, N. Heuer, F. Sayas, Hilbert scales and Sobolev spaces defined by associated Legendre functions. J. Comput. Appl. Math. **235**, 3481–3501 (2011)
13. A. Dijksma, P. Kurasov, Y. Shondin, High order singular rank-one perturbations of a positive operator. Integr. Equ. Oper. Theory **53**, 209–245 (2005)
14. N. Dunford, J. Schwartz, *Linear Operators, Part II* (Wiley Classics Library, New York, 1988)
15. W. Everitt, A catalogue of Sturm–Liouville differential equations, in *Sturm–Liouville Theory: Past and Present* (Birkhäuser, Basel, 2001), pp. 271–331
16. W. Everitt, K. Kwon, L. Littlejohn, R. Wellman, G. Yoon, Jacobi–Stirling numbers, Jacobi polynomials, and the left-definite analysis of the classical Jacobi expression. J. Comput. Appl. Math. **208**, 29–56 (2007)
17. W. Everitt, L. Littlejohn, D. Tuncer, Some remarks on classical lagrangian symmetric differential expressions and their composite powers. Adv. Dyn. Syst. Appl. **2**, 187–206 (2007)

18. M. Fleeman, D. Frymark, C. Liaw, Boundary conditions associated with the general left-definite theory for differential operators. J. Approx. Theory **239**, 1–28 (2019)
19. D. Frymark, Boundary triples and Weyl $m$-functions for powers of the Jacobi differential operator. J. Differ. Equ. **269**, 7931–7974 (2020)
20. D. Frymark, C. Liaw, Properties and decompositions of domains for powers of the Jacobi differential operator. J. Math. Anal. Appl. **489**, 124–155 (2020)
21. D. Frymark, C. Liaw, *Spectral Analysis, Model Theory and Applications of Finite-Rank Perturbations*, invited contribution accepted to Proceedings of IWOTA 2018: Operator Theory, Operator Algebras and Noncommutative Topology (Ronald G. Douglas Memorial Volume). Operator Theory: Advances and Applications (Book 278), 1st edn. (Birkhäuser, Basel, 2021) (November 2020). https://doi.org/10.1007/978--3-030-43380-2
22. T. Kato, *Perturbation Theory for Linear Operators*. Classics in Mathematics (Springer, Berlin (1995). Preprint of the 1980 edition
23. S. Krein, I. Petunin, Scales of Banach spaces. Russ. Math. Surv. **21**, 85–159 (1966)
24. P. Kurasov, $\mathcal{H}_{-n}$-perturbations of self-adjoint operators and Krein's resolvent formula. Integr. Equ. Oper. Theory **45**, 437–460 (2003).
25. C. Liaw, S. Treil, Matrix measures and finite rank perturbations of self-adjoint operators. J. Spectral Th. (2020). https://doi.org/10.4171/JST/324
26. C. Liaw, S. Treil, *Singular Integrals, Rank-One Perturbations and Clark Model in General Situation*, ed. by M.C. Pereyra, S. Marcantognini, A. Stokolos, W. Urbina. Harmonic Analysis, Partial Differential Equations, Complex Analysis, Banach Spaces, and Operator Theory, vol. 2. Celebrating Cora Sadosky's life. AWM-Springer Series, vol. vol. 5 (Springer, Berlin, 2017), pp. 86–132
27. L. Littlejohn, R. Wellman, A general left-definite theory for certain self-adjoint operators with applications to differential equations. J. Differ. Equ. **181**, 280–339 (2002)
28. L. Littlejohn, R. Wellman, On the spectra of left-definite operators. Complex Anal. Oper. Theory **7**, 437–455 (2013)
29. L. Littlejohn, Q. Wicks, Glazman–Krein–Naimark Theory, left-definite theory and the square of the Legendre polynomials differential operator. J. Math. Anal. Appl. **444**, 1–24 (2016)
30. M. Marletta, A. Zettl, The Friedrichs extension of singular differential operators. J. Differ. Equ. **160**, 404–421 (2000)
31. M. Naimark, Linear differential operators parts I, II (Frederick Ungar Publishing, New York, 1972)
32. H. Weyl, Über gewöhnliche Differentialgleichungen mit Singularitäten und die zugehörigen Entwicklungen willkürlicher Funktionen. Math. Ann. **68**, 220–269 (1910)
33. A. Zettl, *Sturm–Liouville Theory*. Mathematical Surveys and Monographs, vol. 121 (American Mathematical Society, Providence, 2005)

# Sampling in the Range of the Analysis Operator of a Continuous Frame Having Unitary Structure

**Antonio G. García**

*The author is very pleased to dedicate this work to his friend, the mathematician Lance L. Littlejohn on the occasion of his 70th birthday.*

**Abstract** We establish a regular sampling theory in the range of the analysis operator of a continuous frame having a unitary structure. The unitary structure is related with a unitary representation of a locally compact abelian group on a separable Hilbert space. The samples are defined by means of suitable discrete convolution systems which generalize some usual sampling settings; here regular sampling means that the samples are taken at a countable discrete subgroup.

**Keywords** Unitary representations of LCA groups · Continuous and discrete frames · Stable sampling expansions

## 1 Statement of the Problem

In this work a regular sampling theory in the range space of the analysis operator of a continuous frame is established. To be more specific, the continuous frame has a unitary structure associated with a unitary representation $t \in G \mapsto U(t)$ of an LCA (locally compact abelian) group $G$ on a separable Hilbert space $\mathcal{H}$. The functions to be recovered, at a countable discrete subgroup $H$ of $G$, are associated with functions in a unitary $H$-invariant subspace $\mathcal{H}_\Phi$ of $\mathcal{H}$ (that, eventually, could coincide with $\mathcal{H}$); the subscript $\Phi = \{\varphi_1, \varphi_2, \dots, \varphi_N\} \subset \mathcal{H}$ denotes a set of stable generators for

A. G. García (✉)
Department of Mathematics, Universidad Carlos III de Madrid, Leganés-Madrid, Spain
e-mail: agarcia@math.uc3m.es



F. Gesztesy, A. Martinez-Finkelshtein (eds.), *From Operator Theory to Orthogonal Polynomials, Combinatorics, and Number Theory*, Operator Theory: Advances and Applications 285, https://doi.org/10.1007/978-3-030-75425-9_7

$\mathcal{H}_\Phi$. Thus, the functions to be recovered from a sequence of samples at $H$ look like:

$$F_f(t) = \langle f, U(t)\phi \rangle_{\mathcal{H}}, \; t \in G, \quad \text{where } f \in \mathcal{H}_\Phi .$$

The set of these functions $F$ (we omit the subscript) forms a reproducing kernel Hilbert space (in short RKHS) of continuous and bounded functions contained in $L^2(G)$ that will be denoted as $\mathcal{H}_{U,\phi,\Phi}$. For some examples of these spaces we cite, among others, the Paley–Wiener spaces $PW_{\pi\sigma}$, shift-invariant subspaces $V_\Phi^2$ in $L^2(\mathbb{R}^d)$ and, in the non-abelian case, the range space of the continuous wavelet or Gabor transforms.

The goal in this work is to recover, in a stable way, any $F$ in $\mathcal{H}_{U,\phi,\Phi}$ from a sequence of samples taken at the subgroup $H$. For instance, for the sequence of *pointwise samples* $\{F(t)\}_{t\in H}$ we obtain a suitable expression as the output of a discrete convolution system (see Eq. (4) in Sect. 3)

$$F(t) = \sum_{n=1}^{N} \big(a_n *_H x_n\big)(t), \; t \in H,$$

where $a_n(t) := \langle \varphi_n, U(t)\phi \rangle_{\mathcal{H}}$, $t \in H$, belongs to $\ell^2(H)$ for $n = 1, 2, \ldots, N$, $x_n \in \ell^2(H)$ are the coefficients of the expansion $f = \sum_{n=1}^{N} \sum_{t\in H} x_n(t)U(t)\varphi_n$, and $*_H$ denotes the convolution in $\ell^2(H)$. The same occurs for the *average samples* $\{\mathcal{M}_m F(t)\}_{t\in H}$ defined by

$$\mathcal{M}_m F(t) := \langle f, U(t)\psi_m \rangle_{\mathcal{H}}, \; t \in H, \; \text{for some fixed } \psi_m \in \mathcal{H}.$$

The above examples lead us to define, for any $F$ in the space $\mathcal{H}_{U,\phi,\Phi}$, a sequence of *generalized samples* $\{\mathcal{L}_m F(t)\}_{t\in H;\, m=1,2,\ldots,M}$, at the subgroup $H$, by means of an $M \times N$ matrix $A = [a_{m,n}]$ with entries in $\ell^2(H)$ as

$$\mathcal{L}_m F(t) := \sum_{n=1}^{N} \big(a_{m,n} *_H x_n\big)(t), \quad t \in H, \quad m = 1, 2, \ldots, M.$$

Thus, under appropriate conditions (see Definition 1 in Sect. 3), the main sampling result (see Theorem 1 Sect. 4) proves that there exist $M$ sampling functions $S_m$ in $\mathcal{H}_{U,\phi,\Phi}$, $m = 1, 2, \ldots, M$, such that the sequence $\{S_m(\cdot - t)\}_{t\in H;\, m=1,2,\ldots,M}$ is a frame for $\mathcal{H}_{U,\phi,\Phi}$, and the sampling expansion

$$F(s) = \sum_{m=1}^{M} \sum_{t\in H} \mathcal{L}_m F(t)\, S_m(s-t), \quad s \in G,$$

holds for every $F \in \mathcal{H}_{U,\phi,\Phi}$. In addition, the sampling functions $S_m$, $m = 1, 2, \ldots, M$, can be obtained, via the matrix $A$, by means of an explicit method (see in the end of Sect. 3).

The used mathematical techniques are similar to those in Ref. [14]. They lie in exploiting the relationship between discrete convolution systems and frames of translates in the product Hilbert space $\ell^2_N(H) := \ell^2(H) \times \cdots \times \ell^2(H)$ ($N$ times); this is an auxiliary space isomorphic to $\mathcal{H}_{U,\phi,\Phi}$.

The work is organized as follows: In Sect. 2 we include some needed preliminaries on continuous and discrete frames; on Fourier analysis for a countable discrete group, and on convolution systems in the Hilbert space $\ell^2_N(H)$. It is worth to mention the relationship between convolution systems in $\ell^2_N(H)$ and frames of translates in $\ell^2_N(H)$ showing the equivalence of their properties. The needed results have been borrowed from Refs. [14, 20]. Section 3 is devoted to introduce the subspace of $L^2(G)$ where the sampling theory will be carried out. Finally, Sect. 4 contains the main sampling result along with some pertinent comments and remarks. Although the work deals with abelian groups, an example involving semi-direct products of groups is included; this particular non-abelian case will be treated by using the theory developed in Sect. 4.

It should be noted that working in LCA groups is not just a unified way of dealing with the classical groups $\mathbb{R}^d$, $\mathbb{Z}^d$, $\mathbb{T}^d$, $\mathbb{Z}^d_s$: signal processing often involves products of these groups which are also locally compact abelian groups. For example, multichannel video signal involves the group $\mathbb{Z}^d \times \mathbb{Z}_s$, where $d$ is the number of channels and $s$ the number of pixels of each image. Finally, some companion references in sampling theory are, for instance, Refs. [2, 6, 9, 19, 21, 23].

## 2 Some Preliminaries

### 2.1 Continuous and Discrete Frames

Let $\mathcal{H}$ be a Hilbert space and let $(\Omega, \mu)$ be a measure space. A mapping $\psi : \Omega \to \mathcal{H}$ is a *continuous frame* for $\mathcal{H}$ with respect to $(\Omega, \mu)$ if $\psi$ is weakly measurable, i.e., for each $x \in \mathcal{H}$ the function $w \mapsto \langle x, \psi(w)\rangle$ is measurable, and there exist constants $0 < A \leq B$ such that

$$A\|x\|^2 \leq \int_\Omega \left|\langle x, \psi(w)\rangle\right|^2 d\mu(w) \leq B\|x\|^2 \quad \text{for each } x \in \mathcal{H}. \tag{1}$$

The constants $A$ and $B$ are the lower and upper continuous frame bounds respectively. The mapping $\psi$ is a *tight continuous frame* if $A = B$; a *Parseval continuous frame* if $A = B = 1$. The mapping $\psi$ is called a *Bessel family* if only the right-hand inequality holds. Throughout this paper we refer a continuous frame as the mapping $\psi : \Omega \to \mathcal{H}$, or as the family $\{\psi(w)\}_{w\in\Omega}$, or $\{\psi_w\}_{w\in\Omega}$, in the Hilbert space $\mathcal{H}$.

There are a lot of examples of continuous frames in the mathematical/physics literature. For instance: the family $\{k_w\}_{w\in\Omega}$ of reproducing kernels of a RKHS $\mathcal{H}_k$ contained in $L^2(\Omega, \mu)$ is a continuous Parseval frame with respect to $(\Omega, \mu)$; a

*Gabor system* $\{M_\xi T_x g \,:\, (x,\xi) \in \mathbb{R}^d \times \mathbb{R}^d\}$ is a tight continuous frame for $L^2(\mathbb{R}^d)$ with respect to $(\mathbb{R}^d \times \mathbb{R}^d, dx\, d\xi)$, where $g \in L^2(\mathbb{R}^d)$ is a fixed non zero function, and $M_\xi$ and $T_x$ denote the modulation and translation operators in $L^2(\mathbb{R}^d)$ respectively; a *wavelet system* $\{T_b D_a \psi \,:\, (a,b) \in (\mathbb{R}\setminus\{0\}) \times \mathbb{R}\}$ is a tight continuous frame for $L^2(\mathbb{R})$ with respect to $\big((\mathbb{R}\setminus\{0\}) \times \mathbb{R}, \dfrac{da db}{a^2}\big)$, where $T_b$ and $D_a$ denote the translation and dilation operators in $L^2(\mathbb{R})$ respectively, and $\psi \in L^2(\mathbb{R})$ is an admissible function, i.e., a function for which the constant $c_\psi := \int_{\mathbb{R}} \frac{|\widehat{\psi}(w)|^2}{|w|} dw < +\infty$; *coherent states* in physics, etc. (see, for instance, Refs. [4, 8, 10]).

The operator $T_\psi : L^2(\Omega,\mu) \to \mathcal{H}$ weakly defined for each $f \in L^2(\Omega,\mu)$ by

$$\langle T_\psi f, x\rangle = \int_\Omega f(w)\big\langle \psi(w), x\big\rangle d\mu(w)\,, \quad x \in \mathcal{H},$$

is linear and bounded; it is called the *synthesis operator* of $\{\psi_w\}_{w\in\Omega}$. Its adjoint operator $T_\psi^* : \mathcal{H} \to L^2(\Omega,\mu)$ is given by $(T_\psi^* x)(w) = \langle x, \psi(w)\rangle$, $w \in \Omega$, and it is called the *analysis operator* of $\{\psi_w\}_{w\in\Omega}$. The *continuous frame operator* $S_\psi = T_\psi T_\psi^*$ is a bounded, self-adjoint, positive and invertible operator in $\mathcal{H}$. For any $x \in \mathcal{H}$ we have the weak representations

$$x = \int_\Omega \big\langle x, \psi(w)\big\rangle S_\psi^{-1}\psi(w) d\mu(w) = \int_\Omega \big\langle x, S_\psi^{-1}\psi(w)\big\rangle \psi(w) d\mu(w)\,.$$

The counting measure $\mu$ on $\Omega = \mathbb{N}$ gives the classical definition of discrete frame $\{x_n\}_{n=1}^\infty$: there exist two constants $0 < A \le B$ such that

$$A\|x\|^2 \le \sum_{n=1}^{\infty} |\langle x, x_n\rangle|^2 \le B\|x\|^2 \quad \text{for each } x \in \mathcal{H}. \tag{2}$$

Given a discrete frame $\{x_n\}_{n=1}^\infty$ for $\mathcal{H}$, its *synthesis operator* $T : \ell^2(\mathbb{N}) \to \mathcal{H}$ is defined by $T\big(\{c_n\}_{n=1}^\infty\big) = \sum_{n=1}^\infty c_n\, x_n$. Its adjoint operator $T^* : \mathcal{H} \to \ell^2(\mathbb{N})$ is given by $T^* x = \{\langle x, x_n\rangle\}_{n=1}^\infty$, and it is called its *analysis operator*. The *frame operator* $S$ is defined by $S(x) := T\,T^* x = \sum_{n=1}^\infty \langle x, x_n\rangle\, x_n$, $x \in \mathcal{H}$; it is a bounded, invertible, positive and self-adjoint operator in $\mathcal{H}$. The sequence $\{S^{-1}x_n\}_{n=1}^\infty$ is also a frame for $\mathcal{H}$ called the *canonical dual frame*. For each $x \in \mathcal{H}$ we have the expansions

$$x = \sum_{n=1}^{\infty} \langle x, x_n\rangle\, S^{-1}x_n = \sum_{n=1}^{\infty} \langle x, S^{-1}x_n\rangle\, x_n\,.$$

As a consequence, given a frame $\{x_n\}_{n=1}^\infty$ for $\mathcal{H}$ the representation property of any vector $x \in \mathcal{H}$ as a series $x = \sum_{n=1}^\infty c_n\, x_n$ is retained, but, unlike the case of Riesz (orthonormal) bases, the uniqueness of this representation is sacrificed. Suitable frame coefficients $\{c_n\}$ which depend continuously and linearly on $x$ are obtained by using the *dual frames* $\{y_n\}_{n=1}^\infty$ of $\{x_n\}_{n=1}^\infty$, i.e., $\{y_n\}_{n=1}^\infty$ is another frame for $\mathcal{H}$ such that

$$x = \sum_{n=1}^{\infty} \langle x, y_n \rangle x_n = \sum_{n=1}^{\infty} \langle x, x_n \rangle y_n \quad \text{for each } x \in \mathcal{H}.$$

In particular, frames include orthonormal and Riesz bases for $\mathcal{H}$. For more details and proofs on discrete and continuous frames, see, for instance, Refs. [4, 8, 10, 13, 22].

Assume that $\{\psi(w)\}_{w\in\Omega}$ is a continuous frame for a Hilbert space $\mathcal{H}$ with respect to $(\Omega, \mu)$ such that the mapping $w \mapsto \psi(w)$ is weakly continuous, i.e., for each $x \in \mathcal{H}$ the function $w \mapsto \langle x, \psi(w)\rangle$ is continuous. Its analysis operator $T_\psi^* : \mathcal{H} \to L^2(\Omega, \mu)$ is a bounded and boundedly invertible operator on its range denoted as $\mathcal{H}_\psi := \text{Range } T_\psi^*$. This is a closed subspace of $L^2(\Omega, \mu)$ described as the functions $F_x$ such that

$$\begin{aligned} &\mathcal{H} \longrightarrow \mathcal{H}_\psi \\ &x \longmapsto F_x \ : \ F_x(w) = \langle x, \psi(w)\rangle_{\mathcal{H}}, \quad w \in \Omega. \end{aligned}$$

Besides $\mathcal{H}_\psi$ is a RKHS (of continuous functions in $\Omega$) whose reproducing kernel is given by

$$k_\psi(u, v) = \langle \psi(v), S_\psi^{-1}\psi(u)\rangle_{\mathcal{H}}, \quad u, v \in \Omega,$$

where $S_\psi^{-1}$ denotes the inverse of the frame operator $S_\psi$ associated to $\{\psi(w)\}_{w\in\Omega}$. That is, for each $F_x \in \mathcal{H}_\psi$ we have the reproducing property

$$F_x(u) = \int_\Omega F_x(v)\, k_\psi(u, v)\, d\mu(v) = \langle F_x, k_\psi(\cdot, u)\rangle_{L^2(\Omega,\mu)}, \quad u \in \Omega.$$

## 2.2 *Discrete Convolution Systems and Frames of Translates*

Let $(H, +)$ be a countable discrete abelian group and let $\mathbb{T} = \{z \in \mathbb{C} : |z| = 1\}$ be the unidimensional torus. A *character* $\xi$ of $H$ is a homomorphism $\xi : H \to \mathbb{T}$, i.e., $\xi(h + h') = \xi(h)\xi(h')$ for all $h, h' \in H$; we denote $\xi(h) = (h, \xi)$. By defining $(\xi + \xi')(h) = \xi(h)\xi'(h)$, the set of characters $\widehat{H}$ is a group, called the *dual group* of $H$; since $H$ is discrete, the group $\widehat{H}$ is compact [11, Prop. 4.4].

For $x \in \ell^1(H)$ its *Fourier transform* is defined by

$$\widehat{x}(\xi) := \sum_{h\in H} x(h)\overline{(h, \xi)} = \sum_{h\in H} x(h)(-h, \xi), \quad \xi \in \widehat{H}.$$

The Plancherel theorem extends uniquely the Fourier transform on $\ell^1(H) \cap \ell^2(H)$ to a unitary isomorphism from $\ell^2(H)$ to $L^2(\widehat{H})$. For the details see, for instance, Refs. [11, 12].

Let $H$ be a countable discrete group and let consider the product Hilbert space $\ell^2_N(H) := \ell^2(H) \times \cdots \times \ell^2(H)$ ($N$ times). For a matrix $A = [a_{m,n}] \in \mathcal{M}_{M\times N}\big(\ell^2(H)\big)$, i.e., an $M \times N$ matrix with entries in $\ell^2(H)$, and $\mathbf{x} \in \ell^2_N(H)$, the *matrix convolution* $A * \mathbf{x}$ in $H$ is defined by

$$(A * \mathbf{x})(h) := \sum_{h' \in H} A(h - h')\,\mathbf{x}(h'), \quad h \in H\,.$$

Note that the $m$-th entry of $A * \mathbf{x}$ is $\sum_{n=1}^{N}(a_{m,n} * x_n)$, where $x_n$ denotes the $n$-th entry of $\mathbf{x} \in \ell^2_N(H)$. The usual properties of a discrete convolution can be found in Refs. [11, 12].

The discrete *convolution system* $\mathcal{A}$ with associated matrix $A \in \mathcal{M}_{M\times N}\big(\ell^2(H)\big)$ given by

$$\begin{aligned} \mathcal{A} : \ell^2_N(H) &\longrightarrow \ell^2_M(H) \\ \mathbf{x} &\longmapsto \mathcal{A}(\mathbf{x}) = A * \mathbf{x} \end{aligned} \tag{3}$$

is a well defined bounded operator if and only if $\widehat{A} \in \mathcal{M}_{M\times N}\big(L^\infty(\widehat{H})\big)$, where $\widehat{A}(\xi) := \big[\widehat{a}_{m,n}(\xi)\big]$ denotes the *transfer matrix* of $A$ (see Refs. [14, 20]). We identify the matrix $\widehat{A}$ with entries in $L^\infty(\widehat{\Lambda})$ and the essentially bounded matrix-valued function $\widehat{A}(\xi)$, a.e. $\xi \in \widehat{\Lambda}$.

Its adjoint operator $\mathcal{A}^* : \ell^2_M(H) \to \ell^2_N(H)$ is also a bounded convolution system with associated matrix $A^* = \big[a^*_{m,n}\big]^\top \in \mathcal{M}_{N\times M}\big(\ell^2(H)\big)$, where $a^*_{m,n}$ denotes the involution $a^*_{m,n}(h) := \overline{a_{m,n}(-h)}$, $h \in H$ (Refs. [14, 20]). Its transfer matrix is $\widehat{A^*}(\xi)$ is just the transpose conjugate of $\widehat{A}(\xi)$, i.e., $\widehat{A}(\xi)^*$, a.e. $\xi \in \widehat{H}$ (Refs. [14, 20]).

The bounded operator $\mathcal{A}$ is injective with a closed range if and only if the operator $\mathcal{A}^*\mathcal{A}$ is invertible; equivalently (see Ref. [14]), if and only if the constant $\delta_A := \operatorname{ess\,inf}_{\xi\in\widehat{H}} \det[\,\widehat{A}(\xi)^*\widehat{A}(\xi)] > 0$.

Let $\mathbf{a}^*_m$ denote the $m$-th column of the matrix $A^*$, then the $m$-th component of $\mathcal{A}(\mathbf{x})$ is

$$[A * \mathbf{x}]_m(h) = \sum_{n=1}^{N}(a_{m,n} *_H x_n)(h) = \langle \mathbf{x}, T_h\,\mathbf{a}^*_m\rangle_{\ell^2_N(H)}, \quad h \in H\,,$$

where $T_h$ denotes the translation operator by $h \in H$ in $\ell^2_N(H)$, i.e., for $\mathbf{a} \in \ell^2_N(H)$, $T_h\,\mathbf{a}(g) = \mathbf{a}(g-h)$, $g \in H$, and $*_H$ is the convolution indexed by $H$. In other words, the operator $\mathcal{A}$ is the *analysis operator* of the sequence $\big\{T_h\mathbf{a}^*_m\big\}_{h\in H;\, m=1,2,\dots,M}$ in $\ell^2_N(H)$. Since the sequence $\big\{T_h\,\mathbf{a}^*_m\big\}_{h\in H;\, m=1,2,\dots,M}$ is a frame for $\ell^2_N(H)$ if and only if its (bounded) analysis operator is injective with a closed range (see Ref. [8]).

Hence, the sequence $\{T_h \mathbf{a}_m^*\}_{h\in H;\, m=1,2,\dots,M}$ will be a frame for $\ell_N^2(H)$ if and only if the constant $\delta_A > 0$.

Concerning the duals of $\{T_h\, \mathbf{a}_m^*\}_{h\in H;\, m=1,2,\dots,M}$ having the same structure, consider two matrices $\widehat{A} \in \mathcal{M}_{M\times N}(L^\infty(\widehat{H}))$ and $\widehat{B} \in \mathcal{M}_{N\times M}(L^\infty(\widehat{H}))$, and let $\mathbf{b}_m$ denote the $m$-th column of the matrix $B$ associated to $\widehat{B}$. Then, the sequences $\{T_h\, \mathbf{a}_m^*\}_{h\in H;\, m=1,2,\dots,M}$ and $\{T_h\, \mathbf{b}_m\}_{h\in H;\, m=1,2,\dots,M}$ form a pair of dual frames for $\ell_N^2(H)$ if and only if $\widehat{B}(\xi)\,\widehat{A}(\xi) = I_N$, a.e. $\xi \in \widehat{H}$; equivalently, if and only if $\mathcal{B}\mathcal{A} = I_{\ell_N^2(H)}$, i.e., the convolution system $\mathcal{B}$ is a left-inverse of the convolution system $\mathcal{A}$ (see Ref. [14]). Thus in $\ell_N^2(H)$ we have the frame expansion

$$\mathbf{x} = \sum_{m=1}^{M} \sum_{h\in H} \langle \mathbf{x}, T_h \mathbf{a}_m^* \rangle_{\ell_N^2(H)}\, T_h \mathbf{b}_m \quad \text{for each } \mathbf{x} \in \ell_N^2(H)\,.$$

Finally to remind that the convolution system $\mathcal{A}$ in (3) is an isomorphism if and only if $M = N$ and $\operatorname{ess\,inf}_{\xi\in\widehat{H}} \big|\det \widehat{A}(\xi)\big| > 0$ (see Refs. [14, 20]). Thus, for the case $M = N$ the sequence $\{T_h\, \mathbf{a}_m^*\}_{h\in H;\, m=1,2,\dots,N}$ is a Riesz basis for $\ell_N^2(H)$. The square matrix $\widehat{A}(\xi)$ is invertible, a.e. $\xi \in \widehat{H}$, and from the columns of $\widehat{A}(\xi)^{-1}$ we get its dual Riesz basis $\{T_h\, \mathbf{b}_m\}_{h\in H;\, m=1,2,\dots,N}$.

## 3 The Subspace of $L^2(G)$ Where the Sampling Is Carried Out

Let $G \ni t \longmapsto U(t) \in \mathcal{U}(\mathcal{H})$ be a *unitary representation* of a LCA group $(G, +)$ on a separable Hilbert space $\mathcal{H}$. Recall that $\{U(t)\}_{t\in G}$ is a family of unitary operators in $\mathcal{H}$ satisfying: $U(t)\,U(t') = U(t+t')$ for $t, t' \in G$; $U(0) = I_{\mathcal{H}}$; and $\langle U(t)\varphi, \phi\rangle_{\mathcal{H}}$ is a continuous function of $t$ for any $\varphi, \phi \in \mathcal{H}$. Note that $U(t)^{-1} = U(-t)$, and since $U(t)^* = U(t)^{-1}$ we have $U(t)^* = U(-t)$.

Assume that for a fixed $\phi \in \mathcal{H}$ the family $\{U(t)\phi\}_{t\in G}$ is a continuous frame for the Hilbert space $\mathcal{H}$ with respect to $(G, \mu_G)$, where $\mu_G$ denotes the Haar measure on $G$. Let $H$ be a countable discrete subgroup of $G$. For a stable set of generators $\Phi = \{\varphi_1, \varphi_2, \dots, \varphi_N\} \subset \mathcal{H}$ we consider the $U$-invariant subspace in $\mathcal{H}$ generated by $\Phi$:

$$\mathcal{H}_\Phi = \Big\{ \sum_{n=1}^{N} \sum_{h\in H} x_n(h) U(h)\varphi_n \ :\ x_n \in \ell^2(H),\ n = 1, 2, \dots, N \Big\}.$$

We are assuming that the sequence $\{U(h)\varphi_n\}_{h\in H;\, n=1,2,\dots,N}$ is a Riesz sequence in $\mathcal{H}$, i.e., a Riesz basis for $\mathcal{H}_\Phi$. Finally, we define the subspace in $L^2(G)$ given by

$$\mathcal{H}_{U,\phi,\Phi} := \Big\{ F_f : G \longrightarrow \mathbb{C} \ :\ F_f(t) = \langle f, U(t)\phi\rangle_{\mathcal{H}},\ t \in G\,,\ \text{where } f \in \mathcal{H}_\Phi \Big\}.$$

Notice that the mapping $\mathcal{H}_\Phi \ni f \longmapsto F_f \in \mathcal{H}_{U,\phi,\Phi}$ is an isomorphism between Hilbert spaces, and the space $\mathcal{H}_{U,\phi,\Phi}$ is a RKHS of continuous and bounded functions in $L^2(G)$.

Besides, the space $\mathcal{H}_{U,\phi,\Phi}$ is an $H$-shift-invariant subspace of $L^2(G)$; indeed, for each $t \in H$ we have that the function $F_f(s-t) = F_{U(t)f}(s)$, $s \in G$, and consequently it belongs to $\mathcal{H}_{U,\phi,\Phi}$.

For instance, if $\phi = \text{sinc}$ is the cardinal sine function then $\{U(t)\phi = \text{sinc}(\cdot - t)\}_{t\in\mathbb{R}}$ is a continuous frame for $\mathcal{H} = PW_\pi$ with respect to $(\mathbb{R}, dx)$ and the space $\mathcal{H}_{U,\phi,\Phi}$ coincides with $PW_\pi$.

From now on we will omit the subscript $f$ in the notation of $F_f$. The aim of this work is to obtain stable sampling results for the functions $F \in \mathcal{H}_{U,\phi,\Phi}$; for instance, the stable recovery of any $F \in \mathcal{H}_{U,\phi,\Phi}$ from the sequence of its samples $\{F(t)\}_{t\in H}$ taken at the countable subgroup $H$ of $G$ and/or other sequences of samples $\{\mathcal{L}_m F(t)\}_{t\in H}$ introduced in next section.

### 3.1 Sampling Data as a Filtering Process

Let $F$ be a function in $\mathcal{H}_{U,\phi,\Phi}$. For each $t \in H$, we have for the sample $F(t)$ the expression

$$\begin{aligned} F(t) &= \langle f, U(t)\phi\rangle_{\mathcal{H}} = \Big\langle \sum_{n=1}^{N}\sum_{h\in H} x_n(h)U(h)\varphi_n, U(t)\phi\Big\rangle_{\mathcal{H}} \\ &= \sum_{n=1}^{N}\sum_{h\in H} x_n(h)\langle \varphi_n, U(t-h)\phi\rangle_{\mathcal{H}} = \sum_{n=1}^{N}\big(a_n *_H x_n\big)(t), \quad t\in H, \end{aligned} \tag{4}$$

where $a_n(s) := \langle \varphi_n, U(s)\phi\rangle_{\mathcal{H}}$, $s \in H$, and $*_H$ denotes the convolution in $\ell^2(H)$.

Similarly, we can define generalized average samples as follows: Given $M$ fixed elements $\psi_1, \psi_2, \ldots, \psi_M$ in $\mathcal{H}$, for any $F \in \mathcal{H}_{U,\phi,\Phi}$ we define the samples at $H$ as

$$\mathcal{M}_m F(t) := \langle f, U(t)\psi_m\rangle_{\mathcal{H}}, \quad t\in H\,, \text{ and } m = 1, 2, \ldots, M\,. \tag{5}$$

Proceeding as in Eq. (4), for each $m = 1, 2, \ldots, M$ we get

$$\mathcal{M}_m F(t) = \sum_{n=1}^{N}\big(a_{m,n} *_H x_n\big)(t)\,, \quad t\in H\,,$$

where $a_{m,n}(s) := \langle \varphi_n, U(s)\psi_m \rangle_{\mathcal{H}}$, $s \in H$. The above examples together with the results in Sect. 2.2 lead us to define in $\mathcal{H}_{U,\phi,\Phi}$ a *generalized stable sampling procedure* at the subgroup $H$ as follows:

**Definition 1** Let $F(s) = \langle f, U(s)\phi \rangle_{\mathcal{H}}$, $s \in G$, be a function in $\mathcal{H}_{U,\phi,\Phi}$ and suppose that $f = \sum_{n=1}^{N} \sum_{h \in H} x_n(h) U(h)\varphi_n$ in $\mathcal{H}_\Phi$. A generalized stable sampling procedure $\mathcal{L}_A$ at $H$ in $\mathcal{H}_{U,\phi,\Phi}$ is defined for each $F \in \mathcal{H}_{U,\phi,\Phi}$ by

$$\mathcal{L}_A F(t) := \big(A *_H \mathbf{x}\big)(t), \quad t \in H, \tag{6}$$

where $\mathcal{L}_A F(t) := \big(\mathcal{L}_1 F(t), \mathcal{L}_2 F(t), \dots, \mathcal{L}_M F(t)\big)^\top$, $A$ denotes a matrix $[a_{m,n}]$ in $\mathcal{M}_{M \times N}\big(\ell^2(H)\big)$, and $\mathbf{x}(t) = (x_1(t), x_2(t), \dots, x_N(t))^\top$ belongs to $\ell^2_N(H)$, such that:

1. Its tranfer matrix $\widehat{A} \in \mathcal{M}_{M \times N}\big(L^\infty(\widehat{H})\big)$, and
2. the constant $\delta_A := \operatorname{ess\,inf}_{\xi \in \widehat{H}} \det[\widehat{A}(\xi)^* \widehat{A}(\xi)] > 0$.

Definition 1 is, of course, equivalent to the classical one that states the existence of two positive constants $0 < c \le C$ such that

$$c\|F\|^2 \le \sum_{m=1}^{M} \sum_{t \in H} |\mathcal{L}_m F(t)|^2 \le C\|F\|^2 \quad \text{for any } F \in \mathcal{H}_{U,\phi,\Phi}.$$

See Notes 2 and 7 in Sect. 4.

The definition of stable sampling as stated above shows, in a explicit way, the relationship between the stable samples and their associated sampling formulas. Indeed, as it will be proved in Theorem 1 (see Sect. 4), once a generalized stable sampling procedure $\mathcal{L}_A$ at $H$ is defined in $\mathcal{H}_{U,\phi,\Phi}$, there exists a method to obtain stable sampling formulas, associated to $\mathcal{L}_A$, and having the form:

$$F(s) = \sum_{m=1}^{M} \sum_{t \in H} \mathcal{L}_m F(t)\, S_m(s-t), \quad s \in G,$$

for every function $F \in \mathcal{H}_{U,\phi,\Phi}$. Namely:

- Compute a matrix $\widehat{B}(\xi) \in \mathcal{M}_{N \times M}\big(L^\infty(\widehat{H})\big)$ such that $\widehat{B}(\xi)\,\widehat{A}(\xi) = I_N$, a.e. $\xi \in \widehat{H}$.
- Compute the matrix $B \in \mathcal{M}_{N \times M}\big(\ell^2(H)\big)$ such that $\widehat{B}(\xi)$ is its transfer matrix.
- Let $\mathbf{b}_m = \big(b_{1,m}, b_{2,m}, \dots, b_{N,m}\big)^\top$ be the $m$-th column, $m = 1, 2, \dots, M$, of the matrix $B$.
- Finally, we obtain the sampling functions as $S_m(s) = \langle \beta_m, U(s)\phi \rangle_{\mathcal{H}}$, $s \in G$, where $\beta_m = \sum_{n=1}^{N} \sum_{h \in H} b_{n,m}(h) U(h)\varphi_n$, $m = 1, 2, \dots, M$.

Condition (2) in Definition 1 implies necessarily that $M \geq N$, i.e., the number $N$ of stable generators used in the auxiliary space $\mathcal{H}_\Phi$ determines the minimal number $M$ of sequences of samples at $H$ that should be considered. In next section we go into the details.

## 4 The Main Sampling Result and Consequences

In this section we state and prove a general sampling result for $\mathcal{H}_{U,\phi,\Phi}$ associated with a stable sampling procedure $\mathcal{L}_A$ at a subgroup $H$. We will see that other usual sampling results can be deduced from it.

**Theorem 1** *Assume that a generalized sampling procedure $\mathcal{L}_A$ at $H$, as in Definition 1, has been defined in $\mathcal{H}_{U,\phi,\Phi}$. Then, there exist $M$ sampling functions $S_m$, $m = 1, 2, \ldots, M$, in $\mathcal{H}_{U,\phi,\Phi}$ such that the sequence $\{S_m(\cdot - t)\}_{t\in H;\, m=1,2,\ldots,M}$ is a frame for $\mathcal{H}_{U,\phi,\Phi}$, and the sampling expansion*

$$F(s) = \sum_{m=1}^{M} \sum_{t\in H} \mathcal{L}_m F(t)\, S_m(s-t)\,, \quad s \in G\,, \tag{7}$$

*holds for every $F \in \mathcal{H}_{U,\phi,\Phi}$. The pointwise convergence of the above series is absolute in G and uniform on G. It also converges in the $L^2(G)$-norm sense.*

***Proof*** Consider $F(s) = \langle f, U(s)\phi\rangle_{\mathcal{H}}$, $s \in G$, a function in $\mathcal{H}_{U,\phi,\Phi}$ and suppose that for the corresponding $f$ in $\mathcal{H}_\Phi$ we have the expansion $f = \sum_{n=1}^{N} \sum_{h\in H} x_n(h) U(h)\varphi_n$. According to the results in Sect. 2.2, since the matrix $\widehat{A}$ has entries in $L^\infty(\widehat{H})$ and $\delta_A > 0$, the sequence $\{T_t\, \mathbf{a}_m^*\}_{t\in H;\, m=1,2,\ldots,M}$ is a frame for $\ell^2_N(H)$ where $\mathbf{a}_m^* = (a_{m,1}^*, a_{m,2}^*, \ldots, a_{m,N}^*)^\top$ in $\ell^2_N(H)$ denotes the $m$-th column of the matrix $A^* = [a_{m,n}^*]^\top \in \mathcal{M}_{N\times M}(\ell^2(H))$ whose entries are the involutions $a_{m,n}^*(h) = \overline{a_{m,n}(-h)}$, $h \in H$. Moreover, $\mathcal{L}_m F(t) = \langle \mathbf{x}, T_t\, \mathbf{a}_m^*\rangle_{\ell^2_N(H)}$, $t \in H$ and $m = 1, 2, \ldots, M$.

Furthermore, there exists a matrix $\widehat{B}(\xi) \in \mathcal{M}_{N\times M}(L^\infty(\widehat{H}))$ such that $\widehat{B}(\xi)\, \widehat{A}(\xi) = I_N$, a.e. $\xi \in \widehat{H}$; it suffices to take $\widehat{B}(\xi) = \widehat{A}(\xi)^\dagger := \big[\widehat{A}(\xi)^* \widehat{A}(\xi)\big]^{-1} \widehat{A}(\xi)^*$, the Moore-Penrose pseudo-inverse of $\widehat{A}(\xi)$. Besides, the sequence $\{T_t\, \mathbf{b}_m\}_{t\in H;\, m=1,2,\ldots,M}$ in $\ell^2_N(H)$ is a dual frame of $\{T_t\, \mathbf{a}_m^*\}_{t\in H;\, m=1,2,\ldots,M}$, where $\mathbf{b}_m$ is the $m$-th column of the matrix $B \in \mathcal{M}_{N\times M}(\ell^2(H))$ whose transfer matrix is $\widehat{B}(\xi)$. As a consequence, for $\mathbf{x} = (x_1, x_2, \ldots, x_N)^\top \in \ell^2_N(H)$ associated to $f \in \mathcal{H}_\Phi$ we have

$$\mathbf{x} = \sum_{m=1}^{M} \sum_{t\in H} \langle \mathbf{x}, T_t\, \mathbf{a}_m^*\rangle_{\ell^2_N(H)}\, T_t\, \mathbf{b}_m = \sum_{m=1}^{M} \sum_{t\in H} \mathcal{L}_m F(t)\, T_t\, \mathbf{b}_m \quad \text{in } \ell^2_N(H)\,. \tag{8}$$

Now consider the isomorphism $\mathcal{T}_\Phi$ defined as

$$\begin{array}{lll} \mathcal{T}_\Phi : \ell^2_N(H) & \longrightarrow & \mathcal{H}_\Phi \\ \mathbf{x} & \longmapsto & f = \sum_{n=1}^N \sum_{h\in H} x_n(h)U(h)\varphi_n \,, \end{array}$$

which satisfies the shifting property $\mathcal{T}_\Phi(T_t\,\mathbf{b}) = U(t)(\mathcal{T}_\Phi\mathbf{b}), t \in H$ and $\mathbf{b} \in \ell^2_N(H)$. Applying the isomorphism $\mathcal{T}_\Phi$ in Eq. (8), we get

$$f = \sum_{m=1}^M \sum_{t\in H} \mathcal{L}_m F(t)U(t)(\mathcal{T}_\Phi\mathbf{b}_m) = \sum_{m=1}^M \sum_{t\in H} \mathcal{L}_m F(t)U(t)\beta_m\,, \quad \text{in } \mathcal{H}_\Phi\,,$$

where $\beta_m = \mathcal{T}_\Phi\mathbf{b}_m = \sum_{n=1}^N \sum_{h\in H} b_{n,m}(h)U(h)\varphi_n \in \mathcal{H}_\Phi$. Finally, for $F \in \mathcal{H}_{U,\phi,\Phi}$ we obtain

$$\begin{aligned} F(s) &= \Big\langle \sum_{m=1}^M \sum_{t\in H} \mathcal{L}_m F(t)U(t)\beta_m, U(s)\phi\Big\rangle = \sum_{m=1}^M \sum_{t\in H} \mathcal{L}_m F(t)\langle\beta_m, U(s-t)\phi\rangle_{\mathcal{H}} \\ &= \sum_{m=1}^M \sum_{t\in H} \mathcal{L}_m F(t)\, S_m(s-t)\,, \quad s\in G\,, \end{aligned} \tag{9}$$

where $S_m(s) = \langle\beta_m, U(s)\phi\rangle_{\mathcal{H}}$, $s \in G$, and $m = 1, 2, \ldots, M$. The pointwise convergence in (9) is absolute due to the unconditional character of a frame expansion. It is uniform on $G$ due to the inequality $|F(s)| \le \|f\|\,\|U(s)\phi\| \le \|f\|\,\|\phi\|$ for all $s \in G$. The composition of isomorphisms

$$\begin{array}{ccccc} \ell^2_N(H) & \longrightarrow & \mathcal{H}_\Phi & \longrightarrow & \mathcal{H}_{U,\phi,\Phi} \\ \mathbf{x} & \longmapsto & f & \longmapsto & F\,, \end{array} \tag{10}$$

shows that the $\ell^2_N(H)$-convergence in expansion (8) implies the convergence of expansion in (9) in the $L^2(G)$-norm sense; besides, we obtain that the sequence $\{S_m(\cdot - t)\}_{t\in H;\, m=1,2,\ldots,M}$ is a frame for $\mathcal{H}_{U,\phi,\Phi}$. □

As a consequence of the above theorem we obtain a *Shannon-type* sampling formula for the space $\mathcal{H}_{U,\phi,\Phi}$:

**Corollary 1** *In order to recover any $F \in \mathcal{H}_{U,\phi,\Phi}$ from its samples $\{F(t)\}_{t\in H}$, at the subgroup $H$, necessarily $N = 1$. Under conditions in Definition 1, i.e., $\widehat{a} \in L^\infty(\widehat{H})$ and* ess $\inf_{\xi\in\widehat{H}}|\widehat{a}(\xi)| > 0$*, where $a(s) = \langle\varphi, U(s)\phi\rangle_{\mathcal{H}}$, $s \in H$, there exists a unique sampling function $S_a \in \mathcal{H}_{U,\phi,\Phi}$ such that the sampling expansion*

$$F(s) = \sum_{t\in H} F(t)\, S_a(s-t)\,, \quad s\in G\,, \tag{11}$$

*holds in $\mathcal{H}_{U,\phi,\Phi}$. The sequence $\{S_a(\cdot - t)\}_{t\in H}$ is a Riesz basis for $\mathcal{H}_{U,\phi,\Phi}$.*

***Proof*** In this scalar case, there exists a unique $\widehat{b}(\xi) \in L^\infty(\widehat{H})$ such that $\widehat{b}(\xi)\,\widehat{a}(\xi) = 1$, a.e. $\xi \in \widehat{H}$. As a consequence, the associated sampling function $S_a$, given by $S_a(s) = \langle \beta, U(s)\phi \rangle_{\mathcal{H}}$, $s \in G$, where $\beta = \sum_{h\in H} b(h)\, U(h)\varphi$ belongs to $\mathcal{H}_\varphi$, is unique. □

In case $N > 1$ we must add $M - 1$ sequences $\{\mathcal{L}_m F(t)\}_{t\in H}$ of samples, $m = 2, \ldots, M$, with $M \geq N$, to the sequence $\{F(t)\}_{t\in H}$ such that the corresponding matrix $A = [a_{m,n}]$ in $\mathcal{M}_{M\times N}\big(\ell^2(H)\big)$ satisfies the conditions in Definition 1. Thus, by using Theorem 1, there exist $M$ sampling functions $S_m \in \mathcal{H}_{U,\phi,\Phi}$, $m = 1, 2, \ldots, M$, such that $\{S_m(\cdot - t)\}_{t\in H;\, m=1,2,\ldots,M}$ is a frame for $\mathcal{H}_{U,\phi,\Phi}$, and the sampling expansion

$$F(s) = \sum_{t\in H} F(t)\, S_1(s-t) + \sum_{m=2}^{M} \sum_{t\in H} \mathcal{L}_m F(t)\, S_m(s-t)\,, \quad s \in G\,, \tag{12}$$

holds for every $F \in \mathcal{H}_{U,\phi,\Phi}$.

## 4.1 Sampling at a Subgroup R with Finite Index in H

The result in Theorem 1 can be easily modified in order to take just samples at a subgroup $R$ with finite index in $H$. Indeed, let $R$ be a subgroup of $H$ with finite index $L$. We fix a set $\{h_1, h_2, \ldots, h_L\}$ of representatives of the cosets of $R$, i.e., the group $H$ can be decomposed as

$$H = (h_1 + R) \cup (h_2 + R) \cup \cdots \cup (h_L + R) \text{ with } (h_l + R) \cap (h_{l'} + R) = \varnothing \text{ for } l \neq l'\,.$$

The space $\mathcal{H}_\Phi$ can be written as

$$\begin{aligned}
\mathcal{H}_\Phi &= \Big\{ \sum_{n=1}^{N} \sum_{h\in H} x_n(h)\, U(h)\varphi_n \ :\ x_n \in \ell^2(H) \Big\} \\
&= \Big\{ \sum_{n=1}^{N} \sum_{l=1}^{L} \sum_{r\in R} x_n(h_l + r)\, U(h_l + r)\varphi_n \Big\} \\
&= \Big\{ \sum_{n=1}^{N} \sum_{l=1}^{L} \sum_{r\in R} x_{nl}(r)\, U(r)\varphi_{nl} \ :\ x_{nl} \in \ell^2(R) \Big\}\,,
\end{aligned}$$

with $x_{nl}(r) := x_n(h_l + r)$ and $\varphi_{nl} := U(h_l)\varphi_n$, where the new index $nl$ goes from $11$ to $NL$. Thus our subspace $\mathcal{H}_\Phi$ can be rewritten as $\mathcal{H}_{\widetilde{\Phi}}$ with $NL$ generators $\widetilde{\Phi} = \{\varphi_{nl}\}$ and coefficients $x_{nl}$ in $\ell^2(R)$.

For instance, concerning the samples $\{F(r)\}_{r\in R}$ we have

$$F(r) = \langle f, U(r)\phi\rangle_{\mathcal{H}} = \Big\langle \sum_{n=1}^{N}\sum_{l=1}^{L}\sum_{s\in R} x_{nl}(s)U(s)\varphi_{nl}, U(r)\phi\Big\rangle_{\mathcal{H}}$$

$$= \sum_{n=1}^{N}\sum_{l=1}^{L}\sum_{s\in R} x_{nl}(s)\langle\varphi_{nl}, U(r-s)\phi\rangle_{\mathcal{H}} = \sum_{n=1}^{N}\sum_{l=1}^{L}\big(a_{1,nl} *_R x_{nl}\big)(r), \quad r\in R, \tag{13}$$

where $a_{1,nl}(s) = \langle\varphi_{nl}, U(s)\phi\rangle_{\mathcal{H}}$, $s\in R$, and $*_R$ denotes the convolution in $\ell^2(R)$. The new index runs as $nl = 11, 12, \dots, 1L, \dots, N1, N2, \dots, NL$. In general, we could consider a stable sampling procedure $\mathcal{L}_A$ at $R$ with associated matrix $A = [a_{m,nl}] \in \mathcal{M}_{M\times NL}\big(\ell^2(R)\big)$ as in Definition 1. Thus we have:

**Corollary 2** *Let $A = [a_{m,nl}] \in \mathcal{M}_{M\times NL}\big(\ell^2(R)\big)$ be a matrix associated to a stable sampling procedure $\mathcal{L}_A$ at $R$ as in Definition 1. Then, there exist $M \geq NL$ sampling functions $S_m \in \mathcal{H}_{U,\phi,\Phi}$, $m = 1, 2, \dots, M$, such that the sampling expansion*

$$F(s) = \sum_{m=1}^{M}\sum_{r\in R} \mathcal{L}_m F(r)\, S_m(s-r), \quad s\in G. \tag{14}$$

*holds in $\mathcal{H}_{U,\phi,\Phi}$. The sequence $\{S_m(\cdot - r)\}_{r\in R;\, m=1,2,\dots,M}$ is a frame for $\mathcal{H}_{U,\phi,\Phi}$.*

***Proof*** Consider a matrix $B \in \mathcal{M}_{NL\times M}\big(\ell^2(R)\big)$ such that such that its transfer matrix $\widehat{B}(\xi)$ belongs to $\mathcal{M}_{NL\times M}\big(L^\infty(\widehat{R})\big)$ and $\widehat{B}(\xi)\,\widehat{A}(\xi) = I_{NL}$, a.e. $\xi\in\widehat{R}$. The sampling functions are $S_m(s) = \langle\beta_m, U(s)\phi\rangle_{\mathcal{H}}$, $s\in G$, where $\beta_m = \mathcal{T}_{\widetilde{\Phi}}\mathbf{b}_m$ belongs to $\mathcal{H}_{\widetilde{\Phi}}$, and $\mathbf{b}_m \in \ell^2_{NL}(R)$ is the $m$-th column, $m = 1, 2, \dots, M$, of the matrix $B$. □

## 4.2 Additional Notes and Remarks

Next we include some comments and remarks enlightening the above results:

1. In Sect. 3 we are assuming that the sequence $\{U(h)\varphi_n\}_{h\in H;\, n=1,2,\dots,N}$ is a Riesz sequence in $\mathcal{H}$, i.e., a Riesz basis for $\mathcal{H}_\Phi$. Necessary and sufficient conditions for sequences having this unitary structure can be found in Refs. [1, 5, 7, 16, 18, 20].
2. For any stable sampling procedure $\mathcal{L}_A$ at $H$ as in Definition 1 there exist positive constants $0 < c \leq C$ such that

$$c\|F\|^2 \leq \sum_{m=1}^{M}\sum_{t\in H} |\mathcal{L}_m F(t)|^2 \leq C\|F\|^2 \quad \text{for every } F\in\mathcal{H}_{U,\phi,\Phi}. \tag{15}$$

Indeed, it follows since the sequence $\{T_t\, \mathbf{a}^*_m\}_{t\in H;\, m=1,2,\ldots,M}$ is a frame for $\ell^2_N(H)$ and from the isomorphim in Eq. (10).

3. In case $M = N$ in Theorem 1, the sequence $\{T_t\, \mathbf{a}^*_m\}_{t\in H;\, m=1,2,\ldots,N}$ is a Riesz basis for $\ell^2_N(H)$ and the square matrix $\widehat{A}(\xi)$ is invertible; proceeding as in the proof of Theorem 1 its inverse gives the dual Riesz basis $\{T_t\, \mathbf{b}_m\}_{t\in H;\, m=1,2,\ldots,N}$. As a consequence, the sequence $\{S_n(\cdot - t)\}_{t\in H;\, n=1,2,\ldots,N}$ is a Riesz basis for $\mathcal{H}_{U,\phi,\Phi}$. The uniqueness of the coefficients in a Riesz basis expansion gives the *interpolation property*:

$$\mathcal{L}_n S_{n'}(t-t') = \delta_{n,n'}\delta_{t,t'}\,, \quad \text{where } t, t' \in H \text{ and } n, n' = 1, 2, \ldots, N\,.$$

4. The use of multiple generators for the auxiliary space $\mathcal{H}_\Phi$ might have advantages against the single generator case. For instance, in Corollary 1 the conditions of stable sampling in Definition 1 for a single generator $\varphi$ are: $\widehat{a} \in L^\infty(\widehat{H})$ and $\operatorname{ess\,inf}_{\xi\in\widehat{H}}|\widehat{a}(\xi)| > 0$, where $a(s) = \langle \varphi, U(s)\phi\rangle_{\mathcal{H}}$, $s \in H$. A way to overcome the restrictive second condition is to consider a subgroup $R$ of $H$ with finite index $L > 1$, and then consider $M \geq L$ sequences of samples taken at the subgroup $R$.

5. In Theorem 1, whenever $M > N$, there exist infinite sampling functions $S_m$, $m = 1, 2, \ldots, M$, coming from the different dual frames $\{T_t\, \mathbf{b}_m\}_{t\in H;\, m=1,2,\ldots,M}$ of $\{T_t\, \mathbf{a}^*_m\}_{t\in H;\, m=1,2,\ldots,M}$. All these duals come from the left-inverses $\widehat{B}(\xi)$ of $\widehat{A}(\xi)$ which are obtained, from the Moore-Penrose pseudo-inverse $\widehat{A}(\xi)^\dagger = \big[\widehat{A}(\xi)^*\widehat{A}(\xi)\big]^{-1}\widehat{A}(\xi)^*$, by means of the $N \times M$ matrices

$$\widehat{B}(\xi) := \widehat{A}(\xi)^\dagger + C(\xi)\big[I_M - \widehat{A}(\xi)\widehat{A}(\xi)^\dagger\big]\,, \quad \text{a.e. } \xi \in \widehat{H}\,,$$

where $C$ denotes any $N \times M$ matrix with entries in $L^\infty(\widehat{H})$. Indeed, it is straightforward to check that any matrix having this form is a left-inverse of $\widehat{A}(\xi)$. Moreover, any left-inverse $\widehat{B}(\xi)$ of $\widehat{A}(\xi)$ belongs to the above family; it suffices to take $C(\xi) = \widehat{B}(\xi)$.

6. The sequence $\{T_t\, \mathbf{a}^*_m\}_{t\in H;\, m=1,2,\ldots,M}$ is a Bessel sequence in $\ell^2_N(H)$ if and only if the convolution system in (3) is bounded, i.e., if and only if the transfer matrix $\widehat{A}(\xi)$ belongs to $\mathcal{M}_{M\times N}\big(L^\infty(\widehat{H})\big)$. Moreover, having in mind the equivalence between the spectral and Frobenius norms for matrices (see Ref. [17]), it is equivalent to the condition

$$\beta_A := \operatorname*{ess\,sup}_{\xi\in\widehat{H}} \lambda_{\max}[\widehat{A}(\xi)^*\widehat{A}(\xi)] < +\infty\,,$$

where $\lambda_{\max}$ denotes the largest eigenvalue of the positive semidefinite matrix $\widehat{A}(\xi)^*\widehat{A}(\xi)$.

7. Under the hypothesis $\widehat{A} \in \mathcal{M}_{M\times N}(L^\infty(\widehat{H}))$, the condition $\delta_A > 0$ in Definition 1 is equivalent to the condition $\alpha_A := \operatorname{ess\,inf}_{\xi\in\widehat{H}} \lambda_{\min}[\widehat{A}(\xi)^*\widehat{A}(\xi)] > 0$, where $\lambda_{\min}$ denotes the smallest eigenvalue of the positive semidefinite matrix $\widehat{A}(\xi)^*\widehat{A}(\xi)$. Indeed, it comes from the inequalities:
$$\alpha_A^N \le \delta_A \le \alpha_A\, \beta_A^{N-1},$$
where $\beta_A$ was introduced in the above note.
8. Assuming that $\widehat{A} \in \mathcal{M}_{M\times N}(L^\infty(\widehat{H}))$, it is worth to mention that the condition $\delta_A > 0$ is also necessary for the existence of a frame expansion (7) as those in Theorem 1. Indeed, suppose that, for each $F \in \mathcal{H}_{U,\phi,\Phi}$, a frame expansion $F = \sum_{m=1}^M \sum_{t\in H} \mathcal{L}_m F(t)\, S_m(\cdot - t)$ holds, and let $\mathcal{T}$ denote the isomorphism in Eq. (10). Then we have
$$\mathbf{x} = \sum_{m=1}^M \sum_{t\in H} \langle \mathbf{x}, T_t\, \mathbf{a}_m^* \rangle_{\ell^2_N(H)}\, \mathcal{T}^{-1} S_m(\cdot - t) \quad \text{for all } \mathbf{x} \in \ell^2_N(H).$$
In addition, since $\{T_t\, \mathbf{a}_m^*\}_{t\in H;\, m=1,2,\dots,M}$ is a Bessel sequence, both sequences $\{T_t\, \mathbf{a}_m^*\}_{t\in H;\, m=1,2,\dots,M}$ and $\{\mathcal{T}^{-1} S_m(\cdot - t)\}_{t\in H;\, m=1,2,\dots,M}$ form a pair of dual frames in $\ell^2_N(H)$ (see Ref. [8]). In particular, according to a result in Sect. 2.2, since $\{T_t\, \mathbf{a}_m^*\}_{t\in H;\, m=1,2,\dots,M}$ is a frame for $\ell^2_N(H)$ we get $\delta_A > 0$.
9. Whenever the entries $a_{m,n}$ of the matrix $A$ belong to $\ell^1(H)$, their Fourier transforms $\widehat{a}_{m,n}$ are continuous and consequently they belong to $L^\infty(\widehat{H})$. In this case the condition $\delta_A > 0$ is equivalent to $\det[\widehat{A}(\xi)^*\widehat{A}(\xi)] \neq 0$ for all $\xi \in \widehat{H}$.
10. For the samples given in Eqs. (4)–(5), the entries $a_{m,n}$ of the matrix $A$ depend on the unitary representation $U(t)$ of the group $G$, $\psi$ and $\Phi$. For a general stable sampling procedure $\mathcal{L}_A$ at $H$, by changing $\psi$ and $\Phi$ we could recover a function $F$, in different spaces $\mathcal{H}_{U,\phi,\Phi}$, from the same sequence of samples.
11. We can relax the initial assumptions in this section by assuming that $\{U(t)\phi\}_{t\in G}$ is just a complete Bessel family for $\mathcal{H}$ with respect to $(G, \mu_G)$. In this case the mapping $f \in \mathcal{H}_\Phi \longmapsto F_f \in \mathcal{H}_{U,\phi,\Phi}$ is injective and continuous but not necessarily an isomorphism (the subspace $\mathcal{H}_{U,\phi,\Phi}$ is not necessarily closed in $L^2(G)$). Under the hypotheses in Theorem 1, the sampling formula (7) holds but the left-hand inequality in (15) does not hold.

## 4.3 *The Case of a Semi-Direct Product of Groups*

The case where the group $G$ is the *semi-direct product of two groups* can be easily reduced to the situation described in Sect. 4 under appropriate conditions. Let $G = K \rtimes_\sigma H$ be the semi-direct product of the LCA group $(K, +)$ and a not necessarily abelian group $(H, \cdot)$, where $\sigma$ denotes the action of the group $H$ on the group $K$, i.e., a homomorphism $\sigma : H \to Aut(K)$ mapping $h \mapsto \sigma_h$. The composition law in $G$ is $(k_1, h_1)\,(k_2, h_2) := (k_1 + \sigma_{h_1}(k_2), h_1 h_2)$ for $(k_1, h_1),\ (k_2, h_2) \in G$. In general, the group $G = K \rtimes_\sigma H$ is not abelian. In case $\sigma_h \equiv Id_K$ for each $h \in H$ we recover the direct product group $G = K \times H$.

Now we consider a subgroup $G' = K' \rtimes_\sigma H'$ where $(K', +)$ is a countable discrete group and $(H', \cdot)$ is a finite group of order $N$ (we will write $H' = \{h_1 = 1_H, h_2, \cdots, h_N\}$), and such that $\sigma_h(K') = K'$ for each $h \in H'$. Suppose that $(k, h) \longmapsto U(k, h)$ is a unitary representation of the group $G = K \rtimes_\sigma H$ on a separable Hilbert space $\mathcal{H}$. For a fixed $\varphi \in \mathcal{H}$, the corresponding $\mathcal{H}_\varphi$ subspace can be written as

$$\begin{aligned}
\mathcal{H}_\varphi &= \Big\{ \sum_{(k,h)\in G'} x(k,h)\, U(k,h)\varphi \ : \ \{x(k,h)\} \in \ell^2(G') \Big\} \\
&= \Big\{ \sum_{n=1}^{N} \sum_{k\in K'} x(k,h_n)\, U(k,1_H) U(0_K,h_n)\varphi \Big\} \\
&= \Big\{ \sum_{n=1}^{N} \sum_{k\in K'} x_n(k)\, U(k,1_H)\varphi_n \ : \ \{x_n\} \in \ell^2(K') \Big\},
\end{aligned} \tag{16}$$

where $x_n(k) := x(k, h_n)$, $k \in K'$, and $\varphi_n := U(0_K, h_n)\varphi$, $n = 1, 2, \ldots, N$. That is, $\mathcal{H}_\varphi \equiv \mathcal{H}_\Phi$ where $\Phi = \{\varphi_1, \varphi_2, \ldots, \varphi_N\}$ is a set of $N$ generators in $\mathcal{H}$. Assume that $\{U(s,t)\phi\}_{(s,t)\in G}$ is a continuous frame for $\mathcal{H}$ with respect to $(G, \mu_G)$, and consider the corresponding $\mathcal{H}_{U,\phi,\Phi}$ space. For any function $F(s,t) = \langle f, U(s,t)\phi\rangle_{\mathcal{H}}$, $(s,t) \in G$, in $\mathcal{H}_{U,\phi,\Phi}$ we define a sampling procedure at $K'$ by

$$\mathcal{L}_A F(k) = \big(\mathcal{L}_1 F(k), \mathcal{L}_2 F(k), \ldots, \mathcal{L}_M F(k)\big)^\top := \big(A *_{K'} \mathbf{x}\big)(k), \quad k \in K',$$

where $A = [a_{m,n}] \in \mathcal{M}_{M\times N}\big(\ell^2(K')\big)$ and $\mathbf{x} = (x_1, x_2, \ldots, x_N)^\top \in \ell^2_N(K')$. Under conditions in Definition 1, there exist $M$ elements $\beta_m \in \mathcal{H}_\Phi$, $m = 1, 2, \ldots, M$, such that

$$f = \sum_{m=1}^{M} \sum_{k\in K'} \mathcal{L}_m F(k)\, U(k, 1_H)\, \beta_m \quad \text{in } \mathcal{H}_\Phi\,.$$

Remind that the elements $\beta_m \in \mathcal{H}_\Phi$, $m = 1, 2, \ldots, M$, are obtained from a left-inverse $\widehat{B}(\xi)$ of $\widehat{A}(\xi)$, a.e. $\xi \in \widehat{K'}$, as in Theorem 1. Hence, for each $F \in \mathcal{H}_{U,\phi,\Phi}$

we have

$$F(s,t) = \langle f, U(s,t)\phi\rangle_{\mathcal{H}} = \sum_{m=1}^{M}\sum_{k\in K'} \mathcal{L}_m F(k)\langle U(k,1_H)\,\beta_m, U(s,t)\,\phi\rangle_{\mathcal{H}}$$

$$= \sum_{m=1}^{M}\sum_{k\in K'} \mathcal{L}_m F(k)\langle \beta_m, U\big[(k,1_H)^{-1}(s,t)\big]\,\phi\rangle_{\mathcal{H}} \tag{17}$$

$$= \sum_{m=1}^{M}\sum_{k\in K'} \mathcal{L}_m F(k)\, S_m(s-k,t)\,, \quad (s,t)\in G\,,$$

where $S_m(s,t) = \langle \beta_m, U(s,t)\,\phi\rangle_{\mathcal{H}}$, $(s,t) \in G$, $m = 1, 2, \dots, M$. Notice that $(k, 1_H)^{-1}(s,t) = (-k, 1_H)(s,t) = (s-k,t)$, for $s \in K$, $k \in K'$ and $t \in H$.

## Euclidean Motion Group and Crystallographic Subgroups

An example of the above setting is given by *crystallographic groups* as subgroups of the *Euclidean motion group* $E(d)$. This group is the semi-direct product $\mathbb{R}^d \rtimes_\sigma O(d)$ corresponding to the homomorphism $\sigma : O(d) \to Aut(\mathbb{R}^d)$ given by $\sigma_\gamma(x) = \gamma x$, where $\gamma \in O(d)$ and $x \in \mathbb{R}^d$; $O(d)$ denotes the orthogonal group of order $d$. The composition law on $E(d) = \mathbb{R}^d \rtimes_\sigma O(d)$ reads $(x,\gamma)\cdot(x',\gamma') = (x+\gamma x', \gamma\gamma')$.

The subgroup $G'$ would be the crystallographic group $C_{P,\Gamma} := P\mathbb{Z}^d \rtimes_\sigma \Gamma$ where $P$ is a non-singular $d\times d$ matrix and $\Gamma$ is a finite subgroup of $O(d)$ of order $N$ such that $\gamma(P\mathbb{Z}^d) = P\mathbb{Z}^d$ for each $\gamma \in \Gamma$. We will denote $\{\gamma_1 = I, \gamma_2, \dots, \gamma_N\}$ the elements of the group $\Gamma$. In this example we consider the *quasi regular representation* (see Ref. [5]) on $L^2(\mathbb{R}^d)$:

$$U(s,\gamma)f(t) = f[\gamma^\top(t-s)]\,, \quad t, s \in \mathbb{R}^d, \gamma \in O(d) \text{ and } f \in L^2(\mathbb{R}^d)\,.$$

Assume that $\phi \in L^2(\mathbb{R}^d)$ is a function such that the family $\{U(s,\gamma)\phi\}_{(s,\gamma)\in E(d)}$ is a continuous frame for a closed subspace $\mathcal{H}$ of $L^2(\mathbb{R}^d)$ (containing $\mathcal{H}_\varphi$) with respect to $(E(d), ds\,d\mu(\gamma))$, where $d\mu(\gamma)$ denotes the left Haar measure on the group $O(d)$. For instance we could take $\phi$ a bandlimited function to a compact set $\Omega \subset \mathbb{R}^d$ and consider $\mathcal{H} := PW_\Omega$; the details are similar to those in Ref. [15]. See Ref. [5] for the details on the left Haar measure in semi-direct products of groups.

For any function $F(s,\gamma) = \langle f(\cdot), \phi(\gamma^\top(\cdot - s))\rangle_{L^2(\mathbb{R}^d)}$, $(s,\gamma) \in E(d)$, with $f \in \mathcal{H}_\Phi$, we consider a stable sampling procedure $\mathcal{L}_A F(p) := (A *_{P\mathbb{Z}^d} \mathbf{x})(p)$, $p \in P\mathbb{Z}^d$, defined at the lattice $P\mathbb{Z}^d$ as in Definition 1. Remind that in this particular example the space defined in (16) is:

$$\mathcal{H}_\Phi = \Big\{ \sum_{n=1}^{N}\sum_{p\in P\mathbb{Z}^d} x_n(p)\,\varphi_n(t-p) \,:\, \{x_n\} \in \ell^2(P\mathbb{Z}^d) \Big\}\,,$$

with $N$ generators $\varphi_n(t) = \varphi(\gamma_n^\top t), n = 1, 2, \ldots, N$, in $L^2(\mathbb{R}^d)$. The corresponding sampling formula (17) reads:

$$F(s, \gamma) = \sum_{m=1}^{M} \sum_{p \in P\mathbb{Z}^d} \mathcal{L}_m F(p)\, S_m(s - p, \gamma), \quad (s, \gamma) \in E(d),$$

where $S_m(s, \gamma) = \langle \beta_m(\cdot), \phi[\gamma^\top(\cdot - s)] \rangle_{L^2(\mathbb{R}^d)}$, $(s, \gamma) \in E(d)$, for some functions $\beta_m \in \mathcal{H}_\Phi$, $m = 1, 2, \ldots, M$, obtained from a left-inverse of the matrix $\widehat{A}(\xi)$ as in Theorem 1.

Whenever $f = \sum_{n=1}^{N} \sum_{p \in P\mathbb{Z}^d} x_n(p)\, \varphi_n(t-p) \in \mathcal{H}_\Phi$, for the pointwise samples $F(p, I) = \langle f(\cdot), \phi(\cdot - p) \rangle_{L^2(\mathbb{R}^d)}$, $p \in P\mathbb{Z}^d$ we have the expression $F(p, I) = \sum_{n=1}^{N} (a_{1,n} * x_n)(p)$, $p \in P\mathbb{Z}^d$, where, for $n = 1, 2, \ldots, N$, we have

$$a_{1,n}(k) = \langle \varphi_n(\cdot), \phi(\cdot - k) \rangle_{L^2(\mathbb{R}^d)} = \langle \varphi(t), \phi(\gamma_n t - k) \rangle_{L^2(\mathbb{R}^d)}, \quad k \in P\mathbb{Z}^d.$$

## 4.4 Some Final Comments

The theory obtained in this work relies on the use of an LCA group $G$. We have considered non-abelian groups which are semi-direct product of groups; the case treated here can be reduced to the abelian case by increasing the number of generators in the auxiliary space. Formally, the general non-abelian case can be handled in the same way but necessarily it will need other additional mathematical tools (see, for instance, Refs. [5, 24]).

As an example, consider the (positive) *affine group* $G_+ = \{(a, b) : a > 0, b \in \mathbb{R}\}$ with composition law $(a, b) \cdot (a', b') = (aa', b + ab')$, and its unitary representation $(a, b) \mapsto U(a, b)$ on $L^2(\mathbb{R})$ given by

$$\big[U(a, b) f\big](t) = \frac{1}{\sqrt{a}} f\Big(\frac{t - b}{a}\Big), \quad t \in \mathbb{R}, \text{ where } f \in L^2(\mathbb{R}).$$

The non-abelian group $G_+$ is non-unimodular with left Haar measure $d\mu_l = \frac{da\,db}{a^2}$.

Let $\phi$ be a function in $L^2(\mathbb{R})$ such that $\{U(a, b)\phi\}_{(a,b) \in G_+}$ is a continuous frame for $L^2(\mathbb{R})$, and let $\{\psi_{m,n}\}_{m,n \in \mathbb{Z}}$ be an orthonormal basis of wavelets for $L^2(\mathbb{R})$ where we use the notation $\psi_{m,n}(t) = 2^{-m/2} \psi\big(\frac{t-n}{2^m}\big)$, $m, n \in \mathbb{Z}$.

Now we sample any function $F(a, b) = \langle f, U(a, b)\phi \rangle_{L^2(\mathbb{R})}$, $(a, b) \in G_+$, where $f \in L^2(\mathbb{R})$, at the subspace $\Gamma := \{(2^m, n) : m, n \in \mathbb{Z}\}$ of $G_+$. The functions $F$ defined above form a RKHS contained in $L^2(\mathbb{R}^+ \times \mathbb{R}; \frac{da\,db}{a^2})$.

As in Sect. 3.1, the samples $\{F(2^m, n)\}_{m,n\in\mathbb{Z}}$ can be expressed as a discrete convolution in $\ell^2(\Gamma)$. A straightforward computation gives:

$$F(2^m, n) = \sum_{p,q} \langle \psi, U\big[(2^p, q)^{-1}(2^m, n)\big]\phi \rangle_{L^2(\mathbb{R})} \langle f, \psi_{p,q} \rangle_{L^2(\mathbb{R})}$$
$$= \big(\mathbf{a} *_\Gamma \mathbf{b}\big)(2^m, n), \quad (2^m, n) \in \Gamma,$$

where $\mathbf{a}(m, n) = \langle \psi, U\big[(m, n)\big]\phi \rangle_{L^2(\mathbb{R})}$ and $\mathbf{b}(m, n) = \langle f, \psi_{m,n} \rangle_{L^2(\mathbb{R})}$, $(2^m, n) \in \Gamma$. The mathematical techniques used in Sect. 2.2 do not work for the non-abelian group $G_+$, and other mathematical techniques are necessary (see, for instance, Refs. [5, 24]).

Another related classical problem is the following: let $\mathcal{H}_k$ be a RKHS of continuous functions $f : \mathbb{R} \to \mathbb{C}$ contained in $L^2(\mathbb{R})$ with reproducing kernels $\{k_x\}_{k\in\mathbb{R}}$. Assume that there exists a Riesz basis for $\mathcal{H}_k$ having the form $\{\varphi_n(t - kN)\}_{k\in\mathbb{Z};\, n=1,2,\dots,N}$, and we want to recover any function $f \in \mathcal{H}_k$ from the sequences of its samples $\{f(kN + r)\}_{k\in\mathbb{Z};\, r=0,1,\dots,N-1}$ in a stable way. For the sample $f(kN + r) = \langle f, k_{kN+r} \rangle_{L^2(\mathbb{R})}$, $k \in \mathbb{Z}$ and $r = 0, 1, \dots, N - 1$, we have the expression:

$$f(kN + r) = \langle f, k_{kN+r} \rangle_{L^2(\mathbb{R})} = \Big\langle \sum_{n=1}^{N} \sum_{m\in\mathbb{Z}} x_n(mN)\, \varphi_n(\cdot - mN), k_{kN+r} \Big\rangle_{L^2(\mathbb{R})}$$
$$= \sum_{n=1}^{N} \sum_{m\in\mathbb{Z}} x_n(mN)\, \varphi_n(kN + r - mN) = \sum_{n=1}^{N} \big(a_{r,n} *_{N\mathbb{Z}} x_n\big)(kN),$$

where $a_{r,n}(mN) = \varphi_n(mN - r)$, $m \in \mathbb{Z}$, for $n = 1, 2, \dots, N$. Under the conditions in Definition 1 for the $N \times N$ matrix $A = [a_{r,n}]$, there exist $N$ sampling functions $S_r \in \mathcal{H}_k$, $r = 0, 1, \dots, N - 1$, such that the sampling formula

$$f(t) = \sum_{r=0}^{N-1} \sum_{k\in\mathbb{Z}} f(kN + r)\, S_r(t - kN), \quad t \in \mathbb{R},$$

holds in $\mathcal{H}_k$. The sequence $\{S_r(\cdot - kN)\}_{k\in\mathbb{Z};\, r=0,1,\dots,N-1}$ is a Riesz basis for $\mathcal{H}_k$.

Finally to say that there is some affinity of the approach followed in this work with the topic of dynamical sampling (see, for instance, Ref. [3] and references therein). Indeed, from the correlation between a continuous frame $\{U(t)\phi\}_{t\in G}$ and an element $f$ in a suitable Hilbert space $\mathcal{H}_\Phi$ we obtain a function $F(t)$ in $L^2(G)$. In the case studied here, assuming that the space $\mathcal{H}_\Phi$ has a discrete unitary structure and under appropriate conditions, the function $F$ can be recovered, in a stable way, from a finite number of data sequences. An important difference is that dynamical sampling approach relies on a semigroup structure rather than on a group one.

**Acknowledgments** The author thanks *Universidad Carlos III de Madrid* for granting him a sabbatical year in 2020-21. This work has been supported by the grant MTM2017-84098-P from the Spanish *Ministerio de Economía y Competitividad (MINECO).*

## References

1. A. Aldroubi, Oblique proyections in atomic spaces. Proc. Am. Math. Soc. **124**, 2051–2060 (1996)
2. A. Aldroubi, Q. Sun, W.S. Tang, Convolution, average sampling, and a Calderon resolution of the identity for shift-invariant spaces. J. Fourier Anal. Appl. **11**(2), 215–244 (2005)
3. A. Aldroubi, K. Gröchenig, L. Huang, P. Jaming, I. Kryshtal, J.L. Romero, Sampling the flow of a bandlimited function. J. Geom. Anal. (2021). https://doi.org/10.1007/s12220-021-00617-0
4. S.T. Ali, J.P. Antoine, J.P. Gazeau, Continuous frames in Hilbert spaces. Ann. Phys. **222**, 1–37 (1993)
5. D. Barbieri, E. Hernández, J. Parcet, Riesz and frame systems generated by unitary actions of discrete groups. Appl. Comput. Harmon. Anal. **39**(3), 369–399 (2015)
6. A. Bhandari, A.I. Zayed, Shift-invariant and sampling spaces associated with the fractional Fourier transform domain. IEEE Trans. Signal Process. **60**(4), 1627–1637 (2012)
7. C. Cabrelli, V. Paternostro, Shift-invariant spaces on LCA groups. J. Funct. Anal. **258**, 2034–2059 (2010)
8. O. Christensen, *An Introduction to Frames and Riesz Bases*, 2nd edn. (Birkhäuser, Boston, 2016)
9. H.R. Fernández-Morales, A.G. García, M.A. Hernández-Medina, M.J. Muñoz-Bouzo, Generalized sampling: from shift-invariant to $U$-invariant spaces. Anal. Appl. **13**(3), 303–329 (2015)
10. M. Fornasier, H. Rauhut, Continuous frames, function spaces, and the discretization problem. J. Fourier Anal. Appl. **11**(3), 245–287 (2005)
11. G.B. Folland, *A Course in Abstract Harmonic Analysis* (CRC Press, Boca Raton, 1995)
12. H. Führ, *Abstract Harmonic Analysis of Continuous Wavelet Transform* (Springer, Berlin, 2005)
13. J.P. Gabardo, D. Han, Frames associated with measurable spaces. Adv. Compos. Mater. **18**(3), 127–147 (2003)
14. A.G. García, M.A. Hernández-Medina, G. Pérez-Villalón, Convolution systems on discrete abelian groups as a unifying strategy in sampling theory. Results Math. **75**, 40 (2020)
15. A.G. García, M.J. Muñoz-Bouzo, A note on continuous stable sampling. Adv. Oper. Theory **5**(3), 994–1013 (2020)
16. T.N. Goodman, S.L. Lee, W.S. Tang, Wavelet bases for a set of commuting unitary operators. Adv. Comput. Math. **1**(1), 109–126 (1993)
17. R.A. Horn, C.R. Johnson, *Matrix Analysis* (Cambridge University Press, Cambridge, 1999)
18. R.Q. Jia, C.A. Micchelli, Using the refinement equations for the construction of pre-waveles II: Powers of two, in *Curves and Surfaces*, ed. by P.J. Laurent, L. Le Méhauté, L. Schumaker (Academic Press, Boston, 1991), pp.209–246
19. S. Kang, K.H. Kwon, Generalized average sampling in shift-invariant spaces. J. Math. Anal. Appl. **377**, 70–78 (2011)
20. G. Pérez-Villalón, Discrete convolution operators and Riesz systems generated by actions of abelian groups. Ann. Funct. Anal. **11**, 285–297 (2020)
21. V. Pohl, H. Boche, $U$-invariant sampling and reconstruction in atomic spaces with multiple generators. IEEE Trans. Signal Process. **60**(7), 3506–3519 (2012)
22. A. Rahimi, A. Najati, Y.N. Dehghan, Continuous frames in Hilbert spaces. Methods Funct. Anal. Topology **12**(2), 170–182 (2006)

23. Z. Shang, W. Sun, X. Zhou, Vector sampling expansions in shift-invariant subspaces. J. Math. Anal. Appl. **325**, 898–919 (2007)
24. E. Skrettingland, Quantum harmonic analysis on lattices and Gabor multipliers. J. Fourier Anal. Appl. **26**, 48 (2020)

# An Extension of the Coherent Pair of Measures of the Second Kind on the Unit Circle

**Lino G. Garza, F. Marcellán, and A. Sri Ranga**

**Abstract** This paper deals with sequences of monic polynomials $\{\Phi_n(\mu_k; z)\}_{n\geq 0}$, $k = 0, 1$, orthogonal with respect to two nontrivial Borel measures $\mu_k$, $k = 0, 1$, supported on the unit circle, satisfying $(n+1)^{-1}\Phi'_{n+1}(\mu_0; z) = \Phi_n(\mu_1; z) + a_n\Phi_{n-1}(\mu_1; z) + b_n\Phi_{n-2}(\mu_1; z)$, $n \geq 3$, where $b_n \neq 0$. We find examples of pairs of measures $(\mu_0, \mu_1)$ for which this property holds. The analysis of polynomials orthogonal with respect to the Sobolev inner product associated with the pair of measures $(\mu_0, \mu_1)$ is presented. Some properties concerning their connection coefficients are given.

**Keywords** Probability measures on the unit circle · Orthogonal polynomials on the unit circle · Coherent pairs of measures of the second kind · Hessenberg matrices · Sobolev inner products on the unit circle

## 1 Introduction

The concept of coherent pairs of measures on the real line was introduced in the seminal paper [8] by Iserles, Koch, Nørsett and Sanz-Serna. They introduce such pairs $(\nu_0, \nu_1)$ of positive measures in the framework of the theory of orthogonal

---

L. G. Garza
Departamento de Física y Matemáticas, Universidad de Monterrey, San Pedro Garza García, Nuevo León, México
e-mail: lino.garza@udem.edu

F. Marcellán (✉)
Departamento de Matemáticas, Universidad Carlos III de Madrid, Leganés, Spain
e-mail: pacomarc@ing.uc3m.es

A. Sri Ranga
Departamento de Matemática, IBILCE, UNESP – Universidade Estadual Paulista, São José do Rio Preto, São Paulo, Brazil
e-mail: sri.ranga@unesp.br



F. Gesztesy, A. Martinez-Finkelshtein (eds.), *From Operator Theory to Orthogonal Polynomials, Combinatorics, and Number Theory*, Operator Theory: Advances and Applications 285, https://doi.org/10.1007/978-3-030-75425-9_8

polynomials associated with a Sobolev inner product. To be precise, the Sobolev inner product is defined by

$$\langle p, q\rangle_S = \int_{\mathbb{R}} p(x)q(x)d\nu_0(x) + \lambda \int_{\mathbb{R}} p'(x)q'(x)d\nu_1(x), \tag{1}$$

where $p$ and $q$ are polynomials with real coefficients and $\lambda$ is a nonnegative real number. The vector of measures $(\nu_0, \nu_1)$ is said to be coherent if the corresponding sequences of monic orthogonal polynomials $\{P_n(\nu_0; x)\}_{n\geq 0}$ and $\{P_n(\nu_1; x)\}_{n\geq 0}$ satisfy

$$nP_{n-1}(\nu_1; x) = P_n'(\nu_0; x) + a_n P_{n-1}'(\nu_0; x), \quad n \geq 2, \tag{2}$$

with $a_n \neq 0$ for $n \geq 2$.

If (2) holds and if $\{S_n(\nu_0, \nu_1; \lambda; x)\}_{n\geq 0}$ is the sequence of monic Sobolev orthogonal polynomials associated with the inner product (1), then

$$S_n(\nu_0, \nu_1; \lambda; x) + b_n(\lambda)\, S_{n-1}(\nu_0, \nu_1; \lambda; x) = P_n(\nu_0; x) + a_n P_{n-1}(\nu_0; x), \tag{3}$$

for $n \geq 1$. This simple equation connecting the two sequences $\{S_n(\nu_0, \nu_1; \lambda; x)\}_{n\geq 0}$ and $\{P_n(\nu_0; x)\}_{n\geq 0}$ turned out to be a powerful tool to study the properties of such Sobolev orthogonal polynomials. In particular, outer relative asymptotics have been deeply analyzed in the literature (see [13, 14] as well as the survey [12]).

In [15] H. G. Meijer proved that if $(\nu_0, \nu_1)$ is a coherent pair of positive measures supported on the real line, then one of the measures has to be classical (Laguerre or Jacobi). In fact, his result is obtained in a more general framework. That is, he deals with orthogonal polynomials with respect to a pair of quasi-definite linear functionals on the set of polynomials with real coefficients and proves that one of the functionals is the Laguerre, Jacobi or Bessel functional. Observe that positive definite linear functionals are associated with nontrivial probability measures supported on the real line (see [6]). Thus, Meijer [15] also determines all the possible coherent pairs of positive measures supported within the real line.

In [4] the authors show that there are Sobolev inner products of the type (1) where the pair of measures $(\nu_0, \nu_1)$ is not coherent, however the relation (3) still holds [4, Thm. 4.1] or a combination of Sobolev orthogonal polynomials of the form $S_n(\nu_0, \nu_1; \lambda; x) + b_n S_{n-1}(\nu_0, \nu_1; \lambda; x)$ can be represented as a linear combination of at most two orthogonal polynomials $P_n(\nu; x)$, where the measure $\nu$ is closely related to the measures $\nu_0$ and $\nu_1$ [4, Thm. 3.1].

In [5] the concept of coherent pair for Hermitian quasi-definite linear functionals, which can be represented by signed measures supported on the unit circle, is introduced. Notice that in the positive definite case, the linear functional is associated with a nontrivial positive measure supported on the unit circle (see [17]).

Given a non trivial probability measure $\mu$ on the unit circle, let

$$\langle p, q\rangle_\mu = \int_{\mathbb{T}} p(\zeta)\overline{q(\zeta)}d\mu(\zeta) = \int_0^{2\pi} p(e^{i\theta})\overline{q(e^{i\theta})}d\mu(e^{i\theta}).$$

We denote by $\{\Phi_n(\mu; z)\}_{n\geq 0}$ the sequence of monic orthogonal polynomials (or monic OPUC) with respect to the inner product $\langle p, q\rangle_\mu$ and let $A_n^{(\mu)}$ and $\alpha_n^{(\mu)}$ be the quantities

$$A_n^{(\mu)} = \langle \Phi_n(\mu; .), \Phi_n(\mu; .)\rangle_\mu \quad \text{and} \quad \alpha_n^{(\mu)} = -\overline{\Phi_{n+1}(\mu; 0)},$$

for $n \geq 0$. The values $\alpha_n^{(\mu)}$ are called the Verblunsky coefficients with respect to the measure $\mu$ and $A_n^{(\mu)}$ is the square of the norm of $\Phi_n(\mu; z)$ with respect to the measure $\mu$. Clearly, $A_0^{(\mu)} = 1$ and, by convention, we take $A_{-1}^{(\mu)} = 0$.

Following [5], a pair $(\mu_0, \mu_1)$ of positive measures supported on the unit circle is said to be a coherent pair of positive measures on the unit circle if the corresponding sequences of monic orthogonal polynomials $\{\Phi_n(\mu_0; z)\}_{n\geq 0}$ and $\{\Phi_n(\mu_1; z)\}_{n\geq 0}$ satisfy the algebraic relation

$$n\Phi_{n-1}(\mu_1; z) = \Phi_n'(\mu_0; z) + \rho_n\Phi_{n-1}'(\mu_0; z), \quad n \geq 2. \tag{4}$$

Here, $\Phi_n'(\mu; z) = d\Phi_n(\mu; z)/dz$. As established in [5], if $(\mu_0, \mu_1)$ is a coherent pair of positive measures on the unit circle then the following can be stated:

- If $\mu_0$ is the Lebesgue measure, then the companion measure $\mu_1$ is the Bernstein-Szegő measure $d\mu_1(z) = d\mu_0(z)/|z - \alpha|^2$, with $|\alpha| < 1$.
- If $\mu_1$ is the Lebesgue measure then the measure $\mu_0$ is such that $d\mu_0(z) = |z - \alpha|^2 d\mu_1(z)$.

They also prove that the only Bernstein-Szegő measure $\mu_0$ for which $(\mu_0, \mu_1)$ is a coherent pair is the Lebesgue measure. Unfortunately, a full description of all coherent pairs of measures supported on the unit circle is not given and this remains an open problem.

An extension of the concept of coherent pair of measures supported on the unit circle has been introduced in [7] and the connection with Sobolev orthogonal polynomials has been discussed in [10]. Indeed, they deal with $(1, 1)$-coherent pairs of measures such that the corresponding sequences of orthogonal polynomials satisfy

$$n\Phi_{n-1}(\mu_1; z) + \sigma_n\Phi_{n-2}(\mu_1; z) = \Phi_n'(\mu_0; z) + \rho_n\Phi_{n-1}'(\mu_0; z), \quad n \geq 2, \tag{5}$$

with $\rho_n \neq 0$ for $n \geq 2$. In such contribution, the explicit expressions for $\sigma_n$ and $\rho_n, n \geq 2$, are obtained.

The study of sequences of orthogonal polynomials with respect to a pair of measures supported on the unit circle satisfying the coherence property (5), however with $\rho_n = 0, n \geq 2$, was studied in [11]. The set of pairs of measures satisfying such a condition is explored and described. Observe that in this case we have neither a (1,1)-coherent pair nor a coherent pair. The motivation for such a study came from

[18], where a nice example of a family of pairs of measures $(\mu_0, \mu_1)$ such that the relation

$$\frac{1}{n}\Phi'_n(\mu_0; z) = \Phi_{n-1}(\mu_1; z) - \chi_n \Phi_{n-2}(\mu_1; z), \quad n \geq 2, \tag{6}$$

holds has been introduced. The associated sequence of monic Sobolev orthogonal polynomials has also been studied therein.

We will refer to a pair of positive measures $(\mu_0, \mu_1)$ on the unit circle for which the relation (6) holds as a *coherent pair of measures of the second kind*. We will also refer to the constants $\chi_n = \chi_n^{(\mu_0,\mu_1)}$ as the connection coefficients associated with the coherent pair of measures $(\mu_0, \mu_1)$ of the second kind. Notice that very recently, see [16], universality results for polynomial reproducing kernels related to such measures have been obtained.

The aim of the present contribution is the study of pairs of measures $(\mu_0, \mu_1)$ such that the corresponding sequences of monic orthogonal polynomials $\{\Phi_n(\mu_0; z)\}_{n\geq 0}$ and $\{\Phi_n(\mu_1; z)\}_{n\geq 0}$ satisfy

$$\frac{1}{n+1}\Phi'_{n+1}(\mu_0; z) = \Phi_n(\mu_1; z) + a_n \Phi_{n-1}(\mu_1; z) + b_n \Phi_{n-2}(\mu_1; z), \quad n \geq 2, \tag{7}$$

where we restrict ourselves to the case $b_n \neq 0$ for $n \geq 3$. The above pair of measures is a (0,2)-coherent pair of measures of the second kind.

The structure of the manuscript is the following. In Sect. 2 we discuss many choices of (0,2) coherent pairs of measures of second kind. The connection between the corresponding Hessenberg matrices, i.e. the representation of the multiplication operator with respect to such bases, is analyzed in Sect. 3. Section 4 is focused on the orthogonal polynomials associated with a Sobolev inner product defined by such vector of measures. The connection coefficients for this sequence and the sequence of OPUC with respect to the measure $\mu_0$ are determined. Some consequences of these relations are given.

## 2 Coherent Pairs of Measures of the Second Kind

The main goal of this chapter is to describe several illustrative examples of pairs of measures $(\mu_0, \mu_1)$ supported on the unit circle for which (7) holds. In (7), if $b_n = 0$ but $a_n \neq 0$ for every $n \geq 2$, then we have the case studied in [11] by F. Marcellán and A. Sri Ranga.

This work has motivations from the points of view of studying the analytical and numerical properties of the sequence of polynomials orthogonal with respect to the Sobolev inner product (27). If (7) holds and if $\{\Psi_n\}_{n\geq 0}$ is the sequence of monic orthogonal polynomials with respect to (27), then as described in detail in Sect. 4 of this manuscript, one finds

$$\Phi_n(\mu_0; z) = \Psi_n(z) + \alpha_n \Psi_{n-1}(z) + \beta_n \Psi_{n-2}(z), \quad n \geq 2,$$

with $\beta_n \neq 0$, $n \geq 3$.

First, we will consider two particular cases of measures on the unit circle for which (7) holds.

## 2.1 The Case $d\mu_1(z) = \frac{1}{2\pi i z} dz$

The measure $\mu_1$ is the so called Lebesgue measure and it is well known that $\Phi_n(\mu_1; z) = z^n$. Thus, (7) reads

$$\frac{1}{n+1}\Phi'_{n+1}(\mu_0; z) = z^n + a_n z^{n-1} + b_n z^{n-2}, \quad n \geq 2.$$

Hence,

$$\Phi_{n+1}(\mu_0; z) = z^{n+1} + \widetilde{a}_n z^n + \widetilde{b}_n z^{n-1} + \kappa_n, \quad n \geq 2,$$

with $\widetilde{a}_n = \frac{n+1}{n} a_n$, $\widetilde{b}_n = \frac{n+1}{n-1} b_n$, $n \geq 2$. By using the Szegő recurrence relation

$$\Phi_{n+1}(\mu_0; z) = z\Phi_n(\mu_0; z) + \Phi_{n+1}(\mu_0; 0)\, \Phi_n^*(\mu_0; z), \quad n \geq 0, \tag{8}$$

for $\{\Phi_n(\mu_0; .)\}_{n\geq 0}$, we then have

$$z^{n+2} + \widetilde{a}_{n+1} z^{n+1} + \widetilde{b}_{n+1} z^n + \kappa_{n+1} = z^{n+2} + \widetilde{a}_n z^{n+1} + \widetilde{b}_n z^n + \kappa_n z$$
$$+ \Phi_{n+2}(\mu_0; 0)\left(1 + \overline{\widetilde{a}_n}\, z + \overline{\widetilde{b}_n} z^2 + \overline{\kappa_n}\, z^{n+1}\right),$$

for $n \geq 2$. Thus,

$$0 = \kappa_2 + \kappa_3 \overline{\widetilde{a}_2}, \quad \widetilde{b}_3 = \widetilde{b}_2 + \kappa_3 \overline{\widetilde{b}_2}, \quad \widetilde{a}_3 = \widetilde{a}_2 + \kappa_3 \overline{\kappa_2},$$

and for $n \geq 3$,

$$0 = \kappa_n + \kappa_{n+1} \overline{\widetilde{a}_n}, \qquad 0 = \kappa_{n+1} \overline{\widetilde{b}_n},$$
$$\widetilde{b}_{n+1} = \widetilde{b}_n \quad \text{and} \quad \widetilde{a}_{n+1} = \widetilde{a}_n + \kappa_{n+1} \overline{\kappa_n}.$$

Since $\widetilde{b}_n \neq 0$ for $n \geq 3$ observe that $\kappa_{n+1} = 0$ for $n \geq 3$. Consequently, the above set of equations also lead to $\kappa_3 = \kappa_2 = 0$, and further, $\widetilde{a}_n = \widetilde{a}_2$ and $\widetilde{b}_n = \widetilde{b}_2$ for $n \geq 3$. That is,

$$\Phi_{n+1}(\mu_0; z) = z^{n+1} + \widetilde{a}_2 z^n + \widetilde{b}_2 z^{n-1} = z^{n-1}(z^2 + \widetilde{a}_2 z + \widetilde{b}_2), \quad n \geq 2.$$

Using the Szegő recurrence relation backwards one also finds $\Phi_2(\mu_0; z) = z^2 + \widetilde{a}_2 z + \widetilde{b}_2$.

Thus, by setting $\Phi_2(\mu_0; z) = h_2(z) = (z - w_1)(z - w_2)$, where $w_j$ is such that $w_1 w_2 = \widetilde{b}_2 \neq 0$, $|w_j| < 1$, $j = 1, 2$, and then using the Bernstein-Szegő theory we can state the following.

**Proposition 1** *Let $\mu_1$ be the Lebesgue measure on the unit circle. Then $(\mu_0, \mu_1)$ is a $(0, 2)$-coherent pair of measures of the second kind on the unit circle if and only if $\mu_0$ is the Bernstein-Szegő measure given by*

$$d\mu_0(z) = \frac{1}{|h_2(z)|^2} d\mu_1(z),$$

*where $h_2(z) = (z - w_1)(z - w_2)$, with $0 < |w_j| < 1$, $j = 1, 2$. Moreover, the connection coefficients $a_n$ and $b_n$ in (7) are*

$$a_n = -\frac{n}{n+1}(w_1 + w_2), \quad b_n = \frac{n-1}{n+1} w_1 w_2, \quad n \geq 3.$$

Notice that a particular case of this (0,2)-coherent pair of measures has been analyzed in [3]. In particular, asymptotic properties for the corresponding sequences of Sobolev orthogonal polynomials are studied therein.

## 2.2 The Case $d\mu_1(z) = \frac{1}{|z-u|^2} \frac{1}{2\pi i z} dz$, $u \neq 0$

The measure $\mu_1$ is the Bernstein-Szegő measure, where it is assumed that $|u| < 1$ and $u \neq 0$. Therefore, $\Phi_n(\mu_1; z) = z^{n-1}(z - u)$, $n \geq 1$, and from (7),

$$\frac{\Phi'_{n+1}(\mu_0; z)}{n+1} = z^{n-1}(z - u) + a_n z^{n-2}(z - u) + b_n z^{n-3}(z - u), \quad n \geq 3. \tag{9}$$

Thus,

$$\Phi_{n+1}(\mu_0; z) = z^{n+1} + (a_n - u)\frac{n+1}{n} z^n + (b_n - u a_n)\frac{n+1}{n-1} z^{n-1} - u b_n \frac{n+1}{n-2} z^{n-2} + \kappa_{n+1},$$

for $n \geq 3$ and from this we can write

$$\Phi_{n+1}(\mu_0; z) = z^{n+1} + \widetilde{a}_n z^n + \widetilde{b}_n z^{n-1} + \widetilde{c}_n z^{n-2} + \kappa_n, \quad n \geq 3, \tag{10}$$

where

$$\widetilde{a}_n = (a_n - u)\frac{n+1}{n}, \quad \widetilde{b}_n = (b_n - ua_n)\frac{n+1}{n-1} \quad \text{and} \quad \widetilde{c}_n = -ub_n\frac{n+1}{n-2}, \quad n \geq 3.$$

Since we have assumed in (7) that $b_n \neq 0$, $n \geq 3$, we also have $\widetilde{c}_n \neq 0$, $n \geq 3$.

Now from the Szegő recurrence for $\{\Phi_n(\mu_0; .)\}_{n\geq 0}$ we have

$$\begin{aligned} z^{n+2} + \widetilde{a}_{n+1} z^{n+1} + \widetilde{b}_{n+1} z^n + \widetilde{c}_{n+1} z^{n-1} + \kappa_{n+1} \\ = z^{n+2} + \widetilde{a}_n z^{n+1} + \widetilde{b}_n z^n + \widetilde{c}_n z^{n-1} + \kappa_n z \\ + \kappa_{n+1}(1 + \overline{\widetilde{a}_n} z + \overline{\widetilde{b}_n} z^2 + \overline{\widetilde{c}_n} z^3 + \overline{\kappa_n} z^{n+1}), \end{aligned} \tag{11}$$

for $n \geq 3$. Thus,

$$0 = \kappa_3 + \kappa_4 \overline{\widetilde{a}_3}, \quad \widetilde{c}_4 = \widetilde{c}_3 + \kappa_4 \overline{\widetilde{b}_3}, \quad \widetilde{b}_4 = \widetilde{b}_3 + \kappa_4 \overline{\widetilde{c}_3}, \quad \text{and} \quad \widetilde{a}_4 = \widetilde{a}_3 + \kappa_4 \overline{\kappa_3},$$

$$0 = \kappa_4 + \kappa_5 \overline{\widetilde{a}_4}, \quad 0 = \kappa_5 \overline{\widetilde{b}_4}, \quad \widetilde{c}_5 = \widetilde{c}_4 + \kappa_5 \overline{\widetilde{c}_4}, \quad \widetilde{b}_5 = \widetilde{b}_4 \quad \text{and} \quad \widetilde{a}s_5 = \widetilde{a}_4 + \kappa_5 \overline{\kappa_4},$$

and for $n \geq 5$,

$$\begin{aligned} &0 = \kappa_n + \kappa_{n+1} \overline{\widetilde{a}_n}, \quad 0 = \kappa_{n+1} \overline{\widetilde{b}_n}, \quad 0 = \kappa_{n+1} \overline{\widetilde{c}_n}, \\ &\widetilde{c}_{n+1} = \widetilde{c}_n, \quad \widetilde{b}_{n+1} = \widetilde{b}_n \quad \text{and} \quad \widetilde{a}_{n+1} = \widetilde{a}_n + \kappa_{n+1} \overline{\kappa_n}. \end{aligned}$$

Since $\widetilde{c}_n \neq 0$ for $n \geq 5$ we first observe that $\kappa_{n+1} = 0$ for $n \geq 5$. Consequently, $\kappa_5 = \kappa_4 = \kappa_3 = 0$, as well as

$$\widetilde{a}_n = \widetilde{a}_3, \quad \widetilde{b}_n = \widetilde{b}_3 \quad \text{and} \quad \widetilde{c}_n = \widetilde{c}_3 \quad \text{for} \quad n \geq 4.$$

Hence, the polynomials $\Phi_{n+1}(\mu_0; .)$ take the form

$$\Phi_{n+1}(\mu_0; z) = z^{n-2}(z^3 + \widetilde{a}_3 z^2 + \widetilde{b}_3 z + \widetilde{c}_3), \quad n \geq 3.$$

The Szegő recurrence applied backwards also gives the validity of the above formula for $n = 2$. That is,

$$\Phi_3(\mu_0; z) = z^3 + \widetilde{a}_3 z^2 + \widetilde{b}_3 z + \widetilde{c}_3.$$

Note that from the above expressions for $\Phi_{n+1}(\mu_0; .)$, $n \geq 2$, we can easily verify that

$$\Phi'_{n+1}(\mu_0; z) - z^{n-3}\Phi'_4(\mu_0; z) = (n-3)z^{n-3}\Phi_3(\mu_0; z), \quad n \geq 3.$$

Thus, from (9) we have $\Phi_3(\mu_0; u) = 0$ and we can write

$$\Phi_3(\mu_0; z) = (z-u)(z-w_1)(z-w_2),$$

where $|w_j| < 1$, $j = 1, 2$, in order that the zeros of $\Phi_3(\mu_0; z)$ are within the open unit disk. Hence, $\widetilde{a}_3 = -(u+w_1+w_2)$, $\widetilde{b}_3 = u(w_1+w_2)+w_1w_2$ and $\widetilde{c}_3 = -uw_1w_2$. Since, $\widetilde{c}_3 \neq 0$, we must also have $|u| > 0$ and $|w_j| > 0$, $j = 1, 2$.

Now we consider the formulas for $\widetilde{a}_n$, $\widetilde{b}_n$ and $\widetilde{c}_n$ given after (10). From the expressions for $\widetilde{a}_n$ and $\widetilde{b}_n$ we find

$$\left.\begin{aligned} a_n &= u + \frac{n}{n+1}\widetilde{a}_3 = u - \frac{n}{n+1}[u + w_1 + w_2], \\ b_n &= ua_n + \frac{n-1}{n+1}\widetilde{b}_3 = \frac{1}{n+1}u^2 - \frac{1}{n+1}u(w_1+w_2) + \frac{n-1}{n+1}w_1w_2, \end{aligned}\right\} \quad n \geq 3.$$

On the other hand, from the expression for $\widetilde{c}_n$,

$$u\, b_n = \frac{n-2}{n+2}u\, w_1w_2, \quad n \geq 3.$$

For the two sets of formulas for the $b_n$'s lead to the same values we must have $(u-w_1)(u-w_2) = 0$. That is, at least one of $w_1$ and $w_2$ must be equal to $u$. Hence, we set $w_2 = u \neq 0$ and $w_1 = w \neq 0$.

Now applying the Bernstein-Szegő theory we can state the following.

**Proposition 2** *Let $\mu_1$ be the Bernstein-Szegő measure on the unit circle given by $d\mu_1(z) = \frac{1}{|z-u|^2}\frac{1}{2\pi i z}dz$, where $0 < |u| < 1$. Then $(\mu_0, \mu_1)$ is a $(0, 2)$-coherent pair of measures of the second kind on the unit circle if and only if $\mu_0$ is the Bernstein-Szegő measure given by*

$$d\mu_0(z) = \frac{1}{|h_2(z)|^2}d\mu_1(z),$$

*where $h_2(z) = (z-u)(z-w)$, with $0 < |w| < 1$. Moreover, the connection coefficients $a_n$ and $b_n$ in (7) are*

$$a_n = u - \frac{n}{n+1}(2u+w), \quad b_n = \frac{n-2}{n+2}uw, \quad n \geq 3.$$

## 2.3 A General Case

The previous examples lead us to assume that

$$d\mu_0(e^{i\theta}) = \frac{1}{|h_2(e^{i\theta})|^2} d\mu_1(e^{i\theta}),$$

where $h_2(e^{i\theta}) - e^{2i\theta} + Ae^{i\theta} + B$.

We now consider the orthogonality conditions

$$\int_{\mathbb{T}} \zeta^{-k}\, \Phi'_{n+1}(d\mu_0;\zeta) d\mu_1(\zeta) = (n+1) b_n \|\Phi_{n-2}\|^2_{\mu_1}\, \delta_{n-2,k}, \quad 0 \le k \le n-2,$$

for $n \ge 3$, which follow from (7). Hence,

$$\int_0^{2\pi} e^{-i(k+1)\theta}\, \frac{d\Phi_{n+1}(d\mu_0; e^{i\theta})}{d\theta} d\mu_1(e^{i\theta}) = 0, \quad 0 \le k \le n-3, \tag{12}$$

and

$$\int_0^{2\pi} e^{-i(n-1)\theta}\, \frac{d\Phi_{n+1}(d\mu_0; e^{i\theta})}{d\theta} d\mu_1(e^{i\theta}) = i(n+1) b_n \|\Phi_{n-2}\|^2_{\mu_1}, \tag{13}$$

for $n \ge 3$.

From (12), using integration by parts we have

$$\int_0^{2\pi} \big[ -i(k+1) e^{-i(k+1)\theta} \omega_1(\theta) + e^{-i(k+1)\theta} \omega'_1(\theta) \big] \Phi_{n+1}(d\mu_0; e^{i\theta}) d\theta = 0,$$

for $0 \le k \le n-3$. Here, we have imposed the differentiability and continuity conditions $d\mu_1(e^{i\theta}) = \omega_1(\theta)d\theta$, $\omega_1(\theta) \in C^1[0, 2\pi]$ and $\omega_1(0) = \omega_1(2\pi)$.

Thus, from $|h_2(e^{i\theta})|^2 d\mu_0(e^{i\theta}) = d\mu_1(e^{i\theta})$, we can write

$$\omega_1(\theta)d\theta = |h_2(e^{i\theta})|^2 d\mu_0(e^{i\theta}) = |h_2(e^{i\theta})|^2 \omega_0(\theta) d\theta,$$

where $\omega_0(\theta) \in C^1(0, 2\pi)$. Consequently,

$$\int_0^{2\pi} \big[ -i(k+1)|h_2(e^{i\theta})|^2 \omega_0(\theta) + d[|h_2(e^{i\theta})|^2 \omega_0(\theta)]/d\theta \big] \times \\ e^{-i(k+1)\theta}\, \Phi_{n+1}(d\mu_0; e^{i\theta}) d\theta \; = \; 0,$$

for $0 \leq k \leq n-3$, which can also be written as

$$\int_0^{2\pi} \Big[ -i(k+1)|h_2(e^{i\theta})|^2 \omega_0(\theta) + \frac{d[|h_2(e^{i\theta})|^2]}{d\theta} \omega_0(\theta) + |h_2(e^{i\theta})|^2 \omega_0'(\theta) \Big] \times e^{-i(k+1)\theta} \Phi_{n+1}(d\mu_0; e^{i\theta}) d\theta = 0, \tag{14}$$

for $0 \leq k \leq n-3$.

*Remark 1* $|h_2(e^{i\theta})|^2 d\mu_0(e^{i\theta}) = |h_2(e^{i\theta})|^2 \omega_0(\theta) d\theta$ also means that $\mu_0$ does not have any mass points when the zeros of $h_2$ are not on the unit circle. However, $\mu_0$ could have one or two mass points if one or both of the zeros of $h_2$ are on the unit circle.

Now, in order that (14) stays valid, we make the following additional assumption

$$|h_2(e^{i\theta})|^2 \omega_0'(\theta) = \frac{g_2(e^{i\theta}) + \overline{g_2(e^{i\theta})}}{2} \omega_0(\theta),$$

where $g_2(e^{i\theta}) = 2Ce^{2i\theta} + 2De^{i\theta} + E$. Thus, we have

$$\int_0^{2\pi} \Big[ -i(k+1)|h_2(e^{i\theta})|^2 + \frac{d[|h_2(e^{i\theta})|^2]}{d\theta} + (Ce^{2i\theta} + De^{i\theta} + \frac{E + \overline{E}}{2} + \overline{D}e^{-i\theta} + \overline{C}e^{-2i\theta}) \Big] \times e^{-i(k+1)\theta} \Phi_{n+1}(d\mu_0; e^{i\theta}) \omega_0(\theta) d\theta = 0, \tag{15}$$

for $0 \leq k \leq n-3$. Since

$$\begin{aligned} |h_2(e^{i\theta})|^2 &= |(e^{2i\theta} + Ae^{i\theta} + B)|^2 \\ &= \overline{B}e^{2i\theta} + e^{i\theta}(\overline{A} + A\overline{B}) + 1 + |A|^2 + |B|^2 + e^{-i\theta}(A + \overline{A}B) + Be^{-2i\theta} \end{aligned}$$

and

$$\frac{d[|h_2(e^{i\theta})|^2]}{d\theta} = 2i\overline{B}e^{2i\theta} + i(\overline{A} + A\overline{B})e^{i\theta} - i(A + \overline{A}B)e^{-i\theta} - 2iBe^{-2i\theta},$$

observe that, for $k = 0, 1, \ldots, n-3$, the polynomials

$$e^{-i(k+1)\theta} \times \Big[ -i(k+1)|h_2(e^{i\theta})|^2 + \frac{d[|h_2(e^{i\theta})|^2]}{d\theta} + (Ce^{2i\theta} + De^{i\theta} + \frac{E + \overline{E}}{2} + \overline{D}e^{-i\theta} + \overline{C}e^{-2i\theta}) \Big],$$

as powers of $\zeta = e^{i\theta}$, are in the span of $\{\zeta^{-n}, \zeta^{-n+1}, \ldots, \zeta^{-1}, 1\}$ if $i\overline{B} + C = 0$.

Thus, in order to have the orthogonality conditions in (12) we take

$$i\overline{B} + C = 0 \quad \text{and} \quad d\mu_0(e^{i\theta}) = \omega_0(\theta)d\theta.$$

Hence, we have

$$|e^{2i\theta} + Ae^{i\theta} + B|^2\, \omega_0'(\theta) = [-i\overline{B}e^{2i\theta} + De^{i\theta} + \text{Re}(E) + \overline{D}e^{-i\theta} + iBe^{-2i\theta}]\omega_0(\theta),$$

which is

$$\frac{\omega_0'(\theta)}{\omega_0(\theta)} = \frac{[-i\overline{B}e^{2i\theta} + De^{i\theta} + \text{Re}(E) + \overline{D}e^{-i\theta} + iBe^{-2i\theta}]}{|h_2(e^{i\theta})|^2}.$$

Thus, with $z = e^{i\theta}$,

$$\frac{\frac{d}{dz}[\omega_0(\theta)]}{\omega_0(\theta)} = \frac{[-i\overline{B}z^2 + Dz + \text{Re}(E) + \overline{D}z^{-1} + iBz^{-2}]}{iz|h_2(z)|^2}. \tag{16}$$

Now if we consider (13), its left hand side becomes the same expression as in (15) but with $k = n - 2$. Proceeding in the same way as for $k \leq n - 3$, we find that the trigonometric polynomial multiplying $\Phi_{n+1}(e^{i\theta})$ has $-i(n + 1)B + \overline{C}$ as leading coefficient (i.e., $-i(n+1)B + \overline{C}$ is the multiplication factor of $e^{-i(n+1)\theta}$). Since we have already assumed that $i\overline{B} + C = 0$, we find $-i(n + 1)B + \overline{C} = 0$ if and only if $B = 0$. Hence, if $b_n \neq 0$ then we must have $B \neq 0$.

However, the condition $b_n = 0$ also means $C = 0$ and hence $h_2(z) = z(z + A)$ and $g_2(z) = 2Dz + E$. This is exactly the situation considered in the paper [11].

We now discuss the different situations (or cases) that follow from (16) according to the nature of the zeros of $h_2(z) = z^2 + Az + B$:

(i) $h_2(z) = (z - \alpha_1)^2$, with $|\alpha_1| > 1$.
(ii) $h_2(z) = (z - \alpha_1)(z - \alpha_2)$, $\alpha_1 \neq \alpha_2$, with $|\alpha_1| > 1$, $|\alpha_2| > 1$.
(iii) $h_2(z) = (z - \alpha_1)(z - \alpha_2)$, with $|\alpha_1| > 1$, $|\alpha_2| = 1$.
(iv) $h_2(z) = (z - \alpha_1)(z - \alpha_2)$, $\alpha_1 \neq \alpha_2$, $|\alpha_1| = |\alpha_2| = 1$.
(v) $h_2(z) = (z - \alpha_1)^2$, with $|\alpha_1| = 1$.

Observe that in all five of these cases we have chosen the zeros of $h_2$ such that $B \neq 0$. Thus, in all the results that follow from the analysis we perform below, the condition $b_n \neq 0$ is satisfied.

**Case (i)**: In this case from $h_2(z) = z^2 + Az + B = (z - \alpha_1)^2$ we first observe that $B = \alpha_1^2$. Moreover, from (16) where it is assumed $z = e^{i\theta}$, we have

$$\begin{aligned}\frac{\frac{d}{dz}[\omega_0(\theta)]}{\omega_0(\theta)} &= \frac{[-i\overline{B}z^4 + Dz^3 + Re(E)z^2 + \overline{D}z + iB]}{iz(z - \alpha_1)^2(1 - \overline{\alpha}_1 z)^2}, \\ &= \frac{[-\overline{B}z^4 - iDz^3 - iRe(E)z^2 - i\overline{D}z + B]}{\overline{\alpha}_1^2 z(z - \alpha_1)^2(z - 1/\overline{\alpha}_1)^2}.\end{aligned}$$

We can see that the right hand side can be expanded in partial fractions. Hence,

$$\frac{\frac{d}{dz}[\omega_0(\theta)]}{\omega_0(\theta)} = \frac{M_1}{z-\alpha_1} + \frac{M_2}{z-1/\overline{\alpha}_1} + \frac{M_3}{(z-\alpha_1)^2} + \frac{M_4}{(z-1/\overline{\alpha}_1)^2} + \frac{M_5}{z},$$

and integrating with respect to $z$ we have

$$\ln\left[\frac{\omega_0(\theta)}{\omega_0(\vartheta)}\right] = M_1 \ln\left[\frac{e^{i\theta}-\alpha_1}{e^{i\vartheta}-\alpha_1}\right] + M_2 \ln\left[\frac{e^{i\theta}-1/\overline{\alpha}_1}{e^{i\vartheta}-1/\overline{\alpha}_1}\right] + iM_5(\theta-\vartheta)$$
$$-\frac{M_3}{e^{i\theta}-\alpha_1} + \frac{M_3}{e^{i\vartheta}-\alpha_1} - \frac{M_4}{e^{i\theta}-1/\overline{\alpha}_1} + \frac{M_4}{e^{i\vartheta}-1/\overline{\alpha}_1}. \tag{17}$$

Simple computations also yield

$$M_5 = \frac{B}{\alpha_1^2} = 1,$$
$$M_3 = \frac{-\overline{B}\alpha_1^4 - iD\alpha_1^3 - i(\operatorname{Re} E)\alpha_1^2 - i\overline{D}\alpha_1 + B}{\alpha_1(|\alpha_1|^2-1)^2}, \qquad M_4 = -\frac{\overline{M}_3}{\overline{\alpha}_1^2},$$
$$M_1 = \frac{-4\overline{B}\alpha_1^3 - 3iD\alpha_1^2 - 2i(\operatorname{Re} E)\alpha_1 - i\overline{D}}{\alpha_1(|\alpha_1|^2-1)^2} - \frac{(3|\alpha_1|^2-1)}{(|\alpha_1|^2-1)}\frac{M_3}{\alpha_1},$$
$$M_2 = \frac{-4\overline{B} - 3iD\overline{\alpha}_1 - 2i(\operatorname{Re} E)\overline{\alpha}_1^2 - i\overline{D}\overline{\alpha}_1^3}{\overline{\alpha}_1^2(|\alpha_1|^2-1)^2} - \frac{(|\alpha_1|^2-3)}{(|\alpha_1|^2-1)}\overline{\alpha}_1 M_4 \quad \text{and}$$
$$M_2 = -2 - M_1.$$

From the latter equation we see that $\operatorname{Re}(M_1) + \operatorname{Re}(M_2) = -2$ as well as $\operatorname{Im}(M_1) + \operatorname{Im}(M_2) = 0$.

Observe that using the above formulas, we also have

$$\overline{M}_2 = \frac{-4B + 3i\overline{D}\alpha_1 + 2i(\operatorname{Re} E)\alpha_1^2 + iD\alpha_1^3}{\alpha_1^2(|\alpha_1|^2-1)^2} - \frac{(|\alpha_1|^2-3)}{(|\alpha_1|^2-1)}\alpha_1\overline{M}_4$$
$$= \frac{-4B + 3i\overline{D}\alpha_1 + 2i(\operatorname{Re} E)\alpha_1^2 + iD\alpha_1^3}{\alpha_1^2(|\alpha_1|^2-1)^2} + \frac{(|\alpha_1|^2-3)}{(|\alpha_1|^2-1)}\frac{M_3}{\alpha_1}$$

and

$$\overline{M}_1 = -2 - \frac{-4B\alpha_1^{-1} + 3i\overline{D} + 2i(\operatorname{Re} E)\alpha_1 + iD\alpha_1^2}{\alpha_1(|\alpha_1|^2-1)^2} - \frac{(|\alpha_1|^2-3)}{(|\alpha_1|^2-1)}\frac{M_3}{\alpha_1}.$$

Hence, from the above formulas for $M_1$, $\overline{M}_1$ and $M_3$ one can also verify that

$$-2 = M_1 + \overline{M}_1 = 2\,\mathrm{Re}(M_1) = 2\,\mathrm{Re}(M_2)$$

and

$$\mathrm{Im}(M_1) = \frac{M_1 - \overline{M}_1}{2i} = \frac{2D\alpha_1 + \mathrm{Re}(E)(|\alpha_1|^2 + 1) + 2\overline{D}\overline{\alpha}_1}{(|\alpha_1|^2 - 1)^3}.$$

Since,

$$\frac{e^{i\theta} - 1/\overline{\alpha}_1}{e^{i\vartheta} - 1/\overline{\alpha}_1} = e^{i(\theta-\vartheta)}\frac{e^{-i\theta} - \overline{\alpha}_1}{e^{-i\vartheta} - \overline{\alpha}_1} \quad \text{and}$$
$$-\frac{1}{e^{i\theta} - 1/\overline{\alpha}_1} + \frac{1}{e^{i\vartheta} - 1/\overline{\alpha}_1} = \frac{\overline{\alpha}_1^2}{e^{-i\theta} - \overline{\alpha}_1} - \frac{\overline{\alpha}_1^2}{e^{-i\vartheta} - \overline{\alpha}_1},$$

together with $M_4 = -\overline{M}_3/\overline{\alpha}_1^2$ and $M_1+M_2 = -2M_5 = -2$, Eq. (17) can be written as

$$\begin{aligned}
\ln\left[\frac{\omega_0(\theta)}{\omega_0(\vartheta)}\right] &= M_1 \ln\left[\frac{e^{i\theta} - \alpha_1}{e^{i\vartheta} - \alpha_1}\right] + M_2 \ln\left[\frac{e^{-i\theta} - \overline{\alpha}_1}{e^{-i\vartheta} - \overline{\alpha}_1}\right] + i(M_2 + M_5)(\theta - \vartheta) \\
&\quad - \frac{M_3}{e^{i\theta} - \alpha_1} + \frac{M_3}{e^{i\vartheta} - \alpha_1} - \frac{\overline{M}_3}{e^{-i\theta} - \overline{\alpha}_1} + \frac{\overline{M}_3}{e^{-i\vartheta} - \overline{\alpha}_1}, \\
&= (M_1 + M_2) \ln\left|\frac{e^{i\theta} - \alpha_1}{e^{i\vartheta} - \alpha_1}\right| + i(M_1 - M_2)\arg\left[\frac{e^{i\theta/2} - \alpha_1 e^{-i\theta/2}}{e^{i\vartheta/2} - \alpha_1 e^{-i\vartheta/2}}\right] \\
&\quad -2\,\mathrm{Re}\left[\frac{M_3}{e^{i\theta} - \alpha_1}\right] + 2\,\mathrm{Re}\left[\frac{M_3}{e^{i\vartheta} - \alpha_1}\right].
\end{aligned}$$

Hence,

$$\begin{aligned}
\omega_0(\theta) &= const \times |e^{i\theta} - \alpha_1|^{-2} e^{-2\,\mathrm{Im}(M_1)\arg(e^{i\theta/2} - \alpha_1 e^{-i\theta/2})} e^{-2\,\mathrm{Re}[M_3/(e^{i\theta} - \alpha_1)]} \\
&= const \times |e^{i\theta} - \alpha_1|^{-2} e^{\mathrm{Im}(M_1)\theta} e^{-2\,\mathrm{Im}(M_1)\arg(e^{i\theta} - \alpha_1)} e^{-2\,\mathrm{Re}[M_3/(e^{i\theta} - \alpha_1)]}.
\end{aligned}$$

However, since $|\alpha_1| > 1$ and $\omega_1(\theta) = |h_2(e^{i\theta})|^2\omega_0(\theta)$, in order to satisfy $\omega_1(0) = \omega_1(2\pi)$, we also must have $\omega_0(0) = \omega_0(2\pi)$. This is possible only if $\mathrm{Im}(M_1) = 0$, which is achieved if

$$2D\alpha_1 + \mathrm{Re}(E)(|\alpha_1|^2 + 1) + 2\overline{D}\overline{\alpha}_1 = 0.$$

As a conclusion,

**Proposition 3** *Given the arbitrary complex numbers $\alpha_1$ and $D$, where $|\alpha_1| > 1$, set*

$$B = \alpha_1^2, \quad E = -4\frac{\mathrm{Re}(D\alpha_1)}{|\alpha_1|^2+1} \quad \textit{and} \quad M_3 = \frac{-\overline{B}\alpha_1^4 - iD\alpha_1^3 - iE\alpha_1^2 - i\overline{D}\alpha_1 + B}{\alpha_1(|\alpha_1|^2-1)^2}.$$

*With*

$$\omega_0(\theta) = const \times |e^{i\theta} - \alpha_1|^{-2} e^{-2\,\mathrm{Re}[M_3/(e^{i\theta}-\alpha_1)]}$$

$$\textit{and} \quad \omega_1(\theta) = const \times |e^{i\theta} - \alpha_1|^{2} e^{-2\,\mathrm{Re}[M_3/(e^{i\theta}-\alpha_1)]},$$

*let $d\mu_0(e^{i\theta}) = \omega_0(\theta)d\theta$ and $d\mu_1(e^{i\theta}) = \omega_1(\theta)d\theta$.*

*Then $(\mu_0, \mu_1)$ is a (0,2)-coherent pair of measures of the second kind on the unit circle such that the corresponding sequences of monic OPUC satisfy* (7).

**Case (ii)**: In this case we have the following.

From $h_2(z) = z^2 + Az + B = (z-\alpha_1)(z-\alpha_2)$ we first observe that $B = \alpha_1\alpha_2$. Now from (16),

$$\begin{aligned}\frac{\frac{d}{dz}[\omega_0(\theta)]}{\omega_0(\theta)} &= \frac{[-i\overline{B}z^4 + Dz^3 + \mathrm{Re}(E)z^2 + \overline{D}z + iB]}{iz(z-\alpha_1)(1-\overline{\alpha}_1 z)(z-\alpha_2)(1-\overline{\alpha}_2 z)} \\ &= \frac{[-\overline{B}z^4 - iDz^3 - i\,\mathrm{Re}(E)z^2 - i\overline{D}z + B]}{\overline{\alpha}_1\overline{\alpha}_2 z(z-\alpha_1)(z-1/\overline{\alpha}_1)(z-\alpha_2)(z-1/\overline{\alpha}_2)}.\end{aligned} \tag{18}$$

Then, using partial fraction decomposition we obtain

$$\frac{\frac{d}{dz}[\omega_0(\theta)]}{\omega_0(\theta)} = \frac{M_1}{z-\alpha_1} + \frac{M_2}{z-\alpha_2} + \frac{M_3}{z-1/\overline{\alpha}_1} + \frac{M_4}{z-1/\overline{\alpha}_2} + \frac{M_5}{z}, \tag{19}$$

where straightforward computations yield

$$M_5 = \frac{B}{(-\alpha_1)(-\alpha_2)} = 1,$$

$$M_1 = \frac{-\overline{B}\alpha_1^4 - iD\alpha_1^3 - i(\mathrm{Re}(E))\alpha_1^2 - i\overline{D}\alpha_1 + B}{\alpha_1(1-|\alpha_1|^2)(\alpha_1-\alpha_2)(1-\alpha_1\overline{\alpha}_2)}, \qquad M_3 = \overline{M}_1,$$

$$M_2 = \frac{-\overline{B}\alpha_2^4 - iD\alpha_2^3 - i(\mathrm{Re}(E))\alpha_2^2 - i\overline{D}\alpha_2 + B}{\alpha_2(1-|\alpha_2|^2)(\alpha_2-\alpha_1)(1-\alpha_2\overline{\alpha}_1)} \quad \text{and } M_4 = \overline{M}_2.$$

Multiplication by $z$ and letting $z \to \infty$ gives $M_1 + M_2 + M_3 + M_4 + M_5 = \frac{-\overline{B}}{\overline{\alpha}_1\overline{\alpha}_2} = -1$. Which means, for example,

$$Re(M_1) + Re(M_2) = -1. \tag{20}$$

Now from (18) and (19), by integration with respect to $z$,

$$\ln\left[\frac{\omega_0(\theta)}{\omega_0(\vartheta)}\right] = M_1 \ln\left[\frac{e^{i\theta}-\alpha_1}{e^{i\vartheta}-\alpha_1}\right] + \overline{M}_1 \ln\left[\frac{e^{i\theta}-1/\overline{\alpha}_1}{e^{i\vartheta}-1/\overline{\alpha}_1}\right]$$
$$+M_2 \ln\left[\frac{e^{i\theta}-\alpha_2}{e^{i\vartheta}-\alpha_2}\right] + \overline{M}_2 \ln\left[\frac{e^{i\theta}-1/\overline{\alpha}_2}{e^{i\vartheta}-1/\overline{\alpha}_2}\right] + iM_5(\theta-\vartheta),$$

where the fixed value of $\vartheta$ is assumed to be such that $\omega_0(\vartheta) \neq 0$.

Using

$$\frac{e^{i\theta}-1/\overline{\alpha}}{e^{i\vartheta}-1/\overline{\alpha}} = e^{i(\theta-\vartheta)}\frac{e^{-i\theta}-\overline{\alpha}}{e^{-i\vartheta}-\overline{\alpha}}$$

and $M_5 = 1$, we can write

$$\ln\left[\frac{\omega_0(\theta)}{\omega_0(\vartheta)}\right] = M_1 \ln\left[\frac{e^{i\theta}-\alpha_1}{e^{i\vartheta}-\alpha_1}\right] + \overline{M}_1 \ln\left[\frac{e^{-i\theta}-\overline{\alpha}_1}{e^{-i\vartheta}-\overline{\alpha}_1}\right]$$
$$+ M_2 \ln\left[\frac{e^{i\theta}-\alpha_2}{e^{i\vartheta}-\alpha_2}\right] + \overline{M}_2 \ln\left[\frac{e^{-i\theta}-\overline{\alpha}_2}{e^{-i\vartheta}-\overline{\alpha}_2}\right]$$
$$+ i(\overline{M}_1 + \overline{M}_2 + 1)(\theta - \vartheta).$$

Furthermore, using (20) there follows $i(\overline{M}_1+\overline{M}_2+1) = \mathrm{Im}(M_1)+\mathrm{Im}(M_2)$. Hence,

$$\omega_0(\theta) = const \times e^{(\mathrm{Im}(M_1)+\mathrm{Im}(M_2))\theta} e^{[M_1 \ln(e^{i\theta}-\alpha_1)+\overline{M}_1 \ln(e^{-i\theta}-\overline{\alpha}_1)]}$$
$$\times e^{[M_2 \ln(e^{i\theta}-\alpha_2)+\overline{M}_2 \ln(e^{-i\theta}-\overline{\alpha}_2)]}.$$

Since,

$$M \ln(e^{i\theta}-\alpha) + \overline{M}\ln(e^{-i\theta}-\overline{\alpha}) = 2\,\mathrm{Re}(M)\ln|e^{i\theta}-\alpha| - 2\,\mathrm{Im}(M)\arg(e^{i\theta}-\alpha),$$

we can also write

$$\omega_0(\theta) = const \times e^{\mathrm{Im}(M_1)[\theta-2\arg(e^{i\theta}-\alpha_1)]}\, e^{\mathrm{Im}(M_2)[\theta-2\arg(e^{i\theta}-\alpha_2)]}$$
$$\times |e^{i\theta}-\alpha_1|^{2\,\mathrm{Re}(M_1)}\,|e^{i\theta}-\alpha_2|^{2\,\mathrm{Re}(M_2)}$$

and $\omega_1(\theta) = |h_2(e^{i\theta})|^2\omega_0(\theta)$.

However, since $|\alpha_1| > 1$, $|\alpha_2| > 1$ and $\omega_1(\theta) = |h_2(e^{i\theta})|^2\omega_0(\theta)$, in order to satisfy $\omega_1(0) = \omega_1(2\pi)$, we also must have $\omega_0(0) = \omega_0(2\pi)$. This is possible only if $\mathrm{Im}(M_1) + \mathrm{Im}(M_2) = 0$. Since $\alpha_1 \neq \alpha_2$, it is not clear if there is a general way to choose the parameters $E$ and $D$ so that this can be achieved. However, we find that when $E=0$ and $\alpha_2=-\alpha_1$ this is the case. Thus,

$$-1-M_2=M_1=-\frac{|\alpha_1|^4\alpha_1^2-iD\alpha_1^2-i\overline{D}-\alpha_1}{2\alpha_1(|\alpha_1|^4-1)},$$

and we can state the following.

**Proposition 4** *Given the arbitrary complex numbers* $\alpha_1$ *and* $D$*, where* $|\alpha_1|>1$*, set*

$$M_1=-\frac{|\alpha_1|^4\alpha_1^2-iD\alpha_1^2-i\overline{D}-\alpha_1}{2\alpha_1(|\alpha_1|^4-1)}.$$

*With*

$$\omega_0(\theta)=const\times e^{-2\,\mathrm{Im}(M_1)[\arg(e^{i\theta}-\alpha_1)-\arg(e^{i\theta}+\alpha_1)]}\\ \times|e^{i\theta}-\alpha_1|^{2\,\mathrm{Re}(M_1)}\,|e^{i\theta}+\alpha_1|^{-2-2\,\mathrm{Re}(M_1)}$$

$$\text{and}\quad \omega_1(\theta)=const\times|e^{i\theta}-\alpha_1|^2\,|e^{i\theta}+\alpha_1|^2\,\omega_0(\theta),$$

*let* $d\mu_0(e^{i\theta})=\omega_0(\theta)d\theta$ *and* $d\mu_1(e^{i\theta})=\omega_1(\theta)d\theta$.

*Then* $(\mu_0,\mu_1)$ *is a (0,2)-coherent pair of measures of the second kind on the unit circle such that the corresponding sequences of monic OPUC satisfy* (7).

**Case (iii)**: In this case we have $|\alpha_1|>1$, $\alpha_2=e^{i\theta_2}$ and

$$\frac{\frac{d}{dz}[\omega_0(\theta)]}{\omega_0(\theta)}=\frac{\alpha_2[-\overline{B}z^4-iDz^3-i\,\mathrm{Re}(E)z^2-i\overline{D}z+B]}{\overline{\alpha}_1z(z-\alpha_1)(z-1/\overline{\alpha}_1)(z-\alpha_2)^2},$$

with $B=\alpha_1\alpha_2$. Hence, expressing in partial fractions we obtain

$$\frac{\frac{d}{dz}[\omega_0(\theta)]}{\omega_0(\theta)}=\frac{M_1}{z-\alpha_1}+\frac{M_2}{z-1/\overline{\alpha}_1}+\frac{M_3}{z-\alpha_2}+\frac{M_4}{(z-\alpha_2)^2}+\frac{M_5}{z}.$$

Again, simple computations yield

$$M_5=\frac{B}{\alpha_1\alpha_2}=1,\qquad M_4=-\frac{-\overline{B}\alpha_2^4-iD\alpha_2^3-i\,\mathrm{Re}(E)\alpha_2^2-i\overline{D}\alpha_2+B}{\alpha_2|\alpha_2-\alpha_1|^2},$$

$$M_1=\frac{\alpha_2[-\overline{B}\alpha_1^4-iD\alpha_1^3-i\,\mathrm{Re}(E)\alpha_1^2-i\overline{D}\alpha_1+B]}{\alpha_1(\alpha_1-\alpha_2)^2(|\alpha_1|^2-1)},\qquad M_2=\overline{M}_1$$

$$\text{and } M_3=\frac{4\overline{B}\alpha_2^3+3iD\alpha_2^2+2i\,\mathrm{Re}(E)\alpha_2+i\overline{D}}{\alpha_2|\alpha_2-\alpha_1|^2}\\ +\frac{[3\overline{\alpha}_1\alpha_2^2-2|\alpha_1|^2\alpha_2+\alpha_1-2\alpha_2]}{\alpha_2^2|\alpha_2-\alpha_1|^2}M_4.$$

In addition, we also have $M_1+M_2+M_3=-1-M_5=-2$.

Since $M_2 = \overline{M}_1$, clearly $M_3$ is real and from the formula for $M_4$ one can also verify that $M_4/\alpha_2 = -\overline{M}_4/\overline{\alpha}_2$, which means $M_4/\alpha_2$ is purely imaginary.

Now, integration with respect to $z$ gives

$$\ln\left[\frac{\omega_0(\theta)}{\omega_0(\vartheta)}\right] = M_1 \ln\left[\frac{e^{i\theta}-\alpha_1}{e^{i\vartheta}-\alpha_1}\right] + M_2 \ln\left[\frac{e^{i\theta}-1/\overline{\alpha}_1}{e^{i\vartheta}-1/\overline{\alpha}_1}\right] + M_3 \ln\left[\frac{e^{i\theta}-\alpha_2}{e^{i\vartheta}-\alpha_2}\right]$$
$$-\frac{M_4}{e^{i\theta}-\alpha_2} + \frac{M_4}{e^{i\vartheta}-\alpha_2} + iM_5(\theta-\vartheta)$$
$$= M_1 \ln\left[\frac{e^{i\theta}-\alpha_1}{e^{i\vartheta}-\alpha_1}\right] + M_2 \ln\left[\frac{e^{-i\theta}-\overline{\alpha}_1}{e^{-i\vartheta}-\overline{\alpha}_1}\right] + M_3 \ln\left[\frac{e^{i\theta/2}-\alpha_2 e^{-i\theta/2}}{e^{i\vartheta/2}-\alpha_2 e^{-i\vartheta/2}}\right]$$
$$-\frac{M_4}{e^{i\theta}-\alpha_2} + \frac{M_4}{e^{i\vartheta}-\alpha_2} + i\big(M_2 + \frac{1}{2}M_3 + M_5\big)(\theta-\vartheta).$$

Observe that, from $\alpha_2 = e^{i\theta_2}$ and $M_4\overline{\alpha}_2 = -\overline{M}_4\,\alpha_2$, we have

$$\frac{e^{i\theta/2}-\alpha_2 e^{-i\theta/2}}{e^{i\vartheta/2}-\alpha_2 e^{-i\vartheta/2}} = \frac{\sin((\theta-\theta_2)/2)}{\sin((\vartheta-\theta_2)/2)}$$

and

$$\frac{1}{e^{i\theta}-\alpha_2} + \frac{1}{2\alpha_2} = \frac{(e^{i\theta}+\alpha_2)}{2\alpha_2(e^{i\theta}-\alpha_2)} = \frac{\cos((\theta-\theta_2)/2)}{i2\alpha_2\sin((\theta-\theta_2)/2)}.$$

Hence, we can write

$$\omega_0(\theta) = const \times |e^{i\theta}-\alpha_1|^{M_1+M_2}\, e^{i(M_1-M_2)\arg(e^{i\theta}-\alpha_1)}$$
$$\times\ \sin((\theta-\theta_2)/2)^{M_3}\, e^{-\left[\frac{M_4\cos((\theta-\theta_2))}{i2\alpha_2\sin((\theta-\theta_2))}\right]} e^{-i(M_1-M_2)\theta}.$$

Observe that for $\omega_0$ is to be real and bounded, and also non-negative throughout $[0, 2\pi]$, one must have $\alpha_2 = 1$, i. e. $\theta_2 = 0$, and $M_4 = 0$. This also implies, from $\omega_1(\theta) = |h_2(e^{i\theta})|^2\omega_0(\theta)$, that $\omega_1(0) = \omega_1(2\pi) = 0$. Moreover, for the existence of $\int_0^{2\pi} \omega_0(\theta)d\theta$, we also must have $M_3 > -1$.

From $\alpha_2 = 1$ and $M_4 = 0$ we immediately find,

$$\mathrm{Re}(E) = 2\,\mathrm{Im}(\alpha_1) - 2\,\mathrm{Re}(D).$$

With this, using $M_3 > -1$ we find

$$2\,\mathrm{Im}(D) < |\alpha_1 + 1|^2.$$

**Proposition 5** *Given the complex numbers $\alpha_1$ and $D$, such that $|\alpha_1| > 1$ and $2\,\mathrm{Im}(D) < |\alpha_1 + 1|^2$, let*

$$E = 2\,\mathrm{Im}(\alpha_1) - 2\,\mathrm{Re}(D),$$

$$M_3 = \frac{4\,\mathrm{Re}(\alpha_1) - 2\,\mathrm{Im}(D)}{|\alpha_1 - 1|^2}, \quad M_1 = \frac{-|\alpha_1|^2\alpha_1^2 - iD\alpha_1^2 - iE\alpha_1 - i\overline{D} + 1}{(\alpha_1 - 1)^2(|\alpha_1|^2 - 1)}$$

*and $h_2(z) = (z - \alpha_1)(z - 1)$. With*

$$\omega_0(\theta) = const \times |e^{i\theta} - \alpha_1|^{-2-M_3} e^{i(M_1 - \overline{M}_1)\arg(e^{i\theta} - \alpha_1)} e^{-i(M_1 - \overline{M}_1)\theta} \sin(\theta/2)^{M_3},$$

$$\omega_1(\theta) = |h_2(e^{i\theta})|^2 \omega_0(\theta),$$

*let $d\mu_0(e^{i\theta}) = \omega_0(\theta)d\theta$ and $d\mu_1(e^{i\theta}) = \omega_1(\theta)d\theta$.*

*Then $(\mu_0, \mu_1)$ is a (0,2)-coherent pair of measures of the second kind on the unit circle such that the corresponding sequences of monic OPUC satisfy* (7).

**Case (iv)**: In this case we have $B = \alpha_1\alpha_2$, $\alpha_1 = e^{i\theta_1}$, $\alpha_2 = e^{i\theta_2}$, $\theta_1 \neq \theta_2$ and

$$\frac{\frac{d}{dz}[\omega_0(\theta)]}{\omega_0(\theta)} = \frac{[-\overline{B}z^4 - iDz^3 - i\,\mathrm{Re}(E)z^2 - i\overline{D}z + B]\alpha_1\alpha_2}{z(z - \alpha_1)^2(z - \alpha_2)^2}.$$

Using the partial fractions decomposition

$$\frac{\frac{d}{dz}[\omega_0(\theta)]}{\omega_0(\theta)} = \frac{M_1}{z - \alpha_1} + \frac{M_2}{z - \alpha_2} + \frac{M_3}{(z - \alpha_1)^2} + \frac{M_4}{(z - \alpha_2)^2} + \frac{M_5}{z},$$

we obtain

$$M_5 = \frac{B}{\alpha_1\alpha_2} = 1,$$

$$M_3 = \frac{(-\overline{B}\alpha_1^4 - iD\alpha_1^3 - i\,\mathrm{Re}(E)\alpha_1^2 - i\overline{D}\alpha_1 + B)\alpha_2}{(\alpha_2 - \alpha_1)^2},$$

$$M_4 = \frac{(-\overline{B}\alpha_2^4 - iD\alpha_2^3 - i\,\mathrm{Re}(E)\alpha_2^2 - i\overline{D}\alpha_2 + B)\alpha_1}{(\alpha_2 - \alpha_1)^2},$$

$$M_1 = \frac{[-4\overline{B}\alpha_1^3 - 3iD\alpha_1^2 - 2i(\mathrm{Re}\,E)\alpha_1 - i\overline{D}]\alpha_2}{(\alpha_2 - \alpha_1)^2} - \frac{\alpha_2 - 3\alpha_1}{\alpha_1(\alpha_2 - \alpha_1)}M_3 \text{ and}$$

$$M_2 = -2 - M_1.$$

From the formulas for $M_3$ and $M_4$ we also have

$$\overline{M}_3 = -M_3/\alpha_1^2 \quad \text{and} \quad \overline{M}_4 = -M_4/\alpha_2^2.$$

Moreover, from the expression of $M_1$ one can also verify that $\mathrm{Im}(M_1) = 0$ as well as $\mathrm{Im}(M_2) = 0$.

Now, integrating with respect to $z$ we have

$$\ln\left[\frac{\omega_0(\theta)}{\omega_0(\vartheta)}\right] = M_1 \ln\left[\frac{e^{i\theta}-\alpha_1}{e^{i\vartheta}-\alpha_1}\right] + M_2 \ln\left[\frac{e^{i\theta}-\alpha_2}{e^{i\vartheta}-\alpha_2}\right] + iM_5(\theta-\vartheta)$$
$$-\frac{M_3}{e^{i\theta}-\alpha_1} + \frac{M_3}{e^{i\vartheta}-\alpha_1} - \frac{M_4}{e^{i\theta}-\alpha_2} + \frac{M_4}{e^{i\vartheta}-\alpha_2}$$
$$= M_1 \ln\left[\frac{e^{i\theta/2}-\alpha_1 e^{-i\theta/2}}{e^{i\vartheta/2}-\alpha_1 e^{-i\vartheta/2}}\right] + M_2 \ln\left[\frac{e^{i\theta/2}-\alpha_2 e^{-i\theta/2}}{e^{i\vartheta/2}-\alpha_2 e^{-i\vartheta/2}}\right]$$
$$-\frac{M_3}{e^{i\theta}-\alpha_1} + \frac{M_3}{e^{i\vartheta}-\alpha_1} - \frac{M_4}{e^{i\theta}-\alpha_2} + \frac{M_4}{e^{i\vartheta}-\alpha_2}.$$

Hence,

$$\omega_0(\theta) = const \times \sin((\theta-\theta_1)/2)^{M_1} \times \sin((\theta-\theta_2)/2)^{M_2}$$
$$\times e^{-\left[\frac{M_3(e^{i\theta}+\alpha_1)}{2\alpha_1(e^{i\theta}-\alpha_1)}\right]} e^{-\left[\frac{M_4(e^{i\theta}+\alpha_2)}{2\alpha_2(e^{i\theta}-\alpha_2)}\right]},$$

where we have used

$$\frac{M_3(e^{i\theta}+\alpha_1)}{4\alpha_1(e^{i\theta}-\alpha_1)} + \frac{\overline{M}_3(e^{-i\theta}+\overline{\alpha}_1)}{4\overline{\alpha}_1(e^{-i\theta}-\overline{\alpha}_1)} = \frac{M_3(e^{i\theta}+\alpha_1)}{2\alpha_1(e^{i\theta}-\alpha_1)} = \frac{M_3}{e^{i\theta}-\alpha_1} + \frac{M_3}{2\alpha_1}$$

and

$$\frac{M_4(e^{i\theta}+\alpha_2)}{4\alpha_2(e^{i\theta}-\alpha_2)} + \frac{\overline{M}_4(e^{-i\theta}+\overline{\alpha}_2)}{4\overline{\alpha}_2(e^{-i\theta}-\overline{\alpha}_2)} = \frac{M_4(e^{i\theta}+\alpha_2)}{2\alpha_2(e^{i\theta}-\alpha_2)} = \frac{M_4}{e^{i\theta}-\alpha_2} + \frac{M_4}{2\alpha_2}.$$

We need to find $\theta_1$, $\theta_2$ and $M_1$ such that $\theta_1 \neq \theta_2$, $M_1 > -1$, $-M_1 - 2 > -1$ and

$$\sin((\theta-\theta_1)/2)^{M_1} \sin((\theta-\theta_2)/2)^{M_2} = \sin((\theta-\theta_1)/2)^{M_1} \sin((\theta-\theta_2)/2)^{-M_1-2},$$

does not change within $[0, 2\pi]$. As this is not possible, we can conclude that case (iv) is not feasible.

**Proposition 6** *Let $(\mu_0, \mu_1)$ be a pair of positive measures on the unit circle such that*

$$d\mu_0(z) = \frac{1}{|h_2(z)|^2} d\mu_1(z),$$

*where $h_2(z) = (z-\alpha_1)(z-\alpha_2)$. Moreover, let $d\mu_0(e^{i\theta}) = \omega_0(\theta)d\theta$, $d\mu_1(e^{i\theta}) = \omega_1(\theta)d\theta$, with $\omega_1(\theta) \in C^1[0, 2\pi]$, $\omega_1(0) = \omega_1(2\pi)$ and that there exists a $g_2(z) =$*

*$2Cz^2+2Dz+E$ such that*

$$|h_2(e^{i\theta})|^2\omega_0'(\theta)=\frac{g_2(e^{i\theta})+\overline{g_2(e^{i\theta})}}{2}\omega_0(\theta).$$

*Then if $\alpha_1\neq\alpha_2$ $|\alpha_1|=|\alpha_2|=1$ the pair $(\mu_0,\mu_1)$ can not be a (0,2)-coherent pair of measures of the second kind on the unit circle.*

**Case (v)**: Finally, considering case (v) we have from (16)

$$\frac{\frac{d}{dz}[\omega_0(\theta)]}{\omega_0(\theta)}=\frac{[-\overline{B}z^4-iDz^3-iRe(E)z^2-i\overline{D}z+B]\alpha_1^2}{z(z-\alpha_1)^4},$$

where $B=\alpha_1^2$ and $|\alpha_1|=1$. Let us write $\alpha_1=e^{i\theta_1}$.

Using partial fractions we obtain

$$\frac{\frac{d}{dz}[\omega_0(\theta)]}{\omega_0(\theta)}=\frac{M_1}{(z-\alpha_1)^4}+\frac{M_2}{(z-\alpha_1)^3}+\frac{M_3}{(z-\alpha_1)^2}+\frac{M_4}{z-\alpha_1}+\frac{M_5}{z}.$$

Simple computations also yield,

$$M_5=1,\qquad M_4=-2,$$

$$M_1=-i[D\alpha_1^2+\mathrm{Re}(E)\alpha_1+\overline{D}]\alpha_1^2,\quad M_2=-[4+2iD\alpha_1+i\,\mathrm{Re}(E)]\alpha_1^2$$

and $M_3=-[2+iD\alpha_1]\alpha_1$.

Integrating with respect to $z$ gives

$$\ln\left[\frac{\omega_0(\theta)}{\omega_0(\vartheta)}\right]=M_4\ln\left[\frac{e^{i\theta}-\alpha_1}{e^{i\vartheta}-\alpha_1}\right]+iM_5(\theta-\vartheta)-\frac{M_1}{3}\frac{1}{(e^{i\theta}-\alpha_1)^3}+\frac{M_1}{3}\frac{1}{(e^{i\vartheta}-\alpha_1)^3}$$
$$-\frac{M_2}{2}\frac{1}{(e^{i\theta}-\alpha_1)^2}+\frac{M_2}{2}\frac{1}{(e^{i\vartheta}-\alpha_1)^2}-M_3\frac{1}{e^{i\theta}-\alpha_1}+M_3\frac{1}{e^{i\vartheta}-\alpha_1}. \tag{21}$$

Again observe that

$$\ln\left[\frac{e^{i\theta}-\alpha_1}{e^{i\vartheta}-\alpha_1}\right]=\frac{1}{2}(\theta-\vartheta)+\ln\left[\frac{\sin(\theta-\theta_1)}{\sin(\vartheta-\theta_1)}\right],$$

$$\frac{1}{e^{i\theta}-\alpha_1}=\frac{(e^{i\theta}+\alpha_1)}{2\alpha_1(e^{i\theta}-\alpha_1)}-\frac{1}{2\alpha_1}=\frac{\cos((\theta-\theta_1)/2)}{2i\alpha_1\sin((\theta-\theta_1)/2)}-\frac{1}{2\alpha_1}.$$

Consequently, taking exponential on both sides of (21) we obtain as one of the terms on the right hand side $\sin((\theta-\theta_1)/2)^{M2}$ and the remaining terms can be given as exponential of some powers of

$$\frac{\cos((\theta-\theta_1)/2)}{2i\alpha_1\sin((\theta-\theta_1)/2)}.$$

Clearly we have to take $\alpha_1 = 1$, i.e. $\theta_1 = 0$, for the positiveness of the function $\sin((\theta-\theta_1)/2)^{M2}$ in $[0, 2\pi]$. However, $M_2 = -2$ leads to the non existence of the integral $\int_0^{2\pi}\omega_0(\theta)d\theta$. This means case (v) is also not feasible.

**Proposition 7** *Let* $(\mu_0, \mu_1)$ *be a pair of positive measures on the unit circle such that*

$$d\mu_0(z) = \frac{1}{|h_2(z)|^2}d\mu_1(z),$$

*where* $h_2(z) = (z-\alpha_1)^2$. *Moreover, let* $d\mu_0(e^{i\theta}) = \omega_0(\theta)d\theta$, $d\mu_1(e^{i\theta}) = \omega_1(\theta)d\theta$, *with* $\omega_1(\theta) \in C^1[0, 2\pi]$, $\omega_1(0) = \omega_1(2\pi)$ *and that there exists a* $g_2(z) = 2Cz^2 + 2Dz + E$ *such that*

$$|h_2(e^{i\theta})|^2\omega_0'(\theta) = \frac{g_2(e^{i\theta}) + \overline{g_2(e^{i\theta})}}{2}\omega_0(\theta).$$

*Then if* $|\alpha_1| = 1$ *the pair* $(\mu_0, \mu_1)$ *can not be a (0,2)-coherent pair of measures of the second kind on the unit circle.*

## 3 Hessenberg Matrices

Now, let us denote by $\{\varphi_n(\mu_0; z)\}_{n\geq 0}$ and $\{\varphi_n(\mu_1; z)\}_{n\geq 0}$ the sequences of orthonormal polynomials associated with $d\mu_0$ and $d\mu_1$, respectively. In this situation, assuming they constitute a (0,2)-coherent pair of measures, (7) becomes

$$\varphi_{n+1}'(\mu_0; z) = \hat{c}_n\varphi_n(\mu_1; z) + \hat{a}_n\varphi_{n-1}(\mu_1; z) + \hat{b}_n\varphi_{n-2}(\mu_1; z), \quad n \geq 3, \tag{22}$$

for some normalized coefficients $\{\hat{a}_n\}_{n\geq 3}$, $\{\hat{b}_n\}_{n\geq 3}$ and $\{\hat{c}_n\}_{n\geq 3}$, where $\hat{b}_n \neq 0, n \geq 3$. Notice that the above equation also holds for $0 \leq n \leq 2$ since the left hand side is a polynomial of degree $n \leq 2$ and it can be represented as a linear combination of $\varphi_k(\mu_1; z), 0 \leq k \leq 2$. Indeed, by convention, $\hat{a}_0 = \hat{b}_0 = \hat{b}_1 = 0$, and the coefficients $\hat{a}_1, \hat{a}_2, \hat{b}_2, \hat{c}_k, 0 \leq k \leq 2$, can be deduced in a straightforward way taking into account $\varphi_2'(\mu_0; z) = \hat{c}_1\varphi_1(\mu_1; z) + \hat{a}_1\varphi_0(\mu_1; z)$ and $\varphi_3'(\mu_0; z) = \hat{c}_2\varphi_2(\mu_1; z)+\hat{a}_2\varphi_1(\mu_1; z)+\hat{b}_2\varphi_0(\mu_1; z)$ are representations which do not depend on the constraint (22) for polynomials associated with (0,2)-coherent pair of measures.

In matrix form, the above equation can be written as

$$[\boldsymbol{\varphi}_{\mu_0}]' = X^T C \boldsymbol{\varphi}_{\boldsymbol{\mu}_1}.$$

Here $\boldsymbol{\varphi}_{\mu_0} = [\varphi_0(\mu_0; z), \ldots, \varphi_n(\mu_0; z), \ldots]^t$, $\boldsymbol{\varphi}_{\mu_1} = [\varphi_0(\mu_1; z), \ldots, \varphi_n(\mu_1; z), \ldots]^t$ and $C$ is the lower triangular matrix

$$C = \begin{bmatrix} \hat{c}_0 & 0 & 0 & 0 & \ldots \\ \hat{a}_1 & \hat{c}_1 & 0 & 0 & \ldots \\ \hat{b}_2 & \hat{a}_2 & \hat{c}_2 & 0 & \ldots \\ 0 & \hat{b}_3 & \hat{a}_3 & \hat{c}_3 & \ddots \\ \vdots & \ddots & \ddots & \ddots & \ddots \end{bmatrix}.$$

Indeed, since $[\boldsymbol{\varphi}_{\mu_0}]' = D\boldsymbol{\varphi}_{\mu_0}$, we get

$$D\boldsymbol{\varphi}_{\mu_0} = X^T C \boldsymbol{\varphi}_{\mu_1}, \tag{23}$$

where $D$ denotes the derivative operator and $X$ is the shift matrix, i.e. if $X = (x_{j,k})_{j,k=0}^{\infty}$, then $x_{j,j+1} = 1$, $j \geq 0$, and $x_{j,k} = 0$, otherwise.

Recall that the sequences $\{\varphi_n(\mu_0; z)\}_{n\geq 0}$ and $\{\varphi_n(\mu_1; z)\}_{n\geq 0}$ satisfy

$$z\boldsymbol{\varphi}_{\mu_0} = H_{\mu_0}\boldsymbol{\varphi}_{\mu_0}, \tag{24}$$

$$z\boldsymbol{\varphi}_{\mu_1} = H_{\mu_1}\boldsymbol{\varphi}_{\mu_1}, \tag{25}$$

where $H_{\mu_0}$ and $H_{\mu_1}$ are the corresponding Hessenberg matrices associated with $\{\varphi_n(\mu_0; z)\}_{n\geq 0}$ and $\{\varphi_n(\mu_1; z)\}_{n\geq 0}$, respectively. These matrices are called GGT matrices (see Chapter 4 in [17]) and provide a representation of the multiplication operator in terms of the orthonormalized OPUC. We must point out that those Hessenberg matrices are not unitary, up to the measure does not belong to the Szegő class. In a similar way, the Hessenberg matrix is unitary if and only if the linear space of polynomials is dense in the linear space of Laurent polynomials (see [17]).

Taking the derivative with respect to $z$ in (24) we obtain

$$zD\boldsymbol{\varphi}_{\mu_0} + \boldsymbol{\varphi}_{\mu_0} = H_{\mu_0} D\boldsymbol{\varphi}_{\mu_0},$$

and using (23) and (25), we get

$$X^T C H_{\mu_1} \boldsymbol{\varphi}_{\mu_1} + \boldsymbol{\varphi}_{\mu_0} = H_{\mu_0} X^T C \boldsymbol{\varphi}_{\mu_1}.$$

Now, we can say that $\boldsymbol{\varphi}_{\mu_1} = L\boldsymbol{\varphi}_{\mu_0}$, where $L$ is the matrix of change of basis so that

$$X^T C H_{\mu_1} L + I = H_{\mu_0} X^T C L. \tag{26}$$

Then, multiplying by $z$ and using (24), we get

$$\begin{aligned} X^T C H_{\mu_1} L H_{\mu 0} \boldsymbol{\varphi}_{\mu 0} + H_{\mu 0} \boldsymbol{\varphi}_{\mu 0} &= H_{\mu 0} X^T C L H_{\mu 0} \boldsymbol{\varphi}_{\mu 0}, \\ H_{\mu 0} \boldsymbol{\varphi}_{\mu 0} &= (H_{\mu 0} X^T C L H_{\mu 0} - X^T C H_{\mu_1} L H_{\mu 0}) \boldsymbol{\varphi}_{\mu 0}, \\ H_{\mu 0} \boldsymbol{\varphi}_{\mu 0} &= (H_{\mu 0} X^T C - X^T C H_{\mu_1}) L H_{\mu 0} \boldsymbol{\varphi}_{\mu 0}, \end{aligned}$$

and, consequently,

$$H_{\mu 0} = (H_{\mu 0} X^T C - X^T C H_{\mu_1}) L H_{\mu 0}.$$

We know that $H_{\mu 0}$ satisfies $H_{\mu 0} H_{\mu 0}^* = I$. Thus,

$$\begin{aligned} I &= H_{\mu 0} X^T C L - X^T C H_{\mu_1} L, \\ H_{\mu 0} X^T C L &= I + X^T C H_{\mu_1} L, \\ H_{\mu 0} X^T &= X^T C H_{\mu_1} C^{-1} + L^{-1} C^{-1}. \end{aligned}$$

Recalling that $XX^T = I$ we get

$$\begin{aligned} X H_{\mu 0} X^T &= C H_{\mu_1} C^{-1} + X L^{-1} C^{-1}, \\ X H_{\mu 0} X^T &= (C H_{\mu_1} + X L^{-1}) C^{-1}, \end{aligned}$$

which can be written as

$$X H_{\mu 0} X^T = C (H_{\mu_1} + C^{-1} X L^{-1}) C^{-1}.$$

Thus, we have proved the following result

**Proposition 8** *Let* $\{\varphi_n(\mu_0; z)\}_{n \geq 0}$ *and* $\{\varphi_n(\mu_1; z)\}_{n \geq 0}$ *be the sequences of orthonormal polynomials satisfying* (23)*. Then their corresponding Hessenberg matrices satisfy*

$$X H_{\mu 0} X^T = C (H_{\mu_1} + C^{-1} X L^{-1}) C^{-1},$$

*where the matrices L and C are defined as above.*

*Remark 2* Notice that in Chapter 4 of [17], another matrix representation of the multiplication operator but in terms of a basis of orthonormal Laurent polynomials (the so called CMV representation) is studied. Such a matrix is pentadiagonal and unitary and the eigenvalues of the leading principal submatrix of size $n \times n$ are the zeros of the corresponding OPUC. The connection between the CMV matrices associated with a (0,2)-coherent pair of measures of the second kind is an open problem.

## 4 Sobolev OPUC

Contributions on the theory of Sobolev orthogonal polynomials on the real line (see [12]) have been very intense during the past few decades. However, it has not been the same for studies concerning Sobolev orthogonal polynomials on the unit circle. For some of the studies on the unit circle we cite [1–3, 5, 9, 11, 18] and [19].

Hence, the objective here is to establish information regarding the sequence of monic Sobolev orthogonal polynomials with respect to the inner product

$$\langle f, g\rangle_{S^{(\mu_0,\mu_1,s)}} = \langle f, g\rangle_{\mu_0} + s\left\langle f', g'\right\rangle_{\mu_1}, \tag{27}$$

where the measures $(\mu_0, \mu_1)$ constitute a (0,2)-coherent pair of measures of the second kind on the unit circle. Here, $\langle f, g\rangle_{\mu_0} = \int_{\mathbb{T}} f(z)\overline{g(z)}d\mu_0(z)$ and $\langle f, g\rangle_{\mu_1} = \int_{\mathbb{T}} f(z)\overline{g(z)}d\mu_1(z)$.

**Theorem 1** *Let $(\mu_0, \mu_1)$ be an (0,2)-coherent pair of measures of the second kind on the unit circle. Then, the sequence of monic orthogonal polynomials $\{\Psi_n(z)\}_{n\geq 0}$ with respect to the Sobolev inner product (27), where $\Psi_n(z) = \Psi_n(\mu_0, \mu_1, s; z)$, satisfies $\Phi_1(\mu_0; z) = \Psi_1(z)$ and*

$$\Phi_n(\mu_0; z) = \Psi_n(z) + \alpha_n\Psi_{n-1}(z) + \beta_n\Psi_{n-2}(z), \quad n \geq 2. \tag{28}$$

*For the coefficients $\alpha_n = \alpha_n^{(\mu_0,\mu_1,s)}$ and $\beta_n = \beta_n^{(\mu_0,\mu_1,s)}$, there hold*

$$\alpha_n = \frac{q_n}{p_n - \overline{\alpha}_{n-1}q_{n-1} - \overline{\beta}_{n-1}r_n}, \quad n \geq 2,$$

*and*

$$\beta_n = \frac{r_n}{p_{n-1} - \overline{\alpha}_{n-2}q_{n-2} - \overline{\beta}_{n-2}r_{n-1}}, \quad n \geq 3,$$

*with $\alpha_0 = \alpha_1 = \beta_0 = \beta_1 = \beta_2 := 0$. Here, $q_n = q_n{}^{(\mu_0,\mu_1,s)}$, $p_n = p_n^{(\mu_0,\mu_1,s)}$ and $r_n = r_n^{(\mu_0,\mu_1,s)}$ are such that*

$$\begin{aligned}
p_n^{(\mu_0,\mu_1,s)} &= \|\Phi_{n-1}(\mu_0; z)\|^2 \\
&\quad + s(n-1)^2\Big[\|\Phi_{n-2}(\mu_1; z)\|^2 + |a_{n-2}|^2\|\Phi_{n-3}(\mu_1; z)\|^2 \\
&\qquad + |b_{n-2}|^2\|\Phi_{n-4}(\mu_1; z)\|^2\Big], \\
q_n^{(\mu_0,\mu_1,s)} &= sn\Big[(n-1)a_{n-1}\|\Phi_{n-2}(\mu_1; z)\|^2 \\
&\qquad + \big((n-1)b_{n-1}\overline{a}_{n-2} - (n-2)b_{n-1}\overline{\alpha}_{n-1}\big)\|\Phi_{n-3}(\mu_1; z)\|^2\Big], \\
r_n^{(\mu_0,\mu_1,s)} &= sn(n-2)b_{n-1}\|\Phi_{n-3}(\mu_1; z)\|^2,
\end{aligned}$$

*for $n \geq 2$, where one must choose $\|\Phi_{-1}(\mu_1; z)\|^2 = \|\Phi_{-2}(\mu_1; z)\|^2 = 0$.*

***Proof*** For simplicity, we will use the notations $\Phi_{n,0}(z)$ and $\Phi_{n,1}(z)$ for $\Phi_n(\mu_0; z)$ and $\Phi_n(\mu_1; z)$, respectively. We consider the Fourier expansion of $\Phi_{n,0}(z)$ in terms of the sequence of Sobolev orthogonal polynomials $\{\Psi_j(z)\}_{j\geq 0}$:

$$\Phi_{n,0}(z) = \sum_{j=0}^{n-1} h_{j,n}\Psi_j(z) + \Psi_n(z), \quad n \geq 1.$$

For $0 \leq j \leq n-1$, we get

$$\begin{aligned} h_{j,n} &= \frac{\langle \Phi_{n,0}(z), \Psi_j(z)\rangle_{S^{(\mu_0,\mu_1,s)}}}{\langle \Psi_j(z), \Psi_j(z)\rangle_{S^{(\mu_0,\mu_1,s)}}}, \\ &= \frac{\langle \Phi_{n,0}(z), \Psi_j(z)\rangle_{\mu_0} + s\left\langle \Phi'_{n,0}(z), \Psi'_j(z)\right\rangle_{\mu_1}}{\langle \Psi_j(z), \Psi_j(z)\rangle_{S^{(\mu_0,\mu_1,s)}}}, \\ &= \frac{s\left\langle \Phi'_{n,0}(z), \Psi'_j(z)\right\rangle_{\mu_1}}{\langle \Psi_j(z), \Psi_j(z)\rangle_{S^{(\mu_0,\mu_1,s)}}}. \end{aligned}$$

Since, $\Psi'_0(x) = 0$, we have $h_{0,n} = 0$, $n \geq 1$, and

$$\Phi_{1,0}(z) = \Psi_1(z) \quad \text{and} \quad \Phi_{2,0}(z) = h_{1,2}\Psi_1(z) + \Psi_2(z).$$

Moreover, from (7) it is easy to see that $h_{j,n} = 0$ for $j = 0, 1, \ldots, n-3$. By relabeling $h_{n-1,n} = \alpha_n$ and $h_{n-2,n} = \beta_n$ we have (28), with $\beta_2 = 0$.

We will now determine the recurrence formula for the coefficients $\alpha_n = \alpha_n^{(\mu_0,\mu_1,s)}$ and $\beta_n = \beta_n^{(\mu_0,\mu_1,s)}$. For $\alpha_n$ we have

$$\alpha_n = \frac{s\left\langle \Phi'_{n,0}(z), \Psi'_{n-1}(z)\right\rangle_{\mu_1}}{\langle \Psi_{n-1}(z), \Psi_{n-1}(z)\rangle_{S^{(\mu_0,\mu_1,s)}}}.$$

From (7) and using $\Psi'_n(z) = \Phi'_{n,0}(z) - \alpha_n\Psi'_{n-1}(z) - \beta_n\Psi'_{n-2}(z)$, we obtain

$$\begin{aligned} s\left\langle \Phi'_{n,0}(z), \Psi'_{n-1}(z)\right\rangle_{\mu_1} &= sn\Big\langle \Phi_{n-1,1}(z) + a_{n-1}\Phi_{n-2,1}(z) + b_{n-1}\Phi_{n-3,1}(z), \\ &\qquad \Phi'_{n-1,0}(z) - \alpha_{n-1}\Psi'_{n-2}(z) - \beta_{n-1}\Psi'_{n-3}(z)\Big\rangle_{\mu_1} \\ &= sn\Big\langle \Phi_{n-1,1}(z) + a_{n-1}\Phi_{n-2,1}(z) + b_{n-1}\Phi_{n-3,1}(z), \\ &\qquad (n-1)[\Phi_{n-2,1}(z) + a_{n-2}\Phi_{n-3,1}(z) + b_{n-2}\Phi_{n-4,1}(z)] \\ &\qquad -\alpha_{n-1}\Psi'_{n-2}(z) - \beta_{n-1}\Psi'_{n-3}(z)\Big\rangle_{\mu_1}. \end{aligned}$$

Thus, we get

$$
\begin{aligned}
&s\left\langle\Phi'_{n,0}(z), \Psi'_{n-1}(z)\right\rangle_{\mu_1} \\
&\quad = sn\Big[a_{n-1}(n-1)\|\Phi_{n-2,1}(z)\|^2 + b_{n-1}(n-1)\overline{a}_{n-2}\|\Phi_{n-3,1}(z)\|^2 \\
&\qquad\qquad - \overline{\alpha}_{n-1}b_{n-1}(n-2)\|\Phi_{n-3,1}(z)\|^2\Big] \\
&\quad = q_n,
\end{aligned}
$$

for $n \geq 2$.

On the other hand, since $\Psi_n(z)$ is monic, we have $\langle\Psi_0(z), \Psi_0(z)\rangle_{S^{(\mu_0,\mu_1,s)}} = \langle 1, 1\rangle_{\mu_0}$ and

$$
\langle\Psi_1(z), \Psi_1(z)\rangle_{S^{(\mu_0,\mu_1,s)}} = \|\Phi_{1,0}(z)\|^2 + s\,\langle 1, 1\rangle_{\mu_1} = \|\Phi_{1,0}(z)\|^2 + s\|\Phi_{0,1}(z)\|^2.
$$

More generally, for $n \geq 2$,

$$
\begin{aligned}
\langle\Psi_n(z), \Psi_n(z)\rangle_{S^{(\mu_0,\mu_1,s)}} &= \left\langle\Phi_{n,0}(z), \Psi_n(z)\right\rangle_{S^{(\mu_0,\mu_1,s)}}, \\
&= \|\Phi_{n,0}(z)\|^2 + s\left\langle\Phi'_{n,0}(z), \Psi'_n(z)\right\rangle_{\mu_1}.
\end{aligned}
$$

Again, from $\Psi'_{n-1}(z) = \Phi'_{n-1,0}(z) - \alpha_{n-1}\Psi'_{n-2}(z) - \beta_{n-1}\Psi'_{n-3}(z)$ and from (7) the expression for $\langle\Psi_{n-1}(z), \Psi_{n-1}(z)\rangle_{S^{(\mu_0,\mu_1,s)}}$ can be written as

$$
\begin{aligned}
&\langle\Psi_{n-1}(z), \Psi_{n-1}(z)\rangle_{S^{(\mu_0,\mu_1,s)}} \\
&= \|\Phi_{n-1,0}(z)\|^2 \\
&\quad + s(n-1)\Big[(n-1)\|\Phi_{n-2,1}(z)\|^2 + (n-1)|a_{n-2}|^2\|\Phi_{n-3,1}(z)\|^2 \\
&\qquad + (n-1)|b_{n-2}|^2\|\Phi_{n-4,1}(z)\|^2 - \overline{\alpha}_{n-1}a_{n-2}(n-2)\|\Phi_{n-3,1}(z)\|^2 \\
&\qquad - \overline{\alpha}_{n-1}b_{n-2}\overline{\alpha}_{n-2}(n-3)\|\Phi_{n-4,1}(z)\|^2 \\
&\qquad - \overline{\alpha}_{n-1}b_{n-2}(n-2)\overline{a}_{n-3}\|\Phi_{n-4,1}(z)\|^2 \\
&\qquad - \overline{\beta}_{n-1}b_{n-2}(n-3)\|\Phi_{n-4,1}(z)\|^2\Big] \\
&= p_n - \overline{\alpha}_{n-1}q_{n-1} - \overline{\beta}_{n-1}r_n,
\end{aligned}
$$

for $n \geq 3$. Thus, we get the desired recurrence formula for $\alpha_n$.

On the other hand, proceeding as above, for $n \geq 3$ we obtain

$$
\begin{aligned}
\beta_n &= \frac{s\left\langle\Phi'_{n,0}(z), \Psi'_{n-2}(z)\right\rangle_{\mu_1}}{\|\Psi_{n-2}(z)\|^2}, \\
&= \frac{snb_{n-1}(n-2)\|\Phi_{n-3,1}(z)\|^2}{\|\Psi_{n-2}(z)\|^2} = \frac{r_n}{p_{n-1} - \overline{\alpha}_{n-2}q_{n-2} - \overline{\beta}_{n-2}r_{n-1}}.
\end{aligned}
$$

□

As a straightforward consequence of the expressions for $\alpha_n$ and $\beta_n$ we get

$$\alpha_n = \frac{q_n}{r_{n+1}}\beta_{n+1}, n \geq 2.$$

Replacing the corresponding expressions for $\alpha_n$ and $\alpha_{n-1}$ as above, we get

**Corollary 1**

$$d_{n+1}(p_n - |q_{n-1}|^2\bar{d}_n - r_n\bar{r}_{n-1}\bar{d}_{n-1}) = 1, n \geq 2,$$

*where* $d_n = \frac{\beta_n}{r_n}$, $n \geq 2$, *and initial conditions* $d_1 = d_2 = 0$.

Notice that, alternatively, from the expressions of the parameters stated in the above Theorem,

$$\beta_{n+1} = \frac{\alpha_n r_{n+1}}{\overline{\alpha_{n-1}}A_n + B_n}, n \geq 2,$$

with

$$A_n = -sn(n-2)b_{n-1}||\Phi_{n-3}(\mu_1; z)||^2,$$
$$B_n = sn(n-1)(a_{n-1}||\Phi_{n-2}(\mu_1; z)||^2 + b_{n-1}\overline{a}_{n-2}||\Phi_{n-3}(\mu_1; z)||^2).$$

This means that one can obtain $\beta_{n+1}$ in terms of $\alpha_n$ and $\alpha_{n-1}$.

On the other hand,

$$\alpha_{n+1} = \frac{A_{n+1}\overline{\alpha_n} + B_{n+1}}{p_{n+1} - (A_n\overline{\alpha}_{n-1} + B_n)\overline{\alpha}_n - r_{n+1}\overline{\beta}_n}, \quad n \geq 2.$$

Thus, one can deduce $\alpha_{n+1}$ from $\alpha_n$, $\alpha_{n-1}$, and $\alpha_{n-2}$.

We now consider $\Psi_n(z)$ as a linear combination of $\{\Phi_n(\mu_0; z)\}_{n\geq 0}$. Again, for notational convenience, we will use $\Phi_{n,0}(z)$ and $\Phi_{n,1}(z)$ instead of $\Phi_n(\mu_0; z)$ and $\Phi_n(\mu_1; z)$, respectively. We have

$$\Psi_{n+1}(z) = \sum_{j=0}^{n+1} c_{j,n+1}\Phi_{j,0}(z),$$

where $\{c_{j,n+1}\}_{j\geq 0}$ are the Fourier coefficients of $\Psi_{n+1}(z)$ with respect to the orthogonal system $\{\Phi_{n,0}(z)\}_{n\geq 0}$.

Taking into account that $\{\Psi_n(z)\}_{n\geq 0}$ is the SMOP with respect to the Sobolev inner product (27), from the initial conditions $a_0 = b_0 = b_1 = 0$ and the convention

$\|\Phi_{n,1}\|^2 = 0$, $n < 0$, we have

$$0 = \langle \Phi_{0,0}, \Psi_{n+1} \rangle_{S(\mu_0,\mu_1,s)},$$
$$= \left\langle \Phi_{0,0}, \sum_{j=0}^{n+1} c_{j,n+1} \Phi_{j,0}(z) \right\rangle_{\mu_0} + s \left\langle 0, \sum_{j=0}^{n+1} c_{j,n+1} \Phi'_{j,0}(z) \right\rangle_{\mu_1}, \quad n \geq 0,$$

so that $c_{0,n+1} = 0$ for $n \geq 0$.

With $n \geq k \geq 1$ and using the coherence relation (7), from

$$0 = \langle \Phi_{k,0}, \Psi_{n+1} \rangle_{S(\mu_0,\mu_1,s)},$$
$$= \left\langle \Phi_{k,0}, \sum_{j=0}^{n+1} c_{j,n+1} \Phi_{j,0}(z) \right\rangle_{\mu_0}$$
$$+ s \Big\langle k[\Phi_{k-1,1} + a_{k-1}\Phi_{k-2,1} + b_{k-1}\Phi_{k-3,1}],$$
$$\sum_{j=0}^{n+1} j c_{j,n+1} [\Phi_{j-1,1} + a_{j-1}\Phi_{j-2,1} + b_{j-1}\Phi_{j-3,1}] \Big\rangle_{\mu_1},$$

we obtain

$$0 = c_{n-2,n+1} \left[ sn(n-2)\overline{b}_{n-1} \|\Phi_{n-3,1}\|^2 \right]$$
$$+ c_{n-1,n+1} \left[ sn(n-1)\overline{a}_{n-1} \|\Phi_{n-2,1}\|^2 + sn(n-1)\overline{b}_{n-1} a_{n-2} \|\Phi_{n-3,1}\|^2 \right]$$
$$+ c_{n,n+1} \left[ \|\Phi_{n,0}\|^2 + sn^2 \|\Phi_{n-1,1}\|^2 + sn^2 |a_{n-1}|^2 \|\Phi_{n-2,1}\|^2 + |b_{n-1}|^2 \|\Phi_{n-3,1}\|^2 \right]$$
$$+ c_{n+1,n+1} \left[ sn(n+1) a_n \|\Phi_{n-1,1}\|^2 + sn(n+1)\overline{a}_{n-1} b_n \|\Phi_{n-2,1}\|^2 \right],$$

for $k = n$, and

$$0 = c_{k-2,n+1} \left[ sk(k-2)\overline{b}_{k-1} \|\Phi_{k-3,1}\|^2 \right]$$
$$+ c_{k-1,n+1} \left[ sk(k-1)\overline{a}_{k-1} \|\Phi_{k-2,1}\|^2 + sk(k-1)\overline{b}_{k-1} a_{k-2} \|\Phi_{k-3,1}\|^2 \right]$$
$$+ c_{k,n+1} \left[ \|\Phi_{k,0}\|^2 + sk^2 \|\Phi_{k-1,1}\|^2 + sk^2 |a_{k-1}|^2 \|\Phi_{k-2,1}\|^2 + sk^2 |b_{k-1}|^2 \|\Phi_{k-3,1}\|^2 \right]$$
$$+ c_{k+1,n+1} \left[ sk(k+1) a_k \|\Phi_{k-1,1}\|^2 + sk(k+1)\overline{a}_{k-1} b_k \|\Phi_{k-2,1}\|^2 \right]$$
$$+ c_{k+2,n+1} \left[ sk(k+2) b_{k+1} \|\Phi_{k-1,1}\|^2 \right],$$

for $1 \leq k \leq n-1$.

The above relations can be written as

$$c_{k-2,n+1}\widehat{\overline{q}}_{k-1} + c_{k-1,n+1}\overline{r}_{k-1} + c_{k,n+1}\widehat{p}_k + c_{k+1,n+1} r_k + c_{k+2,n+1}\widehat{q}_{k+1} = 0, \tag{29}$$

for $1 \leq k \leq n$, with the convention $c_{-1,n+1} = c_{n+2,n+1} = 0$, and

$$\begin{aligned}
\widehat{q_j} &= s(j+1)(j-1)b_j\|\Phi_{j-2,1}\|^2, \\
r_j &= sj(j+1)\left[a_j\|\Phi_{j-1,1}\|^2 + b_j\overline{a_{j-1}}\|\Phi_{j-2,1}\|^2\right], \\
\widehat{p_j} &= \|\Phi_{j,0}\|^2 + sj^2\left[\|\Phi_{j-1,1}\|^2 + |a_{j-1}|^2\|\Phi_{j-2,1}\|^2 + |b_{j-1}|^2\|\Phi_{j-3,1}\|^2\right].
\end{aligned}$$

Notice that $\widehat{q_j} > 0$, for every $j \geq 2$.

Since $c_{n+1,n+1} = 1$, the above linear system (29) can be represented in matrix form as

$$T_n\boldsymbol{c_n} = -\widehat{q_n}\boldsymbol{e_{n-1}} - r_n\boldsymbol{e_n}, \quad n \geq 1,$$

where $\boldsymbol{e_j}$ is the $j$-th column of the $n \times n$ identity matrix, $\boldsymbol{c_n} = [c_{1,n+1}, \ldots, c_{n,n+1}]^t$ and $T_n = T_n^{(\mu_0,\mu_1,s)}$ is the $n \times n$ Hermitian pentadiagonal matrix

$$T_n = \begin{bmatrix}
\widehat{p_1} & r_1 & \widehat{q_2} & 0 & \cdots & \cdots & 0 \\
\overline{r}_1 & \widehat{p_2} & r_2 & \widehat{q_3} & \ddots & & \vdots \\
\overline{\widehat{q}}_2 & \overline{r}_2 & \ddots & \ddots & \ddots & \ddots & \vdots \\
0 & \overline{\widehat{q}}_3 & \ddots & \ddots & \ddots & \widehat{q}_{n-2} & 0 \\
\vdots & \ddots & \ddots & \ddots & \ddots & r_{n-2} & \widehat{q}_{n-1} \\
\vdots & & \ddots & \overline{\widehat{q}}_{n-3} & \overline{r}_{n-2} & \widehat{p}_{n-1} & r_{n-1} \\
0 & \cdots & \cdots & 0 & \overline{\widehat{q}}_{n-2} & \overline{r}_{n-1} & \widehat{p}_n
\end{bmatrix}.$$

Thus, we have proven the following

**Theorem 2** *Let $(\mu_0, \mu_1)$ be a (0,2)-coherent pair of measures of the second kind supported on the unit circle, then the SMOP $\{\Psi_n(z)\}_{n\geq 0}$ associated with the Sobolev inner product* (27) *satisfies*

$$\Psi_1(z) = \Phi_1(\mu_0; z), \quad \textit{and}$$

$$\Psi_{n+1}(z) = \sum_{j=1}^{n} c_{j,n+1}\Phi_j(\mu_0; z) + \Phi_{n+1}(\mu_0; z), \quad n \geq 1,$$

*where $\{c_{j,n+1}\}_{j\geq 0}$ satisfies $c_{0,n+1} = 0$, $c_{n+1,n+1} = 1$, and*

$$T_n\boldsymbol{c_n} = -\widehat{q_n}\boldsymbol{e_{n-1}} - r_n\boldsymbol{e_n}, \quad n \geq 1,$$

*with the notation used above.*

**Acknowledgments** We thank the careful revision by the referee. The suggestions and remarks of his/her report have contributed to improve the presentation of the manuscript.

This work was done as a part of the doctoral thesis of the author Lino G. Garza at Universidad Carlos III de Madrid, Spain, supervised by Francisco Marcellán and supported by the grant MTM2015–65888-C4-2-P, Ministerio de Economía, Industria y Competitividad of Spain.

The research of the author Francisco Marcellán was supported by the grant PGC2018-096504-B-C33 from the Agencia Estatal de Investigación (AEI) of Spain and Fondo Europeo de Desarrollo Regional (FEDER).

The research of the author A. Sri Ranga was supported by the grant 304087/2018-1 from CNPq of Brazil and by the grant 2016/09906-0 from FAPESP of the state of São Paulo, Brazil.

## References

1. A. Aptekarev, E. Berriochoa, A. Cachafeiro, Strong asymptotics for the continuous Sobolev orthogonal polynomials on the unit circle. J. Approx. Theory **100**, 381–391 (1999)
2. E. Berriochoa, A. Cachafeiro, Lebesgue Sobolev orthogonality on the unit circle. J. Comput. Appl. Math. **96**, 27–34 (1998)
3. E. Berriochoa, A. Cachafeiro, A family of Sobolev orthogonal polynomials on the unit circle. J. Comput. Appl. Math. **105**, 163–173 (1999)
4. A.C. Berti, C.F. Bracciali, A. Sri Ranga, Orthogonal polynomials associated with related measures and Sobolev orthogonal polynomials. Numer. Algorithms **34**, 203–216 (2003)
5. A. Branquinho, A. Foulquié Moreno, F. Marcellán, M.N. Rebocho, Coherent pairs of linear functionals on the unit circle. J. Approx. Theory **153**, 122–137 (2008)
6. T.S. Chihara, *An Introduction to Orthogonal Polynomials* (Gordon and Breach, New York, 1978)
7. L. Garza, F. Marcellán, N.C. Pinzón-Cortés, (1, 1)-coherent pairs on the unit circle. Abstr. Appl. Anal. Art. ID 307974 (2013)
8. A. Iserles, P.E. Koch, S.P. Nørsett, J.M. Sanz-Serna, On polynomials orthogonal with respect to certain Sobolev inner products. J. Approx. Theory **65**, 151–175 (1991)
9. X. Li, F. Marcellán, On polynomials orthogonal with respect to Sobolev inner product on the unit circle. Pac. J. Math. **175**, 127–146 (1996)
10. F. Marcellán, N.C. Pinzón-Cortés, Generalized coherent pairs on the unit circle and Sobolev orthogonal polynomials. Publ. Inst. Math. (Beograd) (N.S.) **96**(110), 193–210 (2014)
11. F. Marcellán, A. Sri Ranga, Sobolev orthogonal polynomials on the unit circle and coherent pair of measures of the second kind. Results Math. **71**, 1127–1149 (2017)
12. F. Marcellán, Y. Xu, On Sobolev orthogonal polynomials. Expo. Math. **33**, 308–352 (2015)
13. A. Martínez-Finkelshtein, Asymptotic properties of Sobolev orthogonal polynomials. J. Comput. Appl. Math. **99**, 491–510 (1998)
14. A. Martínez-Finkelshtein, Analytic aspects of Sobolev orthogonal polynomials revisited. J. Comput. Appl. Math. **127**, 255–266 (2001)
15. H.G. Meijer, Determination of all coherent pairs. J. Approx. Theory **89**, 321–343 (1997)
16. B. Simanek, Two universality results for polynomial reproducing kernels. J. Approx. Theory **216**, 16–37 (2017)
17. B. Simon, *Orthogonal Polynomials on the Unit Circle. Part 1. Classical Theory*. American Mathematical Society Colloquium Publications, vol. 54 (American Mathematical Society, Providence, 2005)
18. A. Sri Ranga, Orthogonal polynomials with respect to a family of Sobolev inner products on the unit circle. Proc. Amer. Math. Soc. **144**, 1129–1143 (2016)
19. S.M. Zagorodnyuk, On a family of hypergeometric Sobolev orthogonal polynomials on the unit circle. Constr. Math. Anal. **3**, 75–84 (2020)

# Bessel-Type Operators and a Refinement of Hardy's Inequality

**Fritz Gesztesy, Michael M. H. Pang, and Jonathan Stanfill**

*Dedicated with great pleasure to Lance Littlejohn on the occasion of his 70th birthday.*

**Abstract** The principal aim of this paper is to employ Bessel-type operators in proving the inequality

$$\int_0^\pi dx\, |f'(x)|^2 \geq \frac{1}{4}\int_0^\pi dx\, \frac{|f(x)|^2}{\sin^2(x)} + \frac{1}{4}\int_0^\pi dx\, |f(x)|^2, \quad f \in H_0^1((0,\pi)),$$

where both constants $1/4$ appearing in the above inequality are optimal. In addition, this inequality is strict in the sense that equality holds if and only if $f \equiv 0$. This inequality is derived with the help of the exactly solvable, strongly singular, Dirichlet-type Schrödinger operator associated with the differential expression

$$\tau_s = -\frac{d^2}{dx^2} + \frac{s^2 - (1/4)}{\sin^2(x)}, \quad s \in [0,\infty),\ x \in (0,\pi).$$

The new inequality represents a refinement of Hardy's classical inequality

$$\int_0^\pi dx\, |f'(x)|^2 \geq \frac{1}{4}\int_0^\pi dx\, \frac{|f(x)|^2}{x^2}, \quad f \in H_0^1((0,\pi)),$$

F. Gesztesy (✉) · J. Stanfill
Department of Mathematics, Baylor University, Waco, TX, USA
e-mail: Fritz_Gesztesy@baylor.edu; Jonathan_Stanfill@baylor.edu
https://www.baylor.edu/math/index.php?id=935340, https://sites.baylor.edu/jonathan-stanfill/

M. M. H. Pang
Department of Mathematics, University of Missouri, Columbia, MO, USA
e-mail: pangm@missouri.edu
https://www.math.missouri.edu/people/pang



F. Gesztesy, A. Martinez-Finkelshtein (eds.), *From Operator Theory to Orthogonal Polynomials, Combinatorics, and Number Theory*, Operator Theory: Advances and Applications 285, https://doi.org/10.1007/978-3-030-75425-9_9

and it also improves upon one of its well-known extensions in the form

$$\int_0^\pi dx\, |f'(x)|^2 \geq \frac{1}{4} \int_0^\pi dx\, \frac{|f(x)|^2}{d_{(0,\pi)}(x)^2}, \quad f \in H_0^1((0,\pi)),$$

where $d_{(0,\pi)}(x)$ represents the distance from $x \in (0,\pi)$ to the boundary $\{0,\pi\}$ of $(0,\pi)$.

**Keywords** Hardy-type inequality · Strongly singular differential operators · Friedrichs extension

# 1 Introduction

*Happy Birthday, Lance! We hope this modest contribution to Hardy-type inequalities will cause some joy.*

In a nutshell, the aim of this note is to employ a Bessel-type operator in deriving the Hardy-type inequality

$$\int_0^\pi dx\, |f'(x)|^2 \geq \frac{1}{4} \int_0^\pi dx\, \frac{|f(x)|^2}{\sin^2(x)} + \frac{1}{4} \int_0^\pi dx\, |f(x)|^2, \quad f \in H_0^1((0,\pi)). \tag{1.1}$$

As is readily verified, (1.1) indeed represents an improvement over the classical Hardy inequality

$$\int_0^\pi dx\, |f'(x)|^2 \geq \frac{1}{4} \int_0^\pi dx\, \frac{|f(x)|^2}{x^2}, \quad f \in H_0^1((0,\pi)), \tag{1.2}$$

while also improving upon one of its well-known extensions in the form

$$\int_0^\pi dx\, |f'(x)|^2 \geq \frac{1}{4} \int_0^\pi dx\, \frac{|f(x)|^2}{d_{(0,\pi)}(x)^2}, \quad f \in H_0^1((0,\pi)). \tag{1.3}$$

Here $d_{(0,\pi)}(x)$ represents the distance from $x \in (0,\pi)$ to the boundary $\{0,\pi\}$ of the interval $(0,\pi)$, that is,

$$d_{(0,\pi)}(x) = \begin{cases} x, & x \in (0,\pi/2], \\ \pi - x, & x \in [\pi/2,\pi). \end{cases} \tag{1.4}$$

We emphasize that all constants $1/4$ in (1.1)–(1.3) are optimal and all inequalities are strict in the sense that equality holds in them if and only if $f \equiv 0$.

Our refinement (1.1) (and the optimality of both constants $1/4$ in (1.1)) rests on the exact solvability of the one-dimensional Schrödinger equation with potential $q_s$, $s \in [0, \infty)$, given by

$$q_s(x) = \frac{s^2 - (1/4)}{\sin^2(x)}, \quad x \in (0, \pi), \tag{1.5}$$

as illustrated by Rosen and Morse [57] in 1932, Pöschl and Teller [54] in 1933, and Lotmar [46] in 1935. These authors are either concerned with the following extension of (1.5)

$$\frac{c_1}{\sin^2(x)} + \frac{c_2}{\cos^2(x)}, \quad x \in (0, \pi/2), \tag{1.6}$$

or its hyperbolic analog of the form

$$\frac{c_1}{\sinh^2(x)} + \frac{c_2}{\cosh^2(x)}, \quad x \in \mathbb{R} \text{ (or } x \in (0, \infty)). \tag{1.7}$$

The upshot of these investigations for the purpose at hand was the realization that such problems are exactly solvable in terms of the hypergeometric function $F(a, b; c; \cdot)$ (frequently denoted by ${}_2F_1(a, b; c; \cdot)$). These types of problems are further discussed by Infeld and Hull [36] and summarized in [21, Sect. 38, 39, 93], and more recently in [17]. A discussion of the underlying singular periodic problem (1.5) on $\mathbb{R}$, including the associated Floquet (Bloch) theory, was presented by Scarf [60]. These investigations exclusively focus on aspects of ordinary differential equations as opposed to operator theory even though Dirichlet problems associated with singular endpoints were formally discussed (in this context see also [59]). The operator theoretic approach to (1.5) and (1.6) over a finite interval bounded by singularities, and a variety of associated self-adjoint boundary conditions including coupled boundary conditions leading to energy bands (Floquet–Bloch theory) in the periodic problem on $\mathbb{R}$, on the basis of generalized boundary values due to Rellich [55] (see also [12]), was first discussed in [23] and [27]. Finally, we briefly mention that the case of $n$-soliton potentials

$$q_{(1/2)+n}(x) = n(n+1)/\cosh^2(x), \quad n \in \mathbb{N}, \ x \in \mathbb{R}, \tag{1.8}$$

has received special attention as it represents a solution of infinitely many equations in the stationary Korteweg–de Vries (KdV) hierarchy (starting from level $n$ upward).

Introducing the differential expression

$$\tau_s = -\frac{d^2}{dx^2} + q_s(x) = -\frac{d^2}{dx^2} + \frac{s^2 - (1/4)}{\sin^2(x)}, \quad x \in (0, \pi), \tag{1.9}$$

the exact solvability of the differential equation $\tau_s y = zy$, $z \in \mathbb{C}$, or a comparison with the well-known Bessel operator case $-(d^2/dx^2) + [s^2 - (1/4)]x^{-2}$ near $x = 0$ and $-(d^2/dx^2) + [s^2 - (1/4)](x - \pi)^{-2}$ near $x = \pi$ then yields the nonoscillatory property of $\tau_s$ if and only if $s \in [0, \infty)$. Very roughly speaking, nonnegativity of the Friedrichs extension associated with the differential expression $\tau_0 - (1/4)$, implying nonnegativity of the underlying quadratic form defined on $H_0^1((0, \pi))$, implies the refinement (1.1) of Hardy's inequality.

In Sect. 2 we briefly discuss (principal and nonprincipal) solutions of the exactly solvable Schrödinger equation $\tau_s y = 0$ (solutions of the general equation $\tau_s y = zy$, $z \in \mathbb{C}$, are discussed in Appendix A.1), and introduce minimal $T_{s,min}$ and maximal $T_{s,max} = T_{s,min}^*$ operators corresponding to $\tau_s$ as well as the Friedrichs extension $T_{s,F}$ of $T_{s,min}$ and the boundary values associated with $T_{s,max}$, following recent treatments in [29, 30]. Section 3 contains the bulk of this paper and is devoted to a derivation of inequality (1.1). We also indicate how two related results by Avkhadiev and Wirths [7, 8], involving Drichlet boundary conditions on both ends and a mixture of Dirichlet and Neumann boundary conditions naturally fits into the framework discussed in this paper. In Appendix A.1 we study solutions of $\tau_s y = zy$, $z \in \mathbb{C}$, in more detail and also derive the singular Weyl–Titchmarsh–Kodaira $m$-function associated with $T_{s,F}$. Finally, Appendix B.1 collects some facts on Hardy-type inequalities.

## 2 An Exactly Solvable, Strongly Singular, Periodic Schrödinger Operator

In this section we examine a slight variation of the example found in Sect. 4 of [23] by implementing the methods found in [29].

Let $a = 0$, $b = \pi$,

$$p(x) = r(x) = 1, \quad q_s(x) = \frac{s^2 - (1/4)}{\sin^2(x)}, \quad s \in [0, \infty), \ x \in (0, \pi). \tag{2.1}$$

We now study the Sturm–Liouville operators associated with the corresponding differential expression given by

$$\tau_s = -\frac{d^2}{dx^2} + q_s(x) = -\frac{d^2}{dx^2} + \frac{s^2 - (1/4)}{\sin^2(x)}, \quad s \in [0, \infty), \ x \in (0, \pi), \tag{2.2}$$

which is in the limit circle case at the endpoints $x = 0, \pi$ for $s \in [0, 1)$, and limit point at both endpoints for $s \in [1, \infty)$. The maximal and preminimal operators, $T_{s,max}$ and $\dot{T}_{s,min}$, associated to $\tau_s$ in $L^2((0, \pi); dx)$ are then given by

$$\begin{aligned} &T_{s,max} f = \tau_s f, \quad s \in [0, \infty), \\ &f \in \text{dom}(T_{s,max}) = \big\{g \in L^2((0, \pi); dx) \,\big|\, g, g' \in AC_{loc}((0, \pi)); \\ &\qquad\qquad\qquad\qquad\qquad\qquad\qquad\qquad \tau_s g \in L^2((0, \pi); dx)\big\}, \end{aligned} \tag{2.3}$$

and

$$\begin{aligned} &\dot{T}_{s,min} f = \tau_s f, \quad s \in [0, \infty), \\ &f \in \text{dom}\big(\dot{T}_{s,min}\big) = \big\{g \in L^2((0, \pi); dx) \,\big|\, g, g' \in AC_{loc}((0, \pi)); \\ &\qquad\qquad \text{supp}\,(g) \subset (0, \pi) \text{ is compact}; \tau_s g \in L^2((0, \pi); dx)\big\}. \end{aligned} \tag{2.4}$$

Since

$$q_s \in L^2_{loc}((0, \pi); dx), \quad s \in [0, \infty), \tag{2.5}$$

one can replace $\dot{T}_{s,min}$ by

$$\ddot{T}_{s,min} = \tau_s\big|_{C_0^\infty((0,\pi))}, \quad s \in [0, \infty). \tag{2.6}$$

For $s \in [0, 1)$, we introduce principal and nonprincipal solutions $u_{0,s}(0, \cdot)$ and $\widehat{u}_{0,s}(0, \cdot)$ of $\tau_s u = 0$ at $x = 0$ by

$$\begin{aligned} &u_{0,s}(0, x) = [\sin(x)]^{(1+2s)/2} F\big((1/4) + (s/2), (1/4) + (s/2); 1 + s; \sin^2(x)\big), \\ &\qquad\qquad\qquad\qquad\qquad\qquad\qquad\qquad\qquad\qquad\qquad\qquad s \in [0, 1), \\ &\widehat{u}_{0,s}(0, x) = \begin{cases} (2s)^{-1}[\sin(x)]^{(1-2s)/2} \\ \quad \times F\big((1/4) - (s/2), (1/4) - (s/2); 1 - s; \sin^2(x)\big), & s \in (0, 1), \\ [\sin(x)]^{1/2} F\big(1/4, 1/4; 1; \sin^2(x)\big) \\ \quad \times \displaystyle\int_x^c dx' \, [\sin(x')]^{-1} \big[F\big(1/4, 1/4; 1; \sin^2(x')\big)\big]^{-2}, & s = 0, \end{cases} \end{aligned} \tag{2.7}$$

and principal and nonprincipal solutions $u_{\pi,s}(0,\,\cdot\,)$ and $\widehat{u}_{\pi,s}(0,\,\cdot\,)$ of $\tau_s u = 0$ at $x = \pi$ by

$$u_{\pi,s}(0,x) = [\sin(x)]^{(1+2s)/2} F\big((1/4)+(s/2), (1/4)+(s/2); 1+s; \sin^2(x)\big), \quad s \in [0,1),$$

$$\widehat{u}_{\pi,s}(0,x) = \begin{cases} -(2s)^{-1}[\sin(x)]^{(1-2s)/2} \\ \times F\big((1/4)-(s/2), (1/4)-(s/2); 1-s; \sin^2(x)\big), & s \in (0,1), \\ -[\sin(x)]^{1/2} F\big(1/4, 1/4; 1; \sin^2(x)\big) \\ \times \int_c^x dx'\, [\sin(x')]^{-1} \big[F\big(1/4, 1/4; 1; \sin^2(x')\big)\big]^{-2}, & s = 0. \end{cases} \tag{2.8}$$

Here $F(\,\cdot\,, \,\cdot\,; \,\cdot\,; \,\cdot\,)$ (frequently written as ${}_2F_1(\,\cdot\,, \,\cdot\,; \,\cdot\,; \,\cdot\,)$) denotes the hypergeometric function (see, e.g., [1, Ch. 15]).

*Remark 2.1* We note that the case $c = 1$ in $F(a,b;c;\xi)$, corresponding to the case $s = 0$ in (2.7), (2.8), is a special one in the sense that linearly independent solutions of the hypergeometric differential equation are then of the form (see, e.g., [1, Nos. 15.5.16, 15.5.17])

$$\begin{aligned} y_1(\xi) &= F(a,b;1;\xi), \\ y_2(\xi) &= F(a,b;1;\xi)\ln(\xi) \\ &+ \sum_{n\in\mathbb{N}} \frac{(a)_n (b)_n}{(n!)^2} [\psi(a+n) - \psi(a) + \psi(b+n) - \psi(b) + 2\psi(1) - 2\psi(n+1)]\xi^n. \end{aligned} \tag{2.9}$$

Here $(d)_n$, $n \in \mathbb{N}_0$, represents Pochhammer's symbol (see (A.8)), and $\psi(\,\cdot\,)$ denotes the Digamma function. Since we wanted to ensure (2.12) and our principal aim in connection with the boundary values (2.14)–(2.17) was the derivation of the asymptotic relations (2.10) and (2.11), our choice of $\widehat{u}_{0,0}$ and $\widehat{u}_{\pi,0}$ in (2.7) and (2.8) is to be preferred over the use of the pair of functions in (2.9). For more details in this connection see Appendix A.1. ◇

Since

$$u_{0,s}(0,x) \underset{x\downarrow 0}{=} x^{(1+2s)/2}\big\{1 + \big[(4s^2-1)/(48+48s)\big]x^2 + O\big(x^4\big)\big\}, \quad s \in [0,1),$$

$$\widehat{u}_{0,s}(0,x) \underset{x\downarrow 0}{=} \begin{cases} (2s)^{-1} x^{(1-2s)/2}\big\{1 + \big[(4s^2-1)/(48-48s)\big]x^2 + O\big(x^4\big)\big\}, \\ \qquad s \in (0,1), \\ \ln(1/x) x^{1/2}\big\{1 + \big[([\ln(x)]^{-1} - 1)/48\big]x^2 + O\big(x^4\big)\big\}, \quad s = 0, \end{cases} \tag{2.10}$$

$$u_{\pi,s}(0,x) \underset{x\uparrow\pi}{=} (\pi-x)^{(1+2s)/2}\big\{1+\big[\big(4s^2-1\big)/\big(48+48s\big)\big](\pi-x)^2 + O\big((\pi-x)^4\big)\big\}, \quad s\in[0,1),$$

$$\widehat{u}_{\pi,s}(0,x) \underset{x\uparrow\pi}{=} \begin{cases} -(2s)^{-1}(\pi-x)^{(1-2s)/2}\big\{1+\big[\big(4s^2-1\big)/\big(48-48s\big)\big] \\ \quad \times(\pi-x)^2+O\big((\pi-x)^4\big)\big\}, \quad s\in(0,1), \\ \ln(\pi-x)(\pi-x)^{1/2}\big\{1+\big[\big([\ln(\pi-x)]^{-1}-1\big)/48\big](\pi-x)^2 \\ \quad +O\big((\pi-x)^4\big)\big\}, \quad s=0, \end{cases} \tag{2.11}$$

one deduces that

$$W(\widehat{u}_{0,s}(0,\,\cdot\,),u_{0,s}(0,\,\cdot\,))(0)=1=W(\widehat{u}_{\pi,s}(0,\,\cdot\,),u_{\pi,s}(0,\,\cdot\,))(\pi), \quad s\in[0,1), \tag{2.12}$$

and

$$\lim_{x\downarrow 0}\frac{u_{0,s}(0,x)}{\widehat{u}_{0,s}(0,x)}=0, \quad \lim_{x\uparrow\pi}\frac{u_{\pi,s}(0,x)}{\widehat{u}_{\pi,s}(0,x)}=0, \quad s\in[0,1). \tag{2.13}$$

The generalized boundary values for $g \in \operatorname{dom}(T_{s,max})$ (the maximal operator associated with $\tau_s$) are then of the form

$$\widetilde{g}(0)=\begin{cases} \lim_{x\downarrow 0} g(x)/\big[(2s)^{-1}x^{(1-2s)/2}\big], & s\in(0,1), \\ \lim_{x\downarrow 0} g(x)/\big[x^{1/2}\ln(1/x)\big], & s=0, \end{cases} \tag{2.14}$$

$$\widetilde{g}'(0)=\begin{cases} \lim_{x\downarrow 0}\big[g(x)-\widetilde{g}(0)(2s)^{-1}x^{(1-2s)/2}\big]/x^{(1+2s)/2}, & s\in(0,1), \\ \lim_{x\downarrow 0}\big[g(x)-\widetilde{g}(0)x^{1/2}\ln(1/x)\big]/x^{1/2}, & s=0, \end{cases} \tag{2.15}$$

$$\widetilde{g}(\pi)=\begin{cases} \lim_{x\uparrow\pi} g(x)/\big[-(2s)^{-1}(\pi-x)^{(1-2s)/2}\big], & s\in(0,1), \\ \lim_{x\uparrow\pi} g(x)/\big[(\pi-x)^{1/2}\ln(\pi-x)\big], & s=0, \end{cases} \tag{2.16}$$

$$\widetilde{g}'(\pi)=\begin{cases} \lim_{x\uparrow\pi}\big[g(x)+\widetilde{g}(\pi)(2s)^{-1}(\pi-x)^{(1-2s)/2}\big]/(\pi-x)^{(1+2s)/2}, \\ \qquad\qquad s\in(0,1), \\ \lim_{x\uparrow\pi}\big[g(x)-\widetilde{g}(0)(\pi-x)^{1/2}\ln(\pi-x)\big]/(\pi-x)^{1/2}, \quad s=0. \end{cases} \tag{2.17}$$

As a result, the minimal operator $T_{s,min}$ associated to $\tau_s$, that is,

$$T_{s,min} = \overline{\dot{T}_{s,min}} = \overline{\ddot{T}_{s,min}}, \quad s \in [0, \infty), \tag{2.18}$$

is thus given by

$$\begin{aligned} &T_{s,min} f = \tau_s f, \\ &f \in \operatorname{dom}(T_{s,min}) = \big\{g \in L^2((0,\pi); dx) \,\big|\, g, g' \in AC_{loc}((0,\pi)); \\ &\qquad \widetilde{g}(0) = \widetilde{g}'(0) = \widetilde{g}(\pi) = \widetilde{g}'(\pi) = 0; \ \tau_s g \in L^2((0,\pi); dx)\big\}, \quad s \in [0,1), \end{aligned} \tag{2.19}$$

and satisfies $T_{s,min}^* = T_{s,max}$, $T_{s,max}^* = T_{s,min}$, $s \in [0, \infty)$. Due to the limit point property of $\tau_s$ at $x = 0$ and $x = \pi$ if and only if $s \in [1, \infty)$, one concludes that

$$T_{s,min} = T_{s,max} \text{ if and only if } s \in [1, \infty). \tag{2.20}$$

The Friedrichs extension $T_{s,F}$ of $T_{s,min}$, $s \in [0, 1)$, permits a particularly simple characterization in terms of the generalized boundary conditions (2.14)–(2.17) and is then given by (cf. [39, 49, 56, 58] and the extensive literature cited in [29], [26, Ch. 13])

$$\begin{aligned} &T_{s,F} f = \tau_s f, \\ &f \in \operatorname{dom}(T_{s,F}) = \big\{g \in \operatorname{dom}(T_{s,max}) \,\big|\, \widetilde{g}(0) = \widetilde{g}(\pi) = 0\big\}, \quad s \in [0,1), \end{aligned} \tag{2.21}$$

moreover,

$$T_{s,F} = T_{s,min} = T_{s,max}, \quad s \in [1, \infty), \tag{2.22}$$

is self-adjoint (resp., $\dot{T}_{s,min}$ and $\ddot{T}_{s,min}$, $s \in [1, \infty)$, are essentially self-adjoint) in $L^2((0, \pi); dx)$. In this case the Friedrichs boundary conditions in (2.21) are automatically satisfied and hence can be omitted.

By (A.21) one has

$$\inf(\sigma(T_{s,F})) = [(1/2) + s]^2, \quad s \in [0, \infty), \tag{2.23}$$

in particular,

$$T_{s,F} \geq [(1/2) + s]^2 I_{(0,\pi)}, \quad s \in [0, \infty), \tag{2.24}$$

with $I_{(0,\pi)}$ abbreviating the identity operator in $L^2((0, \pi); dx)$.

All results on 2nd order differential operators employed in this section can be found in classical sources such as [2, Sect. 129], [14, Chs. 8, 9], [18, Sects. 13.6, 13.9, 13.10], [37, Ch. III], [47, Ch. V], [49], [52, Ch. 6], [61, Ch. 9], [62, Sect. 8.3], [63, Ch. 13], [64, Chs. 4, 6–8]. In addition, [29] and [26, Ch. 13] contain very detailed lists of references in this context.

## 3 A Refinement of Hardy's Inequality

The principal purpose of this section is to derive a refinement of the classical Hardy inequality

$$\int_0^\pi dx\,|f'(x)|^2 \geq \frac{1}{4}\int_0^\pi dx\,\frac{|f(x)|^2}{x^2}, \quad f \in H_0^1((0,\pi)), \tag{3.1}$$

as well as of one of its well-known extensions in the form

$$\int_0^\pi dx\,|f'(x)|^2 \geq \frac{1}{4}\int_0^\pi dx\,\frac{|f(x)|^2}{d_{(0,\pi)}(x)^2}, \quad f \in H_0^1((0,\pi)), \tag{3.2}$$

where $d_{(0,\pi)}(x)$ represents the distance from $x \in (0,\pi)$ to the boundary $\{0,\pi\}$ of the interval $(0,\pi)$, that is,

$$d_{(0,\pi)}(x) = \begin{cases} x, & x \in (0,\pi/2], \\ \pi - x, & x \in [\pi/2,\pi). \end{cases} \tag{3.3}$$

The constant $1/4$ in (3.1) and (3.2) is known to be optimal and both inequalities are strict in the sense that equality holds in them if and only if $f \equiv 0$.

For background on Hardy-type inequalities we refer, for instance, to [9, p. 3–5], [10], [13], [15, p. 104–105], [28], [32], [33], [34, Sect. 7.3, p. 240–243], [42, Sect. 5], [43, Ch. 3], [44, Ch. 1], [45], [50, Ch. 1], [53].

The principal result of this section then can be formulated as follows.

**Theorem 3.1** *Let $f \in H_0^1((0,\pi))$. Then,*

$$\int_0^\pi dx\,|f'(x)|^2 \geq \frac{1}{4}\int_0^\pi dx\,\frac{|f(x)|^2}{\sin^2(x)} + \frac{1}{4}\int_0^\pi dx\,|f(x)|^2, \tag{3.4}$$

*where both constants $1/4$ in (3.4) are optimal. In addition, the inequality is strict in the sense that equality holds in (3.4) if and only if $f \equiv 0$.*

***Proof*** By Sect. 2 for $s = 0$ and by [23, Sect. 4] for $s \in (0, \infty)$, one has

$$\left( -\frac{d^2}{dx^2} + \frac{s^2 - (1/4)}{\sin^2(x)} - [(1/2) + s]^2 I_{(0,\pi)} \right)\Big|_{C_0^\infty((0,\pi))} \geq 0, \quad s \in [0, \infty). \tag{3.5}$$

Thus, setting $s = 0$ in (3.5) yields

$$\int_0^\pi dx\, |f'(x)|^2 \geq \frac{1}{4} \int_0^\pi dx\, \frac{|f(x)|^2}{\sin^2(x)} + \frac{1}{4} \int_0^\pi dx\, |f(x)|^2, \quad f \in C_0^\infty((0, \pi)). \tag{3.6}$$

Now denote by $H_0^1((0, \pi))$ the standard Sobolev space on $(0, \pi)$ obtained upon completion of $C_0^\infty((0, \pi))$ in the norm of $H^1((0, \pi))$. Since $C_0^\infty((0, \pi))$ is dense in $H_0^1((0, \pi))$, given $f \in H_0^1((0, \pi))$, there exists a sequence $\{f_n\}_{n\in\mathbb{N}} \subset C_0^\infty((0, \pi))$ such that $\lim_{n\to\infty} \|f_n - f\|^2_{H_0^1((0,\pi))} = 0$. Hence, one can find a subsequence $\{f_{n_p}\}_{p\in\mathbb{N}}$ of $\{f_n\}_{n\in\mathbb{N}}$ such that $f_{n_p}$ converges to $f$ pointwise almost everywhere on $(0, \pi)$ as $p \to \infty$. Thus an application of Fatou's lemma (cf., e.g., [22, Corollary 2.19]) yields that (3.6) extends to $f \in H_0^1((0, \pi))$, namely,

$$\begin{aligned}
&\frac{1}{4} \int_0^\pi dx\, \frac{|f(x)|^2}{\sin^2(x)} + \frac{1}{4} \int_0^\pi dx\, |f(x)|^2 \\
&\quad \leq \liminf_{p\to\infty} \frac{1}{4} \int_0^\pi dx\, \frac{|f_{n_p}(x)|^2}{\sin^2(x)} \\
&\qquad + \liminf_{p\to\infty} \frac{1}{4} \int_0^\pi dx\, |f_{n_p}(x)|^2 \quad \text{(by Fatou's Lemma)} \\
&\quad \leq \liminf_{p\to\infty} \left\{ \frac{1}{4} \int_0^\pi dx\, \frac{|f_{n_p}(x)|^2}{\sin^2(x)} + \frac{1}{4} \int_0^\pi dx\, |f_{n_p}(x)|^2 \right\} \\
&\quad \leq \liminf_{p\to\infty} \int_0^\pi dx\, |f'_{n_p}(x)|^2 \quad \text{(by (3.6))} \\
&\quad = \lim_{p\to\infty} \int_0^\pi dx\, |f'_{n_p}(x)|^2 \\
&\quad = \int_0^\pi dx\, |f'(x)|^2.
\end{aligned} \tag{3.7}$$

The substitution $s \mapsto is$ in (2.7) results in solutions that have oscillatory behavior due to the factor $[\sin(x)]^{\pm is}$ in (A.3), (A.7), rendering all solutions of $\tau_s y(\lambda, \cdot) = \lambda y(\lambda, \cdot)$ oscillatory for each $\lambda \in \mathbb{R}$ if and only if $s^2 < 0$. Classical oscillation theory results (see, e.g., [29, Theorem 4.2]) prove that $\dot{T}_{s,min}$, and hence $T_{s,min}$ are bounded from below if and only if $s \in [0, \infty)$. This proves that the first constant $1/4$ on the right-hand side of (3.4) is optimal.

Next we demonstrate that also the second constant $1/4$ on the right-hand side of (3.4) is optimal arguing by contradiction as follows: Suppose that for some $\varepsilon > 0$,

$$\int_0^{\pi} dx\, |f'(x)|^2 \geq \frac{1}{4}\int_0^{\pi} dx\, \frac{|f(x)|^2}{\sin^2(x)} + \left(\frac{1}{4} + \varepsilon\right)\int_0^{\pi} dx\, |f(x)|^2, \qquad f \in \operatorname{dom}(T_{0,min}). \tag{3.8}$$

Upon integrating by parts in the left-hand side of (3.8) this implies

$$T_{0,min} \geq \left(\frac{1}{4} + \varepsilon\right) I_{(0,\pi)}, \tag{3.9}$$

implying

$$T_{0,F} \geq \left(\frac{1}{4} + \varepsilon\right) I_{(0,\pi)} \tag{3.10}$$

(as $T_{0,min}$ and $T_{0,F}$ share the same lower bound by general principles), contradicting (2.23) for $s = 0$. Hence also the 2nd constant $1/4$ on the right-hand side of (3.4) is optimal.

It remains to prove strictness of inequality (3.4) if $f \not\equiv 0$: Arguing again by contradiction, we suppose there exists $0 \neq f_0 \in H_0^1((0, \pi))$ such that

$$\int_0^{\pi} dx\, |f_0'(x)|^2 = \frac{1}{4}\int_0^{\pi} dx\, \frac{|f_0(x)|^2}{\sin^2(x)} + \frac{1}{4}\int_0^{\pi} dx\, |f_0(x)|^2. \tag{3.11}$$

Since $H_0^1((0, \pi)) \subseteq \operatorname{dom}\left(T_{s,F}^{1/2}\right)$, $s \in [0, \infty)$ (in fact, one even has the equality $H_0^1((0, \pi)) = \operatorname{dom}\left(T_{s,F}^{1/2}\right)$ for all $s \in (0, \infty)$, see, e.g., [4, 11, 16, 24, 39, 41]), one concludes via (3.11) that

$$\left(T_{0,F}^{1/2} f_0, T_{0,F}^{1/2} f_0\right)_{L^2((0,\pi);dx)} = (1/4)\|f_0\|^2_{L^2((0,\pi);dx)}. \tag{3.12}$$

Moreover, since $T_{0,F}$ is self-adjoint with purely discrete and necessarily simple spectrum, $T_{0,F}$ has the spectral representation

$$T_{0,F} = \sum_{n\in\mathbb{N}_0} \lambda_n P_n, \quad \lambda_0 = 1/4 < \lambda_1 < \lambda_2 < \cdots, \tag{3.13}$$

where $\sigma(T_{0,F}) = \{\lambda_n\}_{n\in\mathbb{N}_0}$ and $P_n$ are the one-dimensional projections onto the eigenvectors associated with the eigenvalues $\lambda_n$, $n \in \mathbb{N}_0$, explicitly listed in (A.21), in particular, $\lambda_0 = 1/4$. Thus,

$$\begin{aligned}\big(T_{0,F}^{1/2} f_0, T_{0,F}^{1/2} f_0\big)_{L^2((0,\pi);dx)} &= \sum_{n\in\mathbb{N}_0} \lambda_n (f_0, P_n f_0)_{L^2((0,\pi);dx)} \\ &> \lambda_0 \sum_{n\in\mathbb{N}_0} (f_0, P_n f_0)_{L^2((0,\pi);dx)} \\ &= \lambda_0 \|f_0\|^2_{L^2((0,\pi);dx)} = (1/4)\|f_0\|^2_{L^2((0,\pi);dx)}\end{aligned} \tag{3.14}$$

contradicting (3.12) unless

$$P_n f_0 = 0, \ n \in \mathbb{N}, \ \text{ and hence, } P_0 f_0 = f_0, \tag{3.15}$$

that is,

$$f_0 \in \operatorname{dom}(T_{0,F}) \ \text{ and } \ T_{0,F} f_0 = (1/4) f_0, \tag{3.16}$$

employing $\lambda_0 = 1/4$. However, (3.16) implies that

$$f_0(x) \underset{x\downarrow 0}{=} c x^{1/2}\big[1 + O\big(x^2\big)\big] \text{ and hence, } f_0 \notin H_0^1((0,\pi)), \tag{3.17}$$

a contradiction. □

*Remark 3.2* (*i*) That inequality (3.4) represents an improvement over the previously well-known cases (3.1) and (3.2) can be shown as follows: Since trivially

$$\sin(x) \le x, \quad x \in [0,\pi], \tag{3.18}$$

inequality (3.4) is obviously an improvement over the classical Hardy inequality (3.1). On the other hand, since also

$$\sin(x) \le \begin{cases} x, & x \in [0,\pi/2], \\ \pi - x, & x \in [\pi/2,\pi], \end{cases} \tag{3.19}$$

that is (cf. (3.3)),

$$\sin(x) \le d_{(0,\pi)}(x), \quad x \in [0,\pi], \tag{3.20}$$

inequality (3.4) also improves upon the refinement (3.2).

$(ii)$ Assuming $a, b \in \mathbb{R}$, $a < b$, the elementary change of variables

$$\begin{aligned} &(0,\pi) \ni x \mapsto \xi(x) = [(b-a)x + a\pi]/\pi \in (a,b), \\ &f(x) = F(\xi), \end{aligned} \tag{3.21}$$

yields

$$\begin{aligned} \int_a^b d\xi\, |F'(\xi)|^2 \geq \frac{\pi^2}{4(b-a)^2} \int_a^b d\xi\, \frac{|F(\xi)|^2}{\sin^2(\pi(\xi-a)/(b-a))} \\ + \frac{\pi^2}{4(b-a)^2} \int_a^b d\xi\, |F(\xi)|^2, \quad F \in H_0^1((a,b)). \end{aligned} \tag{3.22}$$

These scaling arguments apply to all Hardy-type inequalities considered in this paper and hence it suffices to restrict ourselves to convenient fixed intervals such as $(0,\pi)$, etc. ◇

*Remark 3.3* An earlier version of our preprint contained the following factorization of $\tau_s - (s+(1/2))^2$ into

$$\begin{aligned} \delta_s^+\delta_s = \tau_s - (s+(1/2))^2 = -\frac{d^2}{dx^2} + \frac{s^2-(1/4)}{\sin^2(x)} - (s+(1/2))^2, \\ s \in [0,\infty),\ x \in (0,\pi), \end{aligned} \tag{3.23}$$

where the differential expressions $\delta_s$, $\delta_s^+$ are given by

$$\begin{aligned} \delta_s = \frac{d}{dx} - [s+(1/2)]\cot(x), \quad \delta_s^+ = -\frac{d}{dx} - [s+(1/2)]\cot(x), \\ s \in [0,\infty),\ x \in (0,\pi). \end{aligned} \tag{3.24}$$

Thus, $\delta_s^+\delta_s\big|_{C_0^\infty((0,\pi))} \geq 0$ yields $\tau_s\big|_{C_0^\infty((0,\pi))} \geq (s+(1/2))^2 I$ and taking $s=0$ implies inequality (3.4) for $f \in C_0^\infty((0,\pi))$ and hence for $f \in H_0^1((0,\pi))$ by the usual Fatou-type argument. Hence, if one is primarily interested in the refined Hardy inequality (3.4) itself, taking $s=0$ in (3.23) appears to be its quickest derivation. We're indebted to Ari Laptev for independently pointing this out to us which resulted in our reintroducing the factorization (3.23).

Considering

$$y_s(x) = [\sin(x)]^{(1+2s)/2}, \quad s \in [0,\infty),\ x \in (0,\pi), \tag{3.25}$$

one confirms that $\delta_s y_s = 0$, $s \in [0, \infty)$, and hence

$$(\tau_s y_s)(x) = [s + (1/2)]^2 y_s(x), \quad s \in [0, \infty), \ x \in (0, \pi). \tag{3.26}$$

A second linearly independent solution of $\tau_s y = [s + (1/2)]^2 y$ is then given by

$$\begin{aligned} \widehat{y}_s(x) &= [\sin(x)]^{(1+2s)/2} \int_x^{\pi/2} dt\, [\sin(t)]^{-(1+2s)} \\ &\underset{x \downarrow 0}{=} \begin{cases} (2s)^{-1} x^{(1-2s)/2}[1 + o(1)], & s \in (0, \infty), \\ x^{1/2} \ln(1/x)[1 + o(1)], & s = 0. \end{cases} \end{aligned} \tag{3.27}$$

By inspection, $y_s' \in L^2((0, \pi); dx)$ if and only if $s \in (0, \infty)$, and hence there is a cancellation taking place in $\delta_0 y_0 = 0$ for $s = 0$, whereas $\widehat{y}_s \notin L^2((0, \pi); dx)$ for $s \in [1, \infty)$ (in accordance with $\tau_s$ being in the limit point case for $s \in [1, \infty)$) and $\widehat{y}_s' \notin L^2((0, \pi); dx)$ for $s \in [0, \infty)$. ◇

A closer inspection of the proof of Theorem 3.1 reveals that $[\sin(x)]^{-2}$ is just a very convenient choice for a function that has inverse square singularities at the interval endpoints as it leads to explicit optimal constants $1/4$ in (3.4). To illustrate this point, we consider the differential expressions

$$\omega_0 = -\frac{d^2}{dx^2} - \frac{1}{4x^2}, \quad \alpha_0 = \frac{d}{dx} - \frac{1}{2x}, \quad \alpha_0^+ = -\frac{d}{dx} - \frac{1}{2x}, \quad x \in (0, \pi), \tag{3.28}$$

such that

$$\alpha_0^+ \alpha_0 = \omega_0. \tag{3.29}$$

The minimal and maximal $L^2((0, \pi); dx)$-realizations associated with $\omega_0$ are then given by

$$\begin{aligned} &S_{0,min} f = \omega_0 f, \\ &f \in \operatorname{dom}(S_{0,min}) = \big\{g \in L^2((0, \pi); dx) \,\big|\, g, g' \in AC_{loc}((0, \pi)); \\ &\qquad \operatorname{supp}(g) \subset (0, \pi) \text{ is compact}; \, \omega_0 g \in L^2((0, \pi); dx)\big\}. \end{aligned} \tag{3.30}$$

$$\begin{aligned} &S_{0,max} f = \omega_0 f, \\ &f \in \operatorname{dom}(S_{0,max}) = \big\{g \in L^2((0, \pi); dx) \,\big|\, g, g' \in AC_{loc}((0, \pi)); \\ &\qquad \omega_0 g \in L^2((0, \pi); dx)\big\}, \end{aligned} \tag{3.31}$$

implying $S_{0,min}^* = S_{0,max}$, $S_{0,max}^* = S_{0,min}$, and we also introduce the following self-adjoint extensions of $S_{0,min}$, respectively, restrictions of $S_{0,max}$ (see, e.g., [3–5, 11, 16, 20, 24, 29, 39, 41, 58]),

$$\begin{aligned}
&S_{0,D,N} f = \omega_0 f, \\
&f \in \operatorname{dom}(S_{0,D,N}) = \{g \in \operatorname{dom}(S_{0,max}) \,|\, \widetilde{g}(0) = g'(\pi) = 0\} \\
&\qquad\qquad = \big\{g \in \operatorname{dom}(S_{0,max}) \,\big|\, g'(\pi) = 0;\ \alpha_0 g \in L^2((0,\pi); dx)\big\},
\end{aligned} \tag{3.32}$$

$$\begin{aligned}
&S_{0,F} f = \omega_0 f, \\
&f \in \operatorname{dom}(S_{0,F}) = \{g \in \operatorname{dom}(S_{0,max}) \,|\, \widetilde{g}(0) = g(\pi) = 0\} \\
&\qquad\qquad = \big\{g \in \operatorname{dom}(S_{0,max}) \,\big|\, g(\pi) = 0;\ \alpha_0 g \in L^2((0,\pi); dx)\big\},
\end{aligned} \tag{3.33}$$

with $S_{0,F}$ the Friedrichs extension of $S_{0,min}$. The quadratic forms corresponding to $S_{0,D,N}$ and $S_{0,F}$ are of the form

$$\begin{aligned}
&Q_{S_{0,D,N}}(f,g) = (\alpha_0 f, \alpha_0 g)_{L^2((0,\pi);dx)}, \\
&f,g \in \operatorname{dom}(Q_{S_{0,D,N}}) = \big\{g \in L^2((0,\pi); dx) \,\big|\, g \in AC_{loc}((0,\pi));\ g'(\pi) = 0, \\
&\qquad\qquad\qquad\qquad \alpha_0 g \in L^2((0,\pi); dx)\big\},
\end{aligned} \tag{3.34}$$

$$\begin{aligned}
&Q_{S_{0,F}}(f,g) = (\alpha_0 f, \alpha_0 g)_{L^2((0,\pi);dx)}, \\
&f,g \in \operatorname{dom}(Q_{S_{0,F}}) = \big\{g \in L^2((0,\pi); dx) \,\big|\, g \in AC_{loc}((0,\pi));\ g(\pi) = 0, \\
&\qquad\qquad\qquad\qquad \alpha_0 g \in L^2((0,\pi); dx)\big\}.
\end{aligned} \tag{3.35}$$

One verifies (see (B.6)) that for all $\varepsilon > 0$ and $g \in AC_{loc}((0,\varepsilon))$,

$$\alpha_0 g \in L^2((0,\varepsilon); dx) \text{ implies } \widetilde{g}(0) = 0. \tag{3.36}$$

By inspection,

$$\begin{aligned}
f_0(\lambda, x) &= x^{1/2} J_0\big(\lambda^{1/2} x\big), \quad x \in (0,\pi), \\
&\underset{x\downarrow 0}{=} x^{1/2}\big[1 + O\big(x^2\big)\big]
\end{aligned} \tag{3.37}$$

(where $J_\nu(\,\cdot\,)$ denotes the standard Bessel function of order $\nu \in \mathbb{C}$, cf. [1, Ch. 9]), satisfies (cf. (3.32), (3.33))

$$\widetilde{f_0}(0) = 0. \tag{3.38}$$

Thus, introducing Lamb's constant, now denoted by $\lambda_{D,N,0}^{1/2}$, as the first positive zero of

$$(0, \infty) \ni x \mapsto J_0(x) + 2x J_1(x) \tag{3.39}$$

(see the brief discussion in [7]), one infers that $\lambda_{D,N,0}/\pi^2$ is the first positive zero of $f_0'(\,\cdot\,, \pi)$, that is,

$$f_0'(\lambda_{D,N,0}/\pi^2, \pi) = 0. \tag{3.40}$$

In addition denoting by $\lambda_{F,0}/\pi^2$ the first strictly positive zero of $f_0(\,\cdot\,, \pi)$ one has

$$f_0(\lambda_{F,0}/\pi^2, \pi) = 0, \tag{3.41}$$

and hence $\lambda_{D,N,0}/\pi^2$ and $\lambda_{F,0}/\pi^2$ are the first eigenvalue of the mixed Dirichlet/Neumann operator $S_{0,D,N}$ and the Dirichlet operator (the Friedrichs extension of $S_{0,min}$) $S_{0,F}$, respectively.[1] Equivalently,

$$\inf(\sigma(S_{0,D,N})) = \lambda_{D,N,0}\pi^{-2}, \quad \inf(\sigma(S_{0,F})) = \lambda_{F,0}\pi^{-2}, \tag{3.42}$$

in particular,

$$S_{0,D,N} \geq \lambda_{D,N,0}\pi^{-2} I_{L^2((0,\pi);dx)}, \quad S_{0,F} \geq \lambda_{F,0}\pi^{-2} I_{L^2((0,\pi);dx)}, \tag{3.43}$$

$$Q_{S_{0,D,N}}(f, f) \geq \lambda_{D,N,0}\pi^{-2} \|f\|_{L^2((0,\pi);dx)}^2, \quad f \in \operatorname{dom}(Q_{S_{0,D,N}}), \tag{3.44}$$

$$Q_{S_{0,F}}(f, f) \geq \lambda_{F,0}\pi^{-2} \|f\|_{L^2((0,\pi);dx)}^2, \quad f \in \operatorname{dom}(Q_{S_{0,F}}). \tag{3.45}$$

Numerically, one confirms that

$$\lambda_{D,N,0} = 0.885\ldots, \quad \lambda_{F,0} = 5.783\ldots. \tag{3.46}$$

Thus, arguments analogous to the ones in the proof of Theorem 3.1 yield the following variants of (3.1), (3.2),

$$\begin{aligned} \int_0^\pi dx\, |f'(x)|^2 \geq \frac{1}{4} \int_0^\pi dx\, \frac{|f(x)|^2}{x^2} + \frac{\lambda_{D,N,0}}{\pi^2} \int_0^\pi dx\, |f(x)|^2, \\ f \in \operatorname{dom}(Q_{S_{0,D,N}}) \cap H^1((0,\pi)), \end{aligned} \tag{3.47}$$

[1] In particular, $\lambda_{D,N,0}$ and $\lambda_{F,0}$ are the first eigenvalue of the mixed Dirichlet/Neumann and Dirichlet operator on the interval $(0, 1)$.

$$\int_0^\pi dx\, |f'(x)|^2 \geq \frac{1}{4}\int_0^\pi dx\, \frac{|f(x)|^2}{x^2} + \frac{\lambda_{F,0}}{\pi^2}\int_0^\pi dx\, |f(x)|^2, \qquad f \in H_0^1((0,\pi)), \tag{3.48}$$

as well as,

$$\int_0^\pi dx\, |f'(x)|^2 \geq \frac{1}{4}\int_0^\pi dx\, \frac{|f(x)|^2}{d_{(0,\pi)}(x)^2} + \frac{4\lambda_{D,N,0}}{\pi^2}\int_0^\pi dx\, |f(x)|^2, \qquad f \in H_0^1((0,\pi)). \tag{3.49}$$

All constants in (3.47)–(3.49) are optimal and the inequalities are all strict (for $f \not\equiv 0$).

In obtaining (3.47)–(3.49) one makes use of the fact that the domain of a semibounded, self-adjoint operator $A$ in the complex, separable Hilbert space $\mathcal{H}$ is a form core for $A$, equivalently, a core for $|A|^{1/2}$. In addition, we used (cf. [20, Theorem 7.1]) that for $f \in \operatorname{dom}(S_{0,D,N}) \cup \operatorname{dom}(S_{0,F})$, there exists $K_0(f) \in \mathbb{C}$ such that

$$\begin{aligned} &\lim_{x\downarrow 0} x^{-1/2} f(x) = K_0(f), \quad \lim_{x\downarrow 0} x^{1/2} f'(x) = K_0(f)/2, \\ &\lim_{x\downarrow 0} f(x) f'(x) = K_0(f)^2/2. \end{aligned} \tag{3.50}$$

Moreover, if in addition $f' \in L^2((0,1); dx)$, $f(0) = 0$, combining (3.50) with estimate (B.10) yields $K_0(f) = 0$, and hence $\lim_{x\downarrow} f(x)f'(x) = 0$. This permits one to integrate by parts in $Q_{S_{0,D,N}}(f,f)$ and $Q_{S_{0,F}}(f,f)$ and in the process verify (3.47)–(3.49).

We note that inequalities (3.47) and (3.49) were first derived by Avkhadiev and Wirths [7] (also recorded in [8] and [9, Sect. 3.6.3]; see also [6, 35, 48]) following a different approach applicable to the multi-dimensional case. We have not found inequality (3.48) in the literature, but expect it to be known.

*Remark 3.4* The arguments presented thus far might seem to indicate that Hardy-type inequalities are naturally associated with underlying second-order differential operators satisfying boundary conditions of the Dirichlet and/or Neumann type at the interval endpoints. However, this is not quite the case as the following result (borrowed, e.g., from [15, Lemma 5.3.1], [50, Sect. 1.1]) shows: Suppose $b \in (0,\infty)$, $f \in AC_{loc}((0,b))$, $f' \in L^2((0,b); dx)$, $f(0) = 0$, then (with $f$ real-valued without loss of generality),

$$\begin{aligned} \int_0^b dx\, |f'(x)|^2 &= \int_0^b dx\, \big|x^{1/2}[x^{-1/2}f(x)]' + (2x)^{-1} f(x)\big|^2 \\ &= \int_0^b dx\, \Big\{4^{-1}x^{-2}f(x)^2 + x^{-1/2}f(x)[x^{-1/2}f(x)]' + x\big[(x^{-1/2}f(x))'\big]^2\Big\} \end{aligned}$$

$$
\begin{aligned}
&\geq \int_0^b dx \left\{4^{-1}x^{-2}f(x)^2 + x^{-1/2}f(x)\big[x^{-1/2}f(x)\big]'\right\} \\
&= \int_0^b dx\, \frac{|f(x)|^2}{4x^2} + 2^{-1}\big[x^{-1/2}f(x)\big]^2\big|_{x=0}^b \\
&\geq \int_0^b dx\, \frac{|f(x)|^2}{4x^2} - 2^{-1}\lim_{x\downarrow 0}\big[x^{-1/2}f(x)\big]^2 \\
&= \int_0^b dx\, \frac{|f(x)|^2}{4x^2},
\end{aligned}
\tag{3.51}
$$

employing the estimate (B.10) with $f(0) = 0$. In particular, no boundary conditions whatsoever are needed at the right end point $b$. One notes that the hypotheses on $f$ imply that $f \in AC([0, b])$ and hence actually that $f$ behaves like an $H_0^1$-function in a right neighborhood of $x = 0$, equivalently, $f\widetilde{\chi}_{[0,b/2]} \in H_0^1((0, b))$, where

$$
\widetilde{\chi}_{[0,r/2]}(x) = \begin{cases} 1, & x \in [0, r/4], \\ 0, & x \in [3r/4, r], \end{cases} \quad \widetilde{\chi}_{[0,r/2]} \in C^\infty([0, r]), \ r \in (0, \infty). \tag{3.52}
$$

◇

*Remark 3.5* Employing locality of the operators involved, one can show (cf. [31]) that all considerations in the bulk of this paper, extend to the situation where

$$
q(x) = \frac{s^2 - (1/4)}{x^2}, \text{ respectively, } q(x) = \frac{s^2 - (1/4)}{\sin^2(x)}, \quad s \in [0, \infty), \tag{3.53}
$$

is replaced by a potential $q$ satisfying $q \in L^1_{loc}((0, \pi); dx)$ and for some $s_j \in [0, \infty)$, $j = 1, 2$, and some $0 < \varepsilon$ sufficiently small,

$$
q_{s_1,s_2}(x) = \begin{cases} \big[s_1^2 - (1/4)\big]x^{-2}, & x \in (0, \varepsilon), \\ \big[s_2^2 - (1/4)\big](x - \pi)^{-2}, & x \in (\pi - \varepsilon, \pi). \end{cases} \tag{3.54}
$$

As discussed in [39], this can be replaced by $q \geq q_{s_1,s_2}$ a.e.

In addition, we only presented the tip of an iceberg in this section as these considerations naturally extend to more general Sturm–Liouville operators in $L^2((a, b); dx)$ generated by differential expressions of the type

$$
-\frac{d}{dx}p(x)\frac{d}{dx} + q(x), \quad x \in (a, b), \tag{3.55}
$$

as discussed to some extent in [25]. We will return to this and the general three-coefficient Sturm–Liouville operators in $L^2((a,b); r dx)$ generated by

$$\frac{1}{r(x)}\left[-\frac{d}{dx}p(x)\frac{d}{dx}+q(x)\right], \quad x\in(a,b), \tag{3.56}$$

elsewhere. ◇

## A.1 The Weyl–Titchmarsh–Kodaira $m$-Function Associated with $T_{s,F}$

We start by introducing a normalized fundamental system of solutions $\phi_{0,s}(z,\,\cdot\,)$ and $\theta_{0,s}(z,\,\cdot\,)$ of $\tau_s u = zu$, $s\in[0,1)$, $z\in\mathbb{C}$, satisfying (cf. the generalized boundary values introduced in (2.14), (2.15))

$$\widetilde{\theta}_{0,s}(z,0)=1,\quad \widetilde{\theta}'_{0,s}(z,0)=0,\quad \widetilde{\phi}_{0,s}(z,0)=0,\quad \widetilde{\phi}'_{0,s}(z,0)=1, \tag{A.1}$$

with $\phi_{0,s}(\,\cdot\,,x)$ and $\theta_{0,s}(\,\cdot\,,x)$ entire for fixed $x\in(0,\pi)$. To this end, we introduce the two linearly independent solutions to $\tau_s y = zy$ (entire w.r.t. $z$ for fixed $x\in(0,\pi)$) given by

$$\begin{aligned} y_{1,s}(z,x) &= [\sin(x)]^{(1-2s)/2} \\ &\qquad \times F\big([(1/2)-s+z^{1/2}]/2, [(1/2)-s-z^{1/2}]/2; 1/2; \cos^2(x)\big), \\ y_{2,s}(z,x) &= \cos(x)[\sin(x)]^{(1-2s)/2} \\ &\qquad \times F\big([(3/2)-s+z^{1/2}]/2, [(3/2)-s-z^{1/2}]/2; 3/2; \cos^2(x)\big), \\ &\qquad\qquad s\in[0,1),\ z\in\mathbb{C},\ x\in(0,\pi). \end{aligned} \tag{A.2}$$

Using the connection formula found in [1, Eq. 15.3.6] yields the behavior near $x=0,\pi$,

$$\begin{aligned} y_{1,s}(z,x) &= \frac{\pi^{1/2}\Gamma(s)[\sin(x)]^{(1-2s)/2}}{\Gamma\big([(1/2)+s+z^{1/2}]/2\big)\Gamma\big([(1/2)+s-z^{1/2}]/2\big)} \\ &\quad \times F\big([(1/2)-s+z^{1/2}]/2, [(1/2)-s-z^{1/2}]/2; 1-s; \sin^2(x)\big) \\ &\quad + \frac{\pi^{1/2}\Gamma(-s)[\sin(x)]^{(1+2s)/2}}{\Gamma\big([(1/2)-s+z^{1/2}]/2\big)\Gamma\big([(1/2)-s-z^{1/2}]/2\big)} \\ &\quad \times F\big([(1/2)+s+z^{1/2}]/2, [(1/2)+s-z^{1/2}]/2; 1+s; \sin^2(x)\big), \\ &\qquad\qquad s\in(0,1),\ z\in\mathbb{C},\ x\in(0,\pi), \end{aligned} \tag{A.3}$$

$$y_{2,s}(z,x) = \frac{\pi^{1/2}\Gamma(s)\cos(x)[\sin(x)]^{(1-2s)/2}}{2\Gamma([(3/2)+s+z^{1/2}]/2)\Gamma([(3/2)+s-z^{1/2}]/2)}$$
$$\times F([(3/2)-s+z^{1/2}]/2, [(3/2)-s-z^{1/2}]/2; 1-s; \sin^2(x))$$
$$+ \frac{\pi^{1/2}\Gamma(-s)\cos(x)[\sin(x)]^{(1+2s)/2}}{2\Gamma([(3/2)-s+z^{1/2}]/2)\Gamma([(3/2)-s-z^{1/2}]/2)}$$
$$\times F([(3/2)+s+z^{1/2}]/2, [(3/2)+s-z^{1/2}]/2; 1+s; \sin^2(x)),$$
$$s \in (0,1),\ z \in \mathbb{C},\ x \in (0,\pi)\backslash\{\pi/2\}.$$

*Remark A.1.1* Before we turn to the case $s = 0$, we recall Gauss's identity (cf. [1, no. 15.1.20])

$$F(a,b;c;1) = \frac{\Gamma(c)\Gamma(c-a-b)}{\Gamma(c-a)\Gamma(c-b)}, \quad c \in \mathbb{C}\backslash\{-\mathbb{N}_0\},\ \mathrm{Re}(c-a-b) > 0, \tag{A.4}$$

and the differentiation formula (cf. [1, no. 15.2.1])

$$\frac{d}{dz}F(a,b;c;z) = \frac{ab}{c}F(a+1,b+1;c+1;z), \quad a,b,c \in \mathbb{C},\ z \in \{\zeta \in \mathbb{C} \,|\, |\zeta| < 1\}, \tag{A.5}$$

which imply that for $s \in (0,1)$, the two $F(\cdot\,,\,\cdot;\,\cdot;1)$ exist in (A.2) (indeed, for $j = 1,2$ one obtains from (A.2) with $s \in (0,1)$ that $c - a - b = s > 0$) and hence the asymptotic behavior of $y_{j,s}(z,x)$, $j = 1,2$, as $x \downarrow 0$ and $x \uparrow \pi$ is dominated by $x^{(1-2s)/2}$ and $(\pi - x)^{(1-2s)/2}$, respectively. However, the analogous statement fails for $y'_{j,s}(z,x)$, $j = 1,2$, as, taking into account (A.5), the analog of the 2nd condition in (A.4), namely, $\mathrm{Re}[c + 1 - (a+1) - (b+1)] > 0$, is not fulfilled (in this case (A.2) with $s \in (0,1)$ yields $[c+1-(a+1)-(b+1)] = s - 1 < 0$). The situation is similar for the first two $F(\cdot\,,\,\cdot;\,\cdot;x)$ for $y_{1,s}(z,x)$ in (A.3) as $x \to \pi/2$ as in this case the two $F(\cdot\,,\,\cdot;\,\cdot;1)$ exist. Even though for $y_{2,s}(z,x)$ in (A.3) the two $F(\cdot\,,\,\cdot;\,\cdot;1)$ do not exist individually, the limit of each term does exist due to the multiplication by the factor $\cos(x)$. To see this, one can instead consider the limit (cf. [51, no. 15.4.23])

$$\lim_{z\to 1^-} \frac{F(a,b;c;z)}{(1-z)^{c-a-b}} = \frac{\Gamma(c)\Gamma(a+b-c)}{\Gamma(a)\Gamma(b)}, \quad c \in \mathbb{C}\backslash\{-\mathbb{N}_0\},\ \mathrm{Re}(c-a-b) < 0, \tag{A.6}$$

which through the appropriate change of variable reveals that the connection formula for $y_{2,s}(z,x)$ as $x \to \pi/2$ approaches 0 as expected from evaluating $y_{2,s}(z,\pi/2)$ in (A.2). But once again, the analog of the 2nd condition in (A.4), namely, $\operatorname{Re}[c+1-(a+1)-(b+1)] > 0$, fails for the four $F'(\cdot\,,\,\cdot\,;\,\cdot\,;x)$ in (A.3) as $x \to \pi/2$. ◇

Similarly, by Abramowitz and Stegun [1, Eq. 15.3.10] one obtains for the remaining case $s = 0$,

$$\begin{aligned} y_{1,0}(z,x) &= \frac{\pi^{1/2}[\sin(x)]^{1/2}}{\Gamma\big([(1/2)+z^{1/2}]/2\big)\Gamma\big([(1/2)-z^{1/2}]/2\big)} \\ &\quad \times \sum_{n=0}^{\infty} \frac{\big([(1/2)+z^{1/2}]/2\big)_n \big([(1/2)-z^{1/2}]/2\big)_n}{(n!)^2} \big[2\psi(n+1) \\ &\quad - \psi\big(n+[(1/2)+z^{1/2}]/2\big) - \psi\big(n+[(1/2)-z^{1/2}]/2\big) \\ &\quad - \ln(\sin^2(x))\big][\sin(x)]^{2n}, \qquad\qquad (A.7) \\ y_{2,0}(z,x) &= \frac{\pi^{1/2}\cos(x)[\sin(x)]^{1/2}}{2\Gamma\big([(3/2)+z^{1/2}]/2\big)\Gamma\big([(3/2)-z^{1/2}]/2\big)} \\ &\quad \times \sum_{n=0}^{\infty} \frac{\big([(3/2)+z^{1/2}]/2\big)_n \big([(3/2)-z^{1/2}]/2\big)_n}{(n!)^2} \big[2\psi(n+1) \\ &\quad - \psi\big(n+[(3/2)+z^{1/2}]/2\big) - \psi\big(n+[(3/2)-z^{1/2}]/2\big) \\ &\quad - \ln\big(\sin^2(x)\big)\big][\sin(x)]^{2n}, \\ &\quad s = 0,\ z \in \mathbb{C},\ x \in (0,\pi). \end{aligned}$$

Here $\psi(\cdot) = \Gamma'(\cdot)/\Gamma(\cdot)$ denotes the Digamma function, $\gamma_E = -\psi(1) = 0.57721\ldots$ represents Euler's constant, and

$$(\zeta)_0 = 1, \quad (\zeta)_n = \Gamma(\zeta+n)/\Gamma(\zeta),\ n \in \mathbb{N}, \quad \zeta \in \mathbb{C}\backslash(-\mathbb{N}_0), \tag{A.8}$$

abbreviates Pochhammer's symbol (see, e.g., [1, Ch. 6]). Direct computation now yields

$$\begin{aligned} \widetilde{y}_{1,s}(z,0) &= -\widetilde{y}_{1,s}(z,\pi) \\ &= \frac{2\pi^{1/2}\Gamma(1+s)}{\Gamma\big([(1/2)+s+z^{1/2}]/2\big)\Gamma\big([(1/2)+s-z^{1/2}]/2\big)}, \\ \widetilde{y}'_{1,s}(z,0) &= \widetilde{y}'_{1,s}(z,\pi) = \frac{\pi^{1/2}\Gamma(-s)}{\Gamma\big([(1/2)-s+z^{1/2}]/2\big)\Gamma\big([(1/2)-s-z^{1/2}]/2\big)}, \end{aligned}$$

$$\widetilde{y}_{2,s}(z,0) = \widetilde{y}_{2,s}(z,\pi) = \frac{\pi^{1/2}\Gamma(1+s)}{\Gamma\big(\big[(3/2)+s+z^{1/2}\big]/2\big)\Gamma\big(\big[(3/2)+s-z^{1/2}\big]/2\big)},$$

$$\begin{aligned}\widetilde{y}_{2,s}'(z,0) &= -\widetilde{y}_{2,s}'(z,\pi)\\ &= \frac{\pi^{1/2}\Gamma(-s)}{2\Gamma\big(\big[(3/2)-s+z^{1/2}\big]/2\big)\Gamma\big(\big[(3/2)-s-z^{1/2}\big]/2\big)},\\ &\qquad s\in(0,1),\ z\in\mathbb{C},\end{aligned} \tag{A.9}$$

$$\widetilde{y}_{1,0}(z,0) = -\widetilde{y}_{1,0}(z,\pi) = \frac{2\pi^{1/2}}{\Gamma\big(\big[(1/2)+z^{1/2}\big]/2\big)\Gamma\big(\big[(1/2)-z^{1/2}\big]/2\big)},$$

$$\begin{aligned}\widetilde{y}_{1,0}'(z,0) &= \widetilde{y}_{1,0}'(z,\pi)\\ &= \frac{-\pi^{1/2}\big[2\gamma_E+\psi\big(\big[(1/2)+z^{1/2}\big]/2\big)+\psi\big(\big[(1/2)-z^{1/2}\big]/2\big)\big]}{\Gamma\big(\big[(1/2)+z^{1/2}\big]/2\big)\Gamma\big(\big[(1/2)-z^{1/2}\big]/2\big)},\end{aligned}$$

$$\widetilde{y}_{2,0}(z,0) = \widetilde{y}_{2,0}(z,\pi) = \frac{\pi^{1/2}}{\Gamma\big(\big[(3/2)+z^{1/2}\big]/2\big)\Gamma\big(\big[(3/2)-z^{1/2}\big]/2\big)},$$

$$\begin{aligned}\widetilde{y}_{2,0}'(z,0) &= -\widetilde{y}_{2,0}'(z,\pi)\\ &= \frac{-\pi^{1/2}\big[2\gamma_E+\psi\big(\big[(3/2)+z^{1/2}\big]/2\big)+\psi\big(\big[(3/2)-z^{1/2}\big]/2\big)\big]}{2\Gamma\big(\big[(3/2)+z^{1/2}\big]/2\big)\Gamma\big(\big[(3/2)-z^{1/2}\big]/2\big)},\\ &\qquad s=0,\ z\in\mathbb{C}.\end{aligned} \tag{A.10}$$

In particular, one obtains

$$\begin{aligned}\phi_{0,s}(z,x) &= \widetilde{y}_{2,s}(z,0)y_{1,s}(z,x) - \widetilde{y}_{1,s}(z,0)y_{2,s}(z,x),\\ \theta_{0,s}(z,x) &= \widetilde{y}_{1,s}'(z,0)y_{2,s}(z,x) - \widetilde{y}_{2,s}'(z,0)y_{1,s}(z,x),\\ &\qquad s\in[0,1),\ z\in\mathbb{C},\ x\in(0,\pi),\end{aligned} \tag{A.11}$$

since

$$W(y_{1,s}(z,\cdot),y_{2,s}(z,\cdot)) = \widetilde{y}_{1,s}(z,0)\widetilde{y}_{2,s}'(z,0) - \widetilde{y}_{1,s}'(z,0)\widetilde{y}_{2,s}(z,0) = -1, \tag{A.12}$$

with the generalized boundary values given by (A.9), (A.10). To prove (A.12) one recalls Euler's reflection formula (cf. [1, no. 6.1.17])

$$\Gamma(z)\Gamma(1-z) = \frac{\pi}{\sin(\pi z)}, \quad z \in \mathbb{C}\backslash\mathbb{Z}, \tag{A.13}$$

and hence concludes that

$$\begin{aligned} &\Gamma\big([(1/2) + \varepsilon s \pm z^{1/2}]/2\big)\Gamma\big([(3/2) - \varepsilon s \mp z^{1/2}]/2\big) \\ &\quad = \frac{\pi}{\sin\big(\pi\big[(1/2) + \varepsilon \pm z^{1/2}\big]/2\big)}, \quad \varepsilon \in \{-1, 1\}. \end{aligned} \tag{A.14}$$

Thus one computes for $s \in (0,1)$,

$$\begin{aligned} &W(y_{1,s}(z,\,\cdot\,), y_{2,s}(z,\,\cdot\,)) \\ &\quad = -[\sin(\pi s)]^{-1}\big\{\sin\big(\pi\big[(1/2) + s + z^{1/2}\big]/2\big)\sin\big(\pi\big[(1/2) + s - z^{1/2}\big]/2\big) \\ &\qquad\qquad - \sin\big(\pi\big[(1/2) - s + z^{1/2}\big]/2\big)\sin\big(\pi\big[(1/2) - s - z^{1/2}\big]/2\big)\big\} \\ &\quad = -[2\sin(\pi s)]^{-1}\{-\cos(\pi[(1/2) + s]) + \cos(\pi[(1/2) - s])\} = -1. \end{aligned} \tag{A.15}$$

For the case $s = 0$, one recalls the reflection formula for the Digamma function (cf. [1, no. 6.3.7])

$$\psi(1-z) - \psi(z) = \pi\cot(\pi z), \quad z \in \mathbb{C}\backslash\mathbb{Z}, \tag{A.16}$$

and applies trigonometric identities to obtain $W(y_{1,0}(z,\,\cdot\,), y_{2,0}(z,\,\cdot\,)) = -1$. The singular Weyl–Titchmarsh–Kodaira function $m_{0,0,s}(z)$ is then uniquely determined (cf. [30, Eq. (3.18)] and [29] for background on $m$-functions) to be[2]

$$m_{0,0,s}(z) = -\frac{\widetilde{\theta}_{0,s}(z,\pi)}{\widetilde{\phi}_{0,s}(z,\pi)}, \quad s \in [0,1),\ z \in \rho(T_{s,F}). \tag{A.17}$$

[2]Here the subscripts 0, 0 in $m_{0,0,s}$ indicate the Dirichlet (i.e., Friedrichs) boundary conditions at $x = 0, \pi$, a special case of the $m_{\alpha,\beta}$-function discussed in [29] associated with separated boundary conditions at $x = 0, \pi$, indexed by boundary condition parameters $\alpha, \beta \in [0, \pi]$.

Direct calculation once again yields

$$m_{0,0}(z) = -\frac{\widetilde{y}'_{2,s}(z,0)\widetilde{y}_{1,s}(z,\pi) - \widetilde{y}'_{1,s}(z,0)\widetilde{y}_{2,s}(z,\pi)}{2\widetilde{y}_{1,s}(z,0)\widetilde{y}_{2,s}(z,0)}$$

$$= \begin{cases} \dfrac{\pi\Gamma(-s)}{4\Gamma(1+s)}\Bigg[\dfrac{\Gamma([(3/2)+s+z^{1/2}]/2)\Gamma([(3/2)+s-z^{1/2}]/2)}{\Gamma([(3/2)-s+z^{1/2}]/2)\Gamma([(3/2)-s-z^{1/2}]/2)} \\ \qquad +\dfrac{\Gamma([(1/2)+s+z^{1/2}]/2)\Gamma([(1/2)+s-z^{1/2}]/2)}{\Gamma([(1/2)-s+z^{1/2}]/2)\Gamma([(1/2)-s-z^{1/2}]/2)}\Bigg], \\ \hfill s \in (0,1), \\ -\big[4\gamma_E + \psi([(1/2)+z^{1/2}]/2) + \psi([(1/2)-z^{1/2}]/2) \\ \qquad +\psi([(3/2)+z^{1/2}]/2) + \psi([(3/2)-z^{1/2}]/2)\big]/4, \quad s=0, \end{cases}$$

$$z \in \rho(T_{s,F}), \tag{A.18}$$

which has simple poles precisely at the simple eigenvalues of $T_{s,F}$ given by

$$\sigma(T_{s,F}) = \big\{[(1/2)+s+n]^2\big\}_{n\in\mathbb{N}_0}, \quad s \in [0,1). \tag{A.19}$$

*Remark A.1.2* For the limit point case at both endpoints, that is, for $s \in [1,\infty)$, the solutions $y_{j,s}(z,\,\cdot\,)$ in (A.2) remain linearly independent and also the connection formulas (A.3) remain valid for $s \in [1,\infty)\backslash\mathbb{N}$. Moreover, employing once again (A.4) and (A.5) one verifies that the two $F(\,\cdot\,,\,\cdot\,;\,\cdot\,;1)$ as well as $F'(\,\cdot\,,\,\cdot\,;\,\cdot\,;1)$ are well defined in (A.2) and hence for $s \in [1,\infty)$, the asymptotic behavior of $y_{j,s}(z,x)$ and $y'_{j,s}(z,x)$, $j=1,2$, as $x \downarrow 0$ and $x \uparrow \pi$ is dominated by $x^{(1-2s)/2}$ and $x^{-(1+2s)/2}$ and $(\pi-x)^{(1-2s)/2}$ and $(\pi-x)^{-(1+2s)/2}$, respectively. Since in connection with (A.3) one has $c-a-b=\pm 1/2$, independently of the value of $s \in (0,\infty)$, the situation described in Remark A.1.1 for (A.3) and $s \in (0,1)$ applies without change to the current case $s \in [1,\infty)$.

Actually, some of these failures (as $x \to \pi/2$ in $y'_{j,s}(z,x)$, $j=1,2$) are crucial for the following elementary reason: The function

$$[\sin(x)]^{(1+2s)/2}\frac{\pi^{1/2}\Gamma(-s)}{\Gamma([(1/2)-s+z^{1/2}]/2)\Gamma([(1/2)-s-z^{1/2}]/2)}$$
$$\times F\big([(1/2)+s+z^{1/2}]/2, [(1/2)+s-z^{1/2}]/2; 1+s; \sin^2(x)\big), \tag{A.20}$$
$$s \in [1,\infty),\ z \in \mathbb{C},\ x \in (0,\pi),$$

(i.e., the analog of the second part of $y_{1,s}(z,\,\cdot\,)$ on the right-hand side in (A.3)) generates an $L^2((0,\pi);dx)$-element near $x=0,\pi$, and hence if this function and its $x$-derivative were locally absolutely continuous in a neighborhood of $x=\pi/2$

(the only possibly nontrivial point in the interval $(0, \pi)$), the self-adjoint maximal operator $T_{s,max}$, $s \in [1, \infty)$, would have eigenvalues for all $z \in \mathbb{C}$, an obvious contradiction. ◇

Because of the subtlety pointed out in Remark A.1.2 we omit further details on the limit point case $s \in [1, \infty)$ and refer to [23, Sect. 4], instead. In particular, [23, Theorem 4.1 b)] extends (A.19) to $s \in [1, \infty)$ and hence one actually has

$$\sigma(T_{s,F}) = \left\{[(1/2) + s + n]^2\right\}_{n \in \mathbb{N}_0}, \quad s \subset [0, \infty). \tag{A.21}$$

## B.1 Remarks on Hardy-Type Inequalities

In this appendix we recall a Hardy-type inequality useful in Sect. 2.

Introducing the differential expressions $\alpha_s$, $\alpha_s^+$ (cf. (3.28) for $s = 0$),

$$\alpha_s = \frac{d}{dx} - \frac{s + (1/2)}{x}, \quad \alpha_s^+ = -\frac{d}{dx} - \frac{s + (1/2)}{x}, \quad s \in [0, \infty), \ x \in (0, \pi), \tag{B.1}$$

one confirms that

$$\alpha_s^+ \alpha_s = \omega_s = -\frac{d^2}{dx^2} + \frac{s^2 - (1/4)}{x^2}, \quad s \in [0, \infty), \ x \in (0, \pi). \tag{B.2}$$

Following the Hardy inequality considerations in [24, 38, 40], one obtains the following basic facts.

**Lemma B.1.1** *Suppose $f \in AC_{loc}((0, \pi))$, $\alpha_s f \in L^2((0, \pi); dx)$ for some $s \in \mathbb{R}$, and $0 < r_0 < r_1 < \pi < R < \infty$. Then,*

$$\begin{aligned} \int_{r_0}^{r_1} dx\, |(\alpha_s f)(x)|^2 &\geq s^2 \int_{r_0}^{r_1} dx\, \frac{|f(x)|^2}{x^2} + \frac{1}{4} \int_{r_0}^{r_1} dx\, \frac{|f(x)|^2}{x^2[ln(R/x)]^2} \\ &\quad - s\frac{|f(x)|^2}{x}\bigg|_{x=r_0}^{r_1} - \frac{|f(x)|^2}{2x[ln(R/x)]}\bigg|_{x=r_0}^{r_1}, \end{aligned} \tag{B.3}$$

$$\begin{aligned} &\int_{r_0}^{r_1} dx\, x ln(R/x) \left| \left[ \frac{f(x)}{x^{1/2}[ln(R/x)]^{1/2}} \right]' \right|^2 \\ &\quad = \int_{r_0}^{r_1} dx \left[ |f'(x)|^2 - \frac{|f(x)|^2}{4x^2} - \frac{|f(x)|^2}{4x^2[ln(R/x)]^2} \right] \\ &\quad - \frac{|f(x)|^2}{2x}\bigg|_{x=r_0}^{r_1} + \frac{|f(x)|^2}{2x\, ln(R/x)}\bigg|_{x=r_0}^{r_1} \geq 0, \end{aligned} \tag{B.4}$$

$$\begin{aligned}\int_{r_0}^{r_1} dx\,|(\alpha_s f)(x)|^2 &= \int_{r_0}^{r_1} dx\left[|f'(x)|^2 + [s^2-(1/4)]\frac{|f(x)|^2}{x^2}\right] \\ &\quad - [s+(1/2)]\frac{|f(x)|^2}{x}\bigg|_{x=r_0}^{r_1} \geq 0.\end{aligned} \tag{B.5}$$

*If* $s = 0$,

$$\int_0^{r_1} dx\,\frac{|f(x)|^2}{x^2[ln(R/x)]^2} < \infty, \quad \lim_{x\downarrow 0}\frac{|f(x)|}{[x\,ln(R/x)]^{1/2}} = 0. \tag{B.6}$$

*If* $s \in (0,\infty)$, *then*

$$\int_0^{r_1} dx\,|f'(x)|^2 < \infty, \quad \int_0^{r_1} dx\,\frac{|f(x)|^2}{x^2} < \infty, \quad \lim_{x\downarrow 0}\frac{|f(x)|}{x^{1/2}} = 0, \tag{B.7}$$

*in particular,*

$$f\widetilde{\chi}_{[0,r_1/2]} \in H_0^1((0,r_1)), \tag{B.8}$$

*where*

$$\widetilde{\chi}_{[0,r/2]}(x) = \begin{cases}1, & x\in[0,r/4],\\ 0, & x\in[3r/4,r],\end{cases} \quad \widetilde{\chi}_{[0,r/2]} \in C^\infty([0,r]),\ r\in(0,\infty). \tag{B.9}$$

***Proof*** Relations (B.4) and (B.5) are straightforward (yet somewhat tedious) identities; together they yield (B.3). The 1st relation in (B.6) is an instant consequence of (B.3), so is the fact that $\lim_{x\downarrow 0}|f(x)|^2/[x\ln(R/x)]$ exists. Moreover, since $[x\ln(R/x)]^{-1}$ is not integrable at $x = 0$, the 1st relation in (B.6) yields $\liminf_{x\downarrow 0}|f(x)|^2/[x\ln(R/x)] = 0$, implying the 2nd relation in (B.6).

Finally, if $s \in (0,\infty)$, then inequality (B.3) implies the 2nd relation in (B.7); together with $\alpha_s f \in L^2((0,\pi);dx)$, this yields the 1st relation in (B.7). By inequality (B.3), $\lim_{x\downarrow 0}|f(x)|^2/x$ exists, but then the second relation in (B.7) yields $\liminf_{x\downarrow 0}|f(x)|^2/x = 0$ and hence also $\lim_{x\downarrow 0}|f(x)|^2/x = 0$. □

We also recall the following elementary fact.

**Lemma B.1.2** *Suppose* $f \in H^1((0,r))$ *for some* $r \in (0,\infty)$. *Then, for all* $x \in (0,r)$,

$$\begin{aligned}|f(x)-f(0)| &= \left|\int_0^x dt\,f'(t)\right| \leq x^{1/2}\left(\int_0^x dt\,|f'(t)|^2\right)^{1/2} \\ &\leq x^{1/2}\|f'\|_{L^2((0,x);dt)} \underset{x\downarrow 0}{=} o\big(x^{1/2}\big).\end{aligned} \tag{B.10}$$

*Thus, if* $f \in H^1((0,r))$*, then* $\int_0^r dx\, |f(x)|^2/x^2 < \infty$ *if and only if* $f(0) = 0$*, that is, if and only if* $f\widetilde{\chi}_{[0,r/2]} \in H_0^1((0,r))$*.*

*In particular, if* $f \in H^1((0,r))$ *and* $f(0) = 0$*, then actually,*

$$\lim_{x\downarrow 0} \frac{|f(x)|}{x^{1/2}} = 0. \tag{B.11}$$

***Proof*** Since (B.10) is obvious, we briefly discuss the remaining assertions in Lemma B.1.2. If $f \in H^1((0,r))$ and $\int_0^r dx\, |f(x)|^2/x^2 < \infty$, then identity (B.5) for $s < -1/2$, that is,

$$\begin{aligned} \int_{r_0}^{r_1} dx\, |(\alpha_s f)(x)|^2 &= \int_{r_0}^{r_1} dx \left[ |f'(x)|^2 + [s^2 - (1/4)] \frac{|f(x)|^2}{x^2} \right] \\ &\quad - [s + (1/2)] \frac{|f(x)|^2}{x} \bigg|_{x=r_0}^{r_1} \geq 0, \quad s < -1/2, \end{aligned} \tag{B.12}$$

yields the existence of $\lim_{x\downarrow 0} |f(x)|^2/x$. Since $\int_0^r dx\, |f(x)|^2/x^2 < \infty$ implies that $\liminf_{x\downarrow 0} |f(x)|^2/x = 0$, one concludes that $\lim_{x\downarrow 0} |f(x)|^2/x = 0$ and hence $f$ behaves locally like an $H_0^1$-function in a right neighborhood of $x = 0$. Conversely, if $f(0) = 0$, then $\int_0^r dx\, |f(x)|^2/x^2 < \infty$ by Hardy's inequality as discussed in Remark 3.4. Relation (B.11) is clear from (B.10) with $f(0) = 0$. □

*Remark B.1.3* $(i)$ If $f \in AC_{loc}((0,r))$ and $f' \in L^p((0,r); dx)$ for some $p \in [1,\infty)$, the Hölder estimate analogous to (B.10),

$$\begin{aligned} |f(d) - f(c)| = \left| \int_c^d dt\, f'(t) \right| &\leq |d-c|^{1/p'} \left( \int_c^d dt\, |f'(t)|^p \right)^{1/p}, \\ &(c,d) \subset (0,r), \quad \frac{1}{p} + \frac{1}{p'} = 1, \end{aligned} \tag{B.13}$$

implies the existence of $\lim_{c\downarrow 0} f(c) = f(0)$ and $\lim_{d\uparrow r} f(d) = f(r)$ and hence yields $f \in AC([0,r])$.

$(ii)$ The fact that $f \in H^1((0,r))$ and $\int_0^r dx\, |f(x)|^2/x^2 < \infty$ implies $f\widetilde{\chi}_{[0,r/2]} \in H_0^1((0,r))$ is a special case of a multi-dimensional result recorded, for instance, in [19, Theorem 5.3.4].

$(iii)$ When replacing $x^{-2}$, $x \in (0,r)$, by $[\sin(x)]^{-2}$, $x \in (0,\pi)$, due to locality, the considerations in Lemmas B.1.1 and B.1.2 at the left endpoint $x = 0$ apply of course to the right interval endpoint $\pi$. ◇

**Acknowledgments** We are indebted to Jan Derezinski, Aleksey Kostenko, Ari Laptev, and Gerald Teschl for very helpful discussions.

## References


1. M. Abramowitz, I.A. Stegun, *Handbook of Mathematical Functions* (Dover, New York, 1972)
2. N.I. Akhiezer, I.M. Glazman, *Theory of Linear Operators in Hilbert Space*, vol. II (Pitman, Boston, 1981)
3. V.S. Alekseeva, A.Y. Ananieva, On extensions of the Bessel operator on a finite interval and a half-line. J. Math. Sci. **187**, 1–8 (2012)
4. A.Y. Anan'eva, V.S. Budyka, On the spectral theory of the Bessel operator on a finite interval and the half-line. Diff. Equ. **52**, 1517–1522 (2016)
5. A.Y. Ananieva, V.S. Budyika, To the spectral theory of the Bessel operator on finite interval and half-line. J. Math. Sci. **211**, 624–645 (2015)
6. F.G. Avkhadiev, Brezis–Marcus problem and its generalizations. J. Math. Sci. **252**, 291–301 (2021)
7. F.G. Avkhadiev, K.J. Wirths, Unified Poincaré and Hardy inequalities with sharp constants for convex domains. Angew. Math. Mech. **87**, 632–642 (2007)
8. F.G. Avkhadiev, K.J. Wirths, Sharp Hardy-type inequalities with Lamb's constant. Bull. Belg. Math. Soc. Simon Stevin **18**, 723–736 (2011)
9. A.A. Balinsky, W.D. Evans, R.T. Lewis, *The Analysis and Geometry of Hardy's Inequality* (Universitext, Springer, 2015)
10. H. Brezis, M. Marcus, *Hardy's inequalities revisited.* Ann. Scuola Norm. Sup. Pisa Cl. Sci. (4) **25**, 217–237 (1997)
11. L. Bruneau, J. Dereziński, V. Georgescu, Homogeneous Schrödinger operators on half-line. Ann. H. Poincaré **12**, 547–590 (2011)
12. W. Bulla, F. Gesztesy, *Deficiency indices and singular boundary conditions in quantum mechanics*. J. Math. Phys. **26**, 2520–2528 (1985)
13. R.S. Chisholm, W.N. Everitt, L.L. Littlejohn, An integral operator inequality with applications. J. Inequal. Appl. **3**, 245–266 (1999)
14. E.A. Coddington, N. Levinson, *Theory of Ordinary Differential Equations* (Krieger Publ., Malabar, 1985)
15. E.B. Davies, *Spectral Theory and Differential Operators*. Cambridge Studies in Advanced Mathematics, vol. 42 (Cambridge University Press, Cambridge, 1995)
16. J. Dereziński, V. Georgescu, *On the domains of Bessel operators*. Ann. H. Poincaré **22**, 3291–3309 (2021)
17. J. Dereziński, M. Wrochna, Exactly solvable Schrödinger operators. Ann. H. Poincaré **12**, 397–418 (2011). See also the update at arXiv:1009.0541
18. N. Dunford, J.T. Schwartz, *Linear Operators. Part II: Spectral Theory* (Wiley, Interscience, New York, 1988)
19. D.E. Edmunds, W.D. Evans, *Spectral Theory and Differential Operators*, 2nd edn. (Oxford University Press, Oxford, 2018)
20. W.N. Everitt, H. Kalf, The Bessel differential equation and the Hankel transform. J. Comp. Appl. Math. **208**, 3–19 (2007)
21. S. Flügge, *Practical Quantum Mechanics*, vol. I. Reprinted 2nd 1994 edn. (Springer, Berlin, 1999)
22. G.B. Folland, *Real Analysis. Modern Techniques and Their Applications*, 2nd edn. (Wiley-Interscience, New York, 1999)
23. F. Gesztesy, W. Kirsch, One-dimensional Schrödinger operators with interactions singular on a discrete set. J. Reine Angew. Math. **362**, 28–50 (1985)
24. F. Gesztesy, L. Pittner, On the Friedrichs extension of ordinary differential operators with strongly singular potentials. Acta Phys. Austriaca **51**, 259–268 (1979)

25. F. Gesztesy, M. Ünal, Perturbative oscillation criteria and Hardy-type inequalities. Math. Nachr. **189**, 121–144 (1998)
26. F. Gesztesy, M. Zinchenko, *Sturm–Liouville Operators, Their Spectral Theory, and Some Applications*. In preparation
27. F. Gesztesy, C. Macedo, L. Streit, An exactly solvable periodic Schrödinger operator. J. Phys. **A18**, L503–L507 (1985)
28. F. Gesztesy, L.L. Littlejohn, I. Michael, R. Wellman, On Birman's sequence of Hardy–Rellich-type inequalities, J. Differ. Equ. **264**, 2761–2801 (2018)
29. F. Gesztesy, L.L. Littlejohn, R. Nichols, On self-adjoint boundary conditions for singular Sturm–Liouville operators bounded from below. J. Differ. Equ. **269**, 6448–6491 (2020)
30. F. Gesztesy, L.L. Littlejohn, M. Piorkowski, J. Stanfill, The Jacobi operator and its Weyl–Titchmarsh–Kodaira $m$-functions. In preparation
31. F. Gesztesy, M.M.H. Pang, J. Stanfill, On domain properties of Bessel-type operators. arXiv:2107.09271
32. G.R. Goldstein, J.A. Goldstein, R.M. Mininni, S. Romanelli, Scaling and variants of Hardy's inequality. Proc. Amer. Math. Soc. **147**, 1165–1172 (2019)
33. G.H. Hardy, Notes on some points in the integral calculus, LX. An inequality between integrals. Messenger Math. **54**, 150–156 (1925)
34. G.H. Hardy, J.E. Littlewood, G. Pólya, *Inequalities* (Cambridge University Press, Cambridge, 1988). Reprinted
35. M. Hoffmann-Ostenhof, T. Hoffmann-Ostenhof, A. Laptev, A geometrical version of Hardy's inequality. J. Funct. Anal. **189**, 539–548 (2002)
36. L. Infeld, T.E. Hull, The factorization method. Rev. Mod. Phys. **23**, 21–68 (1951)
37. K. Jörgens, F. Rellich, *Eigenwerttheorie Gewöhnlicher Differentialgleichungen* (Springer, Berlin, 1976)
38. H. Kalf, On the characterization of the Friedrichs extension of ordinary or elliptic differential operators with a strongly singular potential. J. Funct. Anal. **10**, 230–250 (1972)
39. H. Kalf, A characterization of the Friedrichs extension of Sturm–Liouville operators. J. London Math. Soc. (2) **17**, 511–521 (1978)
40. H. Kalf, J. Walter, Strongly singular potentials and essential self-adjointness of singular elliptic operators in $C_0^\infty(\mathbb{R}^n\backslash\{0\})$. J. Funct. Anal. **10**, 114–130 (1972)
41. A. Kostenko, G. Teschl, On the singular Weyl–Titchmarsh function of perturbed spherical Schrödinger operators. J. Differ. Equ. **250**, 3701–3739 (2011)
42. A. Kufner, *Weighted Sobolev Spaces* (A Wiley-Interscience Publication, John Wiley & Sons, Hoboken, 1985)
43. A. Kufner, L. Maligranda, L.-E. Persson, *The Hardy Inequality. About its History and Some Related Results* (Vydavatelský Servis, Pilsen, 2007)
44. A. Kufner, L.-E. Persson, N. Samko, *Weighted Inequalities of Hardy Type*, 2nd edn. (World Scientific, Singapore, 2017)
45. E. Landau, A note on a theorem concerning series of positive terms: extract from a letter of Prof. E. Landau to Prof. I. Schur. J. London Math. Soc. **1**, 38–39 (1926)
46. W. Lotmar, Zur Darstellung des Potentialverlaufs bei zweiatomigen Molekülen. Z. Physik **93**, 528–533 (1935)
47. M.A. Naimark, *Linear Differential Operators. Part II: Linear Differential Operators in Hilbert Space*. Transl. by E.R. Dawson, Engl. translation edited by W.N. Everitt (Ungar Publishing, New York, 1968)
48. R.G. Nasibullin, R.V. Makarov, Hardy's inequalities with remainders and Lamb-type equations. Siberian Math. J. **61**, 1102–1119 (2020)
49. H.-D. Niessen, A. Zettl, Singular Sturm–Liouville problems: the Friedrichs extension and comparison of eigenvalues. Proc. London Math. Soc. (3) **64**, 545–578 (1992)
50. B. Opic, A. Kufner, *Hardy-Type Inequalities*. Pitman Research Notes in Mathematics Series, vol. 219 (Longman Scientific & Technical, Harlow, 1990)

51. F.W.J. Olver, D.W. Lozier, R.F. Boisvert, C.W. Clark (eds.), *NIST Handbook of Mathematical Functions*. National Institute of Standards and Technology (NIST), U.S. Dept. of Commerce (Cambridge University Press, Cambridge, 2010)
52. D.B. Pearson, *Quantum Scattering and Spectral Theory* (Academic Press, London, 1988)
53. L.-E. Persson, S.G. Samko, A note on the best constants in some Hardy inequalities. J. Math. Inequal. **9**, 437–447 (2015)
54. G. Pöschl, E. Teller, Bemerkungen zur Quantenmechanik des anharmonischen Oszillators. Zeitschr. Physik **83**, 143–151 (1933)
55. F. Rellich, Die zulässigen Randbedingungen bei den singulären Eigenwertproblemen der mathematischen Physik. (Gewöhnliche Differentialgleichungen zweiter Ordnung.) Math. Z. **49**, 702–723 (1943/44)
56. F. Rellich, Halbbeschränkte gewöhnliche Differentialoperatoren zweiter Ordnung. Math. Ann. **122**, 343–368 (1951) (German)
57. N. Rosen, P.M. Morse, On the vibrations of polyatomic molecules. Phys. Rev. **42**, 210–217 (1932)
58. R. Rosenberger, A new characterization of the Friedrichs extension of semibounded Sturm–Liouville operators. J. London Math. Soc. (2) **31**, 501–510 (1985)
59. F.L. Scarf, Discrete states for singular potential problems. Phys. Rev. **109**, 2170–2176 (1958)
60. F.L. Scarf, New soluble energy band problem. Phys. Rev. **112**, 1137–1140 (1958)
61. G. Teschl, *Mathematical Methods in Quantum Mechanics. With Applications to Schrödinger Operators*, 2nd edn. Graduate Studies in Mathematics, vol. 157 (American Mathematical Society, Providence, 2014)
62. J. Weidmann, *Linear Operators in Hilbert Spaces*. Graduate Texts in Mathematics, vol. 68 (Springer, New York, 1980)
63. J. Weidmann, *Lineare Operatoren in Hilberträumen. Teil II: Anwendungen* (Teubner, Stuttgart, 2003)
64. A. Zettl, *Sturm–Liouville Theory*. Mathematical Surveys and Monographs, vol. 121 (American Mathematical Society, Providence, 2005)

# Spectral Theory of Exceptional Hermite Polynomials

**David Gómez-Ullate, Yves Grandati, and Robert Milson**

*In recognition of Lance Littlejohn's many contributions to the theory of orthogonal polynomials and in honor of his 70th birthday.*

**Abstract** In this paper we revisit exceptional Hermite polynomials from the point of view of spectral theory, following the work initiated by Lance Littlejohn. Adapting a result of Deift, we provide an alternative proof of the completeness of these polynomial families. In addition, using equivalence of Hermite Wronskians we characterize the possible gap sets for the class of exceptional Hermite polynomials.

**Keywords** Exceptional polynomials · Hermite polynomials · Darboux transformations

## 1 Introduction

Consider a Sturm-Liouville problem (SLP) on $(-\infty, \infty)$:

$$-(Wy')' - Ry = \varepsilon Wy, \tag{1}$$

D. Gómez-Ullate
Escuela Superior de Ingeniería, U. Cádiz, Puerto Real, Spain

Departamento de Física Teórica, U. Complutense, Madrid, Spain
e-mail: david.gomezullate@uca.es

Y. Grandati
L. Physique et Chimie Théoriques, U. de Lorraine, Metz, France
e-mail: yves.grandati@univ-lorraine.fr

R. Milson (✉)
Department of Mathematics and Statistics, Dalhousie U., Halifax, NS, Canada
e-mail: rmilson@dal.ca



F. Gesztesy, A. Martinez-Finkelshtein (eds.), *From Operator Theory to Orthogonal Polynomials, Combinatorics, and Number Theory*, Operator Theory: Advances and Applications 285, https://doi.org/10.1007/978-3-030-75425-9_10

where $W, R = O(e^{-x^2})$ as $|x| \to \infty$. Remarkably, there exist a large class of such eigenvalue problems with polynomial eigenfunctions. Classical Hermite polynomials, which correspond to the case of $W = e^{-x^2}, R = 0$, are just one, very particular, example. Families of polynomials that arise as eigenfunctions of such SLP are called exceptional Hermite orthogonal polynomials.

Being the eigenfunctions of an SLP, exceptional Hermite polynomials are complete families of orthogonal polynomials with weight $W(x) > 0,\ x \in (-\infty, \infty)$ [14, 15]. However, unlike the classical Hermite polynomials, the degree sequence of an exceptional family has a finite number of gaps in the degree sequence; there is a finite number of so-called exceptional degrees for which there is no corresponding eigenpolynomial.

The well-known Bochner theorem [1] asserts that if $y_n(x),\ n = 0, 1, 2, \ldots$ with $\deg y_n = n$ is a family of polynomials that satisfies a second-order eigenvalue equation

$$py_n'' + qy_n' + ry_n = \varepsilon_n y,$$

then necessarily $p(x), q(x)$ are polynomials with $\deg p \leq 2, \deg q \leq 1$ and $r$ is a constant. Multiplying (1) by $-W^{-1}$ gives

$$y'' + qy' + ry = -\varepsilon y \tag{2}$$

where

$$q = W^{-1}W', \quad r = W^{-1}R.$$

In order for (2) to have infinitely many polynomial eigenfunctions, it is necessary for $q(x), r(x)$ to be rational functions with $\deg q \leq 1$ and $\deg r \leq 0$. To obtain non-classical polynomials, it is necessary for $q$ and $r$ to have poles. Bochner's theorem then implies that such a non-classical family of polynomial eigenfunctions must have gaps in the degree sequence.

If we assume that there is at most a finite number of such gaps, then it is possible to show (see [11, 15]) that, necessarily, (2) takes the form

$$y'' - 2\left(x + \frac{\eta'}{\eta}\right)y' + \left(\frac{\eta''}{\eta} + 2x\frac{\eta'}{\eta}\right)y = \varepsilon y, \tag{3}$$

where $\eta(x)$ is a real polynomial without any real zeros. Not every choice of $\eta$ results in an eigenvalue relation (3) with polynomial eigenfunctions. When it does, however, one can show that there are precisely $\deg \eta$ "exceptional" degrees.

Some examples of exceptional polynomials were investigated back in the early 90s [4] but their systematic study started about 10 years ago, where a full classification was given for codimension one [14]. The role of Darboux transformations in the construction process was quickly recognized [17, 31, 32] and the next conceptual step involved the generation of exceptional families by multiple-step or higher order Darboux transformations [18].

Exceptional polynomials appear in mathematical physics as bound states of exactly solvable rational extensions [15, 28, 31] and exact solutions to Dirac's equation [33]. They appear also in connection with super-integrable systems [27, 29] and finite-gap potentials [21]. From a mathematical point of view, the main results are concerned with the full classification of exceptional orthogonal polynomials [11], properties of their zeros [19, 22, 24], and recurrence relations [7, 20]. Typically, exceptional polynomial families are constructed as certain Wronskians where all but one of the arguments is fixed. An interesting alternative due to A.J. Dúran begins with a construction of exceptional discrete polynomials in terms of a Casoratian determinant, and then recovers the continuous exceptional families via a limit process [5, 6, 8]. This approach highlights the bispectral nature of exceptional polynomials [9, 23] and also establishes a connection to Krall polynomials [5, 6, 8]. An entirely different approach, based on integral operators and isospectral deformation, was recently used to construct a novel class of exceptional Legendre polynomials [12].

Lance Littlejohn and collaborators wrote a series of papers analyzing the spectral-theoretic properties of 1-step exceptional operators [25, 26]. It is the ambition of the present work to extend this type of analysis to the class of multi-step exceptional Hermite operators. The primary result of this paper is the merger of the approach of Littlejohn with the spectral-theoretic characterization of Darboux transformations obtained by Deift [3] (see [13] for a further extension to the double-commutator method) to provide a novel demonstration of the completeness of the exceptional polynomials. An additional result involves the characterization the possible gap sets of families of exceptional Hermite polynomial in terms the corresponding partition, recovering the combinatorial formula in [5]. Finally the technical results on factorization chains in Sect. 3.1 and the norming constant identity in Sect. 3.2 may also be of interest.

## 2 Some Spectral Theory

Let $\mathcal{R} \subset \mathbb{R}[x]$ be the set of real-valued polynomials that have no real zeros. Consider a SLP on $(-\infty, \infty)$ of the form

$$-(W_\eta y')' - R_\eta y = \varepsilon W_\eta y, \qquad \eta \in \mathcal{R}, \tag{4}$$

where

$$W_\eta := \eta^{-2} e^{-x^2}, \tag{5}$$

$$R_\eta := \eta^{-3}(\eta'' + 2x\eta')e^{-x^2} \tag{6}$$

In this section we will consider Darboux transformations of self-adjoint operators corresponding to SLP belonging to the class shown in (4).

Fix $\xi, \eta \in \mathcal{R}$ and set

$$\tilde{\eta} := \eta e^{x^2}, \tag{7}$$

$$\hat{\eta} := \eta^{-1} e^{-\frac{x^2}{2}}, \tag{8}$$

with $\tilde{\xi}, \hat{\xi}$ defined analogously.

Next, introduce the bilinear differential expression

$$\chi(f, g) = fg'' - 2f'g' + f''g - 2x(fg' - f'g). \tag{9}$$

and define the following first, and second order differential expressions

$$\alpha_{\xi,\eta} y = \eta^{-1} \operatorname{Wr}(\xi, y) \tag{10}$$

$$\beta_{\eta,\xi} = \alpha_{\tilde{\eta},\tilde{\xi}} \tag{11}$$

$$\tau_\eta y = \eta^{-1} \chi(\eta, y). \tag{12}$$

with $\tau_\xi$ defined analogously. Note that $\tau_\eta$ is precisely the second-order differential expression in the left-side of (3).

A direct calculation shows that

$$(\alpha_{\xi,\eta} f) g W_\xi + f(\beta_{\eta,\xi} g) W_\eta = (\hat{\eta} f \, \hat{\xi} g)', \tag{13}$$

with $W_\eta$, $W_\xi$ as defined in (5). Consequently, $\alpha_{\xi,\eta}$ and $-\beta_{\eta,\xi}$ are formally adjoint with respect to weights $W_\xi$, $W_\eta$. To be precise, if $I \subset \mathbb{R}$ is a compact interval, and $f, g$ sufficiently smooth functions defined on $I$, we have

$$\int_I (\alpha_{\xi,\eta} f) \bar{g} \, W_\xi + \int_I f(\beta_{\eta,\xi} \bar{g}), W_\eta = \hat{\eta} f \, \hat{\xi} \bar{g} \Big|_I \tag{14}$$

Let us also recall Lagrange's identity

$$(\tau_\eta f) g \, W_\eta - f(\tau_\eta g) \, W_\eta = W_\eta \operatorname{Wr}(f, g).$$

Consequently, $\tau_\eta$ is symmetric with respect to $W_\eta$ in the sense that

$$\int_I (\tau_\eta f) \bar{g} W_\eta - \int_I f(\tau_\eta \bar{g}) \, W_\eta = W_\eta \operatorname{Wr}(f, \bar{g}) \Big|_I, \tag{15}$$

An analogous relation holds for $\tau_\xi$ and $W_\xi$.

Let $(\mathcal{H}_\eta, \langle \cdot, \cdot \rangle_\eta) = \mathrm{L}^2(\mathbb{R}, W_\eta)$ be the Hilbert space of square-integrable complex-valued functions defined on $(-\infty, \infty)$, with $\mathcal{H}_\xi$ defined analogously. Let $\mathcal{D}_0$ denote the vector space of smooth, functions with compact support, and let

$$\mathcal{D}_\xi := \{f \in C^\infty(\mathbb{R})\colon \ \mathrm{Wr}(f, \xi)\,(x) \equiv 0 \text{ for } |x| \text{ sufficiently large}\}. \tag{16}$$

In other words, an element of $\mathcal{D}_\xi$ behaves like a multiple of $\xi(x)$ outside a compact interval, but with the constant of proportionality for large $x$ not necessarily equal to the constant of proportionality for small $x$. It is well known that $\mathcal{D}_0$ is dense in $\mathcal{H}_\eta$ and $\mathcal{H}_\xi$. Since elements of $\mathcal{D}_\xi$ are square integrable with respect to $W_\eta$ and since $\mathcal{D}_0 \subset \mathcal{D}_\xi$, the latter is also a dense subspace.

Let $a_{\xi,\eta}\colon \mathcal{H}_\eta \to \mathcal{H}_\xi$, $b_{\eta,\xi}\colon \mathcal{H}_\xi \to \mathcal{H}_\eta$ denote densely defined first-order operators with action $\alpha_{\xi,\eta}$, $\beta_{\eta,\xi}$, respectively, and with

$$\operatorname{Dom} a_{\xi,\eta} = \mathcal{D}_\xi, \qquad \operatorname{Dom} b_{\eta,\xi} = \mathcal{D}_0.$$

Since $\alpha_{\xi,\eta}\xi \equiv 0$, it follows that $\operatorname{Ran} a_{\xi,\eta} = \mathcal{D}_0$. For same reason and by (14), we have $\operatorname{Ran} b_{\eta,\xi} \subset \operatorname{Ann}_\eta \xi$, where

$$\operatorname{Ann}_\eta \xi = \{f \in \mathcal{H}_\eta\colon \ \langle f, \xi\rangle_\eta = 0.\}. \tag{17}$$

Next, we strengthen these assertions as follows.

**Proposition 2.1** *We have* $\overline{\operatorname{Ran} a_{\xi,\eta}} = \mathcal{H}_\xi$.

***Proof*** Since $\mathcal{D}_0$ is dense in $\mathcal{H}_\xi$, it suffices to show that $\mathcal{D}_0 \subset \operatorname{Ran} a_{\xi,\eta}$. Let $h \in \mathcal{D}_0$ be given. Set

$$f(x) = \xi(x) \int_{-\infty}^{x} h\eta\xi^{-2}, \quad x \in \mathbb{R}.$$

Since $h \in \mathcal{D}_0$, for $x$ sufficiently small, we have $f(x) = 0$. For $x$ sufficiently large, we have $f(x) = C\xi(x)$ where $C = \int_{\mathbb{R}} h\eta\xi^{-2}$. Hence, $f \in \mathcal{D}_\xi$. By (10) we have,

$$\alpha_{\xi,\eta} f = \eta^{-1}\xi^2 \left(\xi^{-1} f\right)' = h.$$

□

**Proposition 2.2** *We have* $\overline{\operatorname{Ran} b_{\eta,\xi}} = \operatorname{Ann}_\eta \xi$.

***Proof*** Let $\{f_n \in \mathcal{D}_0\}_{n\in\mathbb{N}}$ be a sequence such that $\beta_{\eta,\xi} f_n \xrightarrow{n} h$ for some $h \in \mathcal{H}_\eta$. Since $\alpha_{\xi,\eta}\xi = 0$, by (13), we have

$$\left\langle \beta_{\eta,\xi} f_n, \xi \right\rangle_\eta = 0, \quad n \in \mathbb{N}.$$

Moreover, since

$$\langle h, \xi \rangle_\eta = \left\langle h - \beta_{\eta,\xi} f_n, \xi \right\rangle_\eta \xrightarrow{n} 0,$$

we have $h \in \operatorname{Ann}_\eta \xi$. Hence, $\overline{\operatorname{Ran} b_{\eta,\xi}} \subseteq \operatorname{Ann}_\eta \xi$.

We now prove the converse. Let $h \in \operatorname{Ann}_\eta \xi$ be given. Let $\{h_n \in \mathcal{D}_0\}_{n \in \mathbb{N}}$ be a sequence such that $\|h_n - h\| \xrightarrow{n} 0$. Choose a $p \in \mathcal{D}_0$ such that $\langle p, \xi \rangle \neq 0$, and set

$$g_n = h_n - \frac{\langle h_n, \xi \rangle_\eta}{\langle p, \xi \rangle_\eta} p \in \mathcal{D}_0, \quad n \in \mathbb{N}.$$

By construction, $g_n \in \operatorname{Ann}_\eta \xi,\ n \in \mathbb{N}$. By assumption, $\langle h_n, \xi \rangle_\eta \xrightarrow{n} 0$. Hence,

$$\|g_n - h\|_\eta \leq \|h_n - h\|_\eta + \left| \frac{\langle h_n, \xi \rangle_\eta}{\langle p, \xi \rangle_\eta} \right| \|p\|_\eta \xrightarrow{n} 0.$$

For $n \in \mathbb{N}$, set

$$f_n(x) = \tilde{\eta}(x) \int_{-\infty}^{x} g_n \tilde{\xi} \tilde{\eta}^{-2} = \tilde{\eta}(x) \int_{-\infty}^{x} g_n \xi W_\eta, \quad x \in \mathbb{R}.$$

Since $g_n$ has compact support, $f_n(x) = 0$ for $x$ sufficiently small. Since $g_n \in \operatorname{Ann}_\eta \xi$, we also have $f_n(x) = 0$ for $x$ sufficiently large. Hence, $f_n \in \mathcal{D}_0$. Observe that

$$\beta_{\eta,\xi} f_n = \tilde{\xi}^{-1} \tilde{\eta}^2 (\tilde{\eta}^{-1} f_n)' = g_n.$$

Hence $g_n \in \operatorname{Ran} b_{\eta,\xi},\ n \in \mathbb{N}$. We already showed that $g_n \xrightarrow{n} h$. This proves that

$$\operatorname{Ann}_\eta \xi \subseteq \overline{\operatorname{Ran} b_{\eta,\xi}}.$$

□

Let $A_{\xi,\eta}, B_{\eta,\xi}$ denote the maximal extensions of $a_{\xi,\eta}, b_{\eta,\xi}$, respectively. Formally,

$$\begin{aligned} \operatorname{Dom} A_{\xi,\eta} &= \{f \in \mathcal{H}_\eta : f \in \mathrm{AC}_{\mathrm{loc}}(\mathbb{R}); \alpha_{\xi,\eta} f \in \mathcal{H}_\xi\}, \\ \operatorname{Dom} B_{\eta,\xi} &= \{g \in \mathcal{H}_\xi : g \in \mathrm{AC}_{\mathrm{loc}}(\mathbb{R}); \beta_{\eta,\xi} g \in \mathcal{H}_\eta\}. \end{aligned} \tag{18}$$

**Proposition 2.3** *For $f \in \operatorname{Dom} A_{\xi,\eta}$ and $g \in \operatorname{Dom} B_{\eta,\xi}$, we have*

$$\left\langle A_{\xi,\eta} f, g \right\rangle_\xi = \left\langle f, B_{\eta,\xi} g \right\rangle_\eta. \tag{19}$$

***Proof*** By (14), we have

$$\hat{\eta} f \hat{\xi} g \Big|_{-t}^{t} \to \langle \alpha_{\xi,\eta} f, g \rangle_{\eta} - \langle f, \beta_{\eta,\xi} g \rangle_{\xi} \quad \text{as } t \to \infty,$$

with $\hat{\eta}, \hat{\xi}$ as per (8). Observe that

$$(\hat{\eta} f)^2 = f^2 W_{\eta}, \quad (\hat{\xi} g)^2 = g^2 W_{\xi}.$$

Hence, $\hat{\eta} f, \ \hat{\xi} g \in L^2(\mathbb{R})$. Therefore, the above limit must be zero. □

**Proposition 2.4** *We have* $b^*_{\eta,\xi} \subseteq -A_{\xi,\eta}$ *and* $a^*_{\xi,\eta} \subseteq -B_{\eta,\xi}$.

***Proof*** Let $f \in \operatorname{Dom} b^*_{\eta,\xi} \subset \mathcal{H}_{\eta}$. By definition, there exists an $h \in \mathcal{H}_{\xi}$ such that

$$\langle f, \beta_{\eta,\xi} g \rangle_{\eta} = \langle h, g \rangle_{\xi}, \quad \text{for all } g \in \mathcal{D}_0.$$

Set

$$\tilde{f}(x) := \xi \int_0^x h \eta \xi^{-2}, \quad x \in \mathbb{R}.$$

Hence, by (10), we have $\alpha_{\xi,\eta}[\tilde{f}] = h$ almost everywhere. By (14), it then follows that

$$\int_{\mathbb{R}} \tilde{f} \, (\beta_{\eta,\xi} \bar{g}) \, W_{\eta} = \langle h, g \rangle_{\xi}, \quad \text{for all } g \in \mathcal{D}_0.$$

Hence,

$$\int_{\mathbb{R}} (f - \tilde{f})(\beta_{\eta,\xi} \bar{g}) W_{\eta} = 0 \text{ for all } g \in \mathcal{D}_0.$$

Now, $f - \tilde{f} \in L^2(I, W_{\eta})$ for every compact interval $I \subset \mathbb{R}$. Hence, by Proposition 2.2, $f$ and $\tilde{f}$ differ by a multiple of $\xi$ a.e. on $I$. Since this is true for every $I$, it follows that $f - \tilde{f}$ is a multiple of $\xi$ a.e. on all of $\mathbb{R}$. Therefore, $f \in \operatorname{Dom} A_{\xi,\eta}$ with $\alpha_{\xi,\eta} f = h$.

Next, let $f \in \operatorname{Dom} a^*_{\xi,\eta}$. By definition, there exists an $h \in \mathcal{H}_{\eta}$ such that

$$\langle f, \alpha_{\xi,\eta} g \rangle_{\xi} = \langle h, g \rangle_{\eta}, \quad \text{for all } g \in \mathcal{D}_{\xi} \subset \mathcal{H}_{\eta}.$$

If $g = \xi$, then the above expressions vanish. Hence, $h \in \operatorname{Ann}_{\eta} \xi$, which implies that

$$\tilde{f}(x) := \tilde{\eta}(x) \int_{-\infty}^x h \xi W_{\eta}, \quad x \in \mathbb{R}$$

is well defined. By construction, $\beta_{\eta,\xi}\tilde{f} = h$ a.e. It follows that

$$\int_{\mathbb{R}} \tilde{f}(\alpha_{\xi,\eta}\bar{g})\, W_\xi = \langle h, g\rangle_\eta\,, \quad \text{for all } g \in \mathcal{D}_\xi\,.$$

Hence,

$$\int_{\mathbb{R}} (f - \tilde{f})(\alpha_{\xi,\eta}\bar{g})\, W_\xi = 0 \text{ for all } g \in \mathcal{D}_\xi\,.$$

Therefore, by Proposition 2.1, $f = \tilde{f}$ a.e. on $\mathbb{R}$. □

**Proposition 2.5** *We have* $A_{\xi,\eta} = -B^*_{\eta,\xi}$ *and* $-B_{\eta,\xi} = A^*_{\xi,\eta}$.

***Proof*** By Proposition 2.3, we have $A_{\xi,\eta} \subseteq -B^*_{\eta,\xi}$ and $-B_{\eta,\xi} \subseteq A^*_{\xi,\eta}$. By Proposition 2.4,

$$-B^*_{\eta,\xi} \subseteq -b^*_{\eta,\xi} \subseteq A_{\xi,\eta}, \quad \text{and} \quad A^*_{\xi,\eta} \subseteq a^*_{\xi,\eta} \subseteq -B_{\eta,\xi}\,,$$

as was to be shown. □

Let $T_\eta : \mathcal{H}_\eta \to \mathcal{H}_\eta$ denote the densely defined operator with action $\tau_\eta$ and with maximal domain

$$\mathrm{Dom}(T_\eta) = \{f \in \mathrm{AC}^1_{\mathrm{loc}}(\mathbb{R})\colon \tau_\eta f \in \mathcal{H}_\eta\} \tag{20}$$

Let $T_\xi : \mathcal{H}_\xi \to \mathcal{H}_\xi$ be the analogous operator with action $\tau_\xi$.

**Proposition 2.6** *Let* $\eta, \xi \in \mathcal{R}$ *be such that*

$$\chi(\eta, \xi) = \varepsilon_0 \eta\xi, \quad \varepsilon_0 \in \mathbb{R}. \tag{21}$$

*Then,*

$$\begin{aligned} T_\eta &= B_{\eta,\xi} A_{\xi,\eta} + \varepsilon_0, \\ T_\xi &= A_{\xi,\eta} B_{\eta,\xi} + \varepsilon_0 + 2 \end{aligned} \tag{22}$$

*in the the sense of naturally defined composition; i.e.*

$$\mathrm{Dom}\, B_{\eta,\xi} A_{\xi,\eta} = \{f \in \mathrm{Dom}\, A_{\xi,\eta} \colon A_{\xi,\eta} f \in \mathrm{Dom}\, B_{\eta,\xi}\};$$
$$\mathrm{Dom}\, A_{\xi,\eta} B_{\eta,\xi} = \{f \in \mathrm{Dom}\, B_{\eta,\xi} \colon B_{\eta,\xi} f \in \mathrm{Dom}\, A_{\xi,\eta}\}.$$

**Lemma 2.1** *Relation* (21) *is equivalent to either of the following:*

$$\tau_\eta \xi = \varepsilon_0 \xi, \tag{23}$$

$$\tau_\xi \tilde{\eta} = (\varepsilon_0 + 2)\tilde{\eta}. \tag{24}$$

***Proof*** The equivalence of (21) and (23) follows directly from the definition (12). The following differential relation can be verified by direct calculation:

$$\chi(\xi, \tilde{\eta}) = e^{x^2} \chi(\eta, \xi) + 2\xi\tilde{\eta} \tag{25}$$

Hence,

$$\tilde{\eta}\tau_\eta\xi = e^{x^2} \chi(\eta, \xi) = \chi(\xi, \tilde{\eta}) - 2\xi\tilde{\eta} = \xi(\tau_\xi - 2)\tilde{\eta}.$$

This proves the equivalence of (23) and (24). □

**Lemma 2.2** *Suppose that* (21) *holds. Then,*

$$\tau_\eta = \beta_{\eta,\xi}\alpha_{\xi,\eta} + \varepsilon_0, \tag{26}$$

$$\tau_\xi = \alpha_{\xi,\eta}\beta_{\eta,\xi} + \varepsilon_0 + 2 \tag{27}$$

***Proof*** By definition and direct calculation,

$$\chi(\eta, f)\,\xi - \chi(\eta, \xi)\, f = \tilde{\eta}^{-1}\,\mathrm{Wr}(\eta\tilde{\eta}, \mathrm{Wr}(\xi, f)) \tag{28}$$

Hence, by (10) - (12),

$$\begin{aligned}\tau_\eta f &= (\tilde{\xi}\eta^2)^{-1}\,\mathrm{Wr}(\eta\tilde{\eta}, \mathrm{Wr}(\xi, f)) + \xi^{-1} f\tau_\eta\xi \\ &= \tilde{\xi}^{-1}\,\mathrm{Wr}\Big(\tilde{\eta}, \eta^{-1}\,\mathrm{Wr}(\xi, f)\Big) + \varepsilon_0 f \\ &= \beta_{\eta,\xi}\alpha_{\xi,\eta} f + \varepsilon_0 f\end{aligned}$$

By definition and direct calculation,

$$\begin{aligned}\alpha_{\tilde{\eta},\xi} f &= e^{x^2}\beta_{\eta,\xi} f, \\ \beta_{\xi,\tilde{\eta}} f &= e^{-2x^2}\eta^{-1}\,\mathrm{Wr}\Big(e^{x^2}\xi, f\Big) = \eta^{-1}\,\mathrm{Wr}\Big(\xi, e^{-x^2} f\Big) = \alpha_{\xi,\eta}(e^{-x^2} f)\end{aligned}$$

Applying (26) with $\eta \mapsto \xi,\ \xi \mapsto \tilde{\eta},\ \varepsilon_0 \mapsto \varepsilon_0 + 2$, and using (24) gives

$$\tau_\xi = \beta_{\xi,\tilde{\eta}}\alpha_{\tilde{\eta},\xi} + \varepsilon_0 + 2 = \alpha_{\xi,\eta}\beta_{\eta,\xi} + \varepsilon_0 + 2.$$

□

***Proof*** By Lemma 2.2, the formal factorization relations (26) and (27) hold. The differential expressions $\tau_\eta, \tau_\xi$ are limit point at $\pm\infty$; see [2, Theorem 2.4]. Hence, $T_\eta, T_\xi$ are self-adjoint. By Proposition 2.5, $A_{\xi,\eta}, -B_{\eta,\xi}$ are adjoint. Hence, the natural compositions $A_{\xi,\eta}B_{\eta,\xi}$ and $B_{\eta,\xi}A_{\xi,\eta}$ are self-adjoint by a standard theorem in functional analysis; see for example [30, Theorem 13.13]. Since there are no boundary conditions at $\pm\infty$ the LHS and the RHS of (22) must coincide. □

**Proposition 2.7** *Let $\eta, \xi \in \mathcal{R}$ such that* (21) *holds. Let $\psi \in \mathcal{H}_\eta$ be an eigenfunction of $T_\eta$; i.e. $\tau_\eta \psi = \varepsilon \psi$, $\varepsilon \in \mathbb{R}$. Then, necessarily $\varepsilon_0 > \varepsilon$ and*

$$\hat{\psi} := \alpha_{\xi,\eta} \psi$$

*is an eigenfunction of $T_\xi$ with eigenvalue $\varepsilon + 2$. Moreover,*

$$\|\hat{\psi}\|_\xi^2 = (\varepsilon_0 - \varepsilon)\|\psi\|_\eta^2 \tag{29}$$

***Proof*** By assumption $\psi \in \operatorname{Dom} T_\eta$. Hence, by Proposition 2.6, $\psi \in \operatorname{Dom} A_{\xi,\eta}$, and $\hat{\psi} \in H_\xi$. By (19) and Proposition 2.6, we have

$$\begin{aligned}\left\langle \hat{\psi}, \hat{\psi} \right\rangle_\xi &= -\left\langle B_{\eta,\xi} A_{\xi,\eta} \psi, \psi \right\rangle_\eta \\ &= \left\langle (\varepsilon_0 - T_\eta)\psi, \psi \right\rangle_\eta = (\varepsilon_0 - \varepsilon) \langle \psi, \psi \rangle_\eta .\end{aligned}$$

By construction,

$$\begin{aligned}\beta_{\eta,\xi} \hat{\psi} &= \beta_{\eta,\xi} \alpha_{\xi,\eta} \psi = (\varepsilon - \varepsilon_0)\psi \in \mathcal{H}_\eta, \\ \alpha_{\xi,\eta} \beta_{\eta,\xi} \hat{\psi} &= (\varepsilon - \varepsilon_0)\hat{\psi} \in \mathcal{H}_\xi .\end{aligned}$$

Hence, $\hat{\psi} \in \operatorname{Dom} B_{\eta,\xi}$ and $\beta_{\eta,\xi}\hat{\psi} \in \operatorname{Dom} A_{\xi,\eta}$. It follows that, $\hat{\psi} \in \operatorname{Dom} T_\xi$, with

$$T_\xi \hat{\psi} = (A_{\xi,\eta} B_{\eta,\xi} + \varepsilon_0 + 2)\hat{\psi} = (\varepsilon + 2)\hat{\psi}.$$

□

**Theorem 2.1** *Let $\eta, \xi \in \mathcal{R}$ be such that* (21) *holds. Let $\sigma(T_\eta), \sigma(T_\xi)$ denote the spectral sets of the indicated operators. Then, $\varepsilon_0 = \max \sigma(T_\eta)$. Moreover,*

$$\sigma(T_\xi - 2) = \sigma(T_\eta) \setminus \{\varepsilon_0\}.$$

***Proof*** Both $T_\eta$ and $T_\xi$ have a pure point spectrum; for a proof see Sect. 5.9 of [34] or Problem 1 in Ch. 9 of [2]. By assumption, $\psi_0 := \xi$ has no real zeros. Relation (21) is equivalent to

$$\tau_\eta \psi_0 = \varepsilon_0 \psi_0.$$

Hence, $\varepsilon_0 = \max \sigma(T_\eta)$ by the Sturm oscillation theorem. Let $\psi_n \in \mathcal{H}_\eta$, $n \in \mathbb{N}$ be the other eigenfunctions of $T_\eta$ with

$$\tau_\eta \psi_n = \varepsilon_n \psi_n, \quad n = 1, 2, \ldots$$

and $\varepsilon_0 > \varepsilon_1 > \varepsilon_2 > \cdots$. Set

$$\hat{\psi}_n = \alpha_{\xi,\eta}\psi_n, \quad n = 1, 2, \ldots.$$

By Proposition 2.7, these are all eigenfunctions of $T_\xi$ with eigenvalues $\varepsilon_n$, $n = 1, 2, \ldots$. By Proposition 2.1, Ran $A_{\xi,\eta} = \mathcal{H}_\xi$. Since $\{\psi_n\}_{n\in\mathbb{N}_0}$ is complete in $\mathcal{H}_\eta$, it follows that $\{\hat{\psi}_n\}_{n\in\mathbb{N}}$ is complete in $\mathcal{H}_\xi$. Therefore $\sigma(T_\xi) = \{\varepsilon_n\}_{n\in\mathbb{N}}$. □

Deift [3] proved a more general version of Theorem 2.1 regarding the spectra of general self-adjoint operators related by a factorization/Darboux transformation. Nonetheless, for the sake of comprehensiveness we prefer to state and prove a more restrictive version limited to Sturm-Liouville operators. The above proof of Theorem 2.1 is more accessible than Deift's more abstract argument.

# 3 The Formal Theory of Exceptional Hermite Polynomials

Classical Hermite polynomials are orthogonal polynomials defined by the recurrence relation

$$H_0 = 1, \quad xH_n = \frac{1}{2}H_{n+1} + nH_{n-1}, \quad n = 1, 2, \ldots \tag{30}$$

They satisfy the following orthogonality relation:

$$\int_{-\infty}^{\infty} H_m(x)H_n(x)e^{-x^2}dx = \sqrt{\pi}\, 2^n n!\delta_{n,m} \tag{31}$$

Hermite polynomials are known as classical orthogonal polynomials, because they also arise as solutions of the Hermite differential equation

$$y'' - 2xy' = \varepsilon y \tag{32}$$

with $y = H_n$, $\varepsilon = -2n$, $n \in \mathbb{N}_0$. Note that multiplication of (32) by $-e^{-x^2}$ gives a singular Sturm-Liouville problem on $(-\infty, \infty)$, namely

$$-(e^{-x^2}y)' = -\varepsilon e^{-x^2}y.$$

This SLP is limit-point at $\pm\infty$, so no explicit boundary conditions are required.

Exceptional Hermite polynomials are a generalization of the classical Hermite polynomials because they satisfy a second-order, Hermite-like differential equation. Each family of such polynomials is indexed by a partition, and so we begin by recalling some relevant definitions.

**Definition 3.1** A partition $\lambda$ of a natural number $N$ is a non-increasing, finitely supported sequence of non-negative integers $\lambda_1 \geq \lambda_2 \geq \ldots \geq 0$ that sum to $N$. The length $\ell$ of a partition $\lambda$ is defined to be the smallest $\ell \in \mathbb{N}_0$ such that $\lambda_{\ell+1} = 0$. Thus, $N = \lambda_1 + \cdots + \lambda_\ell$, with $\lambda_\ell > 0$.

Fix a partition $\lambda$, and let

$$C_{\lambda,l} := 2^{l(l-1)/2} \prod_{1 \leq i < j \leq l} (\lambda_i - \lambda_j + j - i) \tag{33}$$

$$\pi_{\lambda,l}(n) := \prod_{i=1}^{l} (n - N - \lambda_i + i), \quad k \in \mathbb{N}, \tag{34}$$

$$m_i := \lambda_i + \ell - i, \quad i \in \mathbb{N}; \tag{35}$$

$$\eta_\lambda := (C_\ell^\lambda)^{-1} \operatorname{Wr}\big(H_{m_\ell}, \ldots, H_{m_1}\big), \tag{36}$$

$$I_\lambda := \{n \in \mathbb{N}_0 \colon n - N + \ell \geq 0 \text{ and } \pi_{\lambda,\ell}(n) \neq 0.\}, \tag{37}$$

$$H_{\lambda,n} := (2^\ell C_\ell^\lambda \pi_{\lambda,\ell}(n))^{-1} \operatorname{Wr}\big(H_{m_\ell}, \ldots, H_{m_1}, H_{n-N+\ell}\big), \quad n \in I_\lambda \tag{38}$$

where Wr denotes the usual Wronskian determinant.

Observe that $n - N + \ell \in \{m_1, \ldots, m_\ell\}$ if and only if $\pi_{\lambda,\ell}(n) = 0$. Thus, $H_{\lambda,n}$, $n \in I_\lambda$ is a non-zero Wronskian of classical Hermite polynomials. Also note that if $N = 0$ and $\lambda$ is the empty partition, then $\eta_\lambda = 1$ and $H_{\lambda,n} = H_n$, is the classical $n$th degree Hermite polynomial.

**Proposition 3.1** *Let $\eta_\lambda$, $H_{\lambda,n}$, $n \in I_\lambda$ be as above. Then*

$$\eta_\lambda(x) = 2^N x^N + \text{ lower degree terms.} \tag{39}$$

$$H_{\lambda,n}(x) = 2^n x^n + \text{ lower degree terms.} \tag{40}$$

***Proof*** Let $p_i$, $i = 1, \ldots, l$ be a polynomial of degree $d_i$ with leading coefficient $c_i$. Suppose that $d_1, \ldots, d_l$ are distinct and let $P = \operatorname{Wr}(p_1, \ldots, p_l)$. Then

$$\deg P = \sum_{i=1}^{l} d_i - \frac{1}{2} l(l-1) = \sum_{i=1}^{l} (d_i - l + i).$$

Consequently,

$$\deg \eta_\lambda = \sum_{i=1}^{\ell} (m_i - \ell + i) = \sum_i \lambda_i = N,$$

$$\deg H_{\lambda,n} = n - N + \ell - (\ell + 1) + \ell + 1 + \sum_{i=1}^{\ell} (m_i - \ell - 1 + i) = n.$$

The leading coefficient of $P$ is given by $\prod_{i=1}^{k} c_i \prod_{1\leq i<j\leq k}(d_j - d_i)$. It follows that $C_{\lambda,\ell}$ and $2^\ell C_{\lambda,\ell}\pi_{\lambda,\ell}(n)$ are precisely the leading coefficients of $\eta_\lambda$ and $H_{\lambda,n}$, respectively. □

The formulation of $\eta_\lambda$ and $H_{\lambda,n}$ in (36) and (38) is based on Wronskians of size $\ell$ and $\ell+1$ respectively. However, it is possible to define these polynomials using Wronskians of any size $l \geq \ell$.

**Proposition 3.2** *Let $\lambda$ be a partition of $N$ with length $\ell$. Fix $l \geq \ell$, and set*

$$m_{i,l} = \lambda_i + l - i, \quad i = 1, \dots, l.$$

*Then,*

$$\eta_\lambda = C_{\lambda,l}^{-1}\,\mathrm{Wr}\big(H_{m_{l,l}}, \dots, H_{m_{1,l}}\big), \tag{41}$$

$$H_{\lambda,n} = (2^l C_l^\lambda \pi_{\lambda,l}(n))^{-1}\,\mathrm{Wr}\big(H_{m_{l,l}}, \dots, H_{m_{l,1}}, H_{n-N+l}\big) \tag{42}$$

***Proof*** The proof is by induction on $l-\ell$. If $l=\ell$, there is nothing to prove. Suppose that (41) holds for a particular $l \geq \ell$. Observe that

$$m_{i,l+1} = m_{i,l} + 1, \quad i = 1, \dots, l,$$

with $m_{l+1,l+1} = 0$. Since

$$H_n' = 2nH_{n-1}, \quad n = 1, 2, \dots,$$

we have

$$\begin{aligned}\mathrm{Wr}(H_{m_{l+1,l+1}}, H_{m_{l+1,l}}, \dots, H_{m_{l+1,1}}) &= \mathrm{Wr}(1, H_{m_{l,l}+1}, \dots, H_{m_{l,1}+1})\\ &= \mathrm{Wr}(H'_{m_{l,l}+1}, \dots, H'_{m_{l,1}+1})\\ &= 2^l \prod_i (m_{l,i}+1)\,\mathrm{Wr}(H_{m_{l,l}}, \dots, H_{m_{l,1}}).\end{aligned}$$

Observe that

$$\sum_{i=1}^{l+1} m_{l+1,i} = N + l(l+1)/2,$$

the leading term coefficient of the left-side Wronskian is $2^N C_{\lambda,l+1}$, it follows that

$$\mathrm{Wr}(H_{m_{l+1,l+1}}, H_{m_{l+1,l}}, \dots, H_{m_{l+1,1}}) = C_{\lambda,l+1}\eta_\lambda.$$

Relation (42) is proved in an anolgous fashion. □

**Corollary 3.1** *For every $l \geq \ell$ and $n \in I_\lambda$, we have*

$$\frac{H_{\lambda,n}}{\eta_\lambda} = \frac{\operatorname{Wr}\big(H_{m_{l,l}},\dots,H_{m_{1,l}},H_{n-N+l}\big)}{2^l \pi_{\lambda,l}(n)\operatorname{Wr}\big(H_{m_{l,l}},\dots,H_{m_{1,l}}\big)}. \tag{43}$$

In other words, the rational function $\eta_\lambda^{-1} H_{\lambda,n}$ can be defined without invoking the normalizing constant $C_{\lambda,l}$ of (33).

A particularly interesting case of this formulation occurs when we take $l = N$. By (40), the set $I_\lambda$ is simply the degree sequence of the polynomial family $\{H_{\lambda,n}\}_n$. Let

$$k_i := m_{i,N} = \lambda_i + N - i, \quad i = 1,\dots,N, \tag{44}$$

$$K_\lambda := \{k_1,\dots,k_N\}; \tag{45}$$

and observe that

$$K_\lambda = \{0,1,\dots,N-\ell-1\} \cup \{m_i + N - \ell \colon i = 1,\dots,\ell\}. \tag{46}$$

Recall that $n - N + \ell \in \{m_1,\dots,m_\ell\}$ if and only if $\pi_{\lambda,\ell}(n) = 0$. It follows that $K_\lambda = \mathbb{N}_0 \setminus I_\lambda$ is the set of exceptional degrees missing from $I_\lambda$, and that there are precisely $N = \deg \eta_\lambda$ such "exceptional" degrees. Moreover, by (41) and (42), we have

$$\begin{aligned} \eta_\lambda &\propto \operatorname{Wr}\big(H_{k_N},\dots,H_{k_1}\big), \\ H_{\lambda,n} &\propto \operatorname{Wr}\big(H_{k_N},\dots,H_{k_1},H_n\big), \quad n \in I_\lambda. \end{aligned} \tag{47}$$

Thus both $\eta_\lambda$ and the exceptional polynomials $H_{\lambda,n}$ can be formulated as Wronskians involving the exceptional degrees. Of course, for most partitions $N > \ell$ and so the formulations in (36) and (38) are more economical. See [16] for a characterization of the smallest determinant that can be used to represent a Wronskian of Hermite polynomials.

Just like their classical counterparts, exceptional Hermite polynomials satisfy a second-order eigenvalue relation. Set

$$\tau_\lambda y := y'' - 2\left(x + \frac{\eta_\lambda'}{\eta_\lambda}\right) y' + \left(\frac{\eta_\lambda''}{\eta_\lambda} + 2x\frac{\eta_\lambda'}{\eta_\lambda}\right) y; \tag{48}$$

i.e., $\tau_\lambda$ is the second-order differential expression in the left-side of (3) with $\eta = \eta_\lambda$.

**Proposition 3.3** *Let $m_1,\dots,m_l, m \in \mathbb{N}_0$ be distinct non-zero integers. Set*

$$\eta = \operatorname{Wr}\big(H_{m_l},\dots,H_{m_1}\big)$$

$$\xi = \operatorname{Wr}\big(H_{m_l},\dots,H_{m_1},H_m\big)$$

*Then,*

$$\chi(\eta, \xi) = 2(l - m)\eta\xi. \tag{49}$$

***Proof*** Relation (49) is equivalent to

$$(\tau_\eta + 2(m - l))\xi = 0 \tag{50}$$

Set

$$\tilde{\tau}_\eta y = y'' - 2xy' + 2(\log \eta)'' y.$$

An elementary calculation shows that

$$\tau_\eta(\eta y) = \eta \tilde{\tau}_\eta y.$$

Thus, it suffices to show that

$$(\tilde{\tau}_\eta + 2(m - l))(\eta^{-1}\xi) = 0. \tag{51}$$

Let

$$\tau_\emptyset y = y'' - 2xy'$$

be the classic Hermite differential expression. Define the $l$th order differential expression

$$\kappa y = \frac{\operatorname{Wr}(H_{m_l}, \ldots, H_{m_1}, y)}{\operatorname{Wr}(H_{m_l}, \ldots, H_{m_1})} = y^{(l)} - (\log \eta)' y^{(l-1)} + \cdots,$$

and observe that

$$\kappa H_m = \eta^{-1}\xi.$$

Since

$$\tau_\emptyset H_m = -2m H_m,$$

it suffices to show that

$$(\tilde{\tau}_\lambda - 2l)\kappa = \kappa \tau_\emptyset,$$

This, in turn, is equivalent to the relation

$$[\tau_\emptyset, \kappa] = (2l - 2(\log \eta)'')\kappa.$$

By construction, $\ker\kappa = \operatorname{span}\{H_{m_1},\ldots,H_{m_l}\}$. Since these are all eigenfunctions of $\tau_\emptyset$, the commutator $[\tau_\emptyset,\kappa]$ annihilates $\ker\kappa$. By inspection, $[\tau_\emptyset,\kappa]$ has order $l$. Hence, $[\tau_\emptyset,\kappa] = f\kappa$ for some function $f(x)$. Observe that

$$[\tau_\emptyset,\kappa]y = (2l - 2(\log\eta)'')y^{(l)} + \cdots$$

Hence, $f = 2l - 2(\log\eta)''$. This proves (51), which is equivalent to (49). □

**Corollary 3.2** *Let $\lambda$ be a partition and $n \in I_\lambda$ an allowed degree. Then,*

$$\chi\left(\eta_\lambda, H_{\lambda,n}\right) = 2(N-n)\eta_\lambda H_{\lambda,n}. \tag{52}$$

***Proof*** Since $\chi$ is bilinear, this follows by (47) and the preceding Proposition. □

The following converse was proved in [15].

**Proposition 3.4** *Let $\eta(x)$ be a non-zero polynomial. Let $\{y_n\colon \deg y = n\}_{n\in I}$ be the family of polynomial solutions of*

$$\chi(\eta, y) = \varepsilon\eta y \quad \varepsilon\in\mathbb{C}.$$

*Suppose that $\mathbb{N}_0 \setminus I$ is finite; i.e., the family of polynomial solutions is missing at most finitely many degrees. Then, there exists a partition $\lambda$ such that $I = I_\lambda$, and such that, up to a multiplicative constant, $\eta = \eta_\lambda$ and $y_n = H_{\lambda,n},\ n\in I_\lambda$.*

### 3.1 Multi-Step Factorization Chains

In this section, we show that every exceptional operator $T_\lambda$ is connected to the classical

$$\tau_\emptyset y = y'' - 2xy'$$

by a finite factorization chain of exceptional Hermite operators. We will describe two such chains: one will connect an exceptional Hermite operator $\tau_\lambda$ to $\tau_\emptyset$ and the other will connect $\tau_\emptyset$ to $\tau_\lambda$.

Let $\lambda$ be a partition of $N$. Let $k_1,\ldots,k_N \in K_\lambda$ be the exceptional degrees as per (44), and let $n_1,\ldots,n_l \in I_\lambda$ be a list of allowed degrees. Set

$$\eta_i = \operatorname{Wr}\left(H_{k_N},\ldots,H_{k_1},H_{n_1},\ldots,H_{n_i}\right), \quad i = 0,1,\ldots,l,$$

and introduce the differential expressions

$$\alpha_i = \alpha_{\eta_{i+1},\eta_i}, \quad i = 0, \ldots, l-1,$$
$$\beta_i = \alpha_{\eta_i,\eta_{i+1}}$$
$$\tau_i = \tau_{\eta_i}, \quad i = 0, \ldots, l.$$

In particular, $\tau_0 = \tau_\lambda$.

**Proposition 3.5** *With the above definitions, we have*

$$\chi(\eta_i, \eta_{i+1}) = 2(N + i - n_{i+1})\eta_i \eta_{i+1}, \quad i = 0, \ldots, l-1, \tag{53}$$
$$\tau_i = \beta_i \alpha_i + 2(N + i - n_{i+1}), \quad i = 0, \ldots, l-1; \tag{54}$$
$$\tau_{i+1} = \alpha_i \beta_i + 2(N + i + 1 - n_{i+1}). \tag{55}$$

***Proof*** Relation (53) follows by Corollary 3.2. Relations (54) and (55) follow by Lemma 2.2. □

Next, we describe a factorization chain that connects $\tau_\lambda$ to $\tau_\emptyset$ using $\lambda_1$ steps. Let $k_1, k_2, \ldots, k_N \in K_\lambda$ be the exceptional degrees, as per (44). Since $k_1 = \lambda_1 + N - 1$, every $n \geq \lambda_1 + N$ is an allowed degree; i.e. an element of $I_\lambda$.

**Definition 3.2** We call the smallest $\lambda_1$ elements of $I_\lambda$ the *sporadic* degrees of $I_\lambda$.

**Proposition 3.6** *An $n \in I_\lambda$ is a sporadic degree if and only if there exists an exceptional degree $k \in K_\lambda$ such that $n < k$.*

***Proof*** By (44), the $N$ exceptional degrees are contained in $\{0, 1, \ldots, \lambda_1 + N - 1\}$, a set of cardinality $\lambda_1 + N$. Hence, $n \in I_\lambda$ is dominated by some exceptional degree if and only if $n < \lambda_1 + N - 1$. By the preceding remark, there are precisely $\lambda_1$ such degrees. □

**Proposition 3.7** *Let $\lambda$ be a partition of $N$, and let $n_1 < n_2 < \ldots$ be the allowed degrees in $I_\lambda$ listed in increasing order. For $j \in \mathbb{N}_0$, set*

$$\lambda_i^{(j)} = \max(\lambda_i - j, 0) \quad i \in \mathbb{N}, \tag{56}$$

*and let $\ell_j$ be the length of $\lambda^{(j)}$. Then,*

$$\ell_j = N - n_{j+1} + j, \quad j \in \mathbb{N}_0. \tag{57}$$

***Proof*** Observe that if $j \geq \lambda_1$, then $\lambda^{(j)} = \emptyset$ is the trivial partition. Thus it suffices to establish (57) for $j = 0, \ldots, \lambda_1 - 1$.

By definition, $i \leq \ell_j$ if and only if $\lambda_i - j \geq 1$. In other words, the sequence $(\ell_0, \ell_1, \ell_2, \ldots)$ and $\lambda$ are dual partitions. Observe that if $\lambda_i - j \geq 1$, then $i - \ell_j \leq 0$. Conversely, $\lambda_i - j \leq 0$ if and only if $i - \ell_j \geq 1$. Hence,

$$\lambda_i - i \neq j - \ell_j, \quad i \in \mathbb{N},\ j \in \mathbb{N}_0.$$

Using $\sqcup$ for disjoint union, it follows that

$$\{\lambda_i - i : i = 1, \ldots, N\} \sqcup \{j - \ell_j : j = 0, \ldots, \lambda_1 - 1\} = \{-N, -N+1, \ldots, \lambda_1 - 1\}.$$

Therefore,

$$\{j - \ell_j + N : j = 0, \ldots, \lambda_1 - 1\}$$

is precisely the set of sporadic degrees in $I_\lambda$. Relation (57) follows. □

**Proposition 3.8** *Let $\lambda$ be a partition, with $\lambda^{(j)}$, $j \in \mathbb{N}_0$ the sequence of partitions defined by* (56)*. Define the differential expressions*

$$\begin{aligned}
\alpha_j &= \alpha_{\eta_{\lambda(j+1)}, \eta_{\lambda(j)}}, \quad j = 0, \ldots, \lambda_1 - 1, \\
\beta_j &= \alpha_{\eta_{\lambda(j)}, \eta_{\lambda(j+1)}}; \\
\tau_j &= \tau_{\eta_{\lambda(j)}}, \quad j = 0, \ldots, \lambda_1.
\end{aligned}$$

*Then,*

$$\tau_j = \beta_j \alpha_j + 2\ell_j, \quad j = 0, \ldots, \lambda_1 - 1; \tag{58}$$

$$\tau_{j+1} = \alpha_j \beta_j + 2(\ell_j + 1). \tag{59}$$

***Proof*** Let $n_1 < \ldots < n_{\lambda_1}$ be the sporadic degrees of $\lambda$ listed in increasing order. Let

$$\begin{aligned}
K_j &= K_\lambda \cup \{n_1, \ldots, n_j\}, \quad j = 0, \ldots, \lambda_1, \\
\eta_j &= \mathrm{Wr}\big(k_N, \ldots, k_1, n_1, \ldots, n_j\big).
\end{aligned}$$

By (57) and (53),

$$\chi\big(\eta_j, \eta_{j+1}\big) = 2(N + j - n_{j+1})\eta_j \eta_{j+1} = 2\ell_j \eta_j \eta_{j+1}.$$

The desired conclusion now follows by Proposition 3.5. □

For example, consider the partition $\lambda = (3, 3, 1, 1, 0, \ldots)$ of $N = 8$. The exceptional degrees are

$$K_\lambda = \{0, 1, 2, 3, 5, 6, 9, 10\}.$$

The degree set is therefore,

$$I_\lambda = \{4, 7, 8, 11, 12, 13, \ldots\},$$

with 4, 7, 8 being the sporadic degrees. In terms of the above terminology,

$$K_1 = \{0, 1, 2, 3, 4, 5, 6, 9, 10\}, \quad \lambda^{(1)} = (2, 2, 0, 0, 0, \ldots), \quad N_1 = 4,\ \ell_1 = 2$$

$$K_2 = \{0, 1, 2, 3, 4, 5, 6, 7, 9, 10\}, \quad \lambda^{(2)} = (1, 1, 0, 0, 0, \ldots), \quad N_2 = 2,\ \ell_2 = 2$$

$$K_3 = \{0, 1, 2, 3, 4, 5, 6, 7, 8, 9, 10\}, \quad \lambda^{(3)} = (0, 0, 0, 0, 0, \ldots), \quad N_3 = 0,\ \ell_3 = 0.$$

## 3.2 The Norm Identity

In this section, we present and prove a certain algebraic identity that will allow us to derive a formula for the norming constants of the exceptional Hermite polynomials.

**Proposition 3.9** *Let $m_1, \ldots, m_l, m \in \mathbb{N}_0$ be distinct non-negative integers. Set*

$$\begin{aligned} \eta_0 &= 1 \\ \xi_0 &= H_m \\ \eta_i &= \mathrm{Wr}\big(H_{m_1}, \ldots, H_{m_i}\big), \quad i = 1, \ldots, l; \\ \xi_i &= \mathrm{Wr}\big(H_{m_1}, \ldots, H_{m_i}, H_m\big) \end{aligned}$$

*Also, let $\rho_0 = 0$, and recursively define*

$$\rho_{i+1} = \frac{\xi_i \xi_{i+1}}{\eta_i \eta_{i+1}} + 2(m - m_{i+1})\rho_i, \quad i = 0, \ldots, l-1.$$

*We then have,*

$$\left(\frac{\xi_l}{\eta_l}\right)^2 e^{-x^2} - 2^l \prod_{i=1}^{l} (m - m_i) \xi_0^2 e^{-x^2} = \left(\rho_l\, e^{-x^2}\right)'. \tag{60}$$

***Proof*** By (49), we have

$$\chi(\eta_i, \eta_{i+1}) = 2(i - m_{i+1})\eta_i \eta_{i+1}, \quad i = 0, \ldots, l-1.$$

Hence, by the argument used in Proposition 3.5, the differential expressions

$$\alpha_i = \alpha_{\eta_{i+1},\eta_i}, \quad i = 0, \dots, l-1$$
$$\beta_i = \beta_{\eta_i,\eta_{i+1}}, \quad i = 0, \dots, l-1$$
$$\tau_i = \tau_{\eta_i}, \quad i = 0, \dots, l.$$

constitute the factorization chain

$$\tau_i = \beta_i\alpha_i + 2(i - m_{i+1}), \quad i = 0, \dots, l-1$$
$$\tau_{i+1} = \alpha_i\beta_i + 2(i + 1 - m_{i+1}).$$

Next, observe that

$$\alpha_i\xi_i = \frac{\mathrm{Wr}\big(\mathrm{Wr}\big(H_{m_1}, \dots, H_{m_i}, H_{m_{i+1}}\big), \mathrm{Wr}\big(H_{m_1}, \dots, H_{m_i}, H_m\big)\big)}{\mathrm{Wr}\big(H_{m_1}, \dots, H_{m_i}\big)}$$
$$= \mathrm{Wr}\big(H_{m_1}, \dots, H_{m_i}, H_{m_{i+1}}, H_m\big) = \xi_{i+1}.$$

Again, by (49), we have

$$\tau_i(\xi_i) = 2(i - m)\xi_i, \quad i = 0, \dots, l.$$

Hence, by (13), we have

$$\frac{\xi_{i+1}^2}{\eta_{i+1}^2}e^{-x^2} - 2(m - m_{i+1})\frac{\xi_i^2}{\eta_i^2}e^{-x^2} = \left(\frac{\xi_i\xi_{i+1}}{\eta_i\eta_{i+1}}e^{-x^2}\right)', \quad i = 0, \dots, l-1.$$

Hence, by induction,

$$\frac{\xi_l^2}{\eta_l^2}e^{-x^2} - 2^l\prod_{i=1}^{l}(m - m_i)\xi_0^2e^{-x^2} = \left(e^{-x^2}\sum_{i=1}^{l}\prod_{j=1+i}^{l}(2m - 2m_j)\frac{\xi_j\xi_{j-1}}{\eta_j\eta_{j-1}}\right)'$$

Again, by induction,

$$\rho_k = \sum_{i=1}^{k}\prod_{j=1+i}^{l}(2m - 2m_j)\frac{\xi_j\xi_{j-1}}{\eta_j\eta_{j-1}}, \quad k = 1, \dots, \ell-1.$$

The desired identity (60) follows immediately. $\square$

# 4 The L$^2$ Theory

The following theorem was proved by Krein for Sturm-Liouville problems on the half-line, and independently by Adler for Sturm-Liouville problems on a bounded interval. The Adler argument can be extended without difficulty to the case of an infinite interval [10].

**Definition 4.1** We say that $\lambda$ is an even partition $\lambda_{2i-1} = \lambda_{2i}$ for every $i \in \mathbb{N}$.

**Theorem 4.1 (Krein-Adler)** *The polynomial $\eta_\lambda(x)$ has no real zeros if and only if $\lambda$ is an even partition.*

Henceforth we assume that $\lambda$ is an even partition. In that case,

$$W_\lambda := W_{\eta_\lambda} = \eta_\lambda^{-2} e^{-x^2}$$

is a non-singular weight on $(-\infty, \infty)$ with finite moments. Let $\mathcal{H}_\lambda = \mathrm{L}^2(\mathbb{R}, W_\lambda)$ denote the corresponding Hilbert space with weighted inner product,

$$\langle f, g\rangle_\lambda = \int_{\mathbb{R}} f\bar{g}W_\lambda = \int_R \frac{f}{\eta_\lambda}\frac{\bar{g}}{\eta_\lambda} e^{-x^2}, \quad f, g \in \mathcal{H}_\lambda.$$

**Proposition 4.1** *Let $\lambda$ be an even partition. Then, the corresponding exceptional Hermite polynomials enjoy the following orthogonality property:*

$$\left\langle H_{\lambda,n}, H_{\lambda,n'}\right\rangle_\lambda = \sqrt{\pi}\, 2^{n-N} \frac{n!}{\pi_{\lambda,N}(n)}\, \delta_{n,n'}, \quad n, n' \in I_\lambda, \tag{61}$$

*with $\pi_{\lambda,N}(n)$ as defined in* (34).

***Proof*** Being the eigenfunctions of a SLP (4), the $H_{\lambda,n},\ n \in I_\lambda$ are orthogonal with respect to $W_\lambda = W_{\eta_\lambda}$. It remains to establish the form of the norming constants in (61). By Corollary 3.1 and (60),

$$\begin{aligned}
\left\langle H_{\lambda,n}, H_{\lambda,n}\right\rangle_\lambda &= \int_{\mathbb{R}} \left(\frac{H_{\lambda,n}(x)}{\eta_\lambda(x)}\right)^2 e^{-x^2} dx, \quad n \in I_\lambda \\
&= \int_{\mathbb{R}} \left(\frac{\mathrm{Wr}\big(H_{k_N}, \dots, H_{k_1}, H_n\big)}{2^N \pi_{\lambda,N}(n)\,\mathrm{Wr}\big(H_{k_N}, \dots, H_{k_1}\big)}\right)^2 e^{-x^2} dx, \\
&= \frac{1}{2^N \pi_{\lambda,N}(n)} \int_R H_n(x)^2 e^{-x^2} dx \\
&= \sqrt{\pi} 2^{n-N} \frac{n!}{2^N \pi_{\lambda,N}(n)}.
\end{aligned}$$

□

**Theorem 4.2** *Let $\lambda$ be an even partition. Then the SLP of form* (5) *with $\eta = \eta_\lambda$ has eigenvalues*

$$\varepsilon_n = 2(n - N), \quad n \in I_\lambda$$

*The corresponding eigenfunctions are $H_{\lambda,n}$, $n \in I_\lambda$, while the norming constants are given by* (61).

**Lemma 4.1** *Let $\lambda$ be a partition and let $\lambda^{(j)}$, $j \in \mathbb{N}$ be the partitions defined in* (56)*. If $\lambda$ is an even partition, then so is every $\lambda^{(j)}$.*

***Proof*** Suppose that $\lambda_{2i-1} = \lambda_{2i}$ for all $i \in \mathbb{N}$. It follows that

$$\lambda^{(j)}_{2i-1} = \max(\lambda_{2i-1} - j, 0) = \max(\lambda_{2i} - j, 0) = \lambda^{(j)}_{2i} \quad i, j \in \mathbb{N}.$$

□

***Proof*** *(of Theorem 4.2)* Multiplication of (4) by $-W_\lambda^{-1}$ and changing $\varepsilon \to -\varepsilon$, gives (3), which may be expressed as

$$\tau_\lambda y = \varepsilon y.$$

By (12) and (52), we have

$$\tau_\lambda H_{\lambda,n} = 2(N - n)H_{\lambda,n}, \quad n \in I_\lambda.$$

Hence, $2(n - N)$ is an eigenvalue of (5) with corresponding eigenfunction $H_{\lambda,n}$, $n \in I_\lambda$.

It remains to show that there are no either eigenvalues; i.e., that $\{H_{\lambda,n}\}_{n \in I_\lambda}$ is a complete orthogonal basis. To show this, we employ the factorization chain described in Proposition 3.8. Let $\lambda^{(j)}$, $j = 0, \ldots, \lambda_1$ be the partitions defined in (56), and let $\eta_j = \eta_{\lambda^{(j)}}$. By the above Lemma, these are all even partitions, and hence each $\eta_j \in \mathcal{R}$.

Let $T_j = T_{\eta_j}$, $j = 0, 1, \ldots, \lambda_1$ be a sequence of self-adjoint operators with action $\tau_{\eta_j}$ and domains as per (20). By Proposition 2.6, the formal factorization chain of Proposition 3.8 corresponds to a factorization chain of operators

$$T_j = B_j A_j + 2\ell_j, \quad j = 0, \ldots, \lambda_1 - 1,$$
$$T_{j+1} = A_j B_j + 2\ell_j + 2, \quad j = 0, \ldots, \lambda_1 - 1,$$

where $\ell_j$ is the length of partition $\lambda^{(j)}$, and where $A_j$, $B_j$ are first order operators with action $\alpha_{\eta_{j+1},\eta_j}$, $\beta_{\eta_{j+1},\eta_j}$, respectively, and domains as given in (18). Hence, $2\ell_j$ is the largest eigenvalue of $T_j$, $j = 0, \ldots, \lambda_1$. Hence, by Theorem 2.1,

$$\sigma(T_{j+1} - 2) = \sigma(T_j) \setminus \{2\ell_j\}.$$

It follows that

$$\sigma(T_0) = \{2\ell_j - 2j\colon j = 0, \ldots, \lambda_1 - 1\} \cup \sigma(T_{\lambda_1}).$$

However, since $\tau_{\lambda_1} = \tau_\emptyset$ is the classical Hermite operator, we have by (57) that

$$\begin{aligned}\sigma(T_0) &= \{2\ell_j - 2j\colon j = 0, \ldots, \lambda_1 - 1\} \cup (-2\mathbb{N}_0)\\ &= \{2(N - n_{j+1})\colon j = 0, \ldots, \lambda_1 - 1\} \cup (-2\mathbb{N}_0)\\ &= \{2(N - n)\colon n \in I_\lambda\}.\end{aligned}$$

□

**Corollary 4.1** *Let $K = \{k_1, \ldots, k_N\} \subset \mathbb{N}_0$ be a list of $N$ non-negative integers such that*

$$\sum_{i=1}^{N} k_i = \frac{1}{2}N(N+1).$$

*Set*

$$\lambda_i = k_i + i - N, \ i \in \mathbb{N}.$$

*Then, $\{H_{\lambda,n}\}_{n \in I_\lambda}$ is a family of exceptional Hermite polynomials such that $K = K_\lambda$; i.e. the exceptional degrees are precisely $K$. Moreover, if $\lambda$ is an even partition, then the corresponding $\{H_{\lambda,n}\}_{n \in I_\lambda}$ are eigenfunctions of the SLP* (4) *whose spectrum, up to a shift with $N$, differs from the spectrum of the classical Hermite SLP by the removal of eigenvalues at position $2k$, $k \in K$.*

**Acknowledgments** This research has been financed in part by the Spanish MICINN under grants PGC2018-096504-B-C33 and RTI2018-100754-B-I00 and the European Union under the 2014-2020 ERDF Operational Programme and by the Department of Economy, Knowledge, Business and University of the Regional Government of Andalusia (project FEDER-UCA18-108393).

## References

1. S. Bochner, On Sturm-Liouville polynomial systems. Math. J. **29** 730–736 (1929)
2. E.A. Coddington, N. Levinson, *Theory of Ordinary Differential Equations* (Tata McGraw-Hill Education, New York, 1955)
3. P.A. Deift, Applications of a commutation formula. Duke Math. J. **45**, 267–310 (1978)
4. S.Y. Dubov, V.M. Eleonskii, N.E. Kulagin, Equidistant spectra of anharmonic oscillators. Chaos **4**, 47–53 (1994)
5. A.J. Durán, Exceptional Charlier and Hermite orthogonal polynomials. J. Approx. Theory **182**, 29–58 (2014)
6. A.J. Durán, Exceptional Meixner and Laguerre orthogonal polynomials. J. Approx. Theory **184**, 176–208 (2014)

7. A.J. Durán, Higher order recurrence relation for exceptional Charlier, Meixner, Hermite and Laguerre orthogonal polynomials. Integral Transform. Spec. Funct. **26**, 357–376 (2015)
8. A.J. Durán, Exceptional Hahn and Jacobi orthogonal polynomials. J. Approx. Theory **214**, 9–48 (2017)
9. A.J. Durán, Bispectrality of Charlier type polynomials. Integral Transform. Spec. Funct. **30**(8), 601–627 (2019)
10. M.A. García-Ferrero, D. Gómez-Ullate, Oscillation theorems for the Wronskian of an arbitrary sequence of eigenfunctions of Schrödinger's equation. Lett. Math. Phys. **105**, 551–573 (2015)
11. M.A. García-Ferrero, D. Gómez-Ullate, R. Milson, A Bochner type characterization theorem for exceptional orthogonal polynomials. J. Math. Anal. Appl. **472**, 584–626 (2019)
12. M.A. García-Ferrero, D. Gómez-Ullate, R. Milson, Exceptional legendre polynomials and confluent darboux transformations. SIGMA **17**, 016 (2021)
13. F. Gesztesy, G. Teschl, On the double commutation method. Proc. Am. Math. Soc. **124**, 1831–1840 (1996)
14. D. Gomez-Ullate, N. Kamran, R. Milson, An extended class of orthogonal polynomials defined by a Sturm-Liouville problem. J. Math. Anal. Appl. **359**, 352–367 (2009)
15. D. Gomez-Ullate, Y. Grandati, R. Milson, Rational extensions of the quantum harmonic oscillator and exceptional Hermite polynomials. J. Phys. A Math. Theor. **47**, 015203 (2014)
16. D. Gomez-Ullate, Y. Grandati, R. Milson, Durfee rectangles and pseudo-Wronskian equivalences for Hermite polynomials. Stud. Appl. Math. **141**, 596–625 (2018)
17. D. Gomez-Ullate, N. Kamran, R. Milson, Exceptional orthogonal polynomials and the Darboux transformation. J. Phys A **43**, 434016 (2010)
18. D. Gomez-Ullate, N. Kamran, R. Milson, Two-step Darboux transformations and exceptional Laguerre polynomials. J. Math. Anal. Appl. **387**, 410–418 (2012)
19. D. Gómez-Ullate, F. Marcellán, R. Milson, Asymptotic and interlacing properties of zeros of exceptional Jacobi and Laguerre polynomials. J. Math. Anal. Appl. **399**, 480–495 (2013)
20. D. Gómez-Ullate, A. Kasman, A.B.J. Kuijlaars, R. Milson, Recurrence relations for exceptional hermite polynomials. J. Approx. Theory **204**, 1–16 (2016)
21. A.D. Hemery, A.P. Veselov, Whittaker-Hill equation and semifinite-gap Schrödinger operators. J. Math. Phys. **51**, 072108 (2010)
22. Á.P. Horváth, The electrostatic properties of zeros of exceptional Laguerre and Jacobi polynomials and stable interpolation. J. Approx. Theory **194**, 87–107 (2015)
23. A. Kasman, R. Milson, The adelic Grassmannian and exceptional hermite polynomials, mathematical physics. Anal. Geom. **23**, 1–51 (2020)
24. A.B.J. Kuijlaars, R. Milson, Zeros of exceptional Hermite polynomials. J. Approx. Theory **200**, 28–39 (2015)
25. C. Liaw, L. Littlejohn, J.S. Kelly, Spectral analysis for the exceptional $X_m$-Jacobi equation. Elect. J. Differ. Equ. **194**, 1–10 (2015)
26. C. Liaw, L. Littlejohn, J.S. Kelly, R. Milson, The spectral analysis of three families of exceptional Laguerre polynomials. J. Approx. Theory **202**, 5–41 (2016)
27. M. Marquette, C. Quesne, New families of superintegrable systems from Hermite and Laguerre exceptional orthogonal polynomials. J. Math. Phys. **54** 042102 (2013)
28. S. Odake, R. Sasaki, Infinitely many shape invariant potentials and new orthogonal polynomials. Phys. Lett. B **679**, 414–417 (2009)
29. S. Post, S. S. Tsujimoto, L. Vinet, Families of superintegrable Hamiltonians constructed from exceptional polynomials. J. Phys. A **45**, 405202 (2012)
30. W. Rudin, *Functional Analysis* (McGraw-Hill, New York, 1991)
31. C. Quesne, Solvable rational potentials and exceptional orthogonal polynomials in supersymmetric quantum mechanics. SIGMA **5** (2009)
32. R. Sasaki, S. Tsujimoto, A. Zhedanov, Exceptional Laguerre and Jacobi polynomials and the corresponding potentials through Darboux-Crum transformations. J. Phys. A **43** 315204 (2010)
33. A. Schulze-Halberg, B. Roy, Darboux partners of pseudoscalar Dirac potentials associated with exceptional orthogonal polynomials. Ann. Phys. **349**, 159–170 (2014)
34. E.C. Titchmarsh, *Eigenfunction Expansions Associated with Second-Order Differential Equations. Part 1* (Oxford University Press, London, 1962)

# Occupation Time for Classical and Quantum Walks

**F. A. Grünbaum, L. Velázquez, and J. Wilkening**

**Abstract** This is a personal tribute to Lance Littlejohn on the occasion of his 70th birthday. It is meant as a present to him for many years of friendship. It is not written in the "Satz-Beweis" style of Edmund Landau or even in the format of a standard mathematics paper. It is rather an invitation to a fairly new, largely unexplored, topic in the hope that Lance will read it some afternoon and enjoy it. If he cares about complete proofs he will have to wait a bit longer; we almost have them but not in time for this volume. We hope that the figures will convince him and other readers that the phenomena displayed here are interesting enough.

**Keywords** Classical walk · Quantum walk · Occupation time · Monitoring · Verblunsky coefficients · CMV matrices · Riesz walk

## 1 Introduction

The origin of our story goes back to a remarkable observation by Paul Levy [24], that one dimensional Brownian motion behaves in rather unexpected ways. To state his result in everyday terms let us switch to a long night at the casino where you play repeatedly with a fair coin: if you get heads you win one dollar, if you get tails you lose one dollar. You play once every 10 s and you keep track of your "fortune" as a function of time: you are happy if your fortune at that time is positive, i.e. you have won more times than you have lost, otherwise you are unhappy. This game has to be played for a fixed and large number of tosses. What proportion of the total time

---

F. A. Grünbaum (✉) · J. Wilkening
Department of Mathematics, University of California, Berkeley, CA, USA
e-mail: grunbaum@math.berkeley.edu; wilkening@berkeley.edu

L. Velázquez
Departamento de Matemática Aplicada, Universidad de Zaragoza, Zaragoza, Spain
e-mail: velazque@unizar.es



F. Gesztesy, A. Martinez-Finkelshtein (eds.), *From Operator Theory to Orthogonal Polynomials, Combinatorics, and Number Theory*, Operator Theory: Advances and Applications 285, https://doi.org/10.1007/978-3-030-75425-9_11

you play do you expect to be happy? It is not unreasonable to guess that the answer should be about $1/2$.

However, and this is P. Levy's observation, a "wild night" is more likely than a "normal night". Take—for instance—two windows of total length $1/10$: then the probability that you are happy somewhere between 45 and 55% of the time is smaller than the probability that you are happy either more than 95 or less than 5% of the time. This discussion can be found in many standard probability books [9, 25, 30]. The distribution of the random variable in question is explicitly known—and is given by an arcsine law—and as evidence of its interest we observe that one of the first applicaction of the Feynman-Kac formula was a rederivation of this result of P. Levy by M. Kac, see [17]. The careful formulation of the discrete version with Brownian motion replaced by a coin was done by K. Chung and W. Feller, see [7].

Just to put things in perspective, consider two-dimensional Brownian motion starting at the origin, run it for time one and inquire about the time spent in the positive quadrant. The analytical form of the result is not known. See [8, 11] and references there.

The rest of this paper is devoted to a discussion of what happens when one replaces a classical random walk on the integers, such as the ordinary coin, with a quantum walk. This is not the place to give a full account of this notion. We just mention the paper by Y. Aharonov et al. [1], where this was first introduced; a very good and early survey paper by Julia Kempe [18]; and the first paper where this was connected with orthogonal polynomials on the unit circle, see [4]. It is also useful to look at [19].

We will study a few quantum walks whose state space is a Hilbert space spanned by an orthonormal set of states $|i\rangle \otimes |\uparrow\rangle$, $|i\rangle \otimes |\downarrow\rangle$ with a definite "site" $i$ along the integers and a definite value of an extra degree of freedom—the "spin"—represented by an up or down arrow, which may be viewed as the quantum counterpart of the two sides of a coin. At each time step, the evolution is driven by some "transition rules"

$$|i\rangle \otimes |\uparrow\rangle \longrightarrow \begin{cases} |i+1\rangle \otimes |\uparrow\rangle & \text{with (complex) probability amplitude } c_{11}^i \\ |i-1\rangle \otimes |\downarrow\rangle & \text{with (complex) probability amplitude } c_{21}^i \end{cases}$$

$$|i\rangle \otimes |\downarrow\rangle \longrightarrow \begin{cases} |i+1\rangle \otimes |\uparrow\rangle & \text{with (complex) probability amplitude } c_{12}^i \\ |i-1\rangle \otimes |\downarrow\rangle & \text{with (complex) probability amplitude } c_{22}^i \end{cases}$$

where, for each $i \in \mathbb{Z}$,

$$C_i = \begin{pmatrix} c_{11}^i & c_{12}^i \\ c_{21}^i & c_{22}^i \end{pmatrix} \tag{1}$$

is an arbitrary unitary matrix which we will call the $i^{th}$ coin. This defines in the Hilbert state space a unitary operator $U$ governing the one-step evolution, so that a walker originally in the state $\psi$ is in the state $U^n\psi$ after $n$ steps.

This kind of "coined" walk is the simplest of all models. In [4] this is extended by allowing a few more "local transitions" so as to obtain the doubly infinite version of a general CMV matrix—see [3, 4]—as the unitary matrix that gives the discrete time evolution in the basis $|i\rangle \otimes |\uparrow\rangle$, $|i\rangle \otimes |\downarrow\rangle$. The most general doubly infinite CMV matrix has the form

$$\left(\begin{array}{cccc|cccc} \dots & \dots & \dots & \dots & \dots & \dots & \dots & \dots \\ \dots & \rho_{-3}\overline{\alpha}_{-2} & -\alpha_{-3}\overline{\alpha}_{-2} & \rho_{-2}\overline{\alpha}_{-1} & \rho_{-2}\rho_{-1} & 0 & 0 & \dots \\ \dots & \rho_{-3}\rho_{-2} & -\alpha_{-3}\rho_{-2} & -\alpha_{-2}\overline{\alpha}_{-1} & -\alpha_{-2}\rho_{-1} & 0 & 0 & \dots \\ \hline \dots & 0 & 0 & \rho_{-1}\overline{\alpha}_{0} & -\alpha_{-1}\overline{\alpha}_{0} & \rho_{0}\overline{\alpha}_{1} & \rho_{0}\rho_{1} & \dots \\ \dots & 0 & 0 & \rho_{-1}\rho_{0} & -\alpha_{-1}\rho_{0} & -\alpha_{0}\overline{\alpha}_{1} & -\alpha_{0}\rho_{1} & \dots \\ \dots & \dots & \dots & \dots & \dots & \dots & \dots & \dots \end{array}\right)$$

where $\rho_j = \sqrt{1 - |\alpha_j|^2}$ and $(\alpha_j)_{j=-\infty}^{\infty}$ is a sequence of complex numbers such that $|\alpha_j| < 1$. The coefficients $\alpha_j$ are known as the Verblunsky coefficients. A quantum walk whose unitary step in the basis $|i\rangle \otimes |\uparrow\rangle$, $|i\rangle \otimes |\downarrow\rangle$ is a doubly infinite CMV will be called a CMV walk.

CMV walks are the quantum analogue of the classical birth–death processes. The Jacobi matrices underlying the latter provide the canonical representations of self-adjoint operators, while a similar role in the unitary case is played by the CMV matrices. Also, the fruitful connection between birth-death processes and orthogonal polynomials on the real line via Jacobi matrices becomes a similarly useful connection between CMV walks and the theory of orthogonal polynomials on the unit circle, the breeding ground where CMV matrices were born.

CMV walks not only constitute simple quantum walk models, but they are universal models for quantum walks, since any unitary operator—as the unitary step governing the evolution of a quantum walk—has a CMV representation. This paves the way to the use of the CMV representation as a resource for the analysis of general quantum walks, a technique nowadays known as the "CGMV method" [5, 21, 22].

Up to a change of phases in the basis, coined walks correspond to CMV walks whose Verblunsky coefficients with odd index are zero [4, 5]. The corresponding coins are given by

$$C_i = \begin{pmatrix} \rho_{2i} & -\alpha_{2i} \\ \overline{\alpha}_{2i} & \rho_{2i} \end{pmatrix}. \tag{2}$$

The examples that we will consider are mostly coined walks: the Hadamard walk of [1, 18] which illustrates the behaviour for an unbiased constant coin, several instances of biased constant coins, and even a case of a site dependent coin as the the one indicated above. The last example that we consider here, the Riesz walk, not only requires going beyond coined walks to consider a CMV walk, but also involves a non-trivial site dependence for the corresponding Verblunsky coefficients.

## 2 A Look at the Classical Discrete Case

In the discrete case we say that the partial sum $S_k$ (i.e. your fortune at time $k$) is positive if $S_k > 0$ or in case that $S_k = 0$ one has $S_{k-1} > 0$. We put $S_0 = 0$. If $N_n$ denotes the number of "positive" terms among $S_1, S_2, \ldots, S_n$ the result of Chung–Feller [7] is that the probability that $N_{2n} = 2r$ is

$$P(N_{2n} = 2r) = \frac{1}{2^{2n}}\binom{2r}{r}\binom{2n-2r}{n-r}.$$

Using the notation $u_{2k} = \frac{1}{2^{2k}}\binom{2k}{k}$ we have

$$P(N_{2n} = 2r) = u_{2r}u_{2n-2r}\,.$$

In [10] one finds an adaptation of the Feynman-Kac method to reprove the results of Chung and Feller in the discrete case. This leads to supplement the results in [7] with

$$P(N_{2n+1} = 2r) = u_{2r}u_{2n+2-2r}\frac{n-r+1}{n+1}, \qquad r = 0, 1, 2, \ldots, n \tag{3}$$

and

$$P(N_{2n+1} = 2r-1) = u_{2r}u_{2n+2-2r}\frac{r}{n+1}, \qquad r = 1, 2, \ldots, n+1\,. \tag{4}$$

Those who have played with Legendre polynomials, such as Lance, may enjoy the fact that they are connected with this coin-tossing problem. If we denote the Legendre polynomials by $P_n$ (with $P_n(1) = 1$) we get

$$\sum_{k=0}^{n} u_{2k}u_{2n-2k}q^{2k} = P_n\left(\frac{q+q^{-1}}{2}\right)q^n.$$

This connection of the Legendre polynomials and the coin-tossing game is essentially in the literature; see [26], formula(10), page 248.

In the classical case, either the case of Brownian motion or of a fair coin (in the limit of an infinite number of tosses), the cumulative distribution and density functions are well-known; see Fig. 1. The numerical simulations reported in [11] agree almost perfectly with this analytical result. Most of the numerical work in [11] deals with the two dimensional case and the occupation time in a wedge, in which case no analytical results are known. The way that Brownian motion is simulated in [11] goes back to P. Levy himself in terms of a random "Fourier series" using Haar functions to approximate white noise.

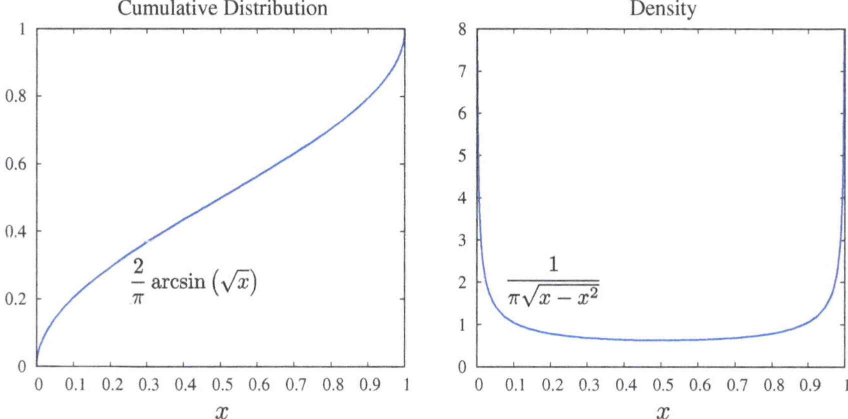


**Fig. 1** Cumulative distribution and density for the occupation time of the positive half line under Brownian motion

## 3 Occupation Times for Quantum Walks

When discussing occupation times for quantum walks one should decide when the walker will be "observed", since this alters the natural evolution due to Born's rule about quantum measurements. Here we consider a "monitoring" approach to occupation times similar to that one proposed in [15] and further considered in [2] to study recurrence in quantum walks. One should also look at [23].

This gives a natural way to look at the times when a coined or CMV quantum walk is on each half of the state space. The positive half consists of the span of the states

$$|0\rangle \otimes |\downarrow\rangle\,,\ |1\rangle \otimes |\uparrow\rangle\,,\ |1\rangle \otimes |\downarrow\rangle\,,\ |2\rangle \otimes |\uparrow\rangle\,,\ |2\rangle \otimes |\downarrow\rangle\,,\ \ldots \tag{5}$$

while the negative half consists of the span of the states

$$\ldots\,,\ |-2\rangle \otimes |\uparrow\rangle\,,\ |-2\rangle \otimes |\downarrow\rangle\,,\ |-1\rangle \otimes |\uparrow\rangle\,,\ |-1\rangle \otimes |\downarrow\rangle\,,\ |0\rangle \otimes |\uparrow\rangle \tag{6}$$

We call these the positive and the negative subspaces of our Hilbert state space respectively. In accordance with the setup in [2, 4], the orthogonal projection $P$ on the positive subspace plays a fundamental role in our results.

The monitoring approach assumes that a measurement is performed after each step to decide whether the walker is on the positive subspace or not. From a mathematical point of view, this results in the application of one of the projections, $P$ or $Q = I - P$, after the unitary step $U$ driving the evolution. While $P$ conditions on the event "the walker has been found on the positive subspace", the complementary projection $Q$ conditions on the event "the walker has not been found on the positive subspace". This gives the probability of any occupation distribution according to Born's rule.

For example, if the walker starts at the state $\psi$, $\|\psi\| = 1$, and runs for 6 steps, the quantity

$$\|PU\,PU\,QU\,PU\,QU\,PU\psi\|^2$$

gives the probability to find the walker on the positive subspace at times 1, 3, 5 and 6, but not at times 2 and 4. As a consequence, the probability of finding the walker, for instance, 2 times on the positive subspace, while running for a total number of 4 steps, is given by the sum

$$\begin{aligned} P(N_4 = 2) \quad = \quad & \|QU\,QU\,PU\,PU\psi\|^2 + \|PU\,QU\,QU\,PU\psi\|^2 \\ & + \|PU\,PU\,QU\,QU\psi\|^2 + \|QU\,PU\,QU\,PU\psi\|^2 \\ & + \|PU\,QU\,PU\,QU\psi\|^2 + \|QU\,PU\,PU\,QU\psi\|^2. \end{aligned}$$

The initial state in all our examples will be

$$\psi = \frac{1}{\sqrt{2}}\Big(|0\rangle \otimes |\uparrow\rangle + \imath\,|0\rangle \otimes |\downarrow\rangle\Big),$$

where $\imath$ is the imaginary unit.

More generally, the probability of finding the walker $r$ times on the positive subspace in the process of taking $n$ steps is given by the sum

$$P(N_n = r) \quad = \quad \sum_{\substack{P_i = P \text{ or } Q \\ P_i = P \text{ for } r \text{ indices}}} \|(P_n U) \cdots (P_2 U)(P_1 U)\psi\|^2.$$

These quantities will be plotted against the relative number of steps $r/n$ in the graphics corresponding to the examples of the next sections. We will compute both the "density" $P(N_n = r)$ and the "cumulative distribution" $P(N_n \le r) = \sum_{k=0}^{r} P(N_n = k)$, which should be compared with Fig. 1. Since we will be dealing with values of $n$ in the range 90–8100 and $r$ between 0 and $n$, it is clear that one needs a smart way to evaluate the sum given above—a direct approach would involve a prohibitive number of terms. Details on reducing the complexity of this calculation from $O(2^n)$ to $O(n^3)$ will be given in [16].

The walker will evolve with a CMV walk with a doubly infinite CMV matrix that treats right and left "just the same", to have a "fair" comparison with the classical case of a fair coin. Restricting for simplicity to real valued Verblunsky coefficients $\alpha_i$, we need to impose

$$\alpha_i = \alpha_{-2-i}, \qquad i = 0, 1, 2, 3, \dots \tag{7}$$

a condition that leaves $\alpha_{-1}$ free. In the case of coined walks this amounts to taking real coins $C_i$ such that $C_i = C_{-1-i}$.

Our results, displayed in the next few sections, illustrate the fact that the quantum case gives us an embarrassment of riches, i.e. widely different behaviours can take place. All of them are unique to the quantum world, and yet some of them mimic the behaviour that most people expect in the classical case.

## 4 A Look at the Hadamard Walk

The first time that the occupation problem was considered for the Hadamard walk is [20]. This is a coined walk with constant coin

$$\frac{1}{\sqrt{2}}\begin{pmatrix} 1 & 1 \\ 1 & -1 \end{pmatrix}. \tag{8}$$

Although this coin is not of the form (2), such a form may be obtained after a change of phases in the basis [4, 5].

Our approach differs from the one in [20], which needs an ad hoc normalization to get true probabilities due to the lack of a connection with a real measurement protocol. Therefore, for a finite number of steps our results do not agree. Nevertheless, we give some evidence that the limiting results apparently—and surprisingly—agree. Clarifying whether this coincidence holds by chance or there are good reasons for it is something that deserves future research.

The monitoring approach to occupation times in the Hadamard walk leads to the results shown in Fig. 2, which plots the density and cumulative distributions for 100, 900 and 8100 time steps. This figure points to a limiting distribution function given by two symmetric deltas at the edges of the interval. The take home message is that, in the limit of an infinite number of steps, the walker is always in the positive subspace or always in the negative subspace, each with probability $1/2$. More precisely, the discrete probabilities $P(N_n = r)$ appear to have a limit as $n \to \infty$ with $r$ held fixed,

$$P_\infty(r) = \lim_{n\to\infty} P(N_n = r) = \lim_{n\to\infty} P(N_n = n - r). \tag{9}$$

As seen in the bottom two panels of Fig. 2, this limit has already nearly been reached for $0 \le r \le 100$ by the time $n = 900$. Indeed the $n = 900$ result lies directly on top of the $n = 8100$ result for $0 \le r \le 100$. The numerical value of $P(N_n \le 100 \text{ or } N_n \ge n - 100)$ with $n = 900$ is 0.99994613 while that of $n = 8100$ is 0.99994530. The reported digits with $n = 8100$ appear to have stabilized, i.e. increasing $n$ further does not change the first eight digits of the probability. So if $n$ is a billion, the probability that the walker will be found on one side all but at most 100 times is 0.99994530, and the fraction of the time it spends on the other side is $\le 100/10^9$, effectively zero.

Using the alternative definitions and approach in [20] we have computed the density and the corresponding cumulative distribution for $n = 100$ and $n = 900$

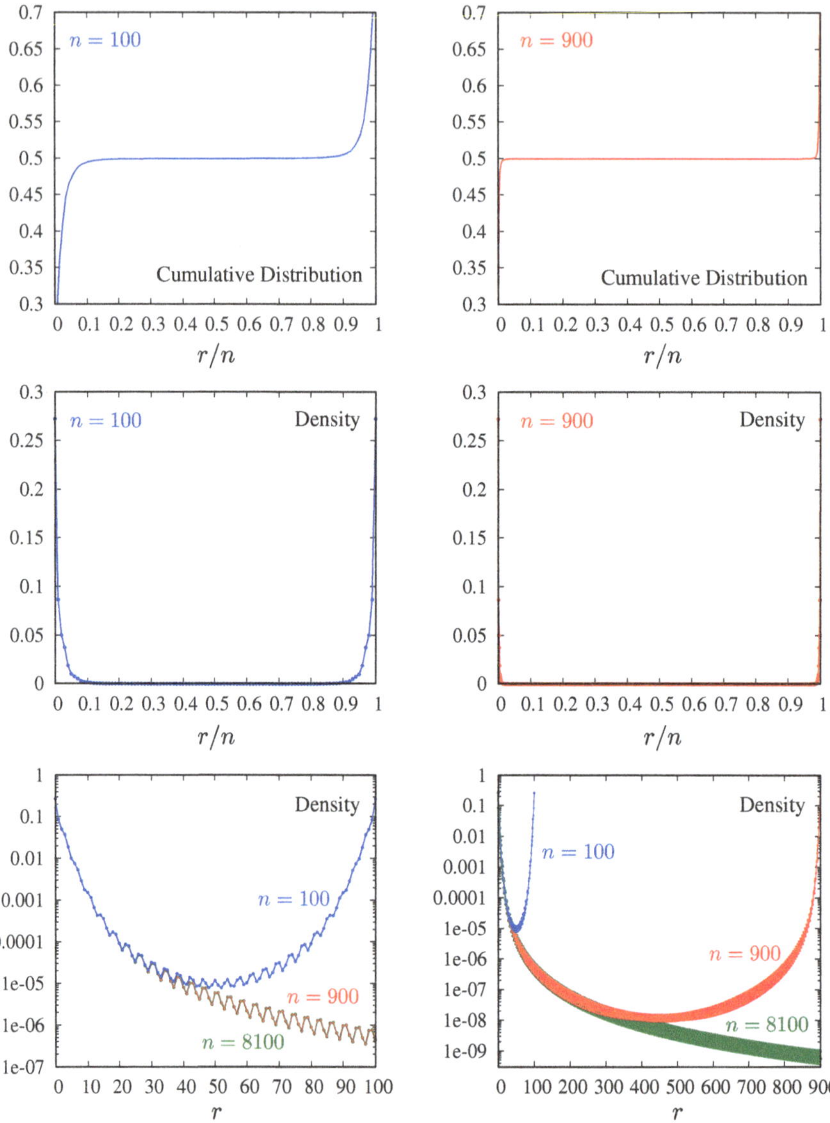


**Fig. 2** Cumulative distribution and density for 100, 900 and 8100 steps of the Hadamard walk (monitoring approach)

steps of the Hadamard walk. The computation was too expensive to proceed past $n = 900$. The setup is different, and only even values of $r$ lead to positive probabilities. Plots of the cumulative distribution and density (i.e. discrete probabilities) are given in Fig. 3. This figure suggests the same limit distribution and behavior as the monitoring approach, though the values of $P_\infty(r)$ will be different.

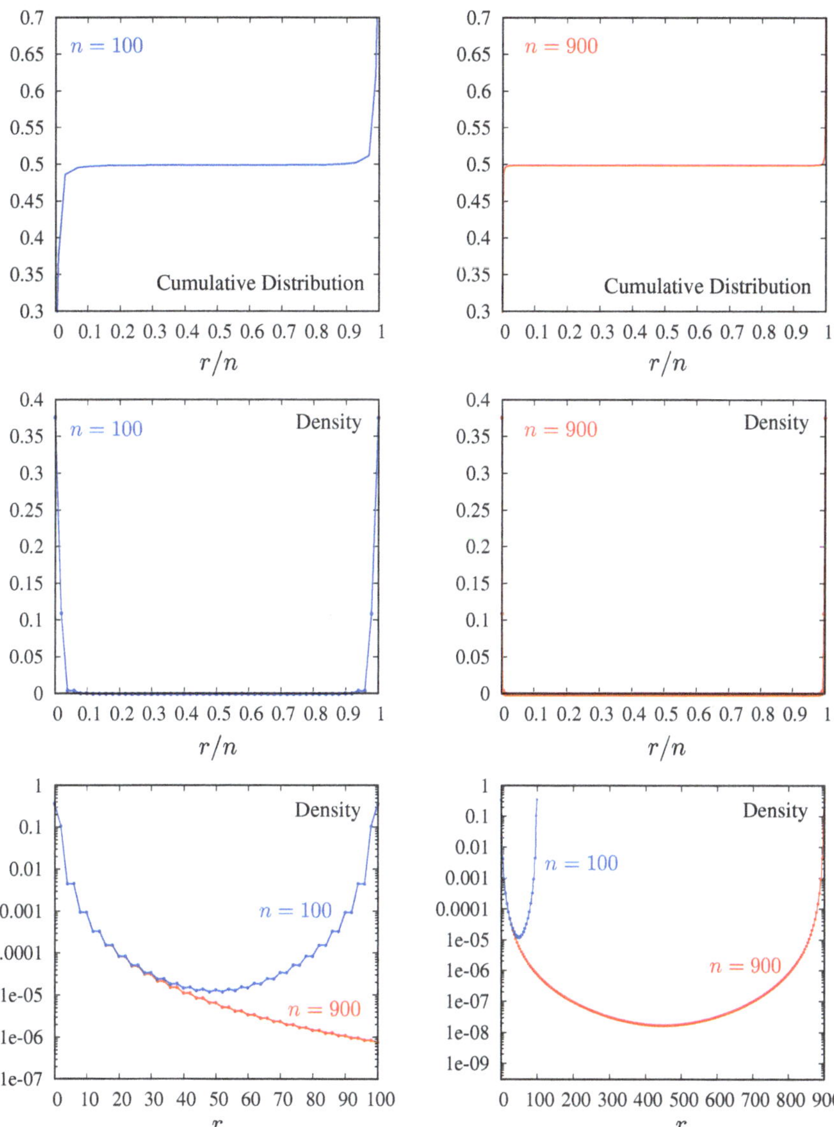


**Fig. 3** Cumulative distribution and density for 100 and 900 steps of the Hadamard walk (following [20])

## 5 The Walk with a Constant Coin

In this section we deal with coined walks with a constant coin

$$C = \begin{pmatrix} \rho & -\alpha \\ \alpha & \rho \end{pmatrix}, \qquad \alpha \in (-1, 1), \qquad \rho = \sqrt{1 - |\alpha|^2}. \tag{10}$$

Here we consider four different cases, corresponding to four choices of $\alpha$, namely $3/5$, $12/13$, $143/145$ and $399/401$. In each case the number of time steps is 120. The plots for the occupation times using the monitoring approach are given in Figs. 4 and 5.

The message here is that for constant coins that are different from the Hadamard one the behaviour can differ substantially from that of the Hadamard case.

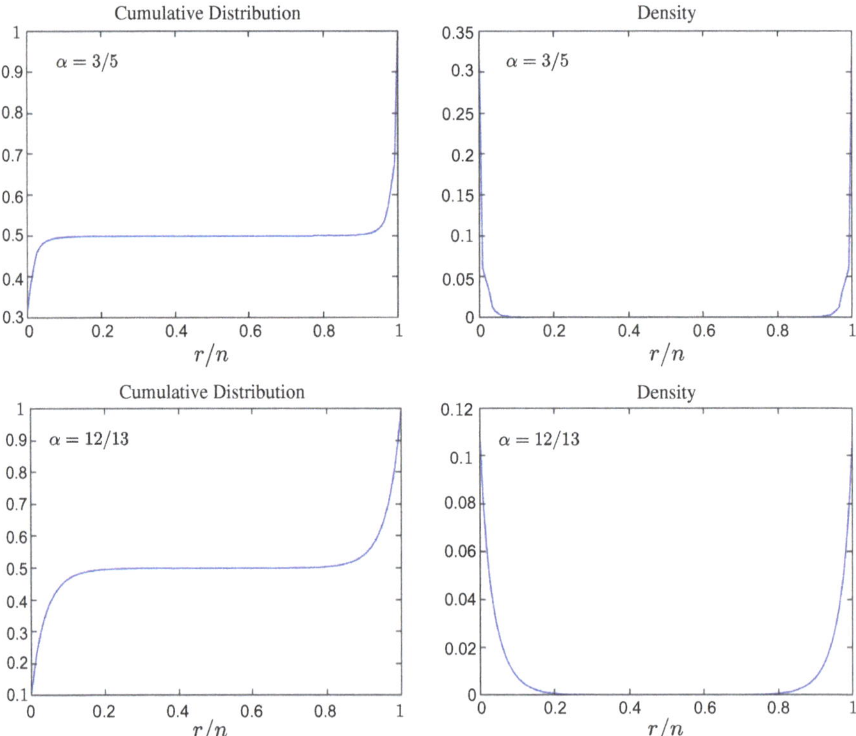


**Fig. 4** Cumulative distribution and density for the coin (10) with $\alpha = 3/5$ and $\alpha = 12/13$, both with 120 steps

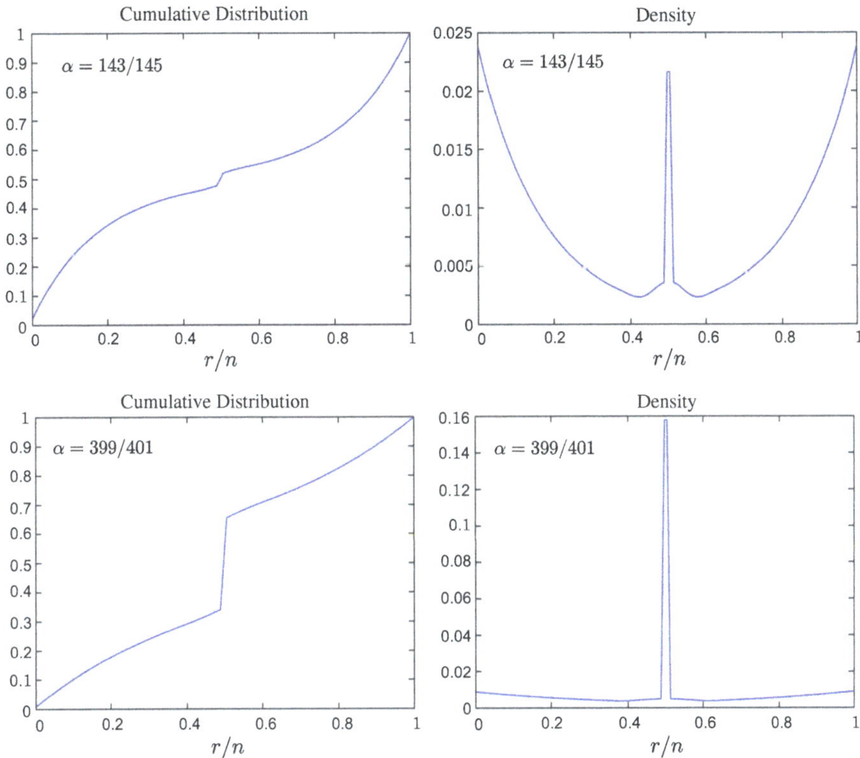


**Fig. 5** Cumulative distribution and density for the coin (10) with $\alpha = 143/145$ and $\alpha = 399/401$, both with 120 steps

## 6 The Even Verblunsky Coefficients Tend to One

In this section we look at a CMV walk whose Verblunsky coefficients $\alpha_{2i} \in (-1, 1)$ depend on the site $i$, approaching the value 1 as $|i|$ tends to infinity. For simplicity we take the odd Verblunsky coefficients to vanish, so as to have a coined walk with site dependent coins $C_i$ of the form (2) such that

$$C_i \quad \xrightarrow{|i|\to\infty} \quad C_\infty = \begin{pmatrix} 0 & -1 \\ 1 & 0 \end{pmatrix}. \tag{11}$$

The three different plots we show differ in the number of time steps. The plots are given in Fig. 6, and the effect of increasing the number of steps is to make the density more and more sharply concentrated around the value $1/2$.

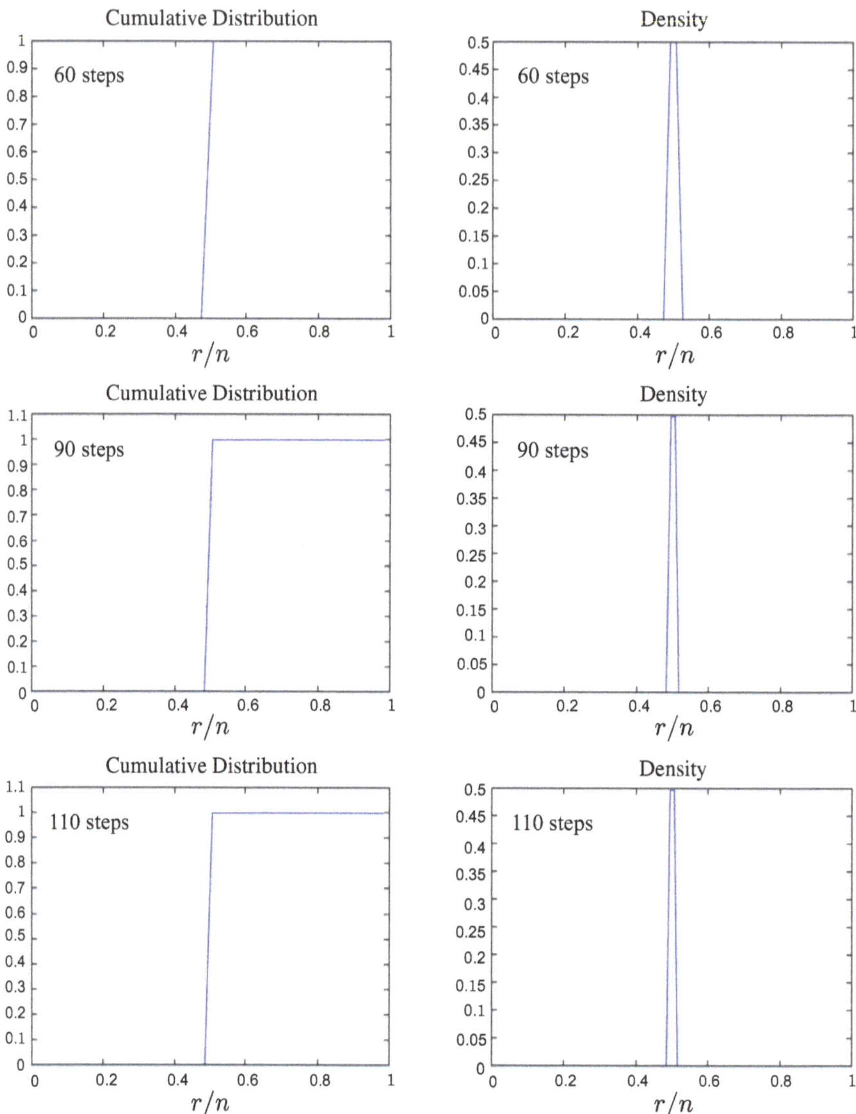


**Fig. 6** Cumulative distribution and density for 60, 90 and 110 steps

The take home message is that in the limit of an infinite number of steps, with probability 1 the walker spends exactly half of the the time on the positive subspace and half on the negative subspace. In spirit, this is what most people feel the classical coin should be doing.

In this example, the Verblunsky coefficients are chosen to satisfy

$$\alpha_{2i-1} = 0, \qquad \alpha_{2i} \xrightarrow{|i|\to\infty} 1.$$

More explicitly, if $i$ is non-negative and even then

$$\alpha_i = \frac{(i+1)^{10}-1}{(i+1)^{10}+1}, \qquad \rho_i = \frac{2(i+1)^5}{(i+1)^{10}+1},$$

and if $i$ is odd then $\alpha_i = 0$ and $\rho_i = 1$.

The corresponding CMV matrix $U$ is therefore a compact perturbation of the unitary matrix obtained by setting $\alpha_{2i-1} = 0$ and $\alpha_{2i} = 1$ in a CMV, which turns out to be a direct sum of the blocks $C_\infty$ in (11). As a consequence of Weyl's theorem on the essential spectrum, the spectrum of $U$ accumulates on the eigenvalues $\pm i$ of $C_\infty$, so that it is pure point.

A couple of natural questions arise: Is the naive behaviour of occupation times observed in this example a common feature of any CMV walk with pure point spectrum? How does the occupation time distribution change when considering a CMV walk with other kind of singular spectrum?

While the first question remains as a challenge, in the next section we explore the second one.

## 7 A Look at the Riesz Walk

In this section we deal with the singular continuous measure constructed by F. Riesz [27], back in 1918. In [12] we study a purely singular continuous quantum walk in the non-negative integers naturally associated to it, by considering the semi-infinite CMV matrix generated by the Riesz measure. In this paper we consider an extension of this walk to the integers.

The measure on the unit circle that F. Riesz built is formally given by the expression

$$d\mu(z) = \prod_{k=1}^{\infty}(1+\cos(4^k\theta))\frac{d\theta}{2\pi} = \prod_{k=1}^{\infty}(1+(z^{4^k}+z^{-4^k})/2)\frac{dz}{2\pi i z}. \tag{12}$$

Here $z = e^{i\theta}$. If one truncates this infinite product the corresponding measure has a nice density. These approximations converge weakly to the Riesz measure.

The construction of this example is based on detailed knowledge of the coefficients $\alpha_i$, $i \geq 0$, for the Riesz measure which are found in [12]. We define a "Riesz walk" on the integers as a a CMV walk defined by extending the Verblunsky coefficients of the Riesz measure to negative indices according to the "fair" rule (7).

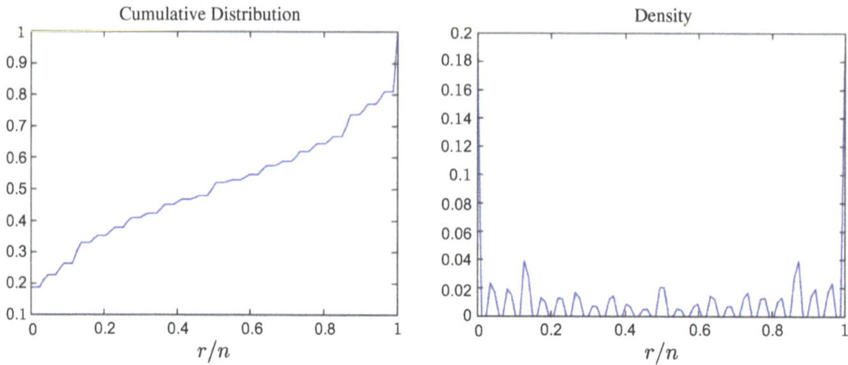


**Fig. 7** Cumulative distribution and density for 90 steps of Riesz's walk

This example is of interest since it is not clear that a "limit law" exists for the "site distribution" of the Riesz walk on the positive half of the integers, see [12]. The same issue arises when one looks at the walk on the integers. The existence of such a limit law for the case of a constant coin is discussed in [19].

Figure 7 shows the density and cumulative distribution for 90 steps of the Riesz walk on the integers.

These examples open up many interesting questions: What can be said in general about the occupation time distribution in CMV walks? Is there any spectral characterization for the different behaviours of such distributions? Is the connection with orthogonal polynomials on the unit circle useful for such a characterization? In particular, is there any important role reserved for the Schur functions in this respect? The last question is motivated by the central role played by Schur functions in both, the theory of orthogonal polynomials on the unit circle [28, 29], and the study of quantum walks [2, 5, 6, 13–15].

**Acknowledgments** The work of the author L. Velázquez has been supported in part by the research project MTM2017-89941-P from Ministerio de Economía, Industria y Competitividad of Spain and the European Regional Development Fund (ERDF), by project UAL18-FQM-B025-A (UAL/CECEU/FEDER) and by projects E26_17R and E48_20R of Diputación General de Aragón (Spain) and the ERDF 2014-2020 "Construyendo Europa desde Aragón."

The work of the author J. Wilkening was supported in part by the US National Science Foundation under award number DMS-1716560 and by the US Department of Energy, Office of Science, Applied Scientific Computing Research, under award number DE-AC02-05CH11231.

## References

1. Y. Aharonov, L. Davidovich, N. Zagury, Quantum random walks. Phys. Rev. A **48**, 1687–1690 (1993)
2. J. Bourgain, F.A. Grünbaum, L. Velázquez, J. Wilkening, Quantum recurrence of a subspace and operator-valued Schur functions. Comm. Math. Phys. **329**, 1031–1067 (2014)

3. M.J. Cantero, L. Moral, L. Velázquez, Five-diagonal matrices and zeros of orthogonal polynomials on the unit circle. Linear Algebra Appl. **362**, 29–56 (2003)
4. M.J. Cantero, F.A. Grünbaum, L. Moral, L. Velázquez, Matrix valued Szegő polynomials and quantum random walks. Commun. Pure Appl. Math. **58**, 464–507 (2010)
5. M.J. Cantero, F.A. Grünbaum, L. Moral, L. Velázquez, The CGMV method for quantum walks. Quantum Inf. Process. **11**, 1149–1192 (2012)
6. C. Cedzich, T. Geib, F.A. Grünbaum, L. Velázquez, A.H. Werner, R.F. Werner, Quantum walks: schur functions meet symmetry protected topological phases (2019). arXiv:1903.07494
7. K.L. Chung, W. Feller, On fluctuations in coin-tossing. PNAS **35**, 605–609 (1949)
8. P. Ernst, L. Shepp, On occupation times of the first and third quadrants for planar Brownian motion. J. Appl. Prob. **54**, 337–342 (2017)
9. W. Feller, *An Introduction to Probability Theory and Its Applications*, vol. 1 (J. Wiley & Sons, Inc., Hoboken, 1968)
10. F.A. Grünbaum, A Feynman-Kac approach to a paper of Chung and Feller on fluctuations in the coin-tossing game. PAMS (2018). https://doi.org/10.1090/proc/14758
11. F.A. Grünbaum, C. McGrouther, Occupation time for two dimensional Brownian motion in a wedge, in *Proceedings of Symposia in Applied Mathematics*, vol. 65 (2007)
12. F.A. Grünbaum, L. Velázquez, The quantum walk of F. Riesz, in *Foundations of Computational Mathematics (Budapest, 2011)*, pp. 93–112. London Mathematical Society. Lecture Note Series 403 (Cambridge University Press, Cambridge, 2013)
13. F.A. Grünbaum, L. Velázquez, A generalization of Schur functions: applications to Nevanlinna functions, orthogonal polynomials, random walks and unitary and open quantum walks. Adv. Math. **326**, 352–464 (2018)
14. F.A. Grünbaum, C. Lardizabal, L. Velázquez, Quantum Markov chains: recurrence, schur functions and splitting rules. Ann. Henri Poincaré **21**, 189–239 (2020)
15. F.A. Grünbaum, L. Velázquez, A.H. Werner, R.F. Werner, Recurrence for discrete time unitary evolutions. Comm. Math. Phys. **320**, 543–569 (2013)
16. F.A. Grünbaum, L. Velázquez, J. Wilkening, Occupation times for quantum walks (Manuscript in preparation)
17. M. Kac, On some connections between probability theory and differential and integral equations, in Proceedings of the Second Berkeley Symposium on Mathematical Statistics and Probability (University of California Press, Berkeley, 1951), pp. 189–215
18. J. Kempe, Quantum random walks-an introductory overview. Contemp. Phys. **44**(4), 307–327 (2003)
19. N. Konno, Quantum walks, in *Quantum Potential Theory*, ed. by U. Franz, M. Schürmann. Lecture Notes in Mathematics 1954 (Springer, Berlin, 2008)
20. N. Konno, Sojourn times of the Hadamard walk in one dimension. Quantum Inf. Process. **11**, 465–480 (2012)
21. N. Konno, E. Segawa, Localization of discrete-time quantum walks on a half line via the CGMV method. Quantum Inf. Comput. **11**, 485–495 (2011)
22. N. Konno, E. Segawa, One-dimensional quantum walks via generating function and the CGMV method. Quantum Inf. Comput. **14**, 1165–1186 (2014)
23. H. Krovi, T. Brun, Hitting time for quantum walks on the hypercube. Phys. Rev. A **73**, 032341 (2006)
24. P. Lévy, Sur certains processus stochastiques homogenes. Compositio Math. **7**, 283–339 (1939)
25. H.P. McKean Jr., *Probability Theory, the Classical Limit Theorems* (Cambridge University Press, Cambridge, 2014)
26. A. Renyi, Legendre polynomials and probability theory. Ann. Univ. Sci. Budapest, Eötvös Sect. Math **3–4**, 247–251 (1960–1961)
27. F. Riesz, Über die Fourierkoeffizienten einer stetigen Funktion von beschränkter Schwankung. Math. Z. **18**, 312–315 (1918)
28. B. Simon, *Orthogonal Polynomials on the Unit Circle, Part 1: Classical Theory*. AMS Colloquium Publications, vol. 54.1 (American Mathematical Society, Providence 2005)

29. B. Simon, *Orthogonal Polynomials on the Unit Circle, Part 2: Spectral Theory*, AMS Colloquium Publications, vol. 54.2 (American Mathematical Society, Providence, 2005)
30. D. Stroock, *Probability Theory, an Analytical View* (Cambridge University Press, Cambridge, 1993)

# On Foci of Ellipses Inscribed in Cyclic Polygons

**Markus Hunziker, Andrei Martinez-Finkelshtein, Taylor Poe, and Brian Simanek**

*To our friend, colleague and mentor Lance Littlejohn in celebration of his distinguished career.*

**Abstract** Given a natural number $n \geq 3$ and two points $a$ and $b$ in the unit disk $\mathbb{D}$ in the complex plane, it is known that there exists a unique elliptical disk having $a$ and $b$ as foci that can also be realized as the intersection of a collection of convex cyclic $n$-gons whose vertices fill the whole unit circle $\mathbb{T}$. What is less clear is how to find a convenient formula or expression for such an elliptical disk. Our main results reveal how orthogonal polynomials on the unit circle provide a useful tool for finding such a formula for some values of $n$. The main idea is to realize the elliptical disk as the numerical range of a matrix and the problem reduces to finding the eigenvalues of that matrix.

**Keywords** Orthogonal polynomials · Poncelet Ellipses · Blaschke products

M. Hunziker · B. Simanek (✉)
Department of Mathematics, Baylor University, Waco, TX, USA
e-mail: Markus_Hunziker@baylor.edu; brian_simanek@baylor.edu

A. Martinez-Finkelshtein
Department of Mathematics, Baylor University, Waco, TX, USA

Department of Mathematics, University of Almería, Almería, Spain
e-mail: a_martinez-finkelshtein@baylor.edu

T. Poe
Department of Mathematics, Mississippi College, Clinton, Mississippi, USA
e-mail: tpoe@mc.edu



F. Gesztesy, A. Martinez-Finkelshtein (eds.), *From Operator Theory to Orthogonal Polynomials, Combinatorics, and Number Theory*, Operator Theory: Advances and Applications 285, https://doi.org/10.1007/978-3-030-75425-9_12

# 1 Introduction

Suppose $n \geq 3$ is a natural number and $E$ is an ellipse in the open unit disk $\mathbb{D}$ in the complex plane. A classical result known as *Poncelet's Theorem* asserts that if there is an $n$-gon $P$ inscribed in the unit circle $\mathbb{T}$ (this is what we call a *cyclic $n$-gon*) with every side of $P$ tangent to $E$, then there are in fact infinitely many such $n$-gons and the union of the vertices of these $n$-gons fills $\mathbb{T}$ (see [6] and Fig. 1). If this property holds and all of the circumscribing $n$-gons are convex, then we will say the ellipse $E$ is a *Poncelet $n$-ellipse*. A simple argument shows that if $a, b \in \mathbb{D}$ and $n \geq 3$ is a natural number, then there exists a unique Poncelet $n$-ellipse with foci at $a$ and $b$ (see [16, Sect. 12]). However, if $n > 3$, then it is not obvious how to write down a formula for this ellipse or deduce any properties of its size (such as area, eccentricity, etc.). Some early relevant formulas for this purpose were found by Cayley [5, Chapter 5], but they are not easy formulas to use. Some of the main results in this paper will show how to write an explicit expression for Poncelet $n$-ellipses when $n = 4$ or $n = 6$.

To accomplish this task, we must frame this problem in a broader context. We start by recalling that the *numerical range* of a matrix $A \in \mathbb{C}^{n \times n}$ is the subset of the complex plane $\mathbb{C}$ given by

$$W(A) = \{\langle x, Ax \rangle : x \in \mathbb{C}^n, \ \|x\| = 1\}.$$

For any matrix $A$, the set $W(A)$ is a compact and convex subset of $\mathbb{C}$ (a fact known as the Toeplitz-Hausdorff Theorem) that contains the eigenvalues of $A$. If $W(A)$ is bounded by an ellipse, then we will say that $W(A)$ is an *elliptical disk*.

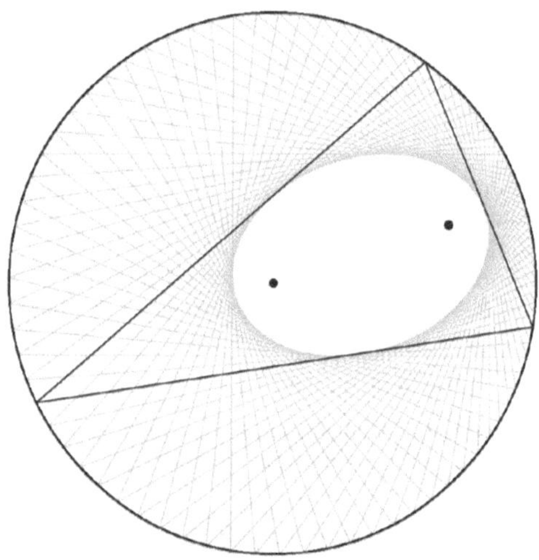

**Fig. 1** A Poncelet 3-ellipse and its two foci

The matrices $A$ that we are most interested in have the following properties:

(i) $\|A\| \leq 1$;
(ii) all eigenvalues of $A$ are in $\mathbb{D}$;
(iii) $\operatorname{rank}(I - AA^*) = \operatorname{rank}(I - A^*A) = 1$.

The set of all $n \times n$ matrices satisfying properties (i-iii) will be denoted by $S_n$. It is known that the numerical range of a matrix in $S_n$ is the convex hull of an algebraic curve of class $n$ and has the $(n + 1)$-Poncelet property, meaning that every point on $\partial W(A)$ is a point of tangency for a convex cyclic $(n + 1)$-gon that is circumscribed about $W(A)$ (see [2, 11, 14] for details).

Our approach to the problem described in the opening paragraph has its roots in work of Gau and Wu [11] and aims to realize the desired ellipse as the numerical range of an appropriate matrix in $S_{n-1}$. This idea is known to have extremely interesting consequences and has inspired a significant amount of recent and ongoing research (see [2–4, 7, 8, 13, 14, 17, 19]). The following theorem (from [19]) is the starting point of our investigation and validates this approach to the problem. We state it using the terminology that we have defined so far.

**Theorem 1** *[19] Suppose $f_1, f_2 \in \mathbb{D}$ and $n \geq 3$. There exists a Poncelet $n$-ellipse with foci at $f_1$ and $f_2$. Furthermore, this ellipse forms the boundary of the numerical range of a matrix $A \in S_{n-1}$ and $f_1, f_2$ are eigenvalues of $A$.*

One simplification of our problem comes from the fact that instead of finding a matrix in $S_{n-1}$ with the desired properties, it suffices to find just the eigenvalues of that matrix. This is because numerical ranges are preserved by unitary conjugation and every matrix in $S_{n-1}$ is unitarily equivalent to a canonical form called a cutoff CMV matrix (see [14, 16]). Such matrices are an important part of the theory of orthogonal polynomials on the unit circle (OPUC) and will play an essential role in our analysis. Further details are presented in Sect. 2.

Returning to our original problem, notice that Theorem 1 states that both $a$ and $b$ must be eigenvalues of the desired matrix. Thus, one really only needs to determine the remaining $n - 3$ eigenvalues, which is why the problem becomes trivial when $n = 3$. When $n = 4$, a concise formula for the one remaining eigenvalue appears in [13], and we give another proof of that formula in Sect. 3 (Fig. 2). In [17], Mirman presented a collection of algebraic relationships that must be satisfied by the eigenvalues we seek. We will state his result as the following theorem, which is a restatement of [17, Equation 30] using the terminology of matrices from $S_n$.

**Theorem 2** *Suppose $E_n$ is a Poncelet $n$-ellipse in $\mathbb{D}$ that is also the boundary of the numerical range of a matrix $A \in S_{n-1}$. Suppose the foci of $E_n$ are $f_1$ and $f_2$. Then the eigenvalues of $A$ can be labeled $\{w_1, \ldots, w_{n-1}\}$ so that $w_1 = f_1$, $w_{n-1} = f_2$, and*

$$w_{j-1}w_{j+1} = \frac{(w_j - f_1)(w_j - f_2)}{(1 - \bar{f}_1 w_j)(1 - \bar{f}_2 w_j)}, \qquad j = 2, 3, \ldots, n-2 \tag{1}$$

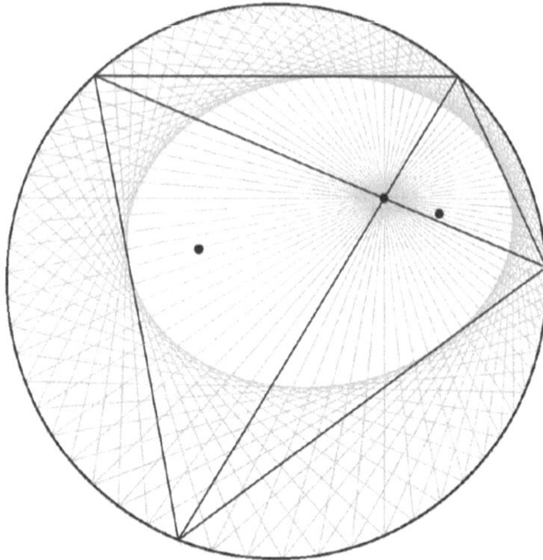

**Fig. 2** The Poncelet 4-ellipse with foci $-0.28+0.12i$, $0.6+0.24i$ and pentagram point $0.4+0.3i$

We will refer to the system of Eqs. (1) as the "Mirman system". It gives us a necessary but not sufficient condition for a set of eigenvalues to be the spectrum of a matrix in $S_{n-1}$ whose numerical range is bounded by an ellipse with foci at $f_1$ and $f_2$. It falls short of presenting a complete solution to our problem because the algebraic relationships admit multiple solutions (even in $\mathbb{D}^{n-3}$), i.e. this system of equations has many solutions and only one of them has the desired interpretation.

Our path forward is now clear. Given $f_1, f_2 \in \mathbb{D}$, we want to find a matrix $A \in S_{n-1}$ such that $\partial W(A)$ is an ellipse with foci at $f_1$ and $f_2$. Theorem 1 tells us that such an $A$ exists, and we know that we can realize it as a cutoff CMV matrix. Such a matrix has $n-1$ eigenvalues in $\mathbb{D}$, two of which must be $f_1$ and $f_2$. A priori, there are no other restrictions on the remaining eigenvalues of $A$ other than they must be in $\mathbb{D}$. In Sect. 3, we will show how to locate the third eigenvalue of $A$ when $n = 4$ and consider an even more general situation in which both $f_1$ and $f_2$ are eigenvalues of $A$, but only one is a focus of the Poncelet 4-ellipse (Fig. 3). In Sect. 4 we will consider the case when $n = 6$ and find a set of algebraic equations in three variables whose unique solution in $\mathbb{D}^3$ marks the locations of the other three eigenvalues that we seek. Section 4 also considers more general problems related to Poncelet ellipses. In Sect. 5, we will consider matrices in $S_4$ and discover some properties of all of the solutions to the Mirman system when $n = 5$. The most significant results currently known for general $n$ come from [17–19].

The next section is a brief review of the relationship between $S_n$ and the theory of OPUC. It includes much of the background, notation, and terminology that will be relevant to the remainder of the paper. Many of these topics are discussed in much greater detail in [4, 14], and a thorough introduction to orthogonal polynomials on the unit circle can be found in [21, 22].

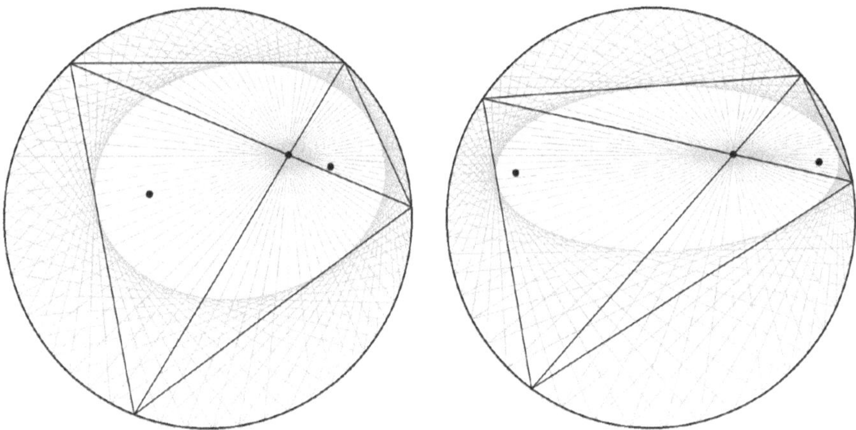

**Fig. 3** Two Poncelet 4-ellipses with the same pentagram point, $0.4 + 0.3i$

## 2 Background and Notation

The class $S_n$ is of such fundamental importance to our analysis that we must begin with a review of some key properties of this class. There are several canonical forms of matrices from $S_n$ (see [14, Sect. 2.3]) and the one that we will use is that of a cutoff CMV matrix. To define a CMV matrix, first define a sequence of $2 \times 2$ matrices $\{\Theta_j\}_{j=0}^{\infty}$ by

$$\Theta_j = \begin{pmatrix} \bar{\alpha}_j & \sqrt{1-|\alpha_j|^2} \\ \sqrt{1-|\alpha_j|^2} & -\alpha_j \end{pmatrix},$$

where $\alpha_j \in \mathbb{D}$. One then defines the operators $\mathcal{L}$ and $\mathcal{M}$ by

$$\mathcal{L} = \Theta_0 \oplus \Theta_2 \oplus \Theta_4 \oplus \cdots, \qquad \mathcal{M} = 1 \oplus \Theta_1 \oplus \Theta_3 \oplus \cdots$$

where the initial 1 in the definition of $\mathcal{M}$ is a $1 \times 1$ identity matrix. The *CMV matrix* corresponding to the sequence $\{\alpha_n\}_{n=0}^{\infty}$ is

$$\mathcal{G} := \mathcal{L}\mathcal{M} = \begin{pmatrix} \overline{\alpha_0} & \overline{\alpha_1}\rho_0 & \rho_1\rho_0 & 0 & 0 & \dots \\ \rho_0 & -\overline{\alpha_1}\alpha_0 & -\rho_1\alpha_0 & 0 & 0 & \dots \\ 0 & \overline{\alpha_2}\rho_1 & -\overline{\alpha_2}\alpha_1 & \overline{\alpha_3}\rho_2 & \rho_3\rho_2 & \dots \\ 0 & \rho_2\rho_1 & -\rho_2\alpha_1 & -\overline{\alpha_3}\alpha_2 & -\rho_3\alpha_2 & \dots \\ 0 & 0 & 0 & \overline{\alpha_4}\rho_3 & -\overline{\alpha_4}\alpha_3 & \dots \\ \dots & \dots & \dots & \dots & \dots & \dots \end{pmatrix}, \quad \rho_n = \sqrt{1-|\alpha_n|^2} \tag{2}$$

(see [21, Sect. 4.2]). Since each of $\mathcal{L}$ and $\mathcal{M}$ is a direct sum of unitary matrices, each of $\mathcal{L}$ and $\mathcal{M}$ is unitary and hence $\mathcal{G}$ is unitary as an operator on $\ell^2(\mathbb{N})$. The principal $n \times n$ submatrix of $\mathcal{G}$ will also be called the *$n \times n$ cut-off CMV matrix*, which we will denote by $\mathcal{G}^{(n)}$. It is easy to see that $\mathcal{G}^{(n)} \in S_n$.

CMV matrices are intimately connected with the theory of orthogonal polynomials on the unit circle. Indeed, if one defines

$$\Phi_n(z) := \det(zI_n - \mathcal{G}^{(n)}), \tag{3}$$

then the polynomial $\Phi_n(z)$ is the degree $n$ monic orthogonal polynomial with respect to the measure $\mu$ that is the spectral measure of $\mathcal{G}$ and the vector $\mathbf{e}_1$. If we need to make explicit reference to the measure $\mu$, then we will write $\Phi_{n,\mu}$. Since $\mathcal{G}$ is unitary, the formula (3) implies that all zeros of $\Phi_n$ are in $\mathbb{D}$. Furthermore, the coefficients $\{\alpha_n\}_{n=0}^{\infty}$ that are used to define $\mathcal{G}$ are related to $\{\Phi_n\}_{n=0}^{\infty}$ by the Szegő recursion:

$$\begin{pmatrix} \Phi_{k+1}(z) \\ \Phi_{k+1}^*(z) \end{pmatrix} = \begin{pmatrix} z & -\overline{\alpha_k} \\ -\alpha_k z & 1 \end{pmatrix} \begin{pmatrix} \Phi_k(z) \\ \Phi_k^*(z) \end{pmatrix}, \tag{4}$$

where if

$$\Phi_n(z) = \sum_{j=0}^{n} c_j z^j,$$

then

$$\Phi_n^*(z) = \sum_{j=0}^{n} \overline{c_j} z^{n-j} = z^n \, \overline{\Phi_n(1/\overline{z})}. \tag{5}$$

$\Phi_n^*$ is called the reversed polynomial of $\Phi_n$. Observe that $\Phi_n^*$ can be of degree strictly less than $n$. It follows from the Szegő recursion that $\alpha_n = -\overline{\Phi_{n+1}(0)}$ and the sequence $\{\alpha_n\}_{n=0}^{\infty}$ is often called the sequence of *Verblunsky coefficients* for the measure $\mu$. For future use, let us define the notation

$$\Phi_n(z) = \mathcal{S}_{\alpha_{n-1}}(\Phi_{n-1}(z)), \qquad \Phi_n^*(z) = \mathcal{T}_{\alpha_{n-1}}(\Phi_{n-1}^*(z)),$$

to say that $\Phi_n$ is related to $\Phi_{n-1}$ by the Szegő recursion and the parameter $\alpha_{n-1}$.

Some of the most important theorems in the study of OPUC come from establishing the following bijections (see [21, Chapter 1]):

- All of the zeros of $\Phi_{n,\mu}(z)$ are in $\mathbb{D}$ and any collection $\{z_j\}_{j=1}^n \in \mathbb{D}^n$ is the zero set of $\Phi_{n,\nu}(z)$ for some $\nu$ supported on $\mathbb{T}$.
- The sequence $\{\Phi_{n,\mu}(0)\}_{n\in\mathbb{N}}$ is a sequence in $\mathbb{D}$ and every sequence $\{\gamma_j\}_{j\in\mathbb{N}} \in \mathbb{D}^{\mathbb{N}}$ satisfies $\Phi_{n,\nu}(0) = \gamma_n$ for all $n \in \mathbb{N}$ and some measure $\nu$ supported on $\mathbb{T}$.

This last fact (known as Verblunsky's Theorem [21, Sect. 1.7]) tells us that the sequence $\{\alpha_n\}_{n=0}^{\infty}$ completely characterizes the measure $\mu$ on $\mathbb{T}$. If it is necessary to make the sequence of Verblunsky coefficients explicit, we will write $\Phi_n(z; \alpha_0, \ldots, \alpha_{n-1})$. We also note here that the Szegő recursion is invertible. This means that if we know $\Phi_n(z; \alpha_0, \ldots, \alpha_{n-1})$, then we can recover $\Phi_j(z; \alpha_0, \ldots, \alpha_{j-1})$ for all $j < n$ and hence we can recover $\alpha_j$ for all $j = 0, \ldots, n-1$ (by evaluation at 0).

Given a family of OPUC, replacing the last Verblunsky coefficient $\alpha_{n-1}$ in the Szegő recursion with $\lambda \in \mathbb{T}$ yields a degree-$n$ *paraorthogonal polynomial* on the unit circle (POPUC)

$$\Phi_n(z; \alpha_0, \ldots, \alpha_{n-2}, \lambda) = z\Phi_{n-1}(z; \alpha_0, \ldots, \alpha_{n-2}) - \overline{\lambda}\Phi_{n-1}^*(z; \alpha_0, \ldots, \alpha_{n-2}).$$

In contrast to OPUC, the zeros of POPUC are on $\mathbb{T}$. Given $\Phi_{n-1}(z)$, we can define $\{\Phi_n^{(\lambda)}\}$ as the set of all degree-$n$ POPUC for $\Phi_{n-1}$ as $\lambda$ varies around $\mathbb{T}$. We have already noted that the OPUC $\Phi_n$ is the characteristic polynomial of a cut-off CMV matrix $\mathcal{G}^{(n)}$. Using this parameter $\lambda \in \mathbb{T}$, we can characterize a family of rank one unitary dilations of $\mathcal{G}^{(n)}$. By adding one row and one column to $\mathcal{G}^{(n)}$, we define a unitary $(n+1) \times (n+1)$ matrix whose characteristic polynomial is $\Phi_{n+1}^{(\lambda)}(z)$. While the numerical range of $\mathcal{G}^{(n)}$ has the $(n+1)$-Poncelet property, the numerical ranges of its unitary dilations are bounded by $(n+1)$-gons inscribed in $\mathbb{T}$ and circumscribed around $W(\mathcal{G}^{(n)})$. The vertices of these $(n+1)$-gons are the eigenvalues of the matrices, or equivalently, the zeros of the POPUC.

We have already mentioned that the boundary of the numerical range of $\mathcal{G}^{(n)}$ has the $(n+1)$-Poncelet property. This phenomenon can be reformulated as saying that $\partial W(\mathcal{G}^{(n)})$ is the envelope of the family of circumscribing $(n+1)$-gons. The precise definition of an envelope and its relationship to numerical ranges is complicated, so we will restrict our attention only to the most relevant facts and refer the reader to [14] for details. The envelope is most easily understood by means of a dual curve. If the matrix is $\mathcal{G}^{(n)}$, then the dual curve is an algebraic curve of degree exactly $n$ and the dual of that dual is an algebraic curve of class $n$ with multiple components. The largest component is the boundary of the numerical range of $\mathcal{G}^{(n)}$ and we denote this component by $C_1$ (to be consistent with notation in [14]). The other components can be numbered $C_2, \ldots, C_{\lceil n/2 \rceil}$ and they too have an interpretation in terms of the $(n+1)$-gons that circumscribe $\partial W(\mathcal{G}^{(n)})$.

To understand this interpretation, let us consider the component $C_2$, which we call the *Pentagram curve* after the pentagram map from [20]. For each $(n+1)$-gon $P$ that circumscribes $\partial W(\mathcal{G}^{(n)})$, let us order the vertices cyclically on $\mathbb{T}$ by $P_1, P_2, \ldots, P_{n+1}$. Consider now the "polygon" obtained by joining $P_j$ to $P_{j+2}$ for every $j = 1, \ldots, n+1$ (where arithmetic is done modulo $n+1$). The resulting shape will be a non-convex $(n+1)$-gon if $n$ is even and it will be two $(n+1)/2$-gons with interlacing vertices if $n$ is odd. In either case, one can define the envelope of this collection of polygons and that will be the curve $C_2$. A similar construction yields the other curves $C_j$. When $n = 6$, we will refer to the curve $C_3$ as the *Brianchon*

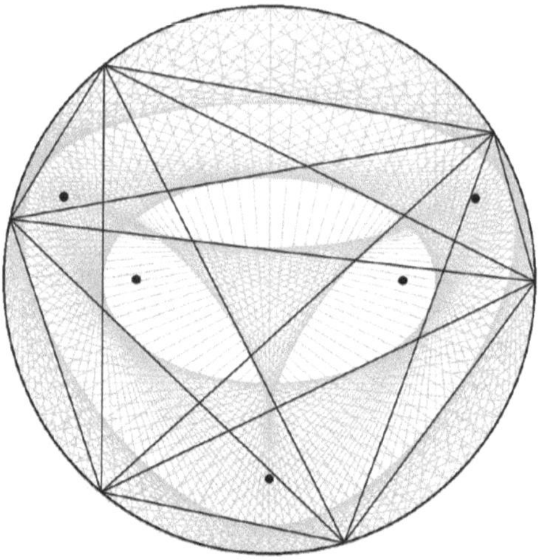

**Fig. 4** A 6-Poncelet curve such that the pentagram curve is an ellipse but the Brianchon curve is not a single point

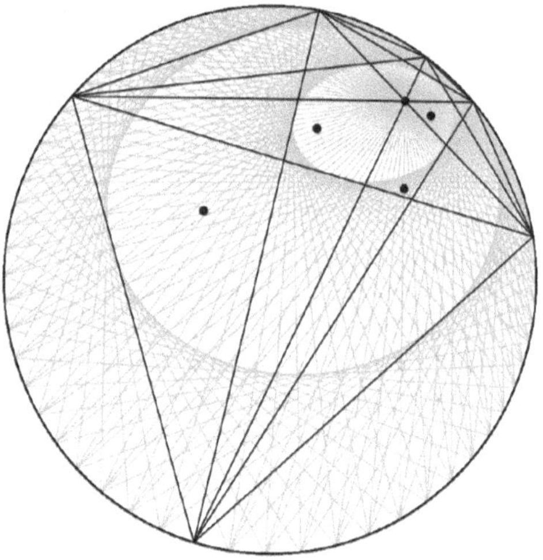

**Fig. 5** A Poncelet 6-ellipse along with the pentagram ellipse and Brianchon point

*curve* (after Brianchon's Theorem [4, Theorem 5.4]) and we note that the curve obtained from this procedure can be a single point (see Figs. 4 and 5).

It was Darboux who first proved that if the curve $C_1$ is an ellipse, then all of the other curves $C_j$ are ellipses, where we consider a single point to be a degenerate ellipse (see [14]). These ellipses all happen to be from the same package (see [17]). In the context of our motivating problem from the opening paragraph, if $C_1$ is an ellipse, then the foci of $C_1$ are eigenvalues of $\mathcal{G}^{(n)}$. Darboux's result implies that the other $C_j$ curves are ellipses, and their foci are also eigenvalues of $\mathcal{G}^{(n)}$. It turns out that one can say something similar even if $C_1$ is not an ellipse. More precisely, if any curve $C_j$ is an ellipse, then the foci of that ellipse are eigenvalues of $\mathcal{G}^{(n)}$. For this reason, finding matrices $\mathcal{G}^{(n)}$ for which some components $C_j$ are ellipses is an interesting problem related to our primary objective, and we will present results of this kind in later sections (see Theorem 3, Theorem 10, and Theorem 11 below).

We will also work with *Blaschke products*

$$B_n(z) := \frac{\Phi_n(z)}{\Phi_n^*(z)}, \tag{6}$$

where

$$\Phi_n(z) = \prod_{j=1}^{n} \left(z - z_j\right), \qquad |z_j| < 1.$$

We will use the notation $B_n(z)$ in this context throughout the rest of the paper and we will say that the Blaschke product $B_n(z)$ has degree $n$. If we need to make the dependence on the zeros explicit, then we will write $B_n(z; z_1, \dots, z_n)$. We will say that a Blaschke product $B_n(z)$ is *regular* if $B_n(0) = 0$.

Our discussion so far shows that the following sets are in bijection with one another:

(i) equivalence classes of matrices in $S_{n-1}$ (where equivalence is defined by unitary conjugation)
(ii) monic polynomials of degree $n - 1$ with all of their zeros in $\mathbb{D}$
(iii) regular degree $n$ Blaschke products
(iv) $\mathbb{D}^{n-1}$ (thought of as collections of Verblunsky coefficients)

Much of what we will do in Sects. 3 and 4 relates properties of the numerical range of a cutoff CMV matrix $\mathcal{G}^{(n-1)}$ to properties of the corresponding Blaschke product $z\Phi_{n-1}(z)/\Phi_{n-1}^*(z)$. The next result will be very helpful in that regard. It is essentially an OPUC version of a result from [3].

**Theorem 3** *Let $n = jk$ and $B_n(z)$ be a regular Blaschke product, i.e. $B_n(z) = \frac{z\Phi_{n-1}(z)}{\Phi_{n-1}^*(z)}$. Then $B_n(z)$ can be expressed as a composition of two regular Blaschke products $B_j(B_k(z))$ (with the degree of $B_m$ equal to $m$) if and only if $\Phi_{n-1}(z)$ factors as*

$$\Phi_{n-1}(z) = \Phi_{k-1}(z) \prod_{m=1}^{j-1} \mathcal{S}_{\bar{a}_m}(\Phi_{k-1}(z))$$

*for some $\Phi_{k-1}$ having all of its zeros in $\mathbb{D}$ and some $\{a_1, \dots, a_{j-1}\} \in \mathbb{D}^{j-1}$. If this factorization holds, then the zeros of $\Phi_{k-1}$ are the zeros of $B_k(z)/z$ and $\{a_1, \dots, a_{j-1}\}$ is the zero set of $B_j(z)/z$.*

***Proof*** Let $B_n(z) = B_j(B_k(z))$. Then

$$B_j(z) = \frac{z(z - a_1)\dots(z - a_{j-1})}{(1 - \overline{a_1}z)\dots(1 - \overline{a_{j-1}}z)}$$

for some $\{a_1, \ldots, a_{j-1}\} \in \mathbb{D}^{j-1}$. If

$$B_k(z) = \frac{z\Phi_{k-1}(z)}{\Phi_{k-1}^*(z)},$$

then

$$\begin{aligned} B_n(z) &= B_j\left(\frac{z\Phi_{k-1}(z)}{\Phi_{k-1}^*(z)}\right) \\ &= \frac{z\Phi_{k-1}(z)(z\Phi_{k-1}(z) - a_1\Phi_{k-1}^*(z)) \cdots (z\Phi_{k-1}(z) - a_{j-1}\Phi_{k-1}^*(z))}{\Phi_{k-1}^*(z)(\Phi_{k-1}^*(z) - \overline{a_1}z\Phi_{k-1}(z)) \cdots (\Phi_{k-1}^*(z) - \overline{a_{j-1}}z\Phi_{k-1}(z))} \end{aligned}$$

It follows that

$$B_n(z) = \frac{z\Phi_{k-1}(z)\mathcal{S}_{\bar{a}_1}(\Phi_{k-1}(z)) \cdots \mathcal{S}_{\bar{a}_{j-1}}(\Phi_{k-1}(z))}{\Phi_{k-1}^*(z)\mathcal{T}_{\bar{a}_1}(\Phi_{k-1}^*(z)) \cdots \mathcal{T}_{\bar{a}_{j-1}}(\Phi_{k-1}^*(z))}$$

and hence $\Phi_{n-1}$ has the desired factorization. Reversing this reasoning shows the converse statement. □

As observed in [3], when the Blaschke product $B_n$ decomposes as in Theorem 3, a particular curve $C_j$ will be the boundary of the numerical range of a matrix in some $S_m$ with $m < n - 1$. When $n = 6$, this will be a key observation that allows us to find Poncelet 5-ellipses. When $n$ is larger, the effected curves will not necessarily be ellipses, but one will observe a certain decoupling of some curves $C_j$ from the other components.

Now that we have the appropriate tools and notation at our disposal, we are ready to explore how the class $S_{n-1}$ can help us find Poncelet $n$-ellipses. The remainder of this paper will focus on small values of $n$.

## 3 The Quadrilateral Case

In this section, we will consider matrices in $S_3$ to show how our approach to Poncelet ellipses using OPUC allows us to reformulate, and in some cases strengthen, existing results in the literature. A classification of those matrices $A \in S_3$ for which $W(A)$ is an elliptical disk is given in [13] (see also [15]). Recall that a Blaschke product is regular if it maps 0 to 0. Gorkin and Wagner [13, Theorem 3.3] showed that $A \in S_3$ has a numerical range that is an elliptical disk if and only if there are regular Blaschke products $B_2, C_2$ of degree 2 such that

$$B_A(z) := \frac{z\det(zI - A)}{[\det(zI - A)]^*} = B_2(C_2(z)) \tag{7}$$

(see also [3, 7]). By Theorem 3, we can state the following, using the notation introduced in Sect. 2.

**Theorem 4** *Suppose $A \in S_3$. The numerical range of $A$ is an elliptical disk if and only if there exist $a, b \in \mathbb{D}$ so that*

$$\det(zI - A) = \Phi_1(z; a)\Phi_2(z; a, b).$$

*If this condition holds, then the pentagram curve is the single point $\bar{a}$ and the foci of $\partial W(A)$ are the zeros of $\Phi_2(z; a, b)$.*

*Remark* Theorem 4 implies that to any Poncelet 4-ellipse $E \subseteq \mathbb{D}$, one can associate a well-defined point in $\mathbb{D}$ that will be the pentagram curve of the matrix $A \in S_3$ that satisfies $\partial W(A) = E$. We will call this point the *pentagram point* of the ellipse $E$. The relationship between Poncelet 4-ellipses and a pentagram point was first observed in [13, Corollary 3.9], where it is stated that this pentagram point is the intersection of all diagonals of all cyclic quadrilaterals that circumscribe $E$.

Theorem 4 gives us a new interpretation of the algorithm for finding a matrix in $S_3$ with elliptic numerical range and two prescribed foci (such an algorithm can be found in [13]). Indeed, one can describe this algorithm as follows:

(i) given $\{f_1, f_2\} \in \mathbb{D}^2$, consider the polynomial $\Phi_2(z) = (z - f_1)(z - f_2)$;
(ii) perform the Inverse Szegő recursion to obtain a degree 1 monic polynomial $\Phi_1(z; \bar{f_3})$ whose zero $f_3$ is in $\mathbb{D}$;
(iii) the $3 \times 3$ cutoff CMV matrix with eigenvalues at $\{f_1, f_2, f_3\}$ has the desired property.

We can even generalize this algorithm to find an $A \in S_3$ with elliptic numerical range having one prescribed focus and a pentagram curve that is a prescribed point.

**Theorem 5** *Given $\{f_1, f_2\} \in \mathbb{D}^2$, there exists a unique $3 \times 3$ cutoff CMV matrix whose numerical range is bounded by an ellipse with one focus at $f_1$ and such that the pentagram curve is the single point $\{f_2\}$.*

***Proof*** By Theorem 4, this amounts to showing that we can find $b \in \mathbb{D}$ such that $\Phi_2(z; \bar{f_2}, b)$ vanishes at $f_1$. It is easy to see that this can be achieved precisely by setting

$$\bar{b} = \frac{f_1 \Phi_1(f_1; \bar{f_2})}{\Phi_1^*(f_1; \bar{f_2})} = \frac{f_1(f_1 - f_2)}{1 - \bar{f_2} f_1}$$

(see also [23]). □

One can also find an $A_\alpha \in S_3$ with $\partial W(A_\alpha)$ an ellipse and whose pentagram curve is a specified point. Since this is a weaker set of conditions than was used in Theorem 5, one expects that we will have many solutions to this problem. Rather than a unique matrix as in Theorem 5, fixing the pentagram point yields a family of matrices in $S_3$ parametrized by the second Verblunsky coefficient of $\Phi_2(z; a, b)$.

**Proposition 1** *Given $\alpha_0 \in \mathbb{D}$, there exists a one-parameter family $A_\alpha \in S_3$, $\alpha \in \mathbb{D}$ such that for each $A_\alpha$, $\partial W(A_\alpha)$ is an ellipse and the pentagram point of $A_\alpha$ is $\alpha_0$.*

***Proof*** Suppose $\alpha_0$ is given and consider the polynomial $\Phi_1(z; \bar{\alpha}_0)$. For any $\alpha \in \mathbb{D}$, consider $\Phi_2(z; \bar{\alpha}_0, \alpha)$ and define

$$\phi_3(z) = \Phi_1(z; \bar{\alpha}_0)\Phi_2(z; \bar{\alpha}_0, \alpha).$$

Thinking of $\phi_3(z)$ as an OPUC and applying the inverse Szegő recursion allows us to recover the Verblunsky coefficients of $\phi_3(z)$ and thus define a cutoff CMV matrix, $A_\alpha \in S_3$, whose characteristic polynomial is $\phi_3(z)$. As $\phi_3(z)$ factors into a degree one and degree two OPUC related by the Szegő recursion, Theorem 4 implies that $\partial W(A_\alpha)$ is an ellipse and the pentagram point of $A_\alpha$ is $\alpha_0$. □

Now suppose we have an $A \in S_3$ with numerical range given by an elliptical disk. Suppose that the foci of the ellipse bounding that elliptical disk are $\{f_1, f_2\}$ and the pentagram point of that ellipse is $\{f_3\}$. Then

$$\det(zI - A) = (z - f_1)(z - f_2)(z - f_3).$$

By Theorem 4, we also have

$$\det(zI - A) = (z - f_3)(z(z - f_3) - b(1 - \bar{f}_3 z))$$

for some $b \in \mathbb{D}$. Evaluating both of these expressions at 0 and equating them shows $b = -f_1 f_2$. It follows that

$$(z-f_1)(z-f_2) = z(z-f_3)+f_1 f_2(1-\bar{f}_3 z) = z^2+z(-f_3-f_1 f_2 \bar{f}_3)+f_1 f_2. \tag{8}$$

Replacing $z$ by $f_3$ in (8) shows

$$(f_3 - f_1)(f_3 - f_2) = f_1 f_2(1 - |f_3|^2). \tag{9}$$

If we look at the reversed polynomials in (8) and replace $z$ by $f_3$, we get

$$(1 - \bar{f}_1 f_3)(1 - \bar{f}_2 f_3) = 1 - |f_3|^2. \tag{10}$$

Dividing (9) by (10) recovers the Mirman system: $f_1 f_2 = B_2(f_3; f_1, f_2)$. Solving for $f_3$ yields the familiar formula (from [13]) for $f_3$ in terms of $f_1, f_2$:

$$f_3 = \frac{f_1 + f_2 - f_2|f_1|^2 - f_1|f_2|^2}{1 - |f_1 f_2|^2}.$$

Notice that we have recovered something that the Mirman system does not give us. One can verify that $f_3 = 0$ is a solution to the Mirman system, but this does not (in general) give us the matrix in $S_3$ with elliptical numerical range. Our calculations using OPUC eliminate this extraneous solution.

# 4 The Hexagon Case

In this section we will consider curves that are realized as the envelope of line segments joining vertices of hexagons inscribed in the unit circle (see [14, Sect. 3] for a rigorous discussion of envelopes of cyclic polygons). We know that such curves have three components: the largest one (outer) formed by connecting adjacent eigenvalues of the unitary dilations is the *Poncelet curve*, the middle component formed by joining alternate eigenvalues is the *pentagram curve*, and the smallest component formed by joining opposite eigenvalues is the *Brianchon curve*. Our first results are the following two theorems.

**Theorem 6** *Let $\mathcal{G}^{(5)}$ be a cut-off CMV matrix and*

$$\Phi_5(z) := \det(zI_5 - \mathcal{G}^{(5)}).$$

*The following are equivalent.*

(i) *the pentagram curve of $\mathcal{G}^{(5)}$ is an ellipse;*
(ii) *there exist regular Blaschke products $\{B_j\}_{j=2}^{3}$ with $\deg(B_j) = j$ such that*

$$\frac{z\Phi_5(z)}{\Phi_5^*(z)} = B_2(B_3(z))$$

(iii) *There exist $\alpha_0, \alpha_1, \alpha_2 \in \mathbb{D}$ so that*

$$\Phi_5(z) = \Phi_2(z; \alpha_0, \alpha_1)\Phi_3(z; \alpha_0, \alpha_1, \alpha_2).$$

*If any of these conditions hold, then the foci of the pentagram curve are the zeros of $B_3(z)/z$ or equivalently, the zeros of $\Phi_2(z; \alpha_0, \alpha_1)$.*

**Theorem 7** *If we retain the notation from Theorem 6, then the following are equivalent:*

(i) *the Brianchon curve of $\mathcal{G}^{(5)}$ is a single point;*
(ii) *there exist regular Blaschke products $\{B_j\}_{j=2}^{3}$ with $\deg(B_j) = j$ such that*

$$\frac{z\Phi_5(z)}{\Phi_5^*(z)} = B_3(B_2(z))$$

(iii) *There exist $\alpha_0, \alpha_1, \gamma_1 \in \mathbb{D}$ so that*

$$\Phi_5(z) = \Phi_1(z; \alpha_0)\Phi_2(z; \alpha_0, \alpha_1)\Phi_2(z; \alpha_0, \gamma_1)$$

*If any of these conditions hold, then the Brianchon point is the zero of $B_2(z)/z$ and the zero of $\Phi_1(z; \alpha_0)$.*

The equivalence of (i) and (ii) in both theorems is proven in [3]. The equivalence of (ii) and (iii) in both theorems follows from Theorem 3. By comparing Theorem 7 with Theorem 4 (and the remark after it), one arrives at the following result.

**Corollary 1** *Let $A \in S_5$ have eigenvalues $\{f_j\}_{j=1}^5$. Suppose the Brianchon curve of $A$ is the point $f_5$. Then $\{f_j\}_{j=1}^4$ can be labelled in such a way that both of the following conditions hold:*

(i) *$f_5$ is the pentagram point of the Poncelet 4-ellipse with foci at $f_1$ and $f_4$;*
(ii) *$f_5$ is the pentagram point of the Poncelet 4-ellipse with foci at $f_2$ and $f_3$.*

Using ideas from [14], we can prove the following.

**Theorem 8** *If we retain the notation from Theorem 6, then the following are equivalent*

(i) *The Poncelet curve associated with $\mathcal{G}^{(5)}$ is an ellipse.*
(ii) *There exist regular Blaschke products $\{B_j\}_{j=2}^3$ with $\deg(B_j) = j$ and regular Blaschke products $\{D_j\}_{j=2}^3$ with $\deg(D_j) = j$ such that*

$$\frac{z\Phi_5(z)}{\Phi_5^*(z)} = B_3(B_2(z)) = D_2(D_3(z)).$$

(iii) *The pentagram curve of $\mathcal{G}^{(5)}$ is an ellipse and the Brianchon curve of $\mathcal{G}^{(5)}$ is a single point.*

***Proof*** The equivalence of (ii) and (iii) is an immediate consequence of Theorems 6 and 7. The fact that (i) implies (iii) is a result of Darboux (see [14, Theorem B]). To see that (iii) implies (i), we will make use of the dual curve described in Sect. 2. Notice that [14, Sect. 3] implies that if $C_1$, $C_2$, or $C_3$ is an algebraic curve of degree 1 or 2, then the same is true of the corresponding component of the dual curve and vice versa (counting a single point as having degree 1). The dual curve in this case has degree 5 (see [9, Sect. 3]). Thus, if the Brianchon curve has degree 1 and the pentagram curve has degree 2, then the Poncelet curve has degree 2, which means it is an ellipse. □

Theorem 7 reveals an interesting phenomenon. Suppose we are given an $A \in S_5$ whose Brianchon curve is a point and whose pentagram curve is not an ellipse (it is easy to find such $A$). Take a rank one unitary dilation $U_1$ of $A$ and look at the hexagon with vertices at the eigenvalues of $U_1$. The diagonals of this hexagon meet in a single point (which is the Brianchon curve of $A$), so by Brianchon's Theorem there exists an ellipse $E_1$ inscribed in this hexagon. By construction, the ellipse $E_1$ is a Poncelet 6-ellipse, so there is in fact an infinite family of hexagons $\{H_\lambda\}$ inscribed in $\partial\mathbb{D}$ and circumscribed about $E_1$. On the other hand, one can look at any other rank one unitary dilation of $A$ and repeat this process. This gives a second infinite family of hexagons, each of which is circumscribed in $\partial\mathbb{D}$ and has its diagonals meeting at the point that is the Brianchon curve of $A$. But these two families of hexagons

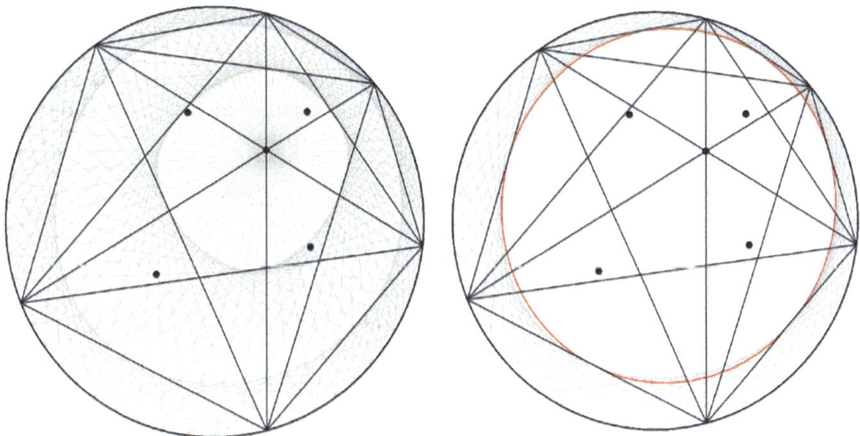

**Fig. 6** A 6-Poncelet curve that is not an ellipse having the property that its Brianchon curve is a single point (Brianchon point). The picture on the right shows the ellipse that is inscribed in one of the Poncelet hexagons. This ellipse exists by Brianchon's theorem. Notice that the 6-Poncelet curve contains points in the exterior as well as points in the interior of this ellipse

cannot be the same, for that would imply that the numerical range of $A$ is bounded by an ellipse and by Theorem 8 we know it is not (Fig. 6).

The next step in our analysis will be very much analogous to that performed in Sect. 3. If we are given $\{f_1, f_2\} \in \mathbb{D}^2$, we want to find $A \in S_5$ whose numerical range is bounded by an ellipse with foci at $\{f_1, f_2\}$. Theorem 1 tells us that such an ellipse exists and is the unique Poncelet 5-ellipse with foci at $f_1$ and $f_2$. We will find a cutoff CMV matrix $\mathcal{G}^{(5)}$ whose numerical range is bounded by an ellipse with foci at $\{f_1, f_2\}$, whose pentagram curve is an ellipse (whose foci will be called $\{f_3, f_4\}$), and whose Brianchon curve is a single point (that we will call $f_5$).

**Theorem 9** *A matrix $A \in S_5$ with eigenvalues $\{f_j\}_{j=1}^5$ satisfies all of the following conditions*

(i) *$W(A)$ is an elliptical disk*
(ii) *the foci of $\partial W(A)$ are $\{f_1, f_2\}$*
(iii) *the foci of the pentagram curve are $\{f_3, f_4\}$*
(iv) *the Brianchon curve is the single point $f_5$*

*if and only if the complex numbers $\{f_j\}_{j=1}^5$ satisfy*

$$\{f_3, f_4\} = \left\{ \frac{f_5 - f_2}{1 - f_2 \bar{f_5}}, \frac{f_5 - f_1}{1 - f_1 \bar{f_5}} \right\} \tag{11}$$

*as sets and*

$$\begin{aligned} f_3 + f_4 + f_1 f_2 f_5 \bar{f_4} \bar{f_3} &= f_1 + f_2 + f_5 \\ f_3 f_4 + f_1 f_2 f_5 (\bar{f_4} + \bar{f_3}) &= f_1 f_2 + f_2 f_5 + f_1 f_5 \end{aligned} \tag{12}$$

Recall that Theorem 7 and Theorem 8 show that if $W(A)$ is bounded by an ellipse, then the characteristic polynomial factors in a certain way (in terms of OPUC) and the degree 1 polynomial in that factorization vanishes at the Brianchon point. We require the following refinement of that result, which tells us about the zeros of the remaining polynomials in that factorization.

**Lemma 1** *Suppose $A \in S_5$ is such that $W(A)$ is an elliptical disk and the eigenvalues of $A$ are $\{f_j\}_{j=1}^5$. Suppose the foci of $\partial W(A)$ are $\{f_1, f_2\}$, the foci of the pentagram curve are $\{f_3, f_4\}$, and the Brianchon point is $f_5$. Write*

$$\det(zI_5 - A) = \Phi_1(z;\alpha_0)\Phi_2(z;\alpha_0,\alpha_1)\Phi_2(z;\alpha_0,\gamma_1)$$

*as in Theorem 7. If $\Phi_2(f_1;\alpha_0,\alpha_1) = 0$, then $\Phi_2(f_2;\alpha_0,\gamma_1) = 0$.*

***Proof*** Suppose $\Phi_2(f_1;\alpha_0,\alpha_1) = 0$ and $\Phi_2(f_2;\alpha_0,\gamma_1) \neq 0$. Then $\Phi_2(f_2;\alpha_0,\alpha_1) = 0$ and hence $\Phi_2(f_3;\alpha_0,\gamma_1) = \Phi_2(f_4;\alpha_0,\gamma_1) = 0$.

Since $\Phi_2(f_2;\alpha_0,\alpha_1) = 0$, we can write

$$(z - f_5)(z - f_1)(z - f_2) = \Phi_1(z;\alpha_0)\Phi_2(z;\alpha_0,\alpha_1).$$

This means $\{f_1, f_2, f_5\}$ are the eigenvalues of some $A \in S_3$ that satisfies the hypotheses of Theorem 4. Applying the Mirman system in the $N = 4$ case shows

$$f_1 f_2 = B_2(f_5; f_1, f_2)$$

The Mirman system in the case $n = 6$ shows shows $f_3 f_4 = B_2(f_5; f_1, f_2)$ and hence $f_1 f_2 = f_3 f_4$. Since $f_3, f_4$ are the zeros of $\Phi_2(z;\alpha_0,\gamma_1)$ and $\gamma_1 = -\overline{\Phi_2(0;\alpha_0,\gamma_1)}$, we conclude that $\gamma_1 = \alpha_1$, which implies $\Phi_2(z;\alpha_0,\gamma_1) = \Phi_2(z;\alpha_0,\alpha_1)$. It follows that $\Phi_2(f_2;\alpha_0,\gamma_1) = 0$, which gives us a contradiction.

□

*Proof of Theorem 9* In Theorem 6 it is stated that the foci of the pentagram ellipse will be the zeros of $\Phi_2(z;\alpha_0,\alpha_1)$. Thus, the foci of the Poncelet curve and the Brianchon point must be the zeros of $\Phi_3(z;\alpha_0,\alpha_1,\alpha_2)$. The product of these zeros is then $\bar{\alpha}_2$ and hence we have

$$z(z - f_3)(z - f_4) - f_1 f_2 f_5(1 - \bar{f}_3 z)(1 - \bar{f}_4 z) = (z - f_1)(z - f_2)(z - f_5) \quad (13)$$

Equating coefficients of $z$ and $z^2$ in (13) tells us that (12) must hold. One can perform a similar calculation invoking Theorem 7 and Lemma 1. By equating coefficients of the appropriate polynomials, we find that (11) must hold.

For the converse statement, the above calculations show that if (11) and (12) hold, then the conditions (iii) in Theorems 6 and 7 are satisfied (by equating coefficients of polynomials). Theorem 8 then implies that the cutoff CMV matrix $\mathcal{G}^{(5)}$ with eigenvalues $\{f_j\}_{j=1}^5$ has numerical range that is bounded by an ellipse, has pentagram curve that is an ellipse with foci $\{f_3, f_4\}$, and has Brianchon point $\{f_5\}$.

The foci of $\partial W(\mathcal{G}^{(5)})$ are eigenvalues of $\mathcal{G}^{(5)}$ (see [14, Sect. 5]). By [17, Corollary 4], we know that one can partition the eigenvalues of $\mathcal{G}^{(5)}$ into the Brianchon point, the foci of the pentagram curve, and the foci of the boundary of the numerical range. Thus, by elimination it must be that the foci of $\partial W(\mathcal{G}^{(5)})$ are $\{f_1, f_2\}$. □

To illustrate Theorem 9, consider the case in which $f_1 = -1/4 + i/4$ and $f_2 = 1/2 + i/3$. Notice that the system of Eqs. (12) is symmetric in $f_3$ and $f_4$. Therefore, we may substitute (11) into (12) and find that there is a unique value of $f_5 \in \mathbb{D}$ that solves the resulting system of two equations, namely $f_5 \approx 0.502489979 \ldots + i0.6539789 \ldots$. One then calculates $f_3 \approx 0.170805399 \ldots + i0.5527964 \ldots$ and $f_4 \approx 0.601612 \ldots + i0.600663814 \ldots$. The resulting ellipses and foci are shown in Fig. 5.

Recall that the Mirman system in the $n = 6$ case is

$$f_1 f_5 = B_2(f_3; f_1, f_2), \qquad f_3 f_4 = B_2(f_5; f_1, f_2), \qquad f_2 f_5 = B_2(f_4; f_1, f_2). \tag{14}$$

The conditions (11) and (12) allow us to recover these relations. Substitute $z$ for $f_3$ in (13) to obtain

$$(f_3 - f_1)(f_3 - f_2) = \frac{f_1 f_2 f_5 (1 - |f_3|^2)(1 - f_3 \bar{f_4})}{f_5 - f_3} \tag{15}$$

If we replace $z$ by $f_3$ in the reversed polynomials from (13), then we obtain

$$(1 - \bar{f_1} f_3)(1 - \bar{f_2} f_3) = \frac{(1 - |f_3|^2)(1 - f_3 \bar{f_4})}{1 - \bar{f_5} f_3} \tag{16}$$

If we divide (15) by (16), and use (11), we find $B_2(f_3; f_1, f_2) = f_1 f_5$. Similar reasoning can be used to derive $B_2(f_4; f_1, f_2) = f_2 f_5$. Replacing $z$ by $f_5$ in (13) tells us that

$$\frac{f_5(f_5 - f_4)(f_5 - f_3)}{(1 - \bar{f_4} f_5)(1 - \bar{f_3} f_5)} = f_1 f_2 f_5$$

If we assume $f_1 f_2 f_5 \neq 0$ and we substitute the relations (11) for $f_3$ and $f_4$ (for an appropriate choice of which to call $f_3$ and which to call $f_4$), then we find

$$(1 - \bar{f_2} f_5)(1 - \bar{f_1} f_5) = (1 - \bar{f_5} f_2)(1 - \bar{f_5} f_1)$$

In other words $(1 - \bar{f_2} f_5)(1 - \bar{f_1} f_5) \in \mathbb{R}$. If we use this fact, then multiplying the expressions in (11) gives $B_2(f_5; f_1, f_2) = f_3 f_4$.

If one is given $(f_1, f_2) \in \mathbb{D}^2$, the construction above shows how to find an $A \in S_5$ so that $\partial W(A)$ is an ellipse with foci at $f_1$ and $f_2$. Our next result reveals that one can similarly find $A \in S_5$ with $\partial W(A)$ an ellipse and whose pentagram curve has prescribed foci (Theorem 8 assures us that if the Poncelet curve of $A$ is an ellipse, then so is its pentagram curve).

**Theorem 10** *Given* $(f_3, f_4) \in \mathbb{D}^2$*, there exists a unique (up to unitary conjugation)* $A \in S_5$ *so that* $\partial W(A)$ *is an ellipse and the pentagram curve of* $A$ *is an ellipse with foci at* $f_3$ *and* $f_4$*.*

Our proof of Theorem 10 requires the following two lemmas.

**Lemma 2** *Given two triangles inscribed in* $\partial\mathbb{D}$ *with interlacing vertices, there is a unique ellipse that is inscribed in both of them.*

***Proof*** By [12, Theorem 3.1], we know that there exists a unique cutoff CMV matrix $\mathcal{G}^{(2)}$ whose numerical range is circumscribed by the two triangles. By the Elliptical Range Theorem [4, Theorem 2.3], we know that this numerical range is bounded by an ellipse with the desired properties. Uniqueness follows from the uniqueness statement in [12, Theorem 3.1] and the fact that every Poncelet 3-ellipse can be realized as boundary of the numerical range of some $2 \times 2$ cutoff CMV matrix (see Theorem 1 above). □

*Remark* Note that Lemma 2 is related to Wendroff's Theorem for POPUC (see [16, Theorem 8] or [1]).

**Lemma 3** *Let* $\{\Phi_3^{(\lambda)}\}_{\lambda\in\partial\mathbb{D}}$ *be the collection of degree 3 POPUC for the same degree 2 OPUC. Label the zeros of* $\Phi_3^{(\lambda)}$ *counterclockwise and cyclically as* $\{z_j^{(\lambda)}\}_{j=1}^3$*. For each* $\lambda \in \partial\mathbb{D}$ *there exists a unique* $\tau \in \partial\mathbb{D}$ *so that the line segments joining* $z_j^{(\lambda)}$ *to* $z_j^{(\tau)}$ *all meet in a single point independent of* $j$ *(this assumes an appropriate labeling of the zeros* $\{z_j^{(\lambda)}\}_{j=1}^3$*).*

***Proof*** For each $\tau$, let $z_1^{(\tau)}$ be the zero that lies between $z_2^{(\lambda)}$ and $z_3^{(\lambda)}$, let $z_2^{(\tau)}$ be the zero that lies between $z_3^{(\lambda)}$ and $z_1^{(\lambda)}$, and let $z_3^{(\tau)}$ be the zero that lies between $z_1^{(\lambda)}$ and $z_2^{(\lambda)}$. Let $L_j^{(\tau)}$ be the line segment that joins $z_j^{(\lambda)}$ to $z_j^{(\tau)}$ for $j = 1, 2, 3$.

Given $\{z_j^{(\lambda)}\}_{j=1}^3$, start with $\tau = \lambda$ and move $\tau$ around $\mathbb{T}$ clockwise.[1] As this happens, consider $\eta_j(\tau) := L_1^{(\tau)} \cap L_j^{(\tau)}$ for $j = 2, 3$ and notice that both of these points are in $L_1^{(\tau)}$. Define

$$d_j(\tau) = \frac{|z_1^{(\lambda)} - \eta_j(\tau)|}{|L_1^{(\tau)}|}, \qquad j = 2, 3.$$

Initially, $d_3(\tau)$ is close to 0 and $d_2(\tau)$ is close to 1. As $\tau$ nears the end of its trip around $\mathbb{T}$, it holds that $d_2(\tau)$ is close to 0 and $d_3(\tau)$ is close to 1. Thus, by the

[1] We rotate $\tau$ clockwise because of the complex conjugation in the Szegő recursion.

Intermediate Value Theorem, there must be a value of $\tau$ such that $d_2(\tau) = d_3(\tau)$ as desired. One can see by inspection that this choice of $\tau$ is unique. □

*Proof of Theorem 10* Suppose $(f_3, f_4) \in \mathbb{D}^2$ are given. Consider the Poncelet 3-ellipse with foci at $f_3$ and $f_4$ (call it $E$). Pick any triangle $T^{(\lambda)}$ that is inscribed in $\partial\mathbb{D}$ and circumscribed about $E$. By Lemma 3, there exists a unique second such triangle $T^{(\tau)}$ such that if $H(\tau, \lambda)$ is the hexagon whose vertices are the vertices of $T^{(\tau)}$ and $T^{(\lambda)}$, then the line segments joining opposite vertices of $H(\tau, \lambda)$ meet in a single point. By Brianchon's Theorem, there is an ellipse inscribed in $H(\tau, \lambda)$. Call this ellipse $E'$ and suppose its foci are $f_1$ and $f_2$.

From what we already know, the ellipse $E'$ is the unique Poncelet 6-ellipse with foci at $f_1$ and $f_2$, and Theorem 1 tells us that it is associated to a matrix in $S_5$. For this matrix, the associated pentagram curve must be an ellipse and the associated Brianchon curve must be a single point. This pentagram ellipse is a Poncelet 3-ellipse and must be inscribed in the triangles $T^{(\lambda)}$ and $T^{(\tau)}$. By Lemma 2, that ellipse must be $E$. □

Our next result is an analog of Theorem 10 for the Brianchon curve. Specifically, we will show that one can find $A \in S_5$ so that $\partial W(A)$ is an ellipse and the Brianchon curve is a single predetermined point. The main difference between Theorem 11 and Theorem 10 is the lack of uniqueness.

**Theorem 11** *Given $f_5 \in \mathbb{D}$, there exists a $5\times5$ cutoff CMV matrix $A$ so that $\partial W(A)$ is an ellipse and the Brianchon curve of $A$ is the single point $f_5$. Furthermore, the set of all possible $5\times5$ cutoff CMV matrices with this property is naturally parametrized by an open triangle.*

***Proof*** Suppose $f_5 \in \mathbb{D}$ is given. Consider the set of all lines passing through $f_5$. Each line intersects $\mathbb{T}$ in two places. One can choose three distinct lines, thus specifying 6 distinct points of $\mathbb{T}$, labeled cyclically as $v_j$ for $j = 1, 2, \ldots, 6$. By Brianchon's Theorem, there is an ellipse inscribed in the hexagon whose vertices are $\{v_j\}_{j=1}^6$. Call this ellipse $E$ and its foci $f_1$ and $f_2$. By Theorem 9, $E$ is the boundary of $W(A)$ for some $A \in S_5$. Thus, the Poncelet curve and pentagram curve of $A$ are ellipses and the Brianchon curve of $A$ is a single point, which we see must be $f_5$ as desired.

Note that the above procedure yields a well-defined map from collections of distinct triples of lines through $f_5$ to matrices in $S_5$ whose numerical range is an elliptical disk and whose Brianchon point is $f_5$. To see this, suppose one starts with three distinct lines through $f_5$. This produces six points on the unit circle by the above procedure. Connecting alternate points forms two triangles that must be tangent to the pentagram ellipse of the matrix we seek. By Lemma 2, there is only one possible choice for such an ellipse, so its foci $f_3$, $f_4$ are a well-defined output. By Theorem 9 and Eq. (11), we can calculate the foci $f_1$, $f_2$ of the Poncelet curve of this matrix. Thus, the eigenvalues $\{f_j\}_{j=1}^5$ of the matrix we seek are a computable quantity from any three distinct lines through $f_5$. Since the eigenvalues determine the matrix in $S_5$, this means the map is well-defined.

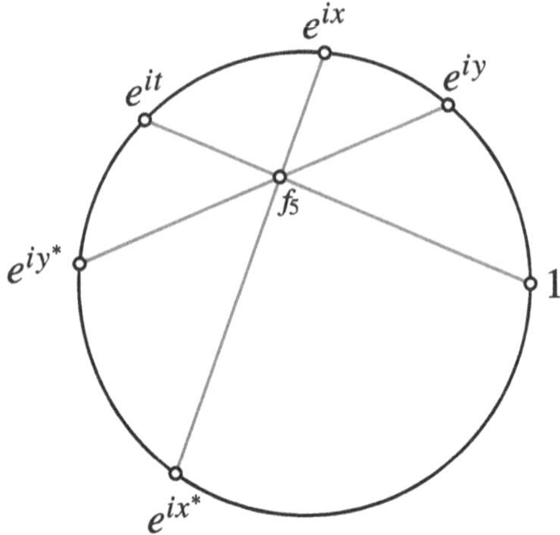


**Fig. 7** Given $f_5$ and the line through $1, f_5$, varying $x, y$ such that $0 < x < y < t$ parametrizes the set of all $A \in S_5$ with elliptic numerical range and Brianchon curve $f_5$

To make this map a bijection, we restrict it to triples of lines through $f_5$ for which one of them also passes through 1. Given any $A \in S_5$ with $W(A)$ an elliptical disk and Brianchon point $f_5$, there exists a hexagon that circumscribes $\partial W(A)$ for which 1 is a vertex, so the restricted map is onto. To show that it is injective, suppose $H$ is a hexagon inscribed in $\mathbb{T}$ with one vertex at 1 and $E$ is an ellipse that is tangent to every edge of $H$. From any given point on $\mathbb{T}$ (in particular, the point 1), there are only two tangents to $E$ through that point. This implies $H$ is the unique hexagon that includes 1 and has the required tangency properties. Thus, our restricted map uniquely determines the numerical range of the matrix and hence uniquely determines the matrix itself.

Suppose that the line through 1 and $f_5$ also intersects $\mathbb{T}$ at $e^{it}$. We have shown that the space of all $5 \times 5$ CMV matrices with the desired property is naturally parametrized by $\{(x, y) : 0 < x < y < t\}$ (see Fig. 7), which is an open triangle. □

## 5 The Pentagon Case

The situation when $n = 5$ is considerably different than the cases $n = 4, 6$. One can see that there exist matrices $A \in S_4$ for which $W(A)$ is an elliptical disk by considering a single $4 \times 4$ Jordan block with eigenvalue 0. However, our approach to finding all such matrices cannot be the same as in the previous two sections. In particular, it will not be possible to consider compositions of Blaschke products as in the previous sections, which partially explains why this case has been studied less often in the literature (see [10]). Instead, we will revisit the Mirman system and prove a structure theorem about the set of possible solutions. In this setting, the system of equations has multiple solutions and our next result describes their relative placement in the plane (Fig. 8).

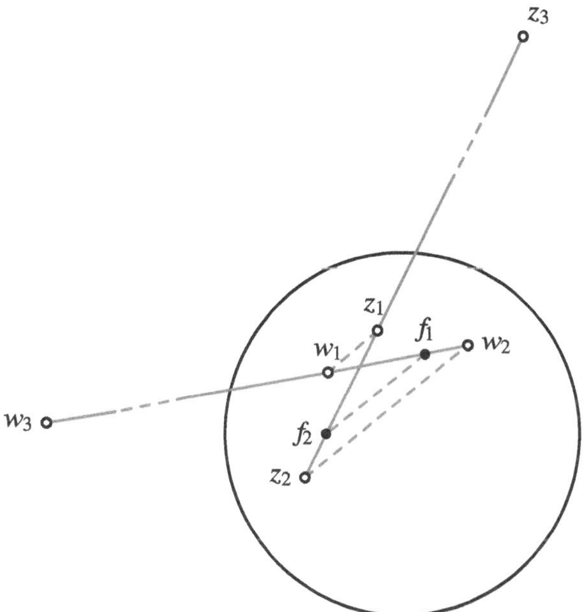


**Fig. 8** Collinearity of the points $f_1$, $f_2$, $z_j$'s and $w_j$'s, as explained in Theorem 12

**Theorem 12** *Let $f_1, f_2 \in \mathbb{D}$, $f_1 f_2 \neq 0$ and set $\Phi_2(z) = (z - f_1)(z - f_2)$. The system*

$$\begin{aligned} \frac{\Phi_2(z)}{\Phi_2^*(z)} &= w f_1, \\ \frac{\Phi_2(w)}{\Phi_2^*(w)} &= z f_2 \end{aligned} \tag{17}$$

*has exactly 5 distinct solution pairs $(z, w) \in \mathbb{C}^2$: four of them are in $\mathbb{D}^2$: $(0, f_2)$, $(f_1, 0)$, $(z_1, w_1)$, $(z_2, w_2)$, and exactly one solution $(z_3, w_3)$ satisfies $|z_3| > 1$, $|w_3| > 1$.*

*Moreover,*

*(i) It holds that*

$$\frac{z_1 - f_2}{w_1 - f_1} = \frac{z_2 - f_2}{w_2 - f_1} = \frac{z_3 - f_2}{w_3 - f_1} = \frac{f_1 \Phi_2^*(f_2)}{f_2 \Phi_2^*(f_1)},$$

*(ii) The points $f_1$, $w_1$, $w_2$ and $w_3$ are collinear, i.e.*

$$\frac{w_i - f_1}{w_j - f_1} \in \mathbb{R}, \quad i, j \in \{1, 2, 3\}, \quad i \neq j.$$

*and the points $f_2$, $z_1$, $z_2$ and $z_3$ are collinear, i.e.*

$$\frac{z_i - f_2}{z_j - f_2} \in \mathbb{R}, \quad i, j \in \{1, 2, 3\}, \quad i \neq j.$$

***Proof*** Let us denote

$$g_1(z) = \frac{1}{f_1} \frac{(z - f_1)(z - f_2)}{(1 - z\bar{f_1})(1 - z\bar{f_2})}, \quad g_2(z) = \frac{1}{f_2} \frac{(z - f_1)(z - f_2)}{(1 - z\bar{f_1})(1 - z\bar{f_2})},$$

so that (17) takes the form $g_1(z) = w$, $g_2(w) = z$. From here, we get that (17) implies that

$$\begin{aligned} g_2 \circ g_1(z) &= z, \\ g_1 \circ g_2(w) &= w. \end{aligned} \tag{18}$$

We claim that both $g_2 \circ g_1$ and $g_1 \circ g_2$ have 4 fixed points inside $\mathbb{D}$. Indeed, consider $g_2 \circ g_1$ (same analysis for $g_1 \circ g_2$). Recall that a Blaschke product is analytic in $\mathbb{D}$, maps $\mathbb{T} \to \mathbb{T}$, $\mathbb{D} \to \mathbb{D}$ and exterior of $\mathbb{D}$ onto its exterior. So,

$$|z| = 1 \quad \Rightarrow \quad |g_1(z)| > 1 \quad \Rightarrow \quad |g_2(g_1(z))| > 1.$$

By Rouché's theorem, $g_2 \circ g_1(z) - z$ and $g_2 \circ g_1(z)$ have the same number of zeros in $\mathbb{D}$; it is straightforward to check that $g_2 \circ g_1$ vanishes at 4 points in $\mathbb{D}$.

Finally, we claim that the statement of the theorem is equivalent to the just established fact that both $g_2 \circ g_1$ and $g_1 \circ g_2$ have 4 fixed points inside $\mathbb{D}$.

Indeed, the 4 pairs of solutions of (17) satisfy (18). Reciprocally, let $z$ be a fixed point of $g_2 \circ g_1$ and denote $\tau = g_1(z)$. Then $g_2(\tau) = g_2(g_1(z)) = z$, and in consequence, $g_1(g_2(\tau)) = g_1(z) = \tau$, meaning that $\tau = g_1(z)$ is a fixed point of $g_1 \circ g_2$. This shows that the fixed points $z$ and $w$ of $g_2 \circ g_1$ and $g_1 \circ g_2$ can be paired $(z, w)$ in such a way that (17) holds.

Now we prove the statement about the remaining solution, this time in $(\mathbb{C} \setminus \overline{\mathbb{D}})^2$. The identity

$$\overline{\frac{(1/\bar{z} - f_1)(1/\bar{z} - f_2)}{(1 - \bar{f_1}/\bar{z})(1 - \bar{f_2}/\bar{z})}} = \frac{1}{\frac{(z - f_1)(z - f_2)}{(1 - z\bar{f_1})(1 - z\bar{f_2})}} \tag{19}$$

allows us to reduce the analysis of (17) in $(\mathbb{C} \setminus \overline{\mathbb{D}})^2$ to the equivalent system

$$\begin{aligned} \frac{(z - f_1)(z - f_2)}{(1 - z\bar{f_1})(1 - z\bar{f_2})} &= \frac{w}{\bar{f_1}}, \\ \frac{(w - f_1)(w - f_2)}{(1 - w\bar{f_1})(1 - w\bar{f_2})} &= \frac{z}{\bar{f_2}} \end{aligned} \tag{20}$$

for $(z, w) \in \mathbb{D}^2$ (we return to the actual solutions outside by the mapping $z \mapsto 1/\overline{z}$, $w \mapsto 1/\overline{w}$). The advantage of working in $\mathbb{D}$ is that again we can use the fixed point argument and Rouché's Theorem. Indeed, as before, define

$$h_1(z) = \overline{f_1}\,\frac{(z - f_1)(z - f_2)}{(1 - z\bar{f_1})(1 - z\bar{f_2})}, \qquad h_2(z) = \overline{f_2}\,\frac{(z - f_1)(z - f_2)}{(1 - z\bar{f_1})(1 - z\bar{f_2})},$$

and look for fixed points of $h_1 \circ h_2$ and $h_2 \circ h_1$. This time

$$|z| = 1 \quad \Rightarrow \quad |h_1(z)| < 1 \quad \Rightarrow \quad |h_2(h_1(z))| < 1,$$

and by Rouché's Theorem, $h_2 \circ h_1(z) - z$ and $f(z) = z$ have the same number of zeros in $\mathbb{D}$, that is, exactly one.

To prove the statements about collinearity, define the linear functions

$$\begin{aligned} z(t) &= \frac{f_2 - f_1}{1 - f_1\bar{f_2}} + \frac{4f_1}{(1 - |f_1|^2)(1 - f_1\bar{f_2})}t \\ w(t) &= \frac{f_1 - f_2}{1 - f_2\bar{f_1}} + \frac{4f_2}{(1 - |f_2|^2)(1 - f_2\bar{f_1})}t \end{aligned} \tag{21}$$

It is a straightforward calculation to verify that the two polynomials (in the variable $t$)

$$\Phi_2(z(t)) - w(t) f_1 \Phi_2^*(z(t)), \qquad \Phi_2(w(t)) - z(t) f_2 \Phi_2^*(w(t))$$

are scalar multiples of each other so any zero of one is a zero of the other. Notice that both of these polynomials have degree 3. One can also check by hand that if $t_z$ is such that $z(t_z) = 0$, then $w(t_z) \neq f_2$. Similarly, if $t_w$ is such that $w(t_w) = 0$, then $z(t_w) \neq f_1$. Thus, the three zeros $\{t_j\}_{j=1}^3$ of

$$Y(t) := \Phi_2(z(t)) - w(t) f_1 \Phi_2^*(z(t)) \tag{22}$$

will be such that $(z(t_j), w(t_j))$ is a solution to the system (17) other than $(0, f_2)$ and $(f_1, 0)$. Thus we can say $(z_j, w_j) = (z(t_j), w(t_j))$. If we set $t_0 = \frac{1}{4}(1 - |f_1|^2)(1 - |f_2|^2)$ and observe that $z(t_0) = f_2$ and $w(t_0) = f_1$, then a short calculation shows

$$\frac{z_j - f_2}{w_j - f_1} = \frac{z(t_j) - z(t_0)}{w(t_j) - w(t_0)} = \frac{f_1 \Phi_2^*(f_2)}{f_2 \Phi_2^*(f_1)}.$$

This proves claim (i) of the theorem.

To prove claim (ii), it suffices to show that each $t_j \in \mathbb{R}$. To this end, a calculation reveals that if we divide the polynomial $Y(t)$ in (22) by its leading coefficient, then we obtain a monic degree 3 polynomial with real coefficients.

Define

$$q(x, y; t) := \left(y + \frac{x - f_1}{1 - \overline{f_1}x}\right)\left(y + \frac{x - f_2}{1 - \overline{f_2}x}\right) - \frac{4txy}{(1 - \overline{f_1}x)(1 - \overline{f_2}x)}$$

There exist two distinguished matrices $A_1, A_2 \in S_4$ such that $A_1$ has numerical range bounded by an ellipse with foci at $\{f_1, f_2\}$ and the pentagram curve of $A_2$ is an ellipse with foci at $\{f_1, f_2\}$. Let us denote the eigenvalues of $A_j$ by $\{f_1, f_2, z_j, w_j\}$ for $j = 1, 2$. Then from [14, Sect. 5] (and also [17, Equations 29–32]) we know that there exist positive real numbers $b_1$ and $b_2$ so that

$$q(f_1, z_j; b_j^2) = q(w_j, f_2; b_j^2) = q(z_j, w_j; b_j^2) = 0$$

for $j = 1, 2$. In fact, $b_j$ is the length of the minor semiaxis of the ellipse associated with $A_j$ whose foci are $\{f_1, f_2\}$ (see [17]).

Using $q(f_1, z_j; b_j^2) = q(w_j, f_2; b_j^2) = 0$ and the formulas (21), we find that $z_j = z(b_j^2)$ and $w_j = w(b_j^2)$. A short calculation shows (recall $t_0 = (1 - |f_1|^2)(1 - |f_2|^2)/4$)

$$q(z(t), w(t); t) = \frac{Y(t)(t - t_0)}{C_{f_1, f_2}\Phi_2^*(z(t))},$$

where $C_{f_1, f_2} = \dfrac{-f_1 t_0}{f_2}$. Thus the zeros of $Y(t)$ are also zeros of $q(z(t), w(t); t)$.

If $t_0$ were a zero of $Y(t)$, then $z = z(t_0)$ and $w = w(t_0)$ would satisfy (17). However, we have seen that $z(t_0) = f_2$ and $w(t_0) = f_1$, giving $0 = \Phi_2(f_2) = f_1^2$, which is nonzero by assumption, so $Y(t_0) \neq 0$.

Thus the zeros of $Y(t)$ are zeros of $q(z(t), w(t); t)$ distinct from $t_0$. We know that $q(z(t), w(t); t)$ has zeros at $t = b_1^2$ and $t = b_2^2$ so these must be zeros of $Y(t)$ as well. This gives us two real zeros of $Y(t)$. Our earlier observation implies that $Y(t)$ has either one or three real zeros, so all zeros of $Y(t)$ are real as desired. □

The solutions to the Mirman system in this setting have a geometric interpretation. Given $\{f_1, f_2\} \in \mathbb{D}^2$, we have just seen that there are three solution pairs $(z, w)$ to the Mirman system and we denote them by $(z_1, w_1)$, $(z_2, w_2)$, and $(z_3, w_3)$. One of the solution pairs in $\mathbb{D}^2$ (say $(z_1, w_1)$) will be such that the $4 \times 4$ cutoff CMV matrix with eigenvalues $\{f_1, f_2, z_1, w_1\}$ will have numerical range bounded by an ellipse with foci at $f_1$ and $f_2$ and the pentagram curve of this matrix will be an ellipse with foci at $z_1$ and $w_1$ (see Fig. 9, left). For the other solution pair in $\mathbb{D}^2$ (say $(z_2, w_2)$), it will be true that the $4 \times 4$ cutoff CMV matrix with eigenvalues $\{f_1, f_2, z_2, w_2\}$ has numerical range bounded by an ellipse with foci at $z_2$ and $w_2$ and the pentagram curve of this matrix will be an ellipse with foci at $f_1$ and $f_2$, see Fig. 9, right. The geometric interpretation of the solution $(z_3, w_3) \in (\mathbb{C} \setminus \bar{\mathbb{D}})^2$ is not clear.

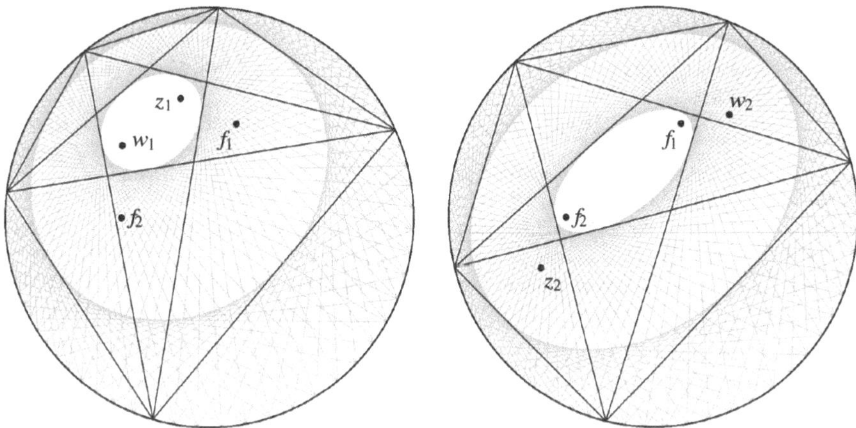


**Fig. 9** Pairs $(z_1, w_1)$ and $(z_2, w_2)$ as foci of the pentagram and the Poncelet ellipses, respectively

Notice that Theorem 12 implies that each line $\ell_j$ passing through $\{z_j, w_j\}$ for $j = 1, 2, 3$ is parallel to the line through $\{f_1, f_2\}$. For $j = 1, 2$, one can obtain this same conclusion from the fact that the ellipses $C_1$ and $C_2$ corresponding to a matrix $A \in S_4$ are in the same package (see [19]). The fact that this same conclusion applies to $\ell_3$ is a new result.

**Acknowledgments** Author Andrei Martinez-Finkelshtein was partially supported by Simons Foundation Collaboration Grants for Mathematicians (grant 710499) and by the Spanish Government–European Regional Development Fund (grant MTM2017-89941-P), Junta de Andalucía (research group FQM-229 and Instituto Interuniversitario Carlos I de Física Teórica y Computacional), and by the University of Almería (Campus de Excelencia Internacional del Mar CEIMAR).

Author Brian Simanek graciously acknowledges support from Simons Foundation Collaboration Grant 707882.

## References

1. Y. Arlinskii, L. Golinskii, E. Tsekanovskii, Contractions with rank one defect operators and truncated CMV matrices. J. Funct. Anal. **254**(1), 154–195 (2008)
2. U. Daepp, P. Gorkin, K. Voss, Poncelet's theorem, Sendov's conjecture, and Blaschke products. J. Math. Anal. Appl. **365**(1), 93–102 (2010)
3. U. Daepp, P. Gorkin, A. Shaffer, B. Sokolowsky, K. Voss, Decomposing finite Blaschke products. J. Math. Anal. Appl. **426**(2), 1201–1216 (2015)
4. U. Daepp, P. Gorkin, A. Shaffer, K. Voss, *Finding Ellipses: What Blaschke Products, Poncelet's Theorem, and the Numerical Range Know About Each Other*. Carus Mathematical Monographs, vol. 34 (MAA Press, Providence, 2018)
5. V. Dragović, M. Radnović, *Poncelet Porisms and Beyond. Integrable Billiards, Hyperelliptic Jacobians and Pencils of Quadrics*. Frontiers in Mathematics (Birkhäuser/Springer Basel AG, Basel, 2011)

6. L. Flatto, *Poncelet's Theorem* (American Mathematical Society, Providence, 2009). Chapter 15 by S. Tabachnikov
7. M. Fujimura, Inscribed Ellipses and Blaschke Products. Comput. Meth. Funct. Theory **13**(4), 557–573 (2013)
8. M. Fujimura, Blaschke products and circumscribed conics. Comput. Meth. Funct. Theory **17**(4), 635–652 (2017)
9. M. Fujimura, Interior and exterior curves of finite Blaschke products. J. Math. Anal. Appl. **467**(1), 711–722 (2018)
10. H.-L. Gau, Elliptic numerical ranges of $4 \times 4$ matrices. Taiwanese J. Math. **10**(1), 117–128 (2006)
11. H.-L. Gau, P. Wu, Numerical range of $S(\phi)$. Linear Multilinear Algebra **45**(1), 49–73 (1998)
12. H.-L. Gau, P. Wu, Numerical range circumscribed by two polygons. Linear Algebra Appl. **382**, 155–170 (2004)
13. P. Gorkin, N. Wagner, Ellipses and compositions of finite Blaschke products. J. Math. Anal. Appl. **445**(2), 1354–1366 (2017)
14. M. Hunziker, A. Martinez-Finkelshtein, T. Poe, B. Simanek, Poncelet-Darboux, Kippenhahn, and Szegő: interactions between projective geometry, matrices, and orthogonal polynomials. Submitted for publication (2021), arXiv: math.2101.12165
15. D. Keeler, L. Rodman, I. Spitkovsky, The numerical range of $3 \times 3$ matrices. Linear Algebra Appl. **252**, 115–139 (1997)
16. A. Martinez-Finkelshtein, B. Simanek, B. Simon, Poncelet's Theorem, Paraorthogonal polynomials and the numerical range of compressed multiplication operators. Adv. Math. **349**, 992–1035 (2019)
17. B. Mirman, $UB$-matrices and conditions for Poncelet polygon to be closed. Linear Algebra Appl. **360**, 123–150 (2003)
18. B. Mirman, Sufficient conditions for Poncelet polygons not to close. Amer. Math. Monthly **112**(4), 351–356 (2005)
19. B. Mirman, P. Shukla, A characterization of complex plane Poncelet curves. Linear Algebra Appl. **408**, 86–119 (2005)
20. R.E. Schwartz, The pentagram map. Exp. Math. **1**, 71–81 (1992)
21. B. Simon, *Orthogonal Polynomials on the Unit Circle, Part One: Classical Theory* (American Mathematical Society, Providence, 2005)
22. B. Simon, *Orthogonal Polynomials on the Unit Circle, Part Two: Spectral Theory* (American Mathematical Society, Providence, 2005)
23. B. Simon, V. Totik, Limits of zeros of orthogonal polynomials on the circle. Math. Nachr. **278**(12–13), 1615–1620 (2005)

# A Differential Analogue of Favard's Theorem

**Arieh Iserles and Marcus Webb**

**Abstract** Favard's theorem characterizes bases of functions $\{p_n\}_{n\in\mathbb{Z}_+}$ for which $xp_n(x)$ is a linear combination of $p_{n-1}(x)$, $p_n(x)$, and $p_{n+1}(x)$ for all $n \geq 0$ with $p_0 \equiv 1$ (and $p_{-1} \equiv 0$ by convention). In this paper we explore the differential analogue of this theorem, that is, bases of functions $\{\varphi_n\}_{n\in\mathbb{Z}_+}$ for which $\varphi_n'(x)$ is a linear combination of $\varphi_{n-1}(x)$, $\varphi_n(x)$, and $\varphi_{n+1}(x)$ for all $n \geq 0$ with $\varphi_0(x)$ a given smooth function (and $\varphi_{-1} \equiv 0$ by convention). We answer questions about orthogonality and completeness of such functions (which are not necessarily polynomials), provide characterisation results, and also, of course, give plenty of examples and list challenges for further research. Motivation for this work originated in the numerical solution of differential equations, in particular spectral methods which give rise to highly structured matrices and stable-by-design methods for partial differential equations of evolution. However, we believe this theory to be of interest in its own right, due to the interesting links between orthogonal polynomials, Fourier analysis and Paley–Wiener spaces, and the resulting identities between different families of special functions.

**Keywords** Orthogonal polynomials · Fourier transform · Spectral methods · Paley–Wiener spaces

A. Iserles (✉)
Department of Applied Mathematics and Theoretical Physics, Centre for Mathematical Sciences, University of Cambridge, Cambridge, United Kingdom
e-mail: ai@maths.cam.ac.uk

M. Webb
Department of Mathematics, University of Manchester, Manchester, United Kingdom
e-mail: marcus.webb@manchester.ac.uk



F. Gesztesy, A. Martinez-Finkelshtein (eds.), *From Operator Theory to Orthogonal Polynomials, Combinatorics, and Number Theory*, Operator Theory: Advances and Applications 285, https://doi.org/10.1007/978-3-030-75425-9_13

# 1 Introduction

Favard's theorem [5, p. 21] states that a sequence of univariate functions $P = \{p_n\}_{n\in\mathbb{Z}_+}$ satisfies

$$\begin{aligned} p_0(x) &= 1 \\ xp_0(x) &= \beta_0 p_0(x) + \gamma_0 p_1(x) \\ xp_n(x) &= \alpha_n p_{n-1}(x) + \beta_n p_n(x) + \gamma_n p_{n+1}(x), \qquad n \geq 1, \end{aligned}$$

for some real sequences $\{\alpha_n\}_{n=1}^{\infty}$, $\{\beta_n\}_{n\in\mathbb{Z}_+}$, $\{\gamma_n\}_{n\in\mathbb{Z}_+}$ if and only if $p_n$ is a polynomial of degree $n$ for all $n \geq 0$ and there exists a finite, signed Borel measure $\mu$ such that

$$\int_{-\infty}^{\infty} \mathrm{d}\mu(x) = 1,$$

$$\int_{-\infty}^{\infty} p_m(x)p_n(x)\,\mathrm{d}\mu(x) = 0 \text{ for all } n \neq m.$$

In other words, $P$ are orthogonal polynomials for the measure $\mu$. Favard's theorem was first announced by Favard in 1935 [9], but was discovered independently at about the same time by Shohat and Natanson.

The measure $\mu$ is clearly only unique up to a diagonal scaling of the polynomial sequence $P$, but it is also possible that $\mu$ is non-unique if there exists another measure with the same moments $M_n = \int_{-\infty}^{\infty} x^n\,\mathrm{d}\mu(x)$ (i.e. the moment problem is indeterminate [1]).

The following observations on Favard's theorem can be readily checked [5].

- $\mu$ is a positive measure if and only if $\alpha_n\gamma_{n-1} > 0$ for all $n$.
- $\mu$ is a symmetric measure, i.e. $\mathrm{d}\mu(-x) = \mathrm{d}\mu(x)$, if and only if $\beta_n = 0$ for all $n$.
- $P$ are orthonormal polynomials with respect to a symmetric measure $\mu$ if and only if $\alpha_n = \gamma_{n-1}$ for $n \geq 1$.

The topic of this paper is the *differential analogue* of Favard's theorem, and its consequences. Specifically, this is the characterization of sequences of functions $\Phi = \{\varphi_n\}_{n\in\mathbb{Z}_+}$ which satisfy

$$\begin{aligned} \varphi_0'(x) &= \beta_0\varphi_0(x) + \gamma_0\varphi_1(x) \\ \varphi_n'(x) &= \alpha_n\varphi_{n-1}(x) + \beta_n\varphi_n(x) + \gamma_n\varphi_{n+1}(x), \qquad n \geq 1, \end{aligned}$$

for some real sequences $\{\alpha_n\}_{n=1}^{\infty}$, $\{\beta_n\}_{n\in\mathbb{Z}_+}$, $\{\gamma_n\}_{n\in\mathbb{Z}_+}$, where $\varphi_0$ is taken as a given smooth function. In general these functions will not be polynomials. Despite the fact that differentiation is arguably a purely real-valued affair, we will see that complex coefficients can arise naturally.

The most well known example which satisfies such a differential recurrence is the set of Hermite functions, familiar in mathematical physics:

$$\varphi_n(x) = \frac{(-1)^n}{(2^n n!)^{1/2}\pi^{1/4}} \mathrm{e}^{-x^2/2}\mathrm{H}_n(x), \qquad n \in \mathbb{Z}_+, \quad x \in \mathbb{R}, \tag{1}$$

where $\mathrm{H}_n$ is the $n$th *Hermite polynomial.* It is known that

$$\varphi_n'(x) = -\sqrt{\frac{n}{2}}\varphi_{n-1}(x) + \sqrt{\frac{n+1}{2}}\varphi_{n+1}(x), \qquad n \in \mathbb{Z}_+. \tag{2}$$

Motivation for studying such bases is numerical solution of differential equations. If a solution is approximated using a *spectral method* in the basis $\Phi$, numerical solutions take the form $u_N(x) = \sum_{n=0}^N a_n\varphi_n(x)$, and the derivative (after projecting back onto the span of the first $N+1$ basis elements) is given by $u_N'(x) = \sum_{n=0}^N \tilde{a}_n\varphi_n(x)$, where $\tilde{\mathbf{a}} = D\mathbf{a}$, and

$$D = \begin{pmatrix} \beta_0 & \alpha_1 & & & \\ \gamma_0 & \beta_1 & \alpha_2 & & \\ & \gamma_1 & \beta_2 & \ddots & \\ & & \ddots & \ddots & \alpha_N \\ & & & \gamma_{N-1} & \beta_N \end{pmatrix}. \tag{3}$$

The matrix $D$ is called the *differentiation* matrix for the basis $\Phi$. Sparse and structured matrices such as this can be used to an advantage for the numerical linear algebra involved in solving a differential equation numerically.

Beyond the tridiagonal nature of $D$ which is guaranteed by such an analogue of Favard's theorem, we have the following wish list:

- We wish for $D$ to be skew-symmetric (i.e. $\gamma_n = -\alpha_{n+1}$). This emulates the skew-self-adjointness of the differentiation operator with respect to the standard inner product and zero Dirichlet (or Cauchy) boundary conditions. For numerical solution of time-dependent PDEs this ensures stability in the $\ell_2$ norm on the expansion coefficients. Moreover, solutions of numerous dispersive PDEs, e.g. the linear Schrödinger equation, preserve the $\mathrm{L}_2$ norm, and skew-self-adjointness of $D$ ensures that their discretisation does so in the $\ell_2$ norm as well.
- We wish for $\Phi$ to be complete in the sense that all functions of interest can be approximated by finite linear combinations in the basis.
- We wish for $\Phi$ to be an orthonormal basis, since then the discrete $\ell_2$ norm on the expansion coefficients is precisely equal to the continuous $\mathrm{L}_2$ norm of the numerical solution, and so the stability properties transfer over to the continuous norm.
- We wish for $\varphi_n(x)$ to be easily computable, perhaps with an explicit analytical formula, and for expansions like $u_N(x)$ to be easy to work with computationally.

Most of the results in this paper have been published elsewhere by the present authors [12–14]. Our purposes here are to provide these results in a self-contained fashion, to present the material in a way more suited for experts in classical analysis, and to offer some new examples and insight that generalize the theory.

# 2 The Main Theory

## 2.1 Fundamental Results

At a basic level, the problem of finding a differential analogue of Favard's theorem is solved by considering functions of the form,

$$\varphi_n(x) = p_n\left(\frac{\mathrm{d}}{\mathrm{d}x}\right)\varphi_0(x),$$

where $\{p_n\}_{n\in\mathbb{Z}_+}$ are the orthogonal polynomials with recurrence coefficients $\{\alpha_n\}_{n=1}^{\infty}, \{\beta_n\}_{n\in\mathbb{Z}_+}, \{\gamma_n\}_{n\in\mathbb{Z}_+}$. However, this formulation does not help us design the differentiation matrix to be skew-symmetric unless we allow non-positive measures $\mu$, nor does it elucidate the function space spanned by $\{\varphi_n\}_{n\in\mathbb{Z}_+}$ or when the basis will be orthogonal. Cue the following three theorems and their corollaries, which approach the problem via Fourier analysis [12, 13].

**Theorem 1** *A set of functions $\Phi = \{\varphi_n\}_{n\in\mathbb{Z}_+} \subset \mathrm{L}_2(\mathbb{R})$ satisfies*

$$\begin{aligned}\varphi_0' &= \mathrm{i}c_0\varphi_0 + b_0\varphi_1\\ \varphi_n' &= -\overline{b_{n-1}}\varphi_{n-1} + \mathrm{i}c_n\varphi_n + b_n\varphi_n\end{aligned}$$

*for some sequences $b_n \in \mathbb{C}\setminus\{0\}$, $c_n \in \mathbb{R}$, if and only if*

$$\varphi_n(x) = \frac{\mathrm{e}^{\mathrm{i}\theta_n}}{\sqrt{2\pi}}\int_{-\infty}^{\infty}\mathrm{e}^{\mathrm{i}x\xi}p_n(\xi)g(\xi)\,\mathrm{d}\xi, \tag{4}$$

*where*

- *$P = \{p_n\}_{n\in\mathbb{Z}_+}$ is an orthonormal polynomial system with respect to $\mathrm{d}\mu$, a probability measure on the real line with all moments finite and with infinitely many points of increase.*
- *$g \in \mathrm{L}^2(\mathbb{R})$ is such that $|\xi^n g(\xi)| \to 0$ as $|\xi| \to \infty$ for all $n \geq 0$*
- *$\{\theta_n\}_{n\in\mathbb{Z}_+} \subset [0, 2\pi)$*

**Corollary 1** *A set of real-valued functions $\Phi = \{\varphi_n\}_{n\in\mathbb{Z}_+} \subset \mathrm{L}_2(\mathbb{R})$ satisfies*

$$\begin{aligned}\varphi_0' &= b_0\varphi_1\\ \varphi_n' &= -b_{n-1}\varphi_{n-1} + b_n\varphi_n\end{aligned}$$

*for some sequence $b_n > 0$, if and only if $\Phi$ satisfies the requirements of Theorem 1 but with the extra constraints,*

1. *The measure $\mu$ is symmetric, i.e.* $\mathrm{d}\mu(-\xi) = \mathrm{d}\mu(\xi)$
2. *The function $g$ has even real part and odd imaginary part*
3. $\theta_n = n\pi/2 \bmod 2\pi$, *i.e.* $\mathrm{e}^{\mathrm{i}\theta_n} = \mathrm{i}^n$.

**Theorem 2 (Orthogonal Systems)** *Let $\Phi = \{\varphi_n\}_{n\in\mathbb{Z}_+}$ satisfy the requirements of Theorem 1. Then $\Phi$ is orthogonal in $\mathrm{L}_2(\mathbb{R})$ if and only if $P$ is orthogonal with respect to the measure $|g(\xi)|^2\mathrm{d}\xi$. Furthermore, whenever $\Phi$ is orthogonal, the functions $\varphi_n/\|g\|_2$ are orthonormal.*

***Proof*** *It follows from Parseval's identity for the inner product of Fourier transforms that*

$$\int_{-\infty}^{\infty} \overline{\varphi_n(x)}\varphi_m(x)\,\mathrm{d}x = \mathrm{e}^{\mathrm{i}(\theta_m-\theta_n)} \int_{-\infty}^{\infty} p_m(\xi)p_n(\xi)|g(\xi)|^2\,\mathrm{d}\xi, \qquad n, m \in \mathbb{Z}_+.$$

**Theorem 3 (Orthogonal Bases for a Paley–Wiener Space)** *Let $\Phi = \{\varphi_n\}_{n\in\mathbb{Z}_+}$ satisfy the requirements of Theorem 2 with a measure $\mathrm{d}\mu$ such that polynomials are dense in $\mathrm{L}_2(\mathbb{R}; \mathrm{d}\mu)$. Then $\Phi$ forms an orthogonal basis for the Paley–Wiener space* $\mathrm{PW}_\Omega(\mathbb{R})$,[1] *where $\Omega$ is the support of* $\mathrm{d}\mu$.

The key consequence of Theorem 3 is that for a basis $\Phi$ satisfying the requirements of Theorem 2 to be complete in $\mathrm{L}_2(\mathbb{R})$, it is necessary that the polynomial basis $P$ is orthogonal with respect to a measure which is supported on the whole real line.

## 2.2 *Relation to Existing Work*

In this paper we advocate for the case where $|g(\xi)|^2 = w(\xi)$, where $\mathrm{d}\mu(\xi) = w(\xi)\,\mathrm{d}\xi$. The reason being that by Theorem 2, we obtain an orthogonal system in $\mathrm{L}^2(\mathbb{R})$.

In existing work it is more common to consider the case $g(\xi)\,\mathrm{d}\xi = \mathrm{d}\mu(\xi)$, and by some authors these functions are known as Fourier–Bessel functions [24]. There are some advantages to using this $g(\xi)$, for example, one does not need to assume that $\mathrm{d}\mu(\xi)$ is absolutely continuous with respect to Lebesgue measure. Furthermore, it can be shown that expansions in the resulting basis have coefficients which depend locally on the function (similarly to Taylor series). In this area of research, such series are called *chromatic expansions* [11, 33]. However, this choice of $g(\xi)$ no

[1] $\mathrm{PW}_\Omega(\mathbb{R})$ for $\Omega \subseteq \mathbb{R}$ is the space of all functions $f \in \mathrm{L}_2(\mathbb{R})$ such that the Fourier transform of $f$ is supported on $\Omega$.

longer produces orthonormal systems on the real line with respect to the standard $L^2(\mathbb{R})$ inner product.

Differential recurrences for orthogonal polynomials are known to be restricted to classical measures (Jacobi, Laguerre, Hermite and Bessel) but they cannot produce a skew-symmetric differentiation matrix—in fact, if $\varphi_n$s are polynomials, a differentiation matrix is always lower-triangular. Such recurrences have been extended to the realm of Sobolev orthogonality using the concept of coherent pairs and their generalisation [6, 7] and this presents interesting connections with the work of this paper. However, this connection should not be confused with Sobolev-orthogonal systems on the real line, which are mentioned briefly in Sect. 6.3.

# 3 Examples

For an absolutely continuous measure $\mathrm{d}\mu(\xi) = w(\xi)\mathrm{d}\xi$ with orthonormal polynomials $P = \{p_n\}_{n\in\mathbb{Z}_+}$, we consider the following to be its canonical associated orthonormal basis functions:

$$\varphi_n(x) = \frac{\mathrm{i}^n}{\sqrt{2\pi}} \int_{-\infty}^{\infty} \mathrm{e}^{\mathrm{i}x\xi} p_n(\xi)|w(\xi)|^{\frac{1}{2}}\,\mathrm{d}\xi. \tag{5}$$

With this particular representation, the coefficients $\{b_n\}_{n\in\mathbb{Z}_+}$ have the convenient property that they are all positive real numbers. The coefficients in the differentiation matrix for $\Phi$ also correspond nicely with the three-term recurrence coefficients of the orthonormal polynomials:

$$\begin{aligned} \xi p_n(\xi) &= b_{n-1}p_{n-1}(\xi) + c_n p_n(\xi) + b_n p_{n+1}(\xi) \\ &\Longleftrightarrow \\ \varphi_n'(\xi) &= -b_{n-1}\varphi_{n-1}(\xi) + \mathrm{i}c_n\varphi_n(\xi) + b_n\varphi_{n+1}(\xi) \end{aligned}$$

The terminology we use for these bases is as follows. If the polynomial basis $P$ has a name, such as the *Chebyshev polynomials*, then we call the basis $\Phi$ the *transformed Chebyshev functions*.

## 3.1 Jacobi

Transformed Jacobi functions are related to Bessel functions. The measure of orthogonality is

$$\mathrm{d}\mu(\xi) = \chi_{(-1,1)}(\xi)(1-\xi)^{\alpha}(1+\xi)^{\beta}\,\mathrm{d}\xi, \tag{6}$$

for given $\alpha, \beta > -1$. It is possible to give an analytic formula for the first few transformed basis functions in terms of hypergeometric functions, for example,

$$\begin{aligned}\varphi_0(x) &= \frac{1}{\sqrt{2\pi}} \int_{-1}^{1} (1-\xi)^{\alpha/2}(1+\xi)^{\beta/2} \frac{e^{ix\xi}}{\sqrt{2^{\alpha/2+\beta/2+1} B(\alpha/2+1, \beta/2+1)}} d\xi \\ &= \frac{2^{\alpha/4+\beta/4}}{\sqrt{\pi}} \sqrt{B(\alpha/2+1, \beta/2+1)} e^{ix} {}_1F_1\left[\begin{matrix} 1+\alpha/2; \\ 2+(\alpha+\beta)/2; \end{matrix} \ -2ix\right].\end{aligned}$$

However, we do not believe that any traction can be gained from working out these functions for general $\alpha$ and $\beta$. A special case of the Jacobi polynomials is the family of ultraspherical polynomials, $\beta = \alpha$, whereby

$$b_n = \sqrt{\frac{(n+1)(n+2\alpha+1)}{(2n+2\alpha+1)(2n+2\alpha+3)}}, \qquad n \in \mathbb{Z}_+,$$

and $c_n = 0$ for all $n \in \mathbb{Z}_+$. The first two transformed ultraspherical functions are

$$\begin{aligned}\varphi_0(x) &= C_\alpha \left(\frac{2}{x}\right)^{(1+\alpha)/2} \mathrm{J}_{(1+\alpha)/2}(x) \\ \varphi_1(x) &= C_\alpha \sqrt{3+2\alpha} \left(\frac{2}{x}\right)^{(1+\alpha)/2} \mathrm{J}_{(3+\alpha)/2}(x),\end{aligned}$$

where $\mathrm{J}_\nu(x)$ is the Bessel function of order $\nu > -1$, and

$$C_\alpha^2 = \frac{\Gamma(\alpha+3/2)\Gamma(\alpha/2+1)}{2^{\alpha+1}\Gamma((1+\alpha)/2)}.$$

For higher indices these functions become more complicated expressions involving polynomials in $1/x$ and Bessel functions. However, in the special case $\alpha = 0$ of Legendre polynomials these reduce to a very neat form [12],

$$\varphi_n(x) = \sqrt{\frac{n+\frac{1}{2}}{x}} \mathrm{J}_{n+1/2}(x), \qquad n \in \mathbb{Z}_+. \tag{7}$$

Figure 1 displays the first four transformed Legendre functions.

Of course, all functions $\Phi$ 'seeded' by Jacobi polynomials are necessarily band limited and 'live' in $\mathrm{PW}_{(-1,1)}(\mathbb{R})$. This could be useful in some applications, such as signal processing, but as stated in the introduction, we would like to generate bases which are complete in $\mathrm{L}_2(\mathbb{R})$. By Theorem 3, we will need to take polynomials which are orthonormal on the whole real line order to achieve this.

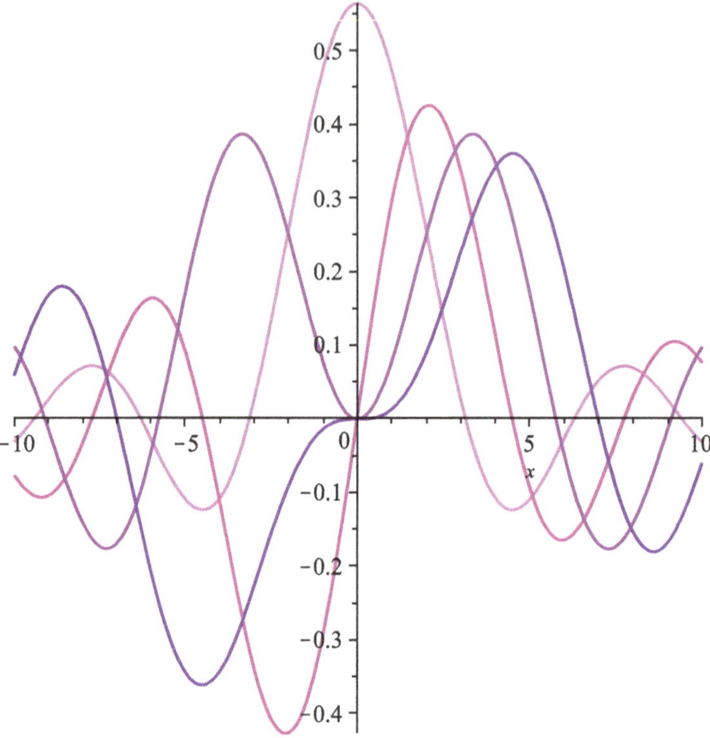


**Fig. 1** The transformed Legendre functions $\varphi_n$ for $n = 0, 1, 2, 3$: darker shade corresponds to higher $n$

## *3.2 Hermite*

Under the transform in Eq. (5), the Hermite polynomials map directly to Hermite functions, so these polynomials hold a special position as a sort of fixed point in this theory. We already mentioned Hermite functions, generated by $\mathrm{d}\mu(\xi) = \mathrm{e}^{-\xi^2}\mathrm{d}\xi$, in (2). They are widely used in computation on the real line and are a univariate case of Hagedorn wave packets [20]. It has been proved in [12] that they are unique among all systems $\Phi$ consistent with Theorems 2–3 with the representation $\varphi_n(x) = h(x)q_n(x)$, where $h \in \mathrm{L}_2(\mathbb{R})$ and each $q_n$ is a polynomial of degree $n$.

Hermite functions obey the Cramér inequality $|\varphi_n(x)| \leq \pi^{-1/4}$, $x \in \mathbb{R}$, $n \in \mathbb{Z}_+$, and several helpful identities, while their generating function is inherited in a straightforward manner from Hermite polynomials.

### 3.3 Generalized Hermite

Let $\eta > -1/2$. The generalized Hermite polynomials $\mathrm{H}_n^{(\eta)}$ are orthogonal with respect to

$$\mathrm{d}\mu(\xi) = |\xi|^{2\eta}\mathrm{e}^{-\xi^2}\mathrm{d}\xi, \qquad \xi \in \mathrm{R}$$

and obey the three-term recurrence relation

$$\mathrm{H}_{n+1}^{(\eta)}(\xi) = 2\xi\mathrm{H}_n^{(\eta)}(\xi) - 2(n+\theta_n)\mathrm{H}_{n-1}^{(\eta)}(\xi),$$

where $\theta_n = 0$ for an even $n$ and $\theta_n = 2\eta$ otherwise [5, p, 156–157]. The explicit form of $\Phi$ has been derived in [12]—the algebra is fairly laborious. Simplifying somewhat the formulæ therein with the first Kummer formula [27, p. 125], we have

$$\varphi_{2n}(x) = c_0\frac{(-1)^n(\frac{1}{2}+\frac{\eta}{2})_n}{2^n}\sum_{\ell=0}^{n}\binom{n}{\ell}\frac{(-1)^\ell(-n-\eta+\frac{1}{2})_\ell}{2^\ell(-n+\frac{1}{2}-\frac{\eta}{2})_\ell}\,{}_1F_1\left[\begin{matrix} n-\ell+\frac{\eta+1}{2}; \\ \frac{1}{2}; \end{matrix}\; -\frac{x^2}{2}\right],$$

$$\varphi_{2n+1}(x) = c_0\frac{(-1)^n(\frac{3}{2}+\frac{\eta}{2})_n}{2^n}x\sum_{\ell=0}^{n}\binom{n}{\ell}\frac{(-1)^\ell(-n-\eta-\frac{1}{2})_\ell}{2^\ell(-n-\frac{1}{2}\eta-\frac{1}{2})_\ell}\,{}_1F_1\left[\begin{matrix} n-\ell+\frac{\eta+3}{2}; \\ \frac{3}{2}; \end{matrix}\; -\frac{x^2}{2}\right],$$

where the normalising constant $c_0$ can be obtained from $c_0^2\int_{-\infty}^{\infty}\varphi_0^2(x)\mathrm{d}x = 1$, $c_0 > 0$.

Figures 2 and 3 display Hermite functions and transformed generalized Hermite system for $\eta = 1$. Hermite functions are, needless to say, the familiar Hermite polynomials scaled by $\mathrm{e}^{-x^2}$ and they demonstrate rapid decay, while their zeros are all real and interlace. Less is known about the case $\eta = 1$ except that the functions evidently decay more sedately.

### 3.4 Laguerre

The transformed Laguerre functions are related to the Fourier basis. The Laguerre weight is

$$\mathrm{d}\mu\xi = \chi_{[0,\infty)}(\xi)\mathrm{e}^{-\xi}\,\mathrm{d}\xi.$$

As described in [13], transformed Laguerre functions have the particularly elegant form,

$$\varphi_n(x) = \sqrt{\frac{2}{\pi}}\mathrm{i}^n\frac{(1+2\mathrm{i}x)^n}{(1-2\mathrm{i}x)^{n+1}}, \qquad n \in \mathbb{Z}_+.$$

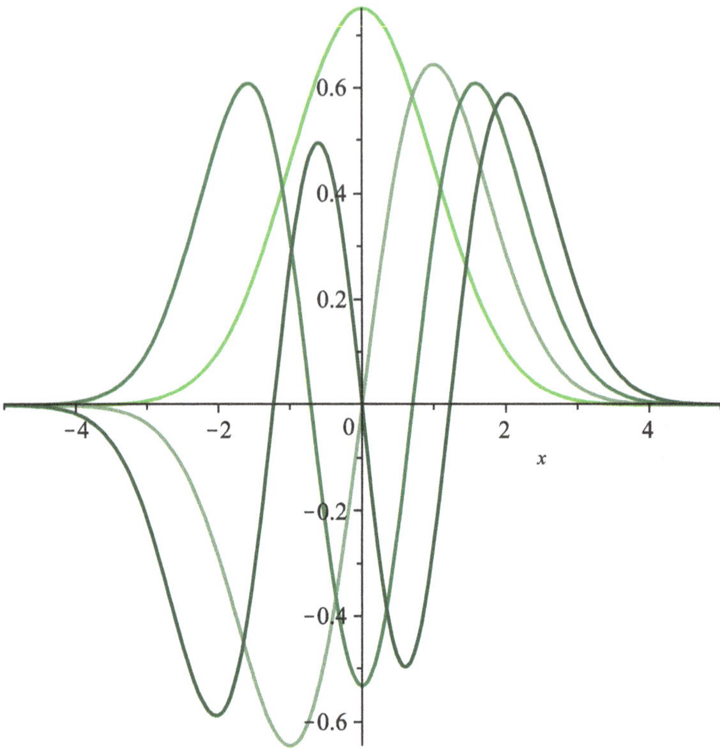


**Fig. 2** Hermite functions $\varphi_n$ for $n = 0, 1, 2, 3$: darker shade corresponds to higher $n$

These functions do not form a complete orthonormal basis for $\mathrm{L}_{(}\mathbb{R})$; by Theorem 3 they are dense in $\mathrm{PW}_{[0,\infty)}(\mathbb{R})$. If we want to obtain a basis which is dense in $\mathrm{L}_2(\mathbb{R})$, all we need to do is take a direct sum of these functions with the functions associated to the Laguerre measure on $(-\infty, 0]$, namely $\mathrm{d}\mu(\xi) = \chi_{(-\infty,0]}(\xi)\mathrm{e}^{\xi}\,\mathrm{d}\xi$. This corresponds to simply taking the above formula and indexing it by $n \in \mathbb{Z}$, resulting in what are known in harmonic analysis as the *Malmquist–Takenaka functions*,

$$\varphi_n(x) = \sqrt{\frac{2}{\pi}}\mathrm{i}^n \frac{(1+2\mathrm{i}x)^n}{(1-2\mathrm{i}x)^{n+1}}, \qquad n \in \mathbb{Z}. \tag{8}$$

These functions have a wealth of beautiful properties and have been discovered and rediscovered over nearly a century since their initial discovery by Malmquist [23] and Takenaka [30], both in 1926 (Fig. 4).

The most notable property of the Malmquist–Takenaka basis is its relation to the Fourier basis. If we make the change of variables $\theta = 2\arctan(2x)$ and

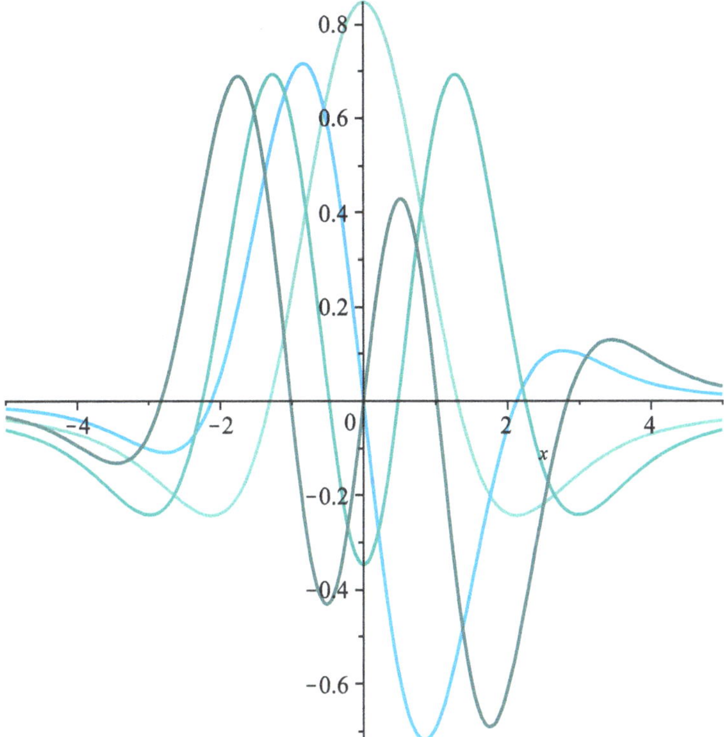


**Fig. 3** Transformed generalized Hermite functions ($\eta = 1$) $\varphi_n$ for $n = 0, 1, 2, 3$: darker shade corresponds to higher $n$

$x = \frac{1}{2}\tan\left(\frac{1}{2}\theta\right)$, then

$$\varphi_n(x) = \sqrt{\frac{2}{\pi}}\mathrm{i}^n \mathrm{e}^{\mathrm{i}(n+\frac{1}{2})\theta}\cos\frac{\theta}{2}. \tag{9}$$

So we see that the Malmquist–Takenaka basis is the Fourier basis in disguise, which leads to a fast FFT-based algorithm to compute the expansion coefficients of a given $f \in \mathrm{L}_2(\mathbb{R})$ (see Sect. 4).

## *3.5 Generalized Laguerre*

The transformed generalized Laguerre functions are related to Szegő–Askey polynomials on the unit circle. The *generalized Laguerre measure* is given by

$$\mathrm{d}\mu(\xi) = \chi_{(0,\infty)}(\xi)\xi^\alpha \mathrm{e}^{-\xi}\,\mathrm{d}\xi,$$

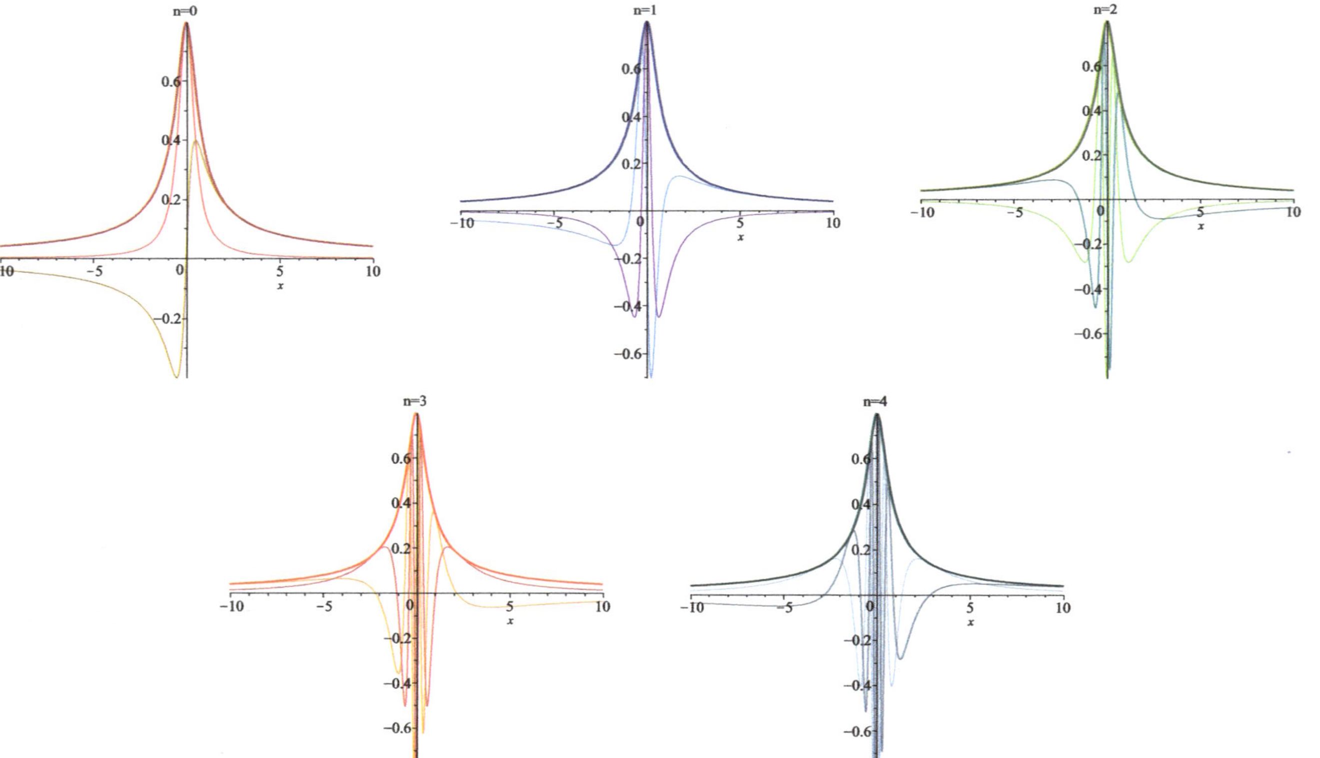


**Fig. 4** The Malmquist–Takenaka system for $n = 0, 1, 2, 34$; Real part in lighter shade than the imaginary part, both within the envelope $(1+4x^2)^{-1/2}$, plotted by a thicker line

where $\alpha > -1$. The case $\alpha = 0$ gives the standard Laguerre polynomials. The corresponding orthogonal polynomials are the *generalized Laguerre polynomials*

$$\mathrm{L}_n^{(\alpha)}(\xi) = \frac{(1+\alpha)_n}{n!} {}_1F_1\left[\begin{matrix}-n; \\ 1+\alpha;\end{matrix} \xi\right] = \frac{(1+\alpha)_n}{n!} \sum_{\ell=0}^{n} (-1)^{\ell} \binom{n}{\ell} \frac{\xi^{\ell}}{(1+\alpha)_{\ell}},$$

where $(z)_m = z(z+1)\cdots(z+m-1)$ is the *Pochhammer symbol* and ${}_1F_1$ is a *confluent hypergeometric function* [27, p. 200]. The Laguerre polynomials obey the recurrence relation

$$(n+1)\mathrm{L}_{n+1}^{(\alpha)}(\xi) = (2n+1+\alpha-\xi)\mathrm{L}_n^{(\alpha)}(\xi) - (n+\alpha)\mathrm{L}_{n-1}^{(\alpha)}(\xi).$$

$$p_n(\xi) = (-1)^n \sqrt{\frac{n!}{\Gamma(n+1+\alpha)}} \mathrm{L}_n^{(\alpha)}(\xi), \qquad n \in \mathbb{Z}_+.$$

We deduce after simple algebra that the normalized polynomials have three-term recurrence coefficients given by,

$$b_n = \sqrt{(n+1)(n+1+\alpha)}, \qquad c_n = 2n+1+\alpha.$$

After lengthy algebra, it was shown in [13] that the transformed generalized Laguerre functions can be expressed by

$$\varphi_n(x) = (-\mathrm{i})^n \sqrt{\frac{2}{\pi}} \left(\frac{1}{1-2\mathrm{i}x}\right)^{1+\frac{\alpha}{2}} \Pi_n^{(\alpha)}\left(\frac{1+2\mathrm{i}x}{1-2\mathrm{i}x}\right), \tag{10}$$

where $\Pi_n^{(\alpha)}$ is a polynomial of degree $n$. Using the substitution $x = \frac{1}{2}\tan\frac{\theta}{2}$ for $\theta \in (-\pi, \pi)$, which implies $(1+2\mathrm{i}x)/(1-2\mathrm{i}x) = \mathrm{e}^{\mathrm{i}\theta}$, the orthonormality of the basis $\Phi$ can be seen to imply that $\{\Pi_n^{(\alpha)}\}_{n\in\mathbb{Z}_+}$ are in fact orthogonal polynomials on the unit circle (OPUC) with respect to the weight

$$W(\theta) = \cos^{\alpha}\frac{\theta}{2}.$$

To be clear, this means that for all $n, m \in \mathbb{Z}_+$,

$$\frac{1}{2\pi}\int_{-\pi}^{\pi} \overline{\Pi_n^{(\alpha)}(\mathrm{e}^{\mathrm{i}\theta})}\Pi_m^{(\alpha)}(\mathrm{e}^{\mathrm{i}\theta}) \cos^{\alpha}\frac{\theta}{2}\,\mathrm{d}\theta = \delta_{n,m}.$$

These polynomials are related to the *Szegő–Askey polynomials* [25, 18.33.13], $\{\phi_n^{(\lambda)}\}_{n\in\mathbb{Z}_+}$, which satisfy

$$\frac{1}{2\pi}\int_{-\pi}^{\pi} \overline{\phi_n^{(\lambda)}(\mathrm{e}^{\mathrm{i}\theta})}\phi_m^{(\lambda)}(\mathrm{e}^{\mathrm{i}\theta}) (1-\cos\theta)^{\lambda}\,\mathrm{d}\theta = \delta_{n,m},$$

by the relation $\Pi_n^{(\alpha)}(z) \propto \phi_n^{(\alpha/2)}(-z)$. In turn, these polynomials are related to the Jacobi polynomials $\mathrm{P}_n^{(\frac{\alpha-1}{2},-\frac{1}{2})}$ and $\mathrm{P}_n^{(\frac{\alpha+1}{2},\frac{1}{2})}$ via the Delsarte–Genin relationship [29].

## 3.6 Continuous Hahn

Continuous Hahn polynomials [17] are orthogonal with respect to the measure

$$\mathrm{d}\mu_{a,b}(\xi) = \frac{1}{2\pi}|\Gamma(a+\mathrm{i}\xi)\Gamma(b-\mathrm{i}\xi)|^2\mathrm{d}\xi, \qquad \xi \in (-\infty,\infty),$$

where $a$ and $b$ are complex parameters with positive real part. The transformed continuous Hahn functions are related to Jacobi polynomials in the following way. Take the following square root of the measure,

$$g_{a,b}(\xi) = \frac{1}{\sqrt{2\pi}}\Gamma(a+\mathrm{i}\xi)\Gamma(b-\mathrm{i}\xi).$$

Note that in general $g_{a,b}$ is a complex valued function, hence it deviates from what we declared as 'canonical' at the beginning of Sect. 3. Note further that when $a$ and $b$ are real, it has an even real part and odd imaginary part, so the resulting transformed functions can be made to be real-valued in this case by Corollary 1.

In [14], the following remarkable identity was shown (in fact an analogous identity was shown for the non-standard continuous Hahn measure $\mathrm{d}\mu_{a,b}(\xi/2)$). Let $p_n$ be the normalized continuous Hahn polynomials (with parameters $a$ and $b$), then

$$\begin{aligned}&\frac{\mathrm{i}^n}{2\pi}\int_{-\infty}^{\infty} p_n(\xi)\Gamma(a+\mathrm{i}\xi)\Gamma(b-\mathrm{i}\xi)\,\mathrm{e}^{\mathrm{i}x\xi}\,\mathrm{d}\xi \qquad (11)\\ &= \left(1-\tfrac{1}{2}\tanh\tfrac{x}{2}\right)^a\left(1+\tfrac{1}{2}\tanh\tfrac{x}{2}\right)^b p_n^{(\alpha,\beta)}\left(\tfrac{1}{2}\tanh\tfrac{x}{2}\right)\end{aligned}$$

where $\alpha = 2a-1$, $\beta = 2b-1$ and $p_n^{(\alpha,\beta)}$ is the $n$th Jacobi polynomial, normalized with respect to the measure (6). It turns out that this relationship between continuous Hahn polynomials and Jacobi polynomials generalizes a little-known identity due to Ramanujan [28], namely

$$\int_{-\infty}^{\infty}|\Gamma(a+\mathrm{i}\xi)|^2\mathrm{e}^{\mathrm{i}x\xi}\mathrm{d}\xi = \frac{\sqrt{\pi}\,\Gamma(a)\Gamma(a+\frac{1}{2})}{\cosh^{2a}\left(\frac{x}{2}\right)}, \qquad a>0.$$

Instead of 'transformed continuous Hahn functions', we call these functions the *tanh–Jacobi* functions [14]. It is convenient to map $\frac{x}{2} \to x$, both for aesthetic

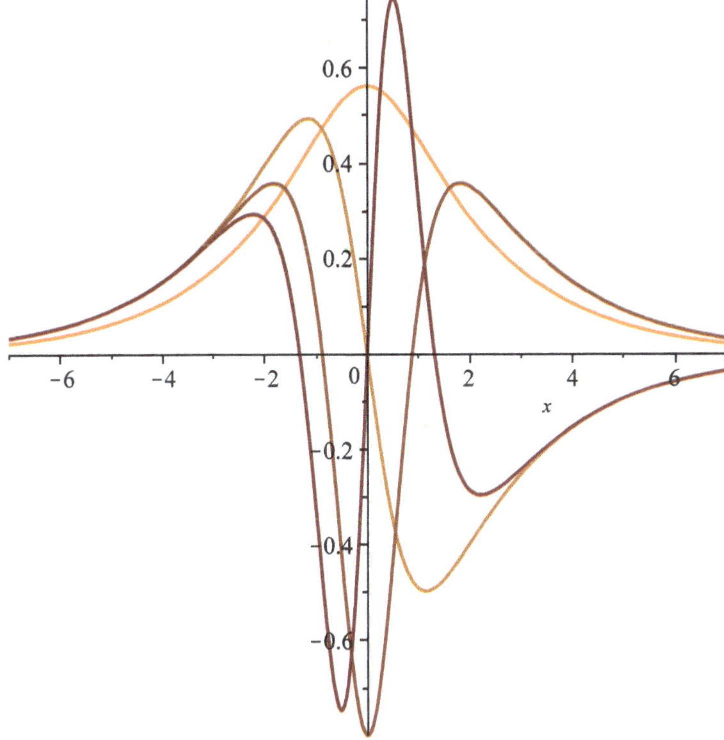


**Fig. 5** The tanh-Chebyshev functions $\varphi_n^{3/4,3/4}$ for $n = 0, 1, 2, 3$: darker shade corresponds to higher $n$

reasons and because this facilitates the computation of expansion coefficients in line with Sect. 4.1. Thus, in place of (11), we have (Fig. 5)

$$\varphi_n^{a,b}(x) = (1 - \tanh x)^a (1 + \tanh x)^b p_n^{(2a-1,2b-1)}(\tanh x), \qquad n \in \mathbb{Z}_+. \tag{12}$$

# 4 Computational Considerations

## 4.1 Computation of Expansion Coefficients

A major consideration in the choice of a practical basis $\Phi$ in the context of spectral methods for PDEs is the speed and ease of the calculation of the first $N$ expansion coefficients $\hat{f}_n$ such that $f(x) = \sum_n \hat{f}_n \varphi_n(x)$ for $f \in L_2(\mathbb{R})$. The simplest algorithm, suitable for all bases $\Phi$, is to compute the coefficients with an $N$-point Gauss–Hermite quadrature. Provided that quadrature nodes and weights are tabulated in advance to allow for repeated calculation, each such computation requires $O(N^2)$ operations if done naively. However, in the case of Hermite functions one can exploit the three-term recurrence relation for Hermite polynomials to derive an $O(N\log^2 N)$ algorithm [21]. The issue, though, is the stability of this procedure. Three-term recurrence relations tend to be unstable in their numerical implementation. While they can be computed stably in a bounded interval using the Clenshaw algorithm [10], once polynomials are orthogonal in $\mathbb{R}$, it is an elementary consequence of standard theory of orthogonal polynomials that their recurrence relations are unbounded (cf. for example (2))—this presents rather delicate implementation issues.

At least five kinds of systems $\Phi$ in the present framework can computed faster (and stably). Letting $\mathrm{e}^{\mathrm{i}\theta}=(1+2\mathrm{i}x)/(1-2\mathrm{i}x)$ in the expression for the coefficients for Malmquist–Takenaka functions (8), we obtain

$$\hat{f}_n=\int_{-\infty}^{\infty} f(x)\overline{\varphi_n(x)}\mathrm{d}x=\frac{(-\mathrm{i})^n}{\sqrt{2\pi}}\int_{-\pi}^{\pi}\left(1-\mathrm{i}\tan\frac{\theta}{2}\right)f\left(\frac{1}{2}\tan\frac{\theta}{2}\right)\mathrm{e}^{-\mathrm{i}n\theta}\mathrm{d}\theta,$$

which can be computed for $-N/2+1\le n\le N/2$ by FFT in $O(N\log N)$ operations. The set of *all* bases $\Phi$ with this feature (namely that the expansion coefficients are equal by a monotone change of variables to the Fourier expansion coefficients of a modified function) is mildly larger:

$$\Phi=\left\{\gamma_n\sqrt{\frac{|\mathrm{Im}\,\lambda|}{\pi}}\mathrm{e}^{\mathrm{i}\omega x}\frac{(\lambda-x)^{n+\delta}}{(\bar{\lambda}-x)^{n+\delta+1}}\;:\;n\in\mathbb{Z}\right\},$$

where $\delta,\omega\in\mathbb{R}$, $\lambda\in\mathbb{C}\setminus\mathbb{R}$ and $\gamma_n\in\mathbb{C}$, $|\gamma_n|=1$, for all $n\in\mathbb{Z}$ [13]. Malmquist–Takenaka corresponds to $\delta=\omega=0$, $\lambda=\mathrm{i}/2$ and $\gamma_n=(-\mathrm{i})^n$.

Four systems (11), corresponding to continuous Hahn polynomials, can be computed in $O(N\log N)$ operations using Fast Cosine Transform—they correspond to

$$(a,b)\in\left\{(\tfrac14,\tfrac14),(\tfrac14,\tfrac34),(\tfrac34,\tfrac14),(\tfrac34,\tfrac34)\right\},$$

whereby the Jacobi polynomials become Chebyshev polynomials of one of four kinds. This follows from (12) by the change of variables $y=\tanh x$ in the integral expression for the coefficients.

We note in passing another approach toward the calculation of expansion coefficients. The expansion coefficients of $f$ with respect to a basis $\Phi$ coincide with the expansion coefficients of its Fourier transform with respect to the basis $P$ of orthogonal polynomials and the inner product induced by their measure. Since computing a Fourier transform with FFT costs $O(N\log N)$ operations for the first $N$ coefficients, we can complement it by a computation of conventional orthogonal expansion. While such $O(N\log N)$ expansions do exist [26], they are unfortunately restricted to Jacobi polynomials. Unless we wish to expand $f$ in $\mathrm{PW}_{(-1,1)}(\mathbb{R})$, this approach—at any rate, in our current state of knowledge—is not competitive.

## 4.2 Approximation Theory on the Real Line

Approximation theory of analytic functions by orthogonal bases is well established in compact intervals, not so in $\mathbb{R}$. The analyticity of $f$ is assumed in a *Bernstein ellipse* surrounding an interval, whereby exponentially-fast convergence of partial sums in the underlying $L_2$ norm is established, at a speed dependent on the eccentricity of the ellipse. This construction fails once the interval is infinite and no alternative overarching theory is available.

The one alternative is the classical method of steepest descent which, with a significant extent of algebraic manipulation, allows the computation of the rate of decay of expansion coefficients. The snag, though, is that each new function calls for new analysis and no general theory is available. And the little we know is baffling!

Take the Malmquist–Takenaka basis, for example. Weideman computed the rate of decay of the coefficients $\hat{f}_n$, $n \in \mathbb{Z}$, for several choices of an analytic $f$ [32], finding

$$\text{for } f(x) = \frac{1}{1+x^4} \quad \text{we have } \hat{f}_n = \mathcal{O}(\rho^{-|n|}), \quad \rho = 1+\sqrt{2}$$

—an exponential rate of decay. Seems like the spectral decay cherished by numerical analysts. Yet,

$$\text{for } f(x) = \frac{\sin x}{1+x^4} \quad \text{we have } \hat{f}_n = \mathcal{O}(|n|^{-9/4})$$

and the convergence is, horrifyingly, little better than quadratic. Thinking naively, $\sin x$ is an entire function, uniformly bounded in magnitude in $\mathbb{R}$: what can go wrong? The cause of the collapse in the speed of convergence is that $\sin x$ has an essential singularity at $\infty$, the North Pole of the Riemann sphere.

Yet, the rules underlying an essential singularity at $\infty$ are hazy as well. If instead of $\sin x/(1+x^4)$ we consider $\sin x/(1+x^2)$, the rate of decay drops to $\mathcal{O}(|n|^{-5/4})$, while

$$\begin{aligned} &\text{for } f(x) = \mathrm{e}^{-x^2} \quad \text{we have } \hat{f}_n = \mathcal{O}(\mathrm{e}^{-3|n|^{2/3}/2}) \\ \text{and } &\text{for } f(x) = \frac{1}{\cosh x} \quad \text{we have } \hat{f}_n = \mathcal{O}(\mathrm{e}^{-2|n|^{1/2}}). \end{aligned}$$

A comprehensive convergence theory for analytic functions on the real line is a significant challenge for approximation theory.

A specific type of functions of significant interest in computational quantum mechanics are *wave packets,* because in the Born–Oppenheimer formulation a wave function of a quantum system can be approximated to high accuracy by a linear combination of such functions. Thus, once we contemplate using a basis $\Phi$ as the 'engine' of a spectral method to discretize PDEs of quantum mechanics, a natural question is how well it does in approximating wave packets. In a univariate setting a wave packet has the form $\cos(\omega x)e^{-\alpha(x-x_0)^2}$, where $\alpha > 0$, $\omega, x_0 \in \mathbb{R}$ and typically $|\omega| \gg 1$. A recent paper [16] analyses, using the method of steepest descent, the performance of different $\Phi$s in this context. It turns out that, once we wish to attain given accuracy, the performance is different for distinct orthonormal systems and that Malmquist–Takenaka functions appear to display the fastest convergence in the large $\omega$ regime.

## 5 Periodic Bases Arising from Discrete Orthogonal Polynomials

Let $\mathcal{Z} \subseteq \mathbb{Z}$ be a set of infinite cardinality. Define the discrete inner product,

$$\langle f, h\rangle = \sum_{k\in\mathcal{Z}} \sigma_k f(k)\overline{h(k)}, \tag{13}$$

where $\sigma_k > 0$, $k \in \mathcal{Z}$, normalized so that $\langle 1, 1\rangle = 1$. Typically we'll choose $\mathcal{Z} = \mathbb{Z}_+$ or $\mathbb{Z}$. The expression in (13) defines an inner product, hence we can form a corresponding orthonormal polynomial system, $P = \{p_n\}_{n\in\mathbb{Z}_+}$.

Consider the $2\pi$-*periodic* functions $\Phi = \{\varphi_n\}_{n\in\mathbb{Z}_+}$ given by,

$$\varphi_n(x) = \mathrm{i}^n \sum_{k\in\mathcal{Z}} \sqrt{\sigma_k}\, p_n(k)e^{\mathrm{i}kx}, \qquad n \in \mathbb{Z}_+, \quad x \in \mathbb{R}.$$

Then, first of all, $P$ being orthonormal with respect to the positive measure $\sum_k \sigma_k \delta(x-k)$, there exist real coefficients $B = \{b_n\}_{n\in\mathbb{Z}_+}$ and $C = \{c_n\}_{n\in\mathbb{Z}_+}$ such that

$$\xi p_n(\xi) = b_{n-1}p_{n-1}(\xi) + c_n p_n(\xi) + b_n p_{n+1}(\xi), \qquad n \in \mathbb{Z}_+,$$

where $b_{-1} = 0$ and $b_n > 0$, $n \in \mathbb{Z}_+$. Differentiating this Fourier series term by term reveals that

$$\varphi_n'(x) = -b_{n-1}\varphi_{n-1}(x) + c_n \mathrm{i}\varphi_n(x) + b_n\varphi_{n+1}(x), \qquad n \in \mathbb{Z}_+.$$

Furthermore, Parseval's identity for a Fourier series gives us

$$\frac{1}{2\pi}\int_{-\pi}^{\pi} \varphi_m(x)\overline{\varphi_n(\xi)}\,\mathrm{d}x = \mathrm{i}^{m-n}\sum_{k\in\mathcal{Z}} \sigma_k p_m(k)p_n(k) = \delta_{m,n}.$$

Therefore, these functions are orthonormal on $\mathrm{L}_2(-\pi, \pi)$. As can be seen, Theorems 1 and 2 appear to generalize naturally to periodic functions in $\mathrm{L}_2(-\pi, \pi)$ via *discrete* orthogonal polynomials. The analogue of Theorem 3 here also holds.

It is of course legitimate to claim that we already have the perfect orthonormal system in $\mathrm{L}_2(-\pi,\pi)\cap C^{\infty}_{\mathrm{per}}(-\pi,\pi)$—the Fourier basis—which has not a tridiagonal differentiation matrix, but a diagonal one. Of course, it also enjoys the added advantage of fast computation with FFT. Thus, it might well be that, in the greater scheme of things, new orthonormal systems of this kind are not of an immediate use. Having said so, at this stage we simply don't know!

Figure 6 displays the first four transformed functions of the bilateral Charlier measure $\mathrm{d}\mu_{1/2}$, where

$$\int_{-\infty}^{\infty} f(\xi)\mathrm{d}\mu_a(\xi) = \sum_{k=-\infty}^{\infty} \frac{a^{|k|}}{|k|!} f(k), \qquad a > -0.$$

(We symmetrize the standard Charlier measure so that the transformed functions are real.) The functions are displayed within a single period.

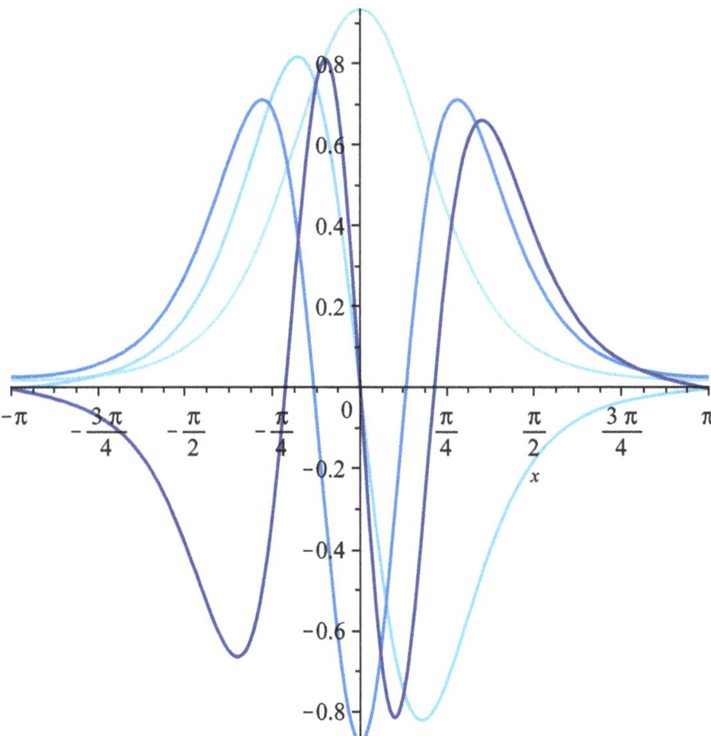


**Fig. 6** Transformed bilateral Charlier functions $\varphi_n$ for $n = 0, 1, 2, 3$ in $[-\pi, \pi]$: darker shade corresponds to higher $n$

# 6 Challenges and Outlook

This paper describes a theory in the making. The results we have uncovered so far are surprisingly elegant, perhaps because they combine the beautiful, rich theories of orthogonal polynomials and Fourier analysis. Standard orthogonal polynomials exhibit deep mathematical structure, and a major challenge is to explore if—and how—this structure is inherited by orthonormal systems $\Phi$.

## 6.1 Transform Pairs

One thing that is particularly interesting is how in certain cases the canonical transformed functions $\Phi$ of a family of orthonormal polynomials $P$ can be expressed in terms of another family of orthonormal polynomials $Q = \{q_n\}_{n\in\mathbb{Z}_+}$ or in terms of known special functions. We summarize the known relationships in the following table.

| Polynomials $p_n$ | Special functions associated to $\varphi_n$ |
|---|---|
| Hermite | Hermite functions/polynomials |
| Laguerre | Malmquist–Takenaka functions, Fourier basis, Chebyshev polynomials |
| Generalized Laguerre | Szegő–Askey polynomials on the unit circle, Jacobi polynomials |
| Ultraspherical | Bessel functions |
| Continuous Hahn | Jacobi polynomials |

The relationship between ultraspherical polynomials and Bessel functions *via* the Fourier transform is well known to those well-versed in special functions. Less so is the relationship between generalized Laguerre polynomials and Szegő–Askey polynomials on the unit circle as in (10), which appears to be noted first by the present authors in [13]. The relationship between continuous Hahn polynomials and Jacobi polynomials as in (11), derived by the present authors in [14], was first noted by Koelink [18], and this follows a long history [2–4, 19, 31].

## 6.2 Location of Zeros

It is well known that zeros of orthogonal polynomials are real, reside in the support of the measure and that zeros of $p_{n-1}$ and $p_n$ interlace. None of these features can be taken for granted for orthonormal bases $\Phi$. Often they are—definitely, and by design, in the case of Hermite and continuous Hahn measures, because each $\varphi$ is a multiple of $p_n$, possibly with a strictly monotone change of argument. Sometimes the problem makes no sense: once $d\mu$ is not symmetric with respect to the origin,

$\varphi_n$ is in general complex-valued. Malmquist–Takenaka functions (8), for example, have no real zeros at all.

In general the picture is more hazy. Transformed Legendre functions (7) have an infinity of real zeros and the zeros of $\varphi_{n-1}$ and $\varphi_n$ interlace: this follows from familiar properties of Bessel functions. The case $\mathrm{d}\mu(\xi) = (1+\xi^2)\mathrm{e}^{-\xi^2}\mathrm{d}\xi$ is displayed in Fig. 7 and all bets are off! $\varphi_0$ has two zeros, $\varphi_1$ has three and $\varphi_2$ four—but, lest a pattern is discerned, $\varphi_3$ has just three, except that the zero at the origin has nontrivial multiplicity, while $\varphi_4$ also has three zeros while the multiplicity of the zero at the origin appears to grow. The analysis of zeros and their locations for general systems $\Phi$ is currently an open problem.

### 6.3 Sobolev Orthogonality

A natural question to ask is on the extension of our theory to more 'exotic' kinds of orthogonality, e.g. the Sobolev $\mathrm{H}^1(\mathbb{R})$ inner product

$$\langle f_1, f_2\rangle = \int_{-\infty}^{\infty} f_1(x) f_2(x)\mathrm{d}x + \int_{-\infty}^{\infty} f_1'(x) f_2'(x)\mathrm{d}x.$$

This will be a subject of a forthcoming paper by the current authors. Spectral methods which are built of such a basis are stable in the $\mathrm{H}_2^1(\mathbb{R})$ Sobolev norm, as opposed to the $\mathrm{L}_2(\mathbb{R})$ norm.

### 6.4 Beyond the Canonical Form

The canonical form given in Eq. (5) guarantees that $\Phi$ is an orthonormal set in $\mathrm{L}_2(\mathbb{R})$ and $b_n > 0$. However, this choice is not unique. Consider a basis of the form

$$\varphi_n(x) = \frac{\mathrm{i}^n}{\sqrt{2\pi}} \int_{-\infty}^{\infty} \mathrm{e}^{\mathrm{i}x\xi} p_n(\xi)\mathrm{e}^{\mathrm{i}\sigma(\xi)}|g(\xi)|^{\frac{1}{2}}\,\mathrm{d}\xi, \tag{14}$$

for a measurable function $\sigma : \mathrm{supp}(\mu) \to \mathbb{R}$. These bases are all orthonormal in $\mathrm{L}_2(\mathbb{R})$ by Theorem 2, and it is readily checked that all of these bases have precisely the same differentiation matrix i.e. the coefficients $b_n$ and $c_n$ do not depend on the function $\sigma$. This is a subtle—yet crucial—distinction between orthogonal polynomials and their transformed functions. Orthogonal monic polynomials are defined uniquely by the Jacobi matrix, while a differentiation matrix is insufficient to define $\Phi$: we also need to specify $\varphi_0$, say, yet not every $\varphi_0$ corresponds to an orthonormal system! The choice of $\sigma$ in (14) captures this added freedom, while ensuring that $\Phi$ is orthonormal.

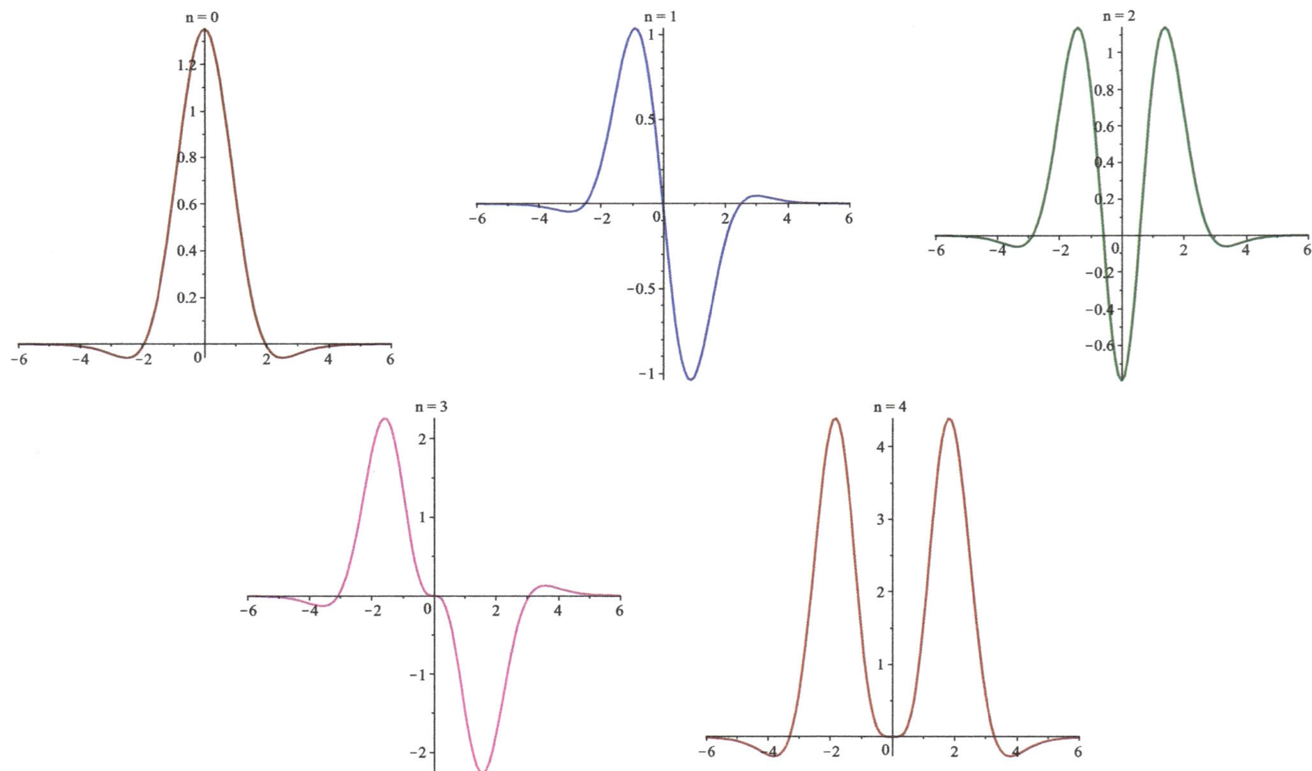


**Fig. 7** The functions $\varphi_n$, $n = 0, \ldots, 4$, for $\mathrm{d}\mu(\xi) = (1+\xi^2)\mathrm{e}^{-\xi^2}\,\mathrm{d}\xi$

This fact has particularly interesting consequences when placed in the context of solving unitary PDEs such as Schrödinger's equation. A Schrödinger equation in one space dimension reads

$$\frac{\partial u}{\partial t} = \mathrm{i}\frac{\partial^2 u}{\partial x^2} + \mathrm{i}F(x,u),$$

where $F$ is the interaction potential. There are good phenomenological reasons to solve it along the entire real line, with the initial condition $u(x,0) = u_0(x)$, $x \in \mathbb{R}$, where $u_0 \in \mathrm{L}_2(\mathbb{R})$. A powerful approach toward the numerical solution of this equation is the concept of *splittings:* the solution is represented as a composition of solutions of the equation

$$\frac{\partial u}{\partial t} = \mathrm{i}\frac{\partial^2 u}{\partial x^2} \tag{15}$$

(the *free Schrödinger equation*) and of the ordinary differential equation $\partial u/\partial t = \mathrm{i}F(x,u)$ [8]. This results in powerful numerical methods that recover many qualitative attributes of the solution. Suppose that we are using a spectral method with the basis $\Phi$, consistent with the theory of Sect. 2 and with a skew-symmetric (or skew-Hermitian) differentiation matrix. Then

$$u_0(x) = \sum_{n\in\mathbb{Z}} \hat{u}_n \varphi_n(x) \qquad \Rightarrow \qquad u(x,t) = \sum_{n\in\mathbb{Z}} \hat{u}_n \psi_n(x,t),$$

where

$$\psi_n(x,t) = \frac{\mathrm{i}^n}{\sqrt{2\pi}} \int_{-\infty}^{\infty} \mathrm{e}^{\mathrm{i}x\xi} p_n(\xi) \mathrm{e}^{\mathrm{i}\xi t^2} |g(\xi)|^{\frac{1}{2}} \,\mathrm{d}\xi.$$

Each $\psi_n$ is the solution of (15) with the initial condition $u(x,0) = \varphi_n(x)$, $x \in \mathbb{R}$ [15]. Cf. Fig. 7 for an example.

## 6.5 A Freudian Slip—Why We Need More Polynomials

The size of the differentiation matrix matters! Numerous applications of spectral methods require the evaluation of the matrix exponential $\mathrm{e}^{\tau D_N}$ or $\mathrm{e}^{\mathrm{i}\tau D_N^2}$, where $D_N$ is the $N \times N$ principal section of $D$ and $\tau$ is the time step. Once $D$ is skew-Hermitian, these exponentials are unitary: this helps to guarantee stability in the sense of Lax. Practical considerations, in particular the wish to use large $\tau$, require the coefficients of $D$ to be as small as possible (more specifically, the spectral radius of $D_N$ should be minimized). However, Sect. 2 and the connection between the entries of $D$ and recurrence coefficients of $P$, imply that the off-diagonal entries cannot be bounded:

$\limsup_{n\to\infty} \alpha_n$, $\limsup_{n\to\infty} \gamma_n = +\infty$. Thus, for Hermite $\alpha_n = \sqrt{(n+1)/2}$ and for Malmquist–Takenaka $\alpha_n = n + 1$.

The objective, thus, is to identify $P$ supported in $\mathbb{R}$ and, for simplicity, with symmetric $\mathrm{d}\mu$, such that the $\alpha_n$s increase at a lower rate. (Symmetry implies $\beta_n \equiv 0$ and $\gamma_n = -\alpha_n$ in (3).) The obvious recourse is to use *Freud polynomials*, orthogonal with respect to $\mathrm{e}^{-|x|^\sigma}$, $\sigma > 0$, $x \in \mathbb{R}$. According to the celebrated *Freud conjecture,*, proved by Lubinsky, Mhaskar, and Saff [22], it is true that $\alpha_n = \mathcal{O}(n^{1/\sigma})$. The larger $\sigma$, the slower the decay! Yet, it is not enough to specify a measure, we also need to have the polynomials in an explicit form in order to construct the set $\Phi$, or at the very least know their recurrence coefficients. Unfortunately, and with the exception of Hermite polynomials ($\sigma = 2$), the explicit form of Freud polynomials is unknown! Although there are known string relations for their recurrence coefficients, they are highly unstable as a numerical means to derive recurrence coefficients. Thus, the challenge is to find orthogonal polynomials supported on the real line (and ideally with a symmetric measure) whose recurrence relations are known explicitly and increase slowly.

The current state of our knowledge of orthogonal polynomials in $\mathrm{L}_2(\mathbb{R})$ is highly incomplete. Few families are known explicitly and their recurrence coefficients grow too fast for our liking. We need more polynomials!

**Acknowledgments** After the online publication of the first version of this paper, several researchers made us aware of related work. We are grateful to Erik Koelink, Doron Lubinsky, Tom Koornwinder, and Enno Diekema for enlightening correspondence and these contributions.

## References

1. N.I. Aheizer, N. Kemmer, *The Classical Moment Problem and Some Related Questions in Analysis* (Oliver and Boyd, Edinburgh, 1965)
2. H. Bateman, Some properties of a certain set of polynomials. Tohoku Math. J. First Series **37**, 23–38 (1933)
3. H. Bateman, An orthogonal property of the hypergeometric polynomial. Proc. Natl. Acad. Sci. USA **28**(9), 374 (1942)
4. L. Carlitz et al., Bernoulli and Euler numbers and orthogonal polynomials. Duke Math. J. **26**(1), 1–15 (1959)
5. T.S. Chihara, *An Introduction to Orthogonal Polynomials* (Gordon and Breach Science Publishers, New York, 1978). Mathematics and its Applications, vol. 13
6. M.N. de Jesus, J. Petronilho, On linearly related sequences of derivatives of orthogonal polynomials. J. Math. Anal. Appl. **347**(2), 482–492 (2008)
7. M.N. de Jesus, F. Marcellán, J. Petronilho, N.C. Pinzón-Cortés, $(M, N)$-coherent pairs of order $(m, k)$ and Sobolev orthogonal polynomials. J. Comput. Appl. Math. **256**, 16–35 (2014)
8. E. Faou, Geometric numerical integration and Schrödinger equations, in *Zurich Lectures in Advanced Mathematics* (European Mathematical Society (EMS), Zürich, 2012)
9. J. Favard, Sur les polynomes de Tchebicheff. C.R. Acad. Sci. Paris **200**, 2052–2053 (1935)
10. L. Fox, I.B. Parker, *Chebyshev Polynomials in Numerical Analysis* (Oxford University Press, London, 1968)

11. A. Ignjatovic, Local approximations based on orthogonal differential operators. J. Fourier Anal. Appl. **13**(3), 309–330 (2007)
12. A. Iserles, M. Webb, Orthogonal systems with a skew-symmetric differentiation matrix. Found. Comput. Math. **19**(6), 1191–1221 (2019)
13. A. Iserles, M. Webb, A family of orthogonal rational functions and other orthogonal systems with a skew-Hermitian differentiation matrix. J. Fourier Anal. Appl. **26**(1), Paper No. 19 (2020)
14. A. Iserles, M. Webb, Fast computation of orthogonal systems with a skew-symmetric differentiation matrix. Commun. Pure Applied Maths **74**(3), 478–506 (2021)
15. A. Iserles, K. Kropielnicka, K. Schratz, M. Webb, Solving the linear Schrödinger equation on the real line. arXiv preprint arXiv:2102.00413 (2021)
16. A. Iserles, K. Luong, M. Webb, Approximation of wave packets on the real line. arXiv preprint arXiv:2101.02566 (2021)
17. R. Koekoek, P.A. Lesky, R.F. Swarttouw, *Hypergeometric Orthogonal Polynomials and their q-Analogues* (Springer, Berlin, 2010)
18. H. Koelink, On Jacobi and continuous Hahn polynomials. Proc. Am. Math. Soc. **124**(3), 887–898 (1996)
19. T.H. Koornwinder, Special orthogonal polynomial systems mapped onto each other by the Fourier-Jacobi transform, in *Polynômes Orthogonaux et Applications* (Springer, Berlin, 1985), pp. 174–183
20. C. Lasser, C. Lubich, Computing quantum dynamics in the semiclassical regime. Acta Numerica **29**, 229–401 (2020)
21. G. Leibon, D.N. Rockmore, W. Park, R. Taintor, G.S. Chirikjian, A fast Hermite transform. Theoret. Comput. Sci. **409**(2), 211–228 (2008)
22. D.S. Lubinsky, H.N. Mhaskar, E.B. Saff, A proof of Freud's conjecture for exponential weights. Constr. Approx. **4**(1), 65–83 (1988)
23. F. Malmquist, Sur la détermination d'une classe de fonctions analytiques par leurs valeurs dans un ensemble donné de points, in *C.R. 6iéme Cong. Mathematical Scand. (Kopenhagen, 1925)*, pp. 253–259. Copenhagen, 1926. Gjellerups
24. G. Mantica, Fourier–Bessel functions of singular continuous measures and their many asymptotics. Electron. Trans. Numer. Anal. **25**, 409–430 (2006)
25. F.W.J. Olver, D.W. Lozier, R.F. Boisvert, C.W. Clark, eds. *NIST Handbook of Mathematical Functions*. U.S. Department of Commerce, National Institute of Standards and Technology, Washington, DC. (Cambridge University Press, Cambridge, 2010). With 1 CD-ROM (Windows, Macintosh and UNIX)
26. S. Olver, R.M. Slevinsky, A. Townsend, Fast algorithms using orthogonal polynomials. Acta Numer. **29**, 573–699 (2020)
27. E.D. Rainville, *Special Functions* (The Macmillan Co., New York, 1960)
28. S. Ramanujan, Some definite integrals connected with Gauss's sums [Messenger Math. **44**, 75–85 (1915)], in *Collected Papers of Srinivasa Ramanujan* (AMS Chelsea Publication, Providence, 2000), pp. 59–67
29. G. Szegő, *Orthogonal Polynomials*, vol. 23 (American Mathematical Society, New York, 1939)
30. S. Takenaka, On the orthogonal functions and a new formula of interpolation. Japanese J. Maths **2**, 129–145 (1926)
31. J. Touchard, Nombres exponentiels et nombres de Bernoulli. Can. J. Math. **8**, 305–320 (1956)
32. J.A.C. Weideman, Theory and applications of an orthogonal rational basis set, in *Proceedings South African of Numerical Mathematics Symposium 1994* (University of KwaZulu-Natal, Glenwood, 1994)
33. A. Zayed, Chromatic expansions in function spaces. Trans. Am. Math. Soc. **366**(8), 4097–4125 (2014)

# Intrinsic Properties of Strongly Continuous Fractional Semigroups in Normed Vector Spaces

**Tiffany Frugé Jones, Joshua Lee Padgett, and Qin Sheng**

**Abstract** Norm estimates for strongly continuous semigroups have been successfully studied in numerous settings, but at the moment there are no corresponding studies in the case of solution operators of singular integral equations. Such equations have recently garnered a large amount of interest due to their potential to model numerous physically relevant phenomena with increased accuracy by incorporating so-called non-local effects. In this article, we provide the first step in the direction of providing such estimates for a particular class of operators which serve as solutions to certain integral equations. The provided results hold in arbitrary normed vector spaces and include the classical results for strongly continuous semigroups as a special case.

**Keywords** Fractional semigroups · Singular integral operators · Logarithmic norm · Non-local operator

T. F. Jones
Department of Mathematics, The University of Arizona, Tucson, AZ, USA
e-mail: tfjones@shsu.edu

J. L. Padgett (✉)
Department of Mathematical Sciences, University of Arkansas, Fayetteville, AR, USA

Center for Astrophysics, Space Physics and Engineering Research, Baylor University, Waco, TX, USA
e-mail: padgett@uark.edu

Q. Sheng
Center for Astrophysics, Space Physics and Engineering Research, Baylor University, Waco, TX, USA

Department of Mathematics, Baylor University, Waco, TX, USA
e-mail: qin_sheng@baylor.edu



F. Gesztesy, A. Martinez-Finkelshtein (eds.), *From Operator Theory to Orthogonal Polynomials, Combinatorics, and Number Theory*, Operator Theory: Advances and Applications 285, https://doi.org/10.1007/978-3-030-75425-9_14

## 1 Introduction

The use of operator theory has been proven to be extremely effective in numerical analysis. It allows for the development of robust tools for applications far outside of the realms of traditional computational mathematics. While these tools have been important and are well understood in cases of classical abstract Cauchy problems—which may be used to represent continuous and discrete problems, alike—there is currently little known about their effectiveness when abstract singular integral problems are presented. In particular, there is no accountable analysis tools available if the generating operators considered are not sectorial.

Researchers in numerical analysis have recently benefited from a surge of activities which overlap with pure mathematical analysis (see, for instance, [4, 7, 18, 19, 28, 30, 31] and publications cited therein). Inspired by this fact and the recent works of Littlejohn and Wellman concerning the investigation of self-adjoint operators in extended Hilbert spaces (cf., e.g., [22–24]), in this article we consider so-called *strongly continuous fractional semigroups* in arbitrary normed vector spaces. The work of Littlejohn and Wellman has added to the existing notion that understanding operators in a more abstract setting can yield extremely insightful results regarding their spectral properties. On the other hand, the theory of orthogonal polynomials is playing an increasingly important role in collocation algorithms for highly oscillatory problems and optimized asymptotic operator splitting (cf., e.g., [9, 37]).

The understanding of the aforementioned properties is integral to the study of numerical algorithms as an entire family, rather than on a case-by-case basis (cf., e.g., [36]). While this task can be difficult in and of itself, the situation is further complicated by the fact that one is also often interested in estimating such families in a non-Hilbert space setting. In this case, it is not so obvious how one can immediately obtain analogous results due to the lack of structure exhibited by many such spaces. In addition to these difficulties, there has been a recent trend in considering operators which exhibit *non-local features* (see, for instance, [3, 5, 14, 30, 32] and references therein). The Cauchy problems associated to these operators no longer exhibit solutions which are semigroups, thus complicating any ensuing analysis. Therefore, even in the classical Hilbert space setting or settings with bounded operators, one cannot employ classical techniques to obtain an understanding of the spectral or norm properties of the solution operator. As an alternative approach, there has been some work in this direction which employs certain orthogonal polynomials (cf., e.g., [2, 12, 27]). However, such endeavors do not yield a meaningful understanding of the qualitative properties of the underlying solution operators. These issues are the primary motivation of the current study.

To this end, herein, we employ techniques which allow for the development of results which mirror the spectral results often derived for local problems in Hilbert spaces for non-local problems in arbitrary normed vector spaces. While such problems have numerous open questions associated to them, we currently focus on the task of developing norm estimates of the solution operators (which will contain classical strongly continuous semigroups as a special case). This goal is accomplished by introducing a particular semi-inner-product on a given normed vector space. This semi-inner-product differs from the classical ones originally studied by Lumer (cf. [11, 15, 26]), but exhibit desirable properties which allow for the development of the desired norm bounds (cf. Lemma 3 and Theorem 1). A nice consequence of these newly derived results is briefly outlined in Corollary 1. An important fact worth mentioning is that Theorem 1 and Corollary 1 demonstrate that while the non-local problems have solution operators which lack certain desired features, it is the case that their norm estimates may be viewed as continuous perturbations of the classical setting with the perturbations being comparable to the amount of non-locality present (e.g., the value of $\alpha \in (0, 1]$ in Corollary 1).

The remainder of this article is organized as follows. In Sect. 2 we introduce a particular notion of a semi-inner-product on arbitrary normed vector spaces and introduce its associated so-called logarithmic norm. In Sect. 2.2 we use these ideas to prove norm growth bounds on strongly continuous semigroups (cf. Definition 3). In Sect. 3 we introduce a notion of fractional semigroups, whose construction depend upon the classical Mittag-Leffler functions (cf. Definitions 5 and 7). Thereafter, these notions are then used to prove analogous growth bounds for strongly continuous fractional semigroups in Sect. 3.2. Finally, in Sect. 4 we provide some concluding remarks on the subject and outline potential future research directions.

## 2 Background

For the purposes of the ensuing analysis, a particular semi-inner-product is defined on arbitrary normed vector spaces. Through it, an associated logarithmic norm is introduced. The necessary concepts of strongly continuous semigroups and their generators will be introduced and studied. An interesting growth bound result of such semigroups employing the logarithmic norm will also be proven (cf. Lemma 3). Our result appears in various forms in the known literature, although, to our knowledge, no formal proof has been provided which is independent of higher regularity in infinite-dimensional spaces (cf., e.g., [8, 25, 39, 40]).

Throughout this article, let $\mathbb{R}$ and $\mathbb{C}$ be the usual real and complex number fields, respectively, and let $i = \sqrt{-1}$. Furthermore, let $\mathbb{N}_0 = \{0, 1, 2, \ldots\}$. In addition, we briefly mention a particular notation used throughout this article which emphasizes how various outside results are applied. If, for example, we cite a result which names a mathematical object $\mathscr{X}$, in order to state results about a family of objects, e.g., $\mathscr{Y}_t$, $t \in \mathbb{R}$, we will write "applied for every $t \in \mathbb{R}$ with $\mathscr{X} \curvearrowleft \mathscr{Y}_t$ in the notation of..." in order to clarify its use.

**Setting 1** *Let $X$ be a vector space defined over the field $\mathbb{R}$ endowed with a proper norm $\|\cdot\|_X$. We denote such a normed $\mathbb{R}$-Banach space as $(X, \|\cdot\|_X)$. We also define the following.*

*(i) Let $B(X) = \{A\colon X \to X\colon \|Ax\|_X < \infty,\ \forall x \in X \text{ with } \|x\|_X = 1\}$.*
*(ii) Let $\mathbb{I}_X\colon X \to X$ satisfy that $\mathbb{I}_X x = x,\ \forall x \in X$.*
*(iii) Let $D(A) \subseteq X$ be the domain of $A\colon X \to X$.*
*(iv) For every $A\colon X \to X$ let $\rho_X(A) = \{\lambda \in \mathbb{C}\colon \exists\, (\lambda\mathbb{I}_X - A)^{-1} \in B(X)\}$ be the resolvent set of $A$.*
*(v) Let $\Re(z) = \frac{1}{2}(z + \bar{z}),\ \forall z \in \mathbb{C}$.*
*(vi) Let $\Gamma\colon \mathbb{C} \to \mathbb{C}$ be Euler's Gamma function endowed with its standard analytic continuation.*
*(vii) Let $R_A\colon \mathbb{C} \to B(X)$ satisfy that $R_A(\lambda) = (\lambda\mathbb{I}_X - A)^{-1},\ \forall A\colon X \to X,\ \forall \lambda \in \rho_X(A)$.*

*Finally, for every $v\colon [0, \infty) \to X$ we define $D_t^+ v(t) = \limsup_{\varepsilon \to 0^+} \frac{v(t+\varepsilon)-v(t)}{\varepsilon} \in [-\infty, \infty],\ t \in [0, \infty)$.*

## 2.1 Logarithmic Norms on Banach Spaces

In Definition 1 we introduce a particular notion of a semi-inner-product on normed vector spaces. Semi-inner-products have long been studied in classical analysis (cf., e.g., [11, 15, 26]) and were originally introduced by Lumer in an attempt to extend classical Hilbert space arguments to more general spaces. While we will not explore all possible properties of semi-inner-products, it is worthwhile to mention that the primary difference between classical inner products and semi-inner-products is the fact that the latter is not necessarily uniquely defined nor bilinear.

**Definition 1 (Right Defined Semi-inner-Product)** Let $(X, \|\cdot\|_X)$ be a normed $\mathbb{R}$-Banach space. We denote by $[\cdot,\cdot]_X\colon X \times X \to \mathbb{R}$ the function which satisfies for all $v, w \in X$ that

$$[v, w]_X = \left[\lim_{\varepsilon \to 0^+} \frac{\|w + \varepsilon v\|_X - \|w\|_X}{\varepsilon}\right] \|w\|_X. \tag{1}$$

We note that the limit in (1) exists due to the fact that the underlying norm on $X$ possesses one-sided Gateaux differentials (cf., e.g., [1]). We also wish to note that the choice of the limit employed in (1) is somewhat arbitrary; in fact, there are numerous other choices which would have demonstrated desirable properties, but the inclusion of these other choices would not have enriched the following discussions.

**Lemma 1** *Let $(X, \|\cdot\|_X)$ be a normed $\mathbb{R}$-Banach space. Then*

*(i) It holds for all $u, v \in X$ that $[u, v]_X \leq \|u\|_X \|v\|_X$,*
*(ii) It holds for all $u \in X$ that $\|u\|_X^2 = [u, u]_X$,*

*(iii) It holds for all* $c \in [0, \infty)$, $u, v \in X$ *that* $[cu, v]_X = c[u, v]_X$, *and*
*(iv) It holds for all* $u, v, w \in X$ *that* $[u + v, w]_X \leq [u, w]_X + [v, w]_X$

*(cf. Definition 1).*

***Proof*** First, note that the triangle inequality ensures that for all $u, v \in X$ it holds that

$$[u, v]_X = \left[\lim_{\varepsilon \to 0^+} \frac{\|v + \varepsilon u\|_X - \|v\|_X}{\varepsilon}\right] \|v\|_X \leq \|u\|_X \|v\|_X \tag{2}$$

(cf. Definition 1). This establishes (i). Next observe that (ii) immediately follows from Definition 1. In addition, note that for all $c \in (0, \infty)$, $u, v \in X$ it holds that

$$\begin{aligned} [cu, v]_X &= \left[\lim_{\varepsilon \to 0^+} \frac{\|v + \varepsilon(cu)\|_X - \|v\|_X}{\varepsilon}\right] \|v\|_X \\ &= c\left[\lim_{c\varepsilon \to 0^+} \frac{\|v + (c\varepsilon)u\|_X - \|v\|_X}{c\varepsilon}\right] \|v\|_X = c[u, v]_X. \end{aligned} \tag{3}$$

Combining this with the fact that Definition 1 implies that for all $u, v \in X$ it holds that $[0u, v]_X = [0, v]_X = 0$ establishes (iii). Moreover, note that the triangle inequality demonstrates that for all $u, v, w \in X$ it holds that

$$\begin{aligned} [u + v, w]_X &= \left[\lim_{\varepsilon \to 0^+} \frac{\|w + \varepsilon(u + v)\|_X - \|w\|_X}{\varepsilon}\right] \|w\|_X \\ &\leq \left[\lim_{\varepsilon \to 0^+} \frac{\|w + 2\varepsilon u\|_X - \|w\|_X}{2\varepsilon}\right] \|w\|_X + \left[\lim_{\varepsilon \to 0^+} \frac{\|w + 2\varepsilon v\|_X - \|w\|_X}{2\varepsilon}\right] \|w\|_X \\ &= [u, w]_X + [v, w]_X. \end{aligned} \tag{4}$$

This establishes (iv). The proof of Lemma 1 is thus completed. □

**Lemma 2** *Assume Setting 1. Then for all differentiable* $v\colon [0, \infty) \to X$, $t \in [0, \infty)$ *it holds that* $\|v(t)\|_X^{-2} D_t^+ \|v(t)\|_X = [\frac{d}{dt} v(t), v(t)]_X \|v(t)\|_X$ *(cf. Definition 1).*

***Proof*** Throughout this proof let $v\colon [0, \infty) \to X$, let $t \in [0, \infty)$, and assume without loss of generality that $\|v(t)\|_X \neq 0$. Observe that the hypothesis that $v$ is differentiable and Taylor's theorem (cf., e.g., Cartan et al. [6, Theorem 5.6.3]) yield that there exists $\delta_t(\varepsilon) \in X$, $\varepsilon \in \mathbb{R}$, such that for all $\varepsilon \in \mathbb{R}$ with $|\varepsilon|$ sufficiently small

(A) it holds that $v(t + \varepsilon) = v(t) + \varepsilon \frac{d}{dt} v(t) + |\varepsilon| \delta_t(\varepsilon)$ and
(B) it holds that $\lim_{\varepsilon \to 0} \delta_t(\varepsilon) = 0$.

Combining (A) and (B) hence shows that

$$
\begin{aligned}
D_t^+ \|v(t)\|_X &= \limsup_{\varepsilon \to 0^+} \frac{\|v(t) + \varepsilon \frac{d}{dt} v(t) + |\varepsilon| \delta_t(\varepsilon)\|_X - \|v(t)\|_X}{\varepsilon} \\
&= \lim_{\varepsilon \to 0^+} \frac{\|v(t) + \varepsilon \frac{d}{dt} v(t)\|_X - \|v(t)\|_X}{\varepsilon} \\
&= \left[ \lim_{\varepsilon \to 0^+} \frac{\|v(t) + \varepsilon \frac{d}{dt} v(t)\|_X - \|v(t)\|_X}{\varepsilon} \right] \frac{\|v(t)\|_X^2}{\|v(t)\|_X^2} = \frac{\left[\frac{d}{dt} v(t), v(t)\right]_X}{\|v(t)\|_X^2} \|v(t)\|_X
\end{aligned}
\tag{5}
$$

(cf. Definition 1). The proof of Lemma 2 is thus completed. □

**Definition 2 (Logarithmic Norm)** Assume Setting 1. Then for every $A\colon X \to X$ we denote by $\mu_X(A) \in \mathbb{R}$ the real number which satisfies

$$
\mu_X(A) = \sup_{\substack{v \in D(A) \subseteq X \\ \|v\|_X \neq 0}} \frac{[Av, v]_X}{\|v\|_X^2}. \tag{6}
$$

We close Sect. 2.1 with a brief discussion of Definition 2. Let $N \in \mathbb{N} = \{1, 2, 3, \ldots\}$, $p \in [1, \infty)$, let $X = L_p([0, \infty); \mathbb{R}^N)$ be the standard $L_p$-space on $\mathbb{R}^N$ (endowed with its typical norm), and let $A\colon \mathbb{R}^N \to \mathbb{R}^N$. Then Definitions 1 and 2 ensure that

$$
\begin{aligned}
\sup_{\substack{v \in D(A) \subseteq X \\ \|v\|_X \neq 0}} \frac{[Av, v]_X}{\|v\|_X^2} &= \sup_{\substack{v \in D(A) \subseteq X \\ \|v\|_X \neq 0}} \frac{\left[\lim_{\varepsilon \to 0^+} \frac{\|v + \varepsilon Av\|_X - \|v\|_X}{\varepsilon}\right] \|v\|_X}{\|v\|_X^2} \\
&= \lim_{\varepsilon \to 0^+} \frac{\|\mathbb{I}_X + \varepsilon A\|_{\mathrm{op}} - 1}{\varepsilon},
\end{aligned}
\tag{7}
$$

where $\|\cdot\|_{\mathrm{op}}$ is the matrix norm induced by $\|\cdot\|_X$, which corresponds to the classical finite-dimensional definition used in, e.g., [40]. The usefulness of such formulations can be seen from the resulting closed-form expressions for (7) for particular choices of $p \in [1, \infty)$. For instance, when $p = 2$, it follows from direct calculation that $\mu_X(A) = \max\{\lambda \in \mathbb{C}\colon \exists x \in \mathbb{R}^N \text{ such that } (A + A^T)x = 2\lambda x\}$.

## 2.2 *Logarithmic Norm Bounds of Classical Semigroups*

We now briefly introduce the classical concepts of strongly continuous semigroups and their associated generators (cf. Definitions 3 and 4). For a more detailed background and applications of these objects, we refer readers to, e.g., [36, 38]. The main result of Sect. 2.2 is Lemma 3, which provides a means to estimate norm bounds for strongly continuous semigroups in a useful manner. In particular, we have that for every generator of a strongly continuous semigroup, $A\colon D(A) \subseteq X \to X$, Lemma 3 implies that if $\mu_X(A) \in (-\infty, 0]$ it holds that $A$ generates a so-called *contraction semigroup* (cf., e.g., [33, Page 10]).

**Definition 3 (Strongly Continuous Semigroup)** Assume Setting 1. Then $\mathbb{T}\colon [0, \infty) \to B(X)$ is a strongly continuous semigroup if

(i) it holds that $\mathbb{T}_0 = \mathbb{I}_X$,
(ii) it holds for all $s, t \in [0, \infty)$ that $\mathbb{T}_{t+s} = \mathbb{T}_t \mathbb{T}_s$, and
(iii) it holds for all $x \in X$ that $\lim_{t\to 0^+} \|\mathbb{T}_t x - x\|_X = 0$.

**Definition 4 (Infinitesimal Generator)** Assume Setting 1 and let $\mathbb{T}\colon [0, \infty) \to B(X)$ be a strongly continuous semigroup (cf. Definition 3). Then $A\colon D(A) \subseteq X \to X$ is the infinitesimal generator of $\mathbb{T}\colon [0, \infty) \to B(X)$ if

(i) it holds that $D(A) = \{x \in X\colon \exists \lim_{h\to 0^+} h^{-1}(\mathbb{T}_h - \mathbb{I}_X)x\}$ and
(ii) it holds for all $x \in D(A)$ that $Ax = \lim_{h\to 0^+} h^{-1}(\mathbb{T}_h - \mathbb{I}_X)x$.

**Lemma 3** *Assume Setting 1 and let $A\colon D(A) \subseteq X \to X$ be the generator of a strongly continuous semigroup $\mathbb{T}_t(A) \in B(X)$, $t \in [0, \infty)$ (cf. Definitions 4 and 3). Then it holds for all $t \in [0, \infty)$, $x \in D(A)$ that $\|\mathbb{T}_t(A)x\|_X \leq \exp(t\mu_X(A))\|x\|_X$ (cf. Definition 2).*

***Proof*** First, observe that the fact that $A$ generates a strongly continuous semigroup and, e.g., Fetahu [13, Theorem 2.12, item c)] (applied for every $t \in [0, \infty)$ with $A \curvearrowleft A$, $x \curvearrowleft x$, $T(t) \curvearrowleft \mathbb{T}_t(A)$ in the notation of [13, Theorem 2.12]) assure that

(I) for all $t \in [0, \infty)$, $x \in D(A)$ it holds that $\mathbb{T}_t(A)x \in D(A)$ and
(II) for all $t \in [0, \infty)$, $x \in D(A)$ it holds that $\frac{d}{dt}\mathbb{T}_t(A)x = A\mathbb{T}_t(A)x$.

Next note that (I) and (II) and Lemma 2 demonstrate that for all $t \in [0, \infty)$, $x \in D(A)$ it holds that

$$\begin{aligned} D_t^+ \|\mathbb{T}_t(A)x\|_X &= \frac{\big[\frac{d}{dt}\mathbb{T}_t(A)x, \mathbb{T}_t(A)x\big]_X}{\|\mathbb{T}_t(A)x\|_X^2} \|\mathbb{T}_t(A)x\|_X \\ &= \frac{\big[A\mathbb{T}_t(A)x, \mathbb{T}_t(A)x\big]_X}{\|\mathbb{T}_t(A)x\|_X^2} \|\mathbb{T}_t(A)x\|_X. \end{aligned} \tag{8}$$

This, (I), the fact that for all $x \in D(A)$ it holds that $\lim_{t\to 0^+} \|\mathbb{T}_t(A)x\|_X = \|x\|_X$, and, e.g., Szarski [41, Theorem 9.6] (applied for every $t \in [0, \infty)$, $x \in D(A)$ with $y(t) \curvearrowleft \|\mathbb{T}_t(A)x\|_X$, $\sigma(t, y) \curvearrowleft [A\mathbb{T}_t(A)x, \mathbb{T}_t(A)x]_X \|\mathbb{T}_t(A)x\|_X^{-2} \|\mathbb{T}_t(A)x\|_X$ in the notation of [41, Theorem 9.6]) prove that for all $t \in [0, \infty)$, $x \in D(A)$ it holds that

$$\|\mathbb{T}_t(A)x\|_X \leq \exp\left(\frac{t\big[A\mathbb{T}_t(A)x, \mathbb{T}_t(A)x\big]_X}{\|\mathbb{T}_t(A)x\|_X^2}\right) \|x\|_X. \tag{9}$$

This and the fact that $\mathbb{R} \ni x \to \exp(x) \in (0, \infty)$ is increasing show that for all $t \in [0, \infty)$, $x \in D(A)$ it holds that

$$\|\mathbb{T}_t(A)x\|_X \leq \exp\left(\frac{t[A\mathbb{T}_t(A)x, \mathbb{T}_t(A)x]_X}{\|\mathbb{T}_t(A)x\|_X^2}\right)\|x\|_X \leq \sup_{\substack{y \in D(A)\\ \|y\|_X \neq 0}} \left[\exp\left(\frac{t[Ay, y]_X}{\|y\|_X^2}\right)\|x\|_X\right]$$

$$\leq \exp\left(\sup_{y \in D(A);\ \|y\|_X \neq 0} \frac{t[Ay, y]_X}{\|y\|_X^2}\right)\|x\|_X = \exp(t\mu_X(A))\|x\|_X \tag{10}$$

(cf. Definition 2). The proof of Lemma 3 is thus completed. □

## 3 Fractional Semigroups

We now introduce the novel solution operators of interest. We note that the use of the term *fractional semigroup* is a bit of a misnomer as the objects in Definition 7 do not satisfy item (ii) of Definition 3 (cf., e.g., [17, Theorem 3.3]). However, we use this term, herein, due to the fact that the classical strongly continuous semigroup is contained as a special case of the operators proposed in Definition 7. In Sect. 3.1 we introduce the concept of the two-parameter Mittag-Leffler and Wright functions. These functions allow for the convenient representation of the fractional semigroups in a fashion which is analogous to the classical functional calculus methods used to represent strongly continuous semigroups. The main result of this article is Lemma 1 of Sect. 3.2. This result can be seen as a generalization of Lemma 3 and, as such, will have analogous useful implications in the study of singular integral problems.

### *3.1 Mittag-Leffler and Wright Functions*

In Definitions 5 and 6 we introduce two families of functions which may be viewed as generalizations of the exponential function and the Bessel functions, respectively. There exists a rich theory behind each of the special functions, but such explorations are tangential to our current goal. As such, we outline their properties which are germane to the current study in Lemmas 4 and 5.

**Definition 5 (Mittag-Leffler Function)** Assume Setting 1 and let $\alpha, \beta \in \mathbb{C}$ satisfy that $\Re(\alpha) \in (0, \infty)$. Then we denote by $E_{\alpha,\beta}\colon \mathbb{C} \to \mathbb{C}$ the function which satisfies for all $z \in \mathbb{C}$ that

$$E_{\alpha,\beta}(z) = \sum_{k=0}^{\infty} \frac{z^k}{\Gamma(\beta + \alpha k)}. \tag{11}$$

**Lemma 4** *Let* $\alpha \in (0, 1]$, $\lambda \in \mathbb{C}$. *Then*

*(i) it holds for all* $z \in \mathbb{C}$ *that* $E_{1,1}(z) = \exp(z)$ *and*
*(ii) it holds for all* $z \in (0, \infty)$ *that* $\frac{d}{dz} E_{\alpha,1}(\lambda z^\alpha) = \lambda z^{\alpha-1} E_{\alpha,\alpha}(\lambda z^\alpha)$

*(cf. Definition 5).*

***Proof*** First, note that Lemma (i) follows directly from Definition 5. Next observe that, e.g., Rudin [35, Theorem 7.17] ensures that for all $z \in (0, \infty)$ it holds that

$$\frac{d}{dz} E_{\alpha,1}(\lambda z^\alpha) = \frac{d}{dz} \sum_{k=0}^{\infty} \frac{(\lambda z^\alpha)^k}{\Gamma(1+\alpha k)} = \sum_{k=0}^{\infty} \frac{d}{dz} \frac{\lambda^k z^{\alpha k}}{\Gamma(1+\alpha k)} = \sum_{k=1}^{\infty} \frac{\alpha k \lambda^k z^{\alpha k-1}}{\Gamma(1+\alpha k)} \tag{12}$$

(cf. Definition 5). This and the fact that for all $z \in \mathbb{C}$ with $\Re(z) \in \mathbb{R}\backslash\{\dots, -2, -1, 0\}$ it holds that $z\Gamma(z) = \Gamma(z+1)$, assure that for all $z \in (0, \infty)$ it holds that

$$\begin{aligned} \frac{d}{dz} E_{\alpha,1}(\lambda z^\alpha) &= \sum_{k=1}^{\infty} \frac{\lambda^k z^{\alpha k-1}}{\Gamma(\alpha k)} = \sum_{k=0}^{\infty} \frac{\lambda^{k+1} z^{\alpha(k+1)-1}}{\Gamma(\alpha(k+1))} \\ &= \lambda z^{\alpha-1} \sum_{k=0}^{\infty} \frac{\lambda^k z^{\alpha k}}{\Gamma(\alpha+\alpha k)} = \lambda z^{\alpha-1} \sum_{k=0}^{\infty} \frac{(\lambda z^\alpha)^k}{\Gamma(\alpha+\alpha k)} = \lambda z^{\alpha-1} E_{\alpha,\alpha}(\lambda z^\alpha). \end{aligned} \tag{13}$$

This establishes Lemma (ii). The proof of Lemma 4 is thus completed. □

**Definition 6 (Wright Function)** Assume Setting 1. Then we denote by $\Psi_{\alpha,\beta} : \mathbb{C} \to \mathbb{C}$ the function which satisfies for all $\alpha \in (-1, \infty)$, $\beta, z \in \mathbb{C}$ that

$$\Psi_{\alpha,\beta}(z) = \sum_{k=0}^{\infty} \frac{(-z)^k}{\Gamma(k+1)\Gamma(\beta+\alpha k)}. \tag{14}$$

**Lemma 5** *Let* $\alpha \in (0, 1)$. *Then*

*(i) it holds for all* $\beta, z \in \mathbb{C}$ *that* $E_{\alpha,\beta}(-z) = \int_0^\infty \Psi_{-\alpha,\beta-\alpha}(t) \exp(-tz)\, dt$,
*(ii) it holds for all* $\beta \in \mathbb{C}$, $n \in \mathbb{N}_0$ *that* $\int_0^\infty z^n \Psi_{-\alpha,\beta}(z)\, dz = \frac{\Gamma(1+n)}{\Gamma(\beta+\alpha(1+n))}$, *and*
*(iii) it holds for all* $\beta, z \in [0, \infty)$ *that* $\Psi_{-\alpha,\beta}(z) \in [0, \infty)$

*(cf. Definitions 5 and 6).*

***Proof*** Throughout this proof for every $f : [0, \infty) \to \mathbb{C}$ let $\mathscr{L}[f; z] \in \mathbb{C}$, $z \in \mathbb{C}$, satisfy for all $z \in \mathbb{C}$ that $\mathscr{L}[f; z] = \int_0^\infty f(t) \exp(-tz)\, dt$. Note that, e.g., Wright [42] ensures that for all $\beta, z \in \mathbb{C}$ it holds that

$$\Psi_{-\alpha,\beta-\alpha}(z) = \frac{1}{2\pi i} \int_{\mathrm{Ha}} \exp(\zeta - z\zeta^\alpha) \zeta^{\alpha-\beta}\, d\zeta \tag{15}$$

(cf. Definition 6), where Ha denotes the Hankel path in the $\zeta$-plane with a cut along the negative real semi-axis $\arg \zeta = \pi$ (cf., e.g., [21, Sect. 13.2.4]). This and Fubini's theorem imply that for all $\beta, z \in \mathbb{C}$ it holds that

$$\int_0^\infty \Psi_{-\alpha,\beta-\alpha}(t) \exp(-tz)\, dt = \int_0^\infty \left[\frac{1}{2\pi i}\int_{\mathrm{Ha}} \exp(\zeta - t\zeta^\alpha)\zeta^{\alpha-\beta}\, d\zeta\right] \exp(-tz)\, dt$$
$$= \frac{1}{2\pi i}\int_{\mathrm{Ha}} \frac{\exp(\zeta)}{\zeta^{\beta-\alpha}}\left[\int_0^\infty \exp\left(-t(z+\zeta^\alpha)\right) dt\right] d\zeta = \frac{1}{2\pi i}\int_{\mathrm{Ha}} \frac{\exp(\zeta)\zeta^{\alpha-\beta}}{z+\zeta^\alpha}\, d\zeta. \tag{16}$$

Combining this and the integral representation of the generalized Mittag-Leffler function as seen in, e.g., Gorenflo and Mainardi [16, Eq. (A.21)] shows that for all $\beta, z \in \mathbb{C}$ it holds that

$$\int_0^\infty \Psi_{-\alpha,\beta-\alpha}(t) \exp(-tz)\, dt = E_{\alpha,\beta}(-z) \tag{17}$$

(cf. Definition 5). This establishes (i). Moreover, observe that the fact that for all $\beta \in \mathbb{C}$, $n \in \mathbb{N}_0$, $f\colon [0,\infty) \to \mathbb{C}$ it holds that $\int_0^\infty t^n f(t)\, dt = \lim_{z\to 0} (-1)^n \frac{d^n}{dz^n}\mathscr{L}[f; z]$ (cf., e.g., Korn and Korn [20, Table 8.3-1]) and (i) guarantee that for all $\beta \in \mathbb{C}$, $n \in \mathbb{N}_0$ it holds that

$$\int_0^\infty z^n \Psi_{-\alpha,\beta}(z)\, dz = \lim_{t\to 0} (-1)^n \frac{d^n \mathscr{L}[\Psi_{-\alpha,\beta}; t]}{dz^n} = \lim_{t\to 0} (-1)^n \frac{d^n E_{\alpha,\alpha+\beta}(-t)}{dt^n}. \tag{18}$$

Next note that induction on $n \in \mathbb{N}_0$ and the dominated convergence theorem show that for all $\beta \in \mathbb{C}$, $n \in \mathbb{N}_0$ it holds that

$$\frac{d^n}{dt^n} E_{\alpha,\alpha+\beta}(-t) = \frac{d^n}{dt^n}\sum_{k=0}^\infty \frac{(-t)^k}{\Gamma((\alpha+\beta)+\alpha k)} = \sum_{k=0}^\infty \frac{d^n}{dt^n}\frac{(-t)^k}{\Gamma((\alpha+\beta)+\alpha k)}$$
$$= \sum_{k=n}^\infty \frac{(-1)^k\big[k\cdot(k-1)\cdot\ldots\cdot(k-n+1)\big]t^{k-n}}{\Gamma(\beta+\alpha(1+k))}. \tag{19}$$

Combining this, e.g., Rudin [35, Theorem 7.17], and (18) ensures that for all $\beta \in \mathbb{C}$, $n \in \mathbb{N}_0$ it holds that

$$\int_0^\infty z^n \Psi_{-\alpha,\beta}(z)\, dz = \lim_{t\to 0} (-1)^n \left[\sum_{k=n}^\infty \frac{(-1)^k\big[k\cdot(k-1)\cdot\ldots\cdot(k-n+1)\big]t^{k-n}}{\Gamma(\beta+\alpha(1+k))}\right]$$
$$= \sum_{k=n}^\infty \left[\lim_{t\to 0} \frac{(-1)^{n+k}\big[k\cdot(k-1)\cdot\ldots\cdot(k-n+1)\big]t^{k-n}}{\Gamma(\beta+\alpha(1+k))}\right] = \frac{\Gamma(1+n)}{\Gamma(\beta+\alpha(1+n))}. \tag{20}$$

This establishes (ii). In addition, observe that, e.g., [10, Theorem 1.3-7] (applied for every $\beta, z \in [0, \infty)$ with $\rho \curvearrowleft -1/\alpha$, $\mu \curvearrowleft \beta - \alpha$, $\Phi \curvearrowleft \Psi$ in the notation of [10, Theorem 1.3-6]) guarantees that (iii) holds. The proof of Lemma 5 is thus completed. □

## 3.2 *Logarithmic Norm Bounds of Fractional Semigroups*

We are now in a position to prove Lemma 1, which is the main result of the article. Note that Definition 7 provides the definition of the two-parameter fractional semigroups which are the focus of the current study. It should be clear that these objects are simply the infinite-dimensional counterparts to the classical Mittag-Leffler functions introduced in Definition 5.

To the authors' knowledge, there have been no result along the lines of Lemma 1 presented in the literature—neither in finite dimensions nor for particular cases of the operator $A$. It is particularly interesting to note that when $\alpha \in (0, 1)$ it holds that Lemma 1 implies the fractional semigroup exhibits sub-exponential growth/decay (which differs from the exponential growth/decay of classical semigroups). We close Sect. 3.2 with Corollary 1 which demonstrates the usefulness of these logarithmic norm bounds in a setting which occurs quite often in applied mathematics. Note that an implication of Corollary 1 is that if $A$ generates a contraction semigroup then it also generates a contraction fractional semigroup.

**Definition 7 (Fractional Semigroup)** Assume Setting 1. Then for every $A \colon X \to X$ we denote by $S_t^{\alpha,\beta}(A) \in B(X)$, $\alpha, \beta \in (0, 1]$, $t \in [0, \infty)$, the function which satisfies for all $\alpha, \beta \in (0, 1]$, $t \in [0, \infty)$ that

$$S_t^{\alpha,\beta}(A) = \frac{1}{2\pi i} \int_{\mathscr{C}} E_{\alpha,\beta}(\lambda t^\alpha) R_A(\lambda)\, d\lambda, \tag{21}$$

where $\mathscr{C} \subseteq \mathbb{C}$ is any contour containing $\sigma(A) = \mathbb{C} \backslash \rho_X(A)$ (cf. Definition 5).

**Theorem 1** *Assume Setting 1 and let $A \colon D(A) \subseteq X \to X$ be the generator of a strongly continuous semigroup $\mathbb{T}_t(A) \in B(X)$, $t \in [0, \infty)$ (cf. Definitions 3 and 4). Then it holds for all $\alpha, \beta \in (0, 1]$, $t \in [0, \infty)$, $x \in D(A)$ that*

$$\|S_t^{\alpha,\beta}(A)x\|_X \leq E_{\alpha,\beta}(t^\alpha \mu_X(A))\|x\|_X \tag{22}$$

*(cf. Definitions 5, 2, and 7).*

***Proof*** Throughout this proof let $\mathscr{C} \subseteq \mathbb{C}$ be a contour containing $\sigma(A) = \mathbb{C} \backslash \rho_X(A)$. Note that the Riesz–Dunford functional calculus (cf., e.g., [34, Theorem 13.5]) and (i) of Lemma 4 guarantee that for all $t \in [0, \infty)$, $x \in D(A)$ it holds that

$$\begin{aligned} \mathbb{T}_t(A)x &= \frac{1}{2\pi i} \int_{\mathscr{C}} \exp(\lambda t) R_A(\lambda) x \, d\lambda \\ &= \frac{1}{2\pi i} \int_{\mathscr{C}} E_{1,1}(\lambda t) R_A(\lambda) x \, d\lambda = S_t^{1,1}(A)x \end{aligned} \tag{23}$$

(cf. Definitions 5 and 7). This, (i) of Lemma 4, (i) of Lemma 5, and Fubini's theorem ensure that for all $\alpha \in (0, 1)$, $\beta \in (0, 1]$, $t \in [0, \infty)$, $x \in D(A)$ it holds that

$$\begin{aligned} S_t^{\alpha,\beta}(A)x &= \frac{1}{2\pi i} \int_{\mathscr{C}} E_{\alpha,\beta}(\lambda t^\alpha) R_A(\lambda) x \, d\lambda \\ &= \frac{1}{2\pi i} \int_{\mathscr{C}} \left[ \int_0^\infty \Psi_{-\alpha,\beta-\alpha}(z) \exp(\lambda t^\alpha z) \, dz \right] R_A(\lambda) x \, d\lambda \\ &= \frac{1}{2\pi i} \int_0^\infty \Psi_{-\alpha,\beta-\alpha}(z) \int_{\mathscr{C}} \exp(\lambda t^\alpha z) R_A(\lambda) x \, d\lambda \, dz \\ &= \frac{1}{2\pi i} \int_0^\infty \Psi_{-\alpha,\beta-\alpha}(z) \int_{\mathscr{C}} E_{1,1}(\lambda t^\alpha z) R_A(\lambda) x \, d\lambda \, dz \\ &= \int_0^\infty \Psi_{-\alpha,\beta-\alpha}(z) S_{t^\alpha z}^{1,1}(A) x \, dz = \int_0^\infty \Psi_{-\alpha,\beta-\alpha}(z) \mathbb{T}_{t^\alpha z}(A) x \, dz. \end{aligned} \tag{24}$$

Combining this, Jensen's inequality, Lemma 3, and (iii) and (i) of Lemma 5 assures that for all $\alpha \in (0, 1)$, $\beta \in (0, 1]$, $t \in [0, \infty)$, $x \in D(A)$ it holds that

$$\begin{aligned} \|S_t^{\alpha,\beta}(A)x\|_X &= \left\| \int_0^\infty \Psi_{-\alpha,\beta-\alpha}(z) \mathbb{T}_{t^\alpha z}(A) x \, dz \right\|_X \\ &\le \int_0^\infty \Psi_{-\alpha,\beta-\alpha}(z) \|\mathbb{T}_{t^\alpha z}(A) x\|_X \, dz \\ &\le \int_0^\infty \Psi_{-\alpha,\beta-\alpha}(z) \exp\big(t^\alpha z \mu_X(A)\big) \|x\|_X \, dz = E_{\alpha,\beta}\big(t^\alpha \mu_X(A)\big) \|x\|_X \end{aligned} \tag{25}$$

(cf. Definition 2). Combining this, (23), and Lemma 3 establishes (22). The proof of Lemma 1 is thus completed. □

An immediate consequence of Lemma 1 is outlined in Corollary 1. While the proof of Corollary 1 is interesting, it is also brief and straightforward; hence, we omit the proof for brevity. Interested readers may note that this result follows by combining Lemma 1, the fact that the Dirichlet Laplace operator on $L^2([0, 1]; \mathbb{R})$, denoted $A$, is negative definite and the generator of a contraction semigroup, and the fact that (in the notation of Corollary 1) it holds for all $t \in [0, \infty)$, $x \in [0, 1]$ that $u(t, x) = S_t^{\alpha,1}(A) u(0, x)$ (cf., e.g., [29, Lemma 3.1]).

**Corollary 1** *Let* $\alpha \in (0, 1]$, *let* $X = L^2([0, 1]; \mathbb{R})$ *be the* $\mathbb{R}$*-Hilbert space of equivalence classes of Lebesgue square-integrable functions from* $(0, 1)$ *to* $\mathbb{R}$ *equipped with its standard norm* $\|\cdot\|_X$, *let* $A \colon D(A) \subseteq X \to X$ *be the Dirichlet Laplace operator, for every* $t \in [0, \infty)$, $x \in [0, 1]$ *let* $u(t, x) \in D(A) \subseteq X$ *satisfy that* $u(t, x) = u(0, x) + \frac{1}{\Gamma(\alpha)} \int_0^t (t - s)^{\alpha-1} Au(s, x)\, ds$ *(cf. Setting 1). Then it holds for all* $t \in [0, \infty)$, $x \in [0, 1]$ *that* $\|u(t, x)\|_X \leq \|u(0, x)\|_X$ *(cf. Definitions 2, 3, and 4).*

## 4 Conclusions and Future Endeavors

In this article we have explored norm bounds of fractional semigroups in arbitrary normed vector spaces. To aid in this study, we introduced a particular semi-inner-product and used its properties to provide norm bounds of strongly continuous semigroups via the logarithmic norm. We then combined these results with representation results for fractional semigroups in order to obtain the main result (i.e., Lemma 1) of the article. This result is the first of its kind and demonstrates that the newly studied fractional semigroups exhibit growth/decay properties which are analogous to the classical setting—the difference being that the exponential function is replaced with a two-parameter Mittag-Leffler function (cf. Definition 5). An immediate consequence of this result is that the contractive property of a semigroup is preserved in the fractional setting—a surprising result which will be quite useful in actual applications (e.g., the setting outlined in Corollary 1).

Future endeavors in this direction will involve both continued abstract analysis and the implementation of actual numerical algorithms. Of particular interest to the authors is the development of operator splitting algorithms for fractional semigroups. Such techniques are local by definition, hence the nonlocality of these new operators (i.e., their lack of a semigroup property) complicates matters considerably. However, one can demonstrate that there exists a semigroup-like property which the fractional semigroups satisfy. From this, we will construct a generalization of the classical Lie-Trotter splitting method (cf., e.g., [36]) and the asymptotic operator splitting (cf., e.g., [37]) and employ the results developed herein to complete the study. Numerical endeavors will pursue the use of the presented operator theoretical representation of singular integral equations for actual simulations. Such methods have yet to be explored and initial results seem to indicate that they will provide a significant improvement in efficiency and accuracy.

**Acknowledgments** The second author acknowledges funding by the National Science Foundation (NSF 1903450). The third author would like to thank the College of Arts and Sciences at Baylor University for partial support through a research leave award. The first and second authors also gratefully acknowledge the impact that Lance Littlejohn has had on this project and their careers—Lance provided their initial exposure to the beauty of operator theory during his graduate functional analysis course at Baylor University. All three authors would like to congratulate Lance Littlejohn on the occasion of his 70th birthday.

# References


1. T.J. Abatzoglou, Norm derivatives on spaces of operators. Math. Ann. **239**(2), 129–135 (1979)
2. M. Abdou, Integral equation of mixed type and integrals of orthogonal polynomials. J. Comput. Appl. Math. **138**(2), 273–285 (2002)
3. W. Arendt, A.F. Ter Elst, M. Warma, Fractional powers of sectorial operators via the Dirichlet-to-Neumann operator. Comm. Partial Differential Equations **43**(1), 1–24 (2018)
4. D. Arnold, R. Falk, R. Winther, Finite element exterior calculus: from Hodge theory to numerical stability. Bull. Am. Math. Soc. **47**(2), 281–354 (2010)
5. L. Caffarelli, L. Silvestre, An extension problem related to the fractional Laplacian. Comm. Partial Differential Equations **32**(8), 1245–1260 (2007)
6. H. Cartan, J. Moore, D. Husemoller, K. Maestro, *Differential Calculus on Normed Spaces: A Course in Analysis* (CreateSpace Independent Publishing Platform, California, 2017)
7. C. Curry, K. Ebrahimi-Fard, H. Munthe-Kaas, What is a post-Lie algebra and why is it useful in geometric integration, in *European Conference on Numerical Mathematics and Advanced Applications* (Springer, Berlin, 2017), pp. 429–437
8. G. Dahlquist, *Stability and Error Bounds in the Numerical Integration of Ordinary Differential Equations*. PhD thesis, (Almqvist and Wiksell, Stockholm, 1958)
9. A. Deaño, D. Huybrechs, A. Iserles, *Computing Highly Oscillatory Integrals* (SIAM, New York, 2017)
10. M.M. Djrbashian, *Harmonic Analysis and Boundary Value Problems in the Complex Domain*, vol. 65 (Springer, Berlin, 1993)
11. S.S. Dragomir, *Semi-inner Products and Applications* (Nova Science, New York, 2004)
12. D. Elliott, Orthogonal polynomials associated with singular integral equations having a Cauchy kernel. SIAM J. Math. Anal. **13**(6), 1041–1052 (1982)
13. E. Fetahu, *On Semigroups of Linear Operators*. PhD thesis, Dissertaçao de Mestrado (Central European University, Budapest-Hungary, 2014)
14. J.E. Galé, P.J. Miana, P.R. Stinga, Extension problem and fractional operators: semigroups and wave equations. J. Evol. Equ. **13**(2), 343–368 (2013)
15. J.R. Giles, Classes of semi-inner-product spaces. Trans. Am. Math. Soc. **129**(3), 436–446 (1967)
16. R. Gorenflo, F. Mainardi, Fractional calculus. In: *Fractals and Fractional Calculus in Continuum Mechanics* (Springer, Berlin, 1997), pp. 223–276
17. B.H. Guswanto, On the properties of solution operators of fractional evolution equations. J. Fractional Calculus Appl. **6**(1), 131–159 (2015)
18. A. Iserles, L.L. Littlejohn, Polynomials orthogonal in a Sobolev space. Linear Complex Anal. Problem Book **3**(Part 2), 190 (2006)
19. A. Iserles, H.Z. Munthe-Kaas, S.P. Nørsett, A. Zanna, Lie-group methods. Acta Numer. **9**, 215–365 (2000)
20. G.A. Korn, T.M. Korn, *Mathematical Handbook for Scientists and Engineers: Definitions, Theorems, and Formulas for Reference and Review*. Dover Civil and Mechanical Engineering Series (Dover Publications, New York, 2000)
21. S.G. Krantz, S.G. Krantz, *Handbook of Complex Variables* (Springer, Berlin, 1999)
22. L.L. Littlejohn, R. Wellman, A general left-definite theory for certain self-adjoint operators with applications to differential equations. J. Differential Equations **181**(1), 280–339 (2002)
23. L.L. Littlejohn, R. Wellman, On the spectra of left-definite operators. Complex Anal. Oper. Theory **7**(1), 437–455 (2016)
24. L.L. Littlejohn, R. Wellman, Self-adjoint operators in extended Hilbert spaces $h \oplus w$: an application of the general GKN-EM theorem. Oper. Matrices **13**(3), 667–704 (2019)
25. S.M. Lozinskii, Error estimate for numerical integration of ordinary differential equations. I. Izvestiya Vysshikh Uchebnykh Zavedenii. Matematika **5**, 52–90 (1958)
26. G. Lumer, Semi-inner-product spaces. Trans. Am. Math. Soc. **100**(1), 29–43 (1961)

27. G.N. Minerbo, M.E. Levy, Inversion of Abel's integral equation by means of orthogonal polynomials. SIAM J. Numer. Anal. **6**(4), 598–616 (1969)
28. H.Z. Munthe-Kaas, K.K. Føllesdal, Lie–Butcher series, geometry, algebra and computation, in *Discrete Mechanics, Geometric Integration and Lie–Butcher Series* (Springer, Berlin, 2018), pp. 71–113
29. J.L. Padgett, The quenching of solutions to time–space fractional Kawarada problems. Comput. Math. Appl. **76**(7), 1583–1592 (2018)
30. J.L. Padgett, Analysis of an approximation to a fractional extension problem. BIT Numer. Math. **60**(3), 715–739 (2020)
31. J.L. Padgett, Q. Sheng, Convergence of an operator splitting scheme for abstract stochastic evolution equations, in *Advances in Mathematical Methods and High Performance Computing* (Springer, Berlin, 2019), pp. 163–179
32. J.L. Padgett, E.G. Kostadinova, C.D. Liaw, K. Busse, L.S. Matthews, T.W. Hyde, Anomalous diffusion in one-dimensional disordered systems: a discrete fractional Laplacian method. J. Phys. A Math. Theor. **53**(13), 135205 (2020)
33. D.W. Robinson, et al. *Basic Theory of One-parameter Semigroups*. (The Australian National University, Mathematical Sciences Institute, Centre, Australia, 1982)
34. W. Rudin, *Real and Complex Analysis* (Tata McGraw-Hill education, New York, 2006)
35. W. Rudin, et al. *Principles of Mathematical Analysis*, vol. 3 (McGraw-Hill, New York, 1964)
36. Q. Sheng, Global error estimate for exponential splitting. IMA J. Numerical Analysis **14**(3), 27–56 (1993)
37. Q. Sheng, R.P. Agarwal, A note on asymptotic splitting and its applications. Math. Comput. Model. **20**(12), 45–58 (1994)
38. K.B. Sinha, S. Srivastava, *Theory of Semigroups and Applications* (Springer, Berlin, 2017)
39. G. Söderlind, The logarithmic norm. History and modern theory. BIT Numer. Math. **46**(3), 631–652 (2006)
40. T. Strom, On logarithmic norms. SIAM J. Numer. Anal. **12**(5), 741–753 (1975)
41. J. Szarski, *Differential Inequalities* (Polish Scientific Publishers, Warszawa, 1965)
42. E.M. Wright, The generalized Bessel function of order greater than one. Q. J. Math. **1**, 36–48 (1940)

# The BFK-gluing Formula for Zeta-determinants and the Conformal Rescaling of a Metric

**Klaus Kirsten and Yoonweon Lee**

*The authors dedicate this work to Lance L. Littlejohn, someone who is defying all odds!*

**Abstract** The Dirichlet-to-Neumann operator plays a central role in the BFK-gluing formula for zeta-determinants of Laplacians, whose homogeneous symbols are invariants with respect to conformal rescaling of Riemannian metrics. We use this property together with the result in Kirsten and Lee (J Math Phys 58(12):123501, 19p, 2015) to recover the main result of Kirsten and Lee (J Spectr Theory 10:1007–1051, 2020), which reduces much of long and tedious computation. We also use this property to prove some relation about the value at zero of the zeta function associated to the Dirichlet-to-Neumann operator, which is obtained in Kirsten and Lee (J Geom Anal 28:3856–3891, 2018) on a warped product manifold.

**Keywords** Gluing formula · Zeta-determinants

## 1 Introduction

Let $(M, g)$ be an $m$-dimensional compact Riemannian manifold with boundary $\partial M$ and $\mathcal{E} \to M$ be a vector bundle of rank $r_0$ on $M$, where $\partial M$ is possibly empty. We denote by $\Delta_M$ a Laplacian acting on smooth sections of $\mathcal{E}$. Then, there exist a

---

K. Kirsten (✉)
Department of Mathematics, Baylor University, Waco, TX, USA

Mathematical Reviews, American Mathematical Society, Ann Arbor, MI, USA
e-mail: Klaus_Kirsten@Baylor.edu

Y. Lee
Department of Mathematics Education, Inha University, Incheon, Korea
e-mail: yoonweon@inha.ac.kr



F. Gesztesy, A. Martinez-Finkelshtein (eds.), *From Operator Theory to Orthogonal Polynomials, Combinatorics, and Number Theory*, Operator Theory: Advances and Applications 285, https://doi.org/10.1007/978-3-030-75425-9_15

connection $\nabla : C^{\infty}(\mathcal{E}) \to C^{\infty}(T^{*}M \otimes \mathcal{E})$ and an endomorphism $E : C^{\infty}(\mathcal{E}) \to C^{\infty}(\mathcal{E})$ [1, 6] such that

$$\Delta_{M} = -\left(\operatorname{Tr} \nabla^{2} + E\right). \tag{1}$$

We choose a closed hypersurface $\mathcal{N}$ satisfying $\mathcal{N} \cap \partial M = \emptyset$ and denote by $M_0$ the closure of $M - \mathcal{N}$. $M_0$ is a compact manifold with boundary $\partial M \cup \mathcal{N} \cup \mathcal{N}$, which may or may not be connected depending on $\mathcal{N}$. We distinguish two copies of $\mathcal{N}$ by $\mathcal{N}^{+}$ and $\mathcal{N}^{-}$ as follows. We choose a unit normal vector field $\partial_{\mathcal{N}}$ along $\mathcal{N}$. We denote $\mathcal{N}$ by $\mathcal{N}^{+}$ on which $\partial_{\mathcal{N}}$ points outward and by $\mathcal{N}^{-}$ on which $\partial_{\mathcal{N}}$ points inward, where $\mathcal{N}^{+} = \mathcal{N}^{-} = \mathcal{N}$.

We denote by $g_0$ and $\Delta_{M_0}$ the extension of $g$ and $\Delta_M$ to $M_0$, respectively, and denote $\mathcal{E}_0 := p^{*}\mathcal{E}$, where $p : M_0 \to M$ is the canonical identification map. We choose a well-posed boundary condition $B$ on $\partial M$ and to simplify notation, we denote by $\Delta_M$ the Laplacian $\Delta_M$ subjected to $B$. We also denote by $\Delta_{M_0,D}$ the Laplacian $\Delta_{M_0}$ subjected to $B$ on $\partial M$ and the Dirichlet boundary condition on $\mathcal{N}^{+} \cup \mathcal{N}^{-}$.

For $\Delta = \Delta_M$ or $\Delta_{M_0,D}$, we define the zeta function $\zeta_{\Delta}(s)$ by

$$\zeta_{\Delta}(s) = \sum_{0 \neq \mu_i \in \operatorname{Spec}(\Delta)} \mu_i^{-s} = \frac{1}{\Gamma(s)} \int_0^{\infty} t^{s-1} \left(\operatorname{Tr} e^{-t\Delta} - \dim \ker \Delta\right) dt. \tag{2}$$

It is known that $\zeta_{\Delta}(s)$ is regular for $\Re s > \frac{m}{2}$ and has an analytic continuation to the whole complex plane having a regular value at $s = 0$ [5]. If $\Delta$ is invertible, we define the zeta-determinant by $\operatorname{Det} \Delta := e^{-\zeta_{\Delta}'(0)}$. When $\Delta$ has a non-trivial null space, we define the modified zeta-determinant by the same formula and denote it by $\operatorname{Det}^{*} \Delta := e^{-\zeta_{\Delta}'(0)}$.

For $0 \leq \lambda \in \mathbb{R}$, we define the Dirichlet-to-Neumann operator $R(\lambda) : C^{\infty}(\mathcal{E}|_{\mathcal{N}}) \to C^{\infty}(\mathcal{E}|_{\mathcal{N}})$ as follows. For $f \in C^{\infty}(\mathcal{E}|_{\mathcal{N}})$, choose $\psi \in C^{\infty}(\mathcal{E}_0)$ such that

$$(\Delta_{M_0} + \lambda)\psi = 0, \qquad \psi|_{\mathcal{N}^{+}} = \psi|_{\mathcal{N}^{-}} = f, \qquad B(\psi) = 0.$$

In fact, $\psi$ is given by $\psi = \widetilde{f} - (\Delta_{M_0,D} + \lambda)^{-1}(\Delta_{M_0} + \lambda)\widetilde{f}$, where $\widetilde{f}$ is an arbitrary smooth extension of $f$ vanishing near $\partial M$. We define

$$R(\lambda) f = \left(\nabla_{\partial_{\mathcal{N}}} \psi\right)\big|_{\mathcal{N}^{+}} - \left(\nabla_{\partial_{\mathcal{N}}} \psi\right)\big|_{\mathcal{N}^{-}}. \tag{3}$$

Then, $R(\lambda)$ is an elliptic pseudodifferential operator ($\Psi$DO) with parameter $\lambda$ of order 1. We define the zeta-determinant $\operatorname{Det} R(\lambda)$ of $R(\lambda)$ in the same way as (2).

The following equality is due to Burghelea, Friedlander and Keppeler [2, 3]. For $\nu = \left[\frac{m-1}{2}\right] + 1$,

$$\frac{d^\nu}{d\lambda^\nu}\left\{\ln \mathrm{Det}\,(\Delta_M + \lambda) - \ln \mathrm{Det}\left(\Delta_{M_0,D} + \lambda\right)\right\} = \frac{d^\nu}{d\lambda^\nu} \ln \mathrm{Det}\, R(\lambda). \tag{4}$$

Equivalently, there exists a real polynomial $P(\lambda) = \sum_{k=0}^{\left[\frac{m-1}{2}\right]} a_k \lambda^k$ such that

$$\ln \mathrm{Det}\,(\Delta_M + \lambda) \;-\; \ln \mathrm{Det}\left(\Delta_{M_0,D} + \lambda\right) = \sum_{k=0}^{\left[\frac{m-1}{2}\right]} a_k \lambda^k \;+\; \ln \mathrm{Det}\, R(\lambda). \tag{5}$$

Let $\{\phi_1, \cdots, \phi_\ell\}$ be an orthonormal basis of $\ker \Delta_M$, where $\ell = \dim \ker \Delta_M$. We define a positive definite $\ell \times \ell$ matrix $A_0$ as follows.

$$d_{ij} = \langle \phi_i|_N,\, \phi_j|_N \rangle_N, \qquad A_0 = \left(d_{ij}\right).$$

It is known [14, 17] that

$$\lim_{\lambda \to 0^+} \ln \mathrm{Det}\left(\Delta_{M_0,D} + \lambda\right) \;=\; \ln \mathrm{Det}\, \Delta_{M_0,D}, \tag{6}$$

$$\lim_{\lambda \to 0^+} \left(\ln \mathrm{Det}\,(\Delta_M + \lambda) - \ln \mathrm{Det}\, R(\lambda)\right) \;=\; \ln \det A_0 + \ln \mathrm{Det}^* \Delta_M - \ln \mathrm{Det}^* R(0),$$

which together with (5) leads to the following result.

$$\ln \mathrm{Det}^* \Delta_M \;-\; \ln \mathrm{Det}\, \Delta_{M_0,D} = a_0 \;-\; \ln \det A_0 \;+\; \ln \mathrm{Det}^* R(0), \tag{7}$$

which shows that the constant term $a_0$ of $P(\lambda)$ plays an important role. We here note that $\ker R(0) = \{\phi|_N \mid \phi \in \ker \Delta_{M,B}\}$, which may be non-trivial. This is called the BFK-gluing formula for the zeta-determinants of Laplacians [2].

We next consider the asymptotic expansions for $\lambda \to \infty$, from which we can compute the coefficients of $P(\lambda)$ precisely. For $\lambda \to \infty$, the asymptotic expansion of the left hand side of (5) is given as follows (Lemma 2.1 in [10], (5.1) in [20]). Suppose that for $t \to 0^+$,

$$\mathrm{Tr}\left(e^{-t\Delta_M} - e^{-t\Delta_{M_0,D}}\right) \sim \sum_{j=0}^{\infty} c_j t^{\frac{-m+j}{2}}, \tag{8}$$

where $c_0 = 0$ (Theorem 3.4.1 in [5]). Then,

$$\ln \mathrm{Det}\,(\Delta_M + \lambda) - \ln \mathrm{Det}\left(\Delta_{M_0,D} + \lambda\right) \sim -\sum_{\substack{j=1\\ j\neq m}}^{N} c_j \frac{d}{ds}\left(\frac{\Gamma(s-\frac{m-j}{2})}{\Gamma(s)}\right)\Big|_{s=0} \lambda^{\frac{m-j}{2}} \tag{9}$$

$$+ c_m \ln\lambda + \sum_{j=1}^{m-1} c_j \left(\frac{\Gamma(s-\frac{m-j}{2})}{\Gamma(s)}\right)\Big|_{s=0} \lambda^{\frac{m-j}{2}} \ln\lambda + O(\lambda^{-\frac{N+1-m}{2}}).$$

Here we note that the constant term does not appear. It is shown in the Appendix of [2] that $R(\lambda)$ is an elliptic ΨDO with parameter of order 1 and weight 2 and for $\lambda \to \infty$, $\ln \mathrm{Det}\, R(\lambda)$ has the following asymptotic expansion,

$$\ln \mathrm{Det}\, R(\lambda) \sim \sum_{j=0}^{\infty} \pi_j \lambda^{\frac{m-1-j}{2}} + \sum_{j=0}^{m-1} q_j \lambda^{\frac{m-1-j}{2}} \ln\lambda, \tag{10}$$

where each $\pi_i$ and $q_j$ can be computed as follows. Setting the homogeneous symbol of the resolvent $(\mu - R(\lambda))^{-1}$ by

$$\sigma\left((\mu - R(\lambda))^{-1}\right)(x,\xi,\lambda,\mu) \sim r_{-1}(x,\xi,\lambda,\mu) + r_{-2}(x,\xi,\lambda,\mu) + r_{-3}(x,\xi,\lambda,\mu) + \cdots, \tag{11}$$

the coefficients $\pi_i$ and $q_j$ are computed as follows:

$$\pi_j = -\frac{\partial}{\partial s}\Big|_{s=0} \int_{\mathcal{N}} \left(\frac{1}{(2\pi)^{m-1}} \int_{T_x^*\mathcal{N}} \frac{1}{2\pi i} \int_\gamma \mu^{-s}\, \mathrm{Tr}\, r_{-1-j}\left(x,\xi,\frac{\lambda}{|\lambda|},\mu\right) d\mu d\xi\right) d\,\mathrm{vol}(\mathcal{N}),$$

$$q_j = \frac{1}{2} \int_{\mathcal{N}} \left(\frac{1}{(2\pi)^{m-1}} \int_{T_x^*\mathcal{N}} \frac{1}{2\pi i} \int_\gamma \mu^{-s}\, \mathrm{Tr}\, r_{-1-j}\left(x,\xi,\frac{\lambda}{|\lambda|},\mu\right) d\mu d\xi\right) d\,\mathrm{vol}(\mathcal{N})\Big|_{s=0}. \tag{12}$$

Comparing the coefficients of (9) with those of (10), we obtain the following relations.

**Lemma 1**

$$a_0 = -\pi_{m-1}, \qquad a_k = -\pi_{m-1-2k} - c_{m-2k}\left(\frac{d}{ds}\Big|_{s=0} \frac{\Gamma(s-k)}{\Gamma(s)}\right), \quad k \geq 1,$$

$$q_{m-1} = c_m, \qquad q_k = c_{k+1}\left(\frac{\Gamma(s-\frac{m-k-1}{2})}{\Gamma(s)}\right)\Big|_{s=0}.$$

It is well known that the density $c_j(x)$ for $x \in \mathcal{N}$ computing $c_j$ in (8) consists of curvature tensors, for example, the Riemann curvature tensor and scalar curvature, etc. [6, 8]. Since $r_{-1-j}(x, \xi, \lambda, \mu)$ consists of the metric tensor, the densities $\pi_j(x)$ and $q_j(x)$ computing $\pi_j$ and $q_j$ consist of some curvature tensors, for example, the Riemann curvature tensor, the scalar curvature and principal curvatures (cf. [6, 8, 18]). Since the heat trace coefficients $c_j(x)$ are well known and the $q_j(x)$ are expressed in terms of $c_j(x)$ by Lemma 1, we are going to concentrate on $\pi_j(x)$. In principle, $\pi_j(x)$ and $q_j(x)$ are computed by the formula (12). However, if $\dim \mathcal{N} \geq 4$, the formula (12) is not very useful since the homogeneous symbol of $(\mu - R(\lambda))^{-1}$ is very complicated. However, when $\mathcal{N}$ is a 2-dimensional manifold, the following is the main result of [12].

**Theorem 1** *Let $(M, g)$ be a compact Riemannian manifold with boundary $\partial M$ and $\mathcal{N}$ be a closed hypersurface of $M$ with $\mathcal{N} \cap \partial M = \emptyset$. We assume that* $\dim M = 3$. *We denote by $\tau_M$, $\tau_{\mathcal{N}}$ the scalar curvatures of $M$, $\mathcal{N}$, respectively. We denote the principal curvatures on $\mathcal{N}$ by $\kappa_1$, $\kappa_2$ and put $H_1 = \frac{1}{2}(\kappa_1 + \kappa_2)$ and $H_2 = \kappa_1\kappa_2$. Then the densities $\pi_i(x)$ ($i = 0, 1, 2$), $x \in \mathcal{N}$, are given as follows:*

$$\pi_0(x) = r_0\left(-\frac{1}{4\pi}\ln 2 + \frac{1}{8\pi}\right), \qquad \pi_1(x) = 0,$$

$$\pi_2(x) = r_0\left\{\left(\frac{\tau_M(x)}{32\pi} + \frac{\tau_{\mathcal{N}}(x)}{96\pi} - \frac{H_1^2(x)}{32\pi} + \frac{H_2(x)}{32\pi}\right)\ln 2 - \frac{1}{64\pi}\left(\tau_M(x) - \tau_{\mathcal{N}}(x)\right)\right.$$
$$\left. - \frac{3H_1^2(x)}{128\pi} + \frac{5H_2(x)}{128\pi}\right\} + \frac{1}{4\pi}\left(\operatorname{Tr} E(x)\right) \cdot \ln 2.$$

Here the computation of $\pi_0(x)$ and $\pi_1(x)$ are simple because the homogeneous symbols $r_{-1}(x, \xi, \lambda, \mu)$ and $r_{-2}(x, \xi, \lambda, \mu)$ are easy to compute. However, $r_{-3}(x, \xi, \lambda, \mu)$ is quite complicated even though $\mathcal{N}$ is a 2-dimensional manifold. In [12], the above result is obtained by long and tedious computations by using the formula (12).

On the other hand, $\pi_2$ is computed on a warped product manifold in terms of a warping function in [9]. More precisely, let $(\mathcal{N}, h)$ be a closed Riemannian manifold and $f : [a, b] \to \mathbb{R}^+$ a smooth function with $f(c) = 1$, where $a < c < b$. We consider a warped product manifold $M := [a, b] \times_f \mathcal{N}$ with the metric $\begin{pmatrix} 1 & 0 \\ 0 & f(r)^2 h_{ij}(y) \end{pmatrix}$, where $h = \left(h_{ij}(y)\right)$ is a metric on $\mathcal{N}$. We denote $M_1 := [a, c] \times_f \mathcal{N}$ and $M_2 := [c, b] \times_f \mathcal{N}$ so that $M = M_1 \cup_{\{c\}\times\mathcal{N}} M_2$, and impose the Dirichlet boundary condition on each boundary. We denote by $\Delta_M$ the Laplacian acting on smooth functions on $M$, which is described by

$$\Delta_M = -\frac{d^2}{dr^2} - (m-1)\frac{f'(r)}{f(r)}\frac{d}{dr} + \frac{1}{f(r)^2}\Delta_{\mathcal{N}}, \tag{13}$$

where $\dim \mathcal{N} = m-1$ and $\Delta_{\mathcal{N}}$ is a Laplacian on $\mathcal{N}$. We denote by $\Delta_{M_i}$ the restriction of $\Delta_M$ to $M_i$ for $i = 1, 2$ and denote by $\Delta_{M,D}$, $\Delta_{M_i,D}$ the Laplacians subject to the Dirichlet boundary condition. When $\dim \mathcal{N} = 2$, it is shown in [9] by using the result of [4] that

$$\begin{aligned} \ln \operatorname{Det} \Delta_{M,D} - \ln \operatorname{Det} \Delta_{M_1,D} - \ln \operatorname{Det} \Delta_{M_2,D} &= -\ln 2 \left(\zeta_{\Delta_{\mathcal{N}}}(0) + \dim \ker \Delta_{\mathcal{N}}\right) \qquad (14) \\ +\left\{\ln 2 \left(\frac{1}{2} f''(c) + \frac{1}{4} f'(c)^2\right) - \frac{1}{4} f''(c) - \frac{3}{16} f'(c)^2\right\} &\operatorname{Res}_{s=1} \zeta_{\Delta_{\mathcal{N}}}(s) + \ln \operatorname{Det} R(0). \end{aligned}$$

This result will be used to simplify the proof of Theorem 1.

We will also comment on the following setting. Let $(W, Y; g)$ be a compact Riemannian manifold with boundary $Y$. We denote the Laplacian acting on smooth functions on $W$ with the Dirichlet and Neumann boundary conditions by $\Delta_{W,D}$ and $\Delta_{W,N}$, respectively. It is shown in [13] that for some real polynomial $S(\lambda) = \sum_{j=0}^{[\frac{m-1}{2}]} b_j \lambda^j$,

$$\ln \operatorname{Det} \left(\Delta_{W,N} + \lambda\right) - \operatorname{Det} \left(\Delta_{W,D} + \lambda\right) = \sum_{j=0}^{[\frac{m-1}{2}]} b_j \lambda^j + \ln \operatorname{Det} Q(\lambda), \qquad (15)$$

where $Q(\lambda)$ is an operator used in the Steklov eigenvalue problem defined in (34) below.

A major argument in our proof will involve conformal scalings. If we rescale the Riemannian metric $g$ to $c^2 g$ for some $c > 0$, then the homogeneous symbol $r_{-1-j}(x, \xi, \lambda, \mu)$ satisfies the following relation ((24) below).

$$r_{-1-j}(x, \xi, \lambda, \mu; c^2 g) = c^{-j} r_{-1-j}\left(x, \frac{1}{c}\xi, \lambda, \mu; g\right). \qquad (16)$$

It is known from the invariance theory (Lemma 1.7.6 in [6]) that $r_{-3}(x, \xi, \lambda, \mu; g)$ is expressed by a linear span of $\{L_{\alpha\alpha} L_{\beta\beta} \operatorname{Id},\ L_{\alpha\beta} L_{\alpha\beta} \operatorname{Id},\ R_{ijji} \operatorname{Id},\ R_{\alpha 33\alpha} \operatorname{Id},\ E\}$, where $L = \left(L_{\alpha\beta}\right)_{1 \leq \alpha, \beta \leq 2}$ is the second fundamental form on $\mathcal{N}$ and $R_{ijkl}$ is the Riemann curvature tensor (Lemma 3.1.11 in [6]). Here $\alpha, \beta$ label variables on $\mathcal{N}$ and $i, j, k, l$ label variables on $M$ so that $1 \leq \alpha, \beta \leq 2$ and $1 \leq i, j, k, l \leq 3$. We use the Einstein convention so that repeated indices are summed over.

In this paper, we use the properties (16) with (5) and (15) to show that

$$\begin{aligned} \zeta_{(\Delta_M(g)+\lambda)}(0) - \zeta_{(\Delta_{M_0,D}(g)+\lambda)}(0) &= \frac{1}{2} \zeta_{R_g(\lambda)}(0), \qquad (17) \\ \zeta_{(\Delta_{W,N}(g)+\lambda)}(0) - \zeta_{(\Delta_{W,D}(g)+\lambda)}(0) &= \frac{1}{2} \zeta_{Q_g(\lambda)}(0). \end{aligned}$$

We also use (16) and (14) to recover Theorem 1, which reduces much of the long and tedious computations performed in [12].

## 2 The Metric Rescaling and Invariance Theory

We begin with the metric rescaling from a Riemannian metric $g$ to $c^2 g$ for $c > 0$ on $M$. For a Laplacian $\Delta_M(g)$ with respect to $g$, it is well known that

$$\Delta_M(c^2 g) = c^{-2} \Delta_M(g), \qquad \Delta_M(c^2 g) + \lambda = c^{-2} \left( \Delta_M(g) + c^2 \lambda \right). \quad (18)$$

The BFK-gluing formula with respect to $c^2 g$ is described as follows.

$$\begin{aligned} \ln \operatorname{Det} \left( \Delta_M(c^2 g) + \lambda \right) - \ln \operatorname{Det} \left( \Delta_{M_0, D}(c^2 g) + \lambda \right) &= P_{c^2 g}(\lambda) + \ln \operatorname{Det} R_{c^2 g}(\lambda) \\ &= \sum_{j=0}^{\nu - 1} a_j(c^2 g) \lambda^j \ + \ \ln \operatorname{Det} R_{c^2 g}(\lambda), \end{aligned} \quad (19)$$

where $P_{c^2 g}(\lambda) = \sum_{j=0}^{\nu-1} a_j(c^2 g) \lambda^j$. For later use, we denote $P_g(\lambda) = \sum_{j=0}^{\nu-1} a_j(g) \lambda^j$. As before, $\nu = \left[ \frac{m-1}{2} \right] + 1$.

The Dirichlet-to-Neumann operator $R_{c^2 g}(\lambda)$ is described as follows. For $f \in C^{\infty}(\mathcal{E}|_{\mathcal{N}})$, we choose $\psi \in C^{\infty}(\mathcal{E}_0)$ such that

$$\left( \Delta_M(c^2 g) + \lambda \right) \psi = c^{-2} \left( \Delta_{M_0}(g) + c^2 \lambda \right) \psi \ = \ 0, \quad \psi|_{\mathcal{N}^+} = \psi|_{\mathcal{N}^-} = f, \quad B(\psi) = 0.$$

$R_{c^2 g}(\lambda) f$ is defined by

$$\begin{aligned} R_{c^2 g}(\lambda) f &= \left( \nabla_{\frac{1}{c} \partial_{x_m}} \psi \right) |_{\mathcal{N}^+} - \left( \nabla_{\frac{1}{c} \partial_{x_m}} \psi \right) |_{\mathcal{N}^-} \ = \ \frac{1}{c} \left\{ \left( \nabla_{\partial_{x_m}} \psi \right) |_{\mathcal{N}^+} - \left( \nabla_{\partial_{x_m}} \psi \right) |_{\mathcal{N}^-} \right\} \\ &= \frac{1}{c} R_g(c^2 \lambda) f, \end{aligned}$$

which shows that

$$R_{c^2 g}(\lambda) = \frac{1}{c} R_g(c^2 \lambda). \quad (20)$$

From (18) and (20), we have

$$\begin{aligned} \ln \operatorname{Det}(\Delta_M(c^2 g) + \lambda) \ &= \ -2 \ln c \cdot \zeta_{(\Delta_M(g) + c^2 \lambda)}(0) \ + \ \ln \operatorname{Det}(\Delta_M(g) + c^2 \lambda), \quad (21) \\ \ln \operatorname{Det} R_{c^2 g}(\lambda) \ &= \ - \ln c \cdot \zeta_{R_g(c^2 \lambda)}(0) \ + \ \ln \operatorname{Det} R_g(c^2 \lambda). \end{aligned}$$

We use the above equalities to rewrite (19) as

$$
\begin{aligned}
&-2\ln c\cdot\left\{\zeta_{(\Delta_M(g)+c^2\lambda)}(0)-\zeta_{(\Delta_{M_0,D}(g)+c^2\lambda)}(0)\right\}\\
&+\left\{\ln\mathrm{Det}(\Delta_M(g)+c^2\lambda)-\ln\mathrm{Det}(\Delta_{M_0,D}(g)+c^2\lambda)\right\}\\
=&-2\ln c\cdot\left\{\zeta_{(\Delta_M(g)+c^2\lambda)}(0)-\zeta_{(\Delta_{M_0,D}(g)+c^2\lambda)}(0)\right\}\ +\ P_g(c^2\lambda)\ +\ \ln\mathrm{Det}\,R_g(c^2\lambda)\\
=&\ P_{c^2g}(\lambda)-\ln c\cdot\zeta_{R_g(c^2\lambda)}(0)\ +\ \ln\mathrm{Det}\,R_g(c^2\lambda),
\end{aligned}
\tag{22}
$$

which leads to the following result.

**Lemma 2**

$$
\begin{aligned}
&-\ln c\cdot\left\{2\left(\zeta_{(\Delta_M(g)+c^2\lambda)}(0)-\zeta_{(\Delta_{M_0,D}(g)+c^2\lambda)}(0)\right)-\zeta_{R_g(c^2\lambda)}(0)\right\}\\
&=P_{c^2g}(\lambda)\ -\ P_g(c^2\lambda)\\
&=\sum_{j=0}^{\nu-1}\left(a_j(c^2g)-a_j(g)c^{2j}\right)\lambda^j.
\end{aligned}
$$

From (20), we have the following.

$$
\sigma\left(\left(\mu-R_{c^2g}(\lambda)\right)^{-1}\right)\ \sim\ \sum_{j=0}^{\infty}r_{-1-j}(x,\xi,\lambda,\mu;c^2g) \tag{23}
$$

$$
\begin{aligned}
&=\sigma\left(\left(\mu-\frac{1}{c}R_g(c^2\lambda)\right)^{-1}\right)\\
&=\sigma\left(c\left(c\mu-R_g(c^2\lambda)\right)^{-1}\right)\ \sim\ c\sum_{j=0}^{\infty}r_{-1-j}(x,\xi,c^2\lambda,c\mu;g)\\
&=c\sum_{j=0}^{\infty}r_{-1-j}(x,c\frac{1}{c}\xi,c^2\lambda,c\mu;g)\ =\ c\sum_{j=0}^{\infty}c^{-1-j}r_{-1-j}\left(x,\frac{1}{c}\xi,\lambda,\mu;g\right)\\
&=\sum_{j=0}^{\infty}c^{-j}r_{-1-j}\left(x,\frac{1}{c}\xi,\lambda,\mu;g\right),
\end{aligned}
$$

which shows that

$$
r_{-1-j}(x,\xi,\lambda,\mu;c^2g)=c^{-j}r_{-1-j}\left(x,\frac{1}{c}\xi,\lambda,\mu;g\right). \tag{24}
$$

We denote the density computing $\pi_j(c^2 g)$ by $\pi_j(x; c^2 g)$. Then, by (12) we get

$$\begin{aligned} \pi_j(x; c^2 g) &= -\frac{\partial}{\partial s}\Big|_{s=0} \left( \frac{1}{(2\pi)^{m-1}} \int_{T_x^* \mathcal{N}} \frac{1}{2\pi i} \int_\gamma \mu^{-s} \right. \\ &\qquad \left. \times \operatorname{Tr} r_{-1-j} \left( x, \xi, \frac{\lambda}{|\lambda|}, \mu; c^2 g \right) d\mu \, d\xi(c^2 g) \right), \\ \pi_j(c^2 g) &= \int_{\mathcal{N}} \pi_j(x; c^2 g) \, d \operatorname{vol}(\mathcal{N}; c^2 g). \end{aligned} \tag{25}$$

Using (24) with the following relations,

$$d\xi(c^2 g) = c^{-(m-1)} d\xi(g), \qquad d \operatorname{vol}(\mathcal{N}; c^2 g) = c^{m-1} d \operatorname{vol}(\mathcal{N}; g), \tag{26}$$

we have the following result.

**Lemma 3**

$$\pi_j(x; c^2 g) = c^{-j} \pi_j(x; g), \qquad \pi_j(c^2 g) = c^{m-1-j} \pi_j(g).$$

Lemma 1 shows that the coefficient $a_k(c^2 g)$ of the polynomial $P_{c^2 g}(\lambda)$ is given by

$$a_k(c^2 g) = -\pi_{m-1-2k}(c^2 g) - c_{m-2k}(c^2 g) \left( \frac{d}{ds}\Big|_{s=0} \frac{\Gamma(s-k)}{\Gamma(s)} \right) = c^{2k} a_k(g), \tag{27}$$

where we used the fact that $c_j(c^2 g) = c^{m-j} c_j(g)$ (Theorem 3.1.9 in [6]). Replacing $\lambda$ with $\frac{1}{c^2}\lambda$, we obtain the following result from Lemma 2, which was shown in [11] on a warped product manifold.

**Theorem 2** *For $\lambda > 0$, the following equality holds.*

$$\zeta_{(\Delta_M(g)+\lambda)}(0) - \zeta_{(\Delta_{M_0,D}(g)+\lambda)}(0) = \frac{1}{2} \zeta_{R_g(\lambda)}(0).$$

*If* $\dim \ker \Delta_M(g) = \ell$, *we obtain the following equality by taking $\lambda \to 0$.*

$$\left( \zeta_{\Delta_M(g)}(0) + \ell \right) - \zeta_{\Delta_{M_0,D}(g)}(0) = \frac{1}{2} \left( \zeta_{R_g(0)}(0) + \ell \right).$$

We next discuss the value of the zeta function $\zeta_{R_g(\lambda)}(s)$ at $s = 0$. We denote the homogeneous symbols of $R_g(\lambda)$ and $(\mu - R_g(\lambda))^{-1}$ by

$$\sigma\left(R_g(\lambda)\right)(x, \xi, \lambda) \sim \sum_{j=0}^{\infty} \widetilde{\theta}_{1-j}(x, \xi, \lambda), \tag{28}$$

$$\sigma\left((\mu - R(\lambda)^{-1}\right)(x, \xi, \lambda, \mu) \sim \sum_{j=0}^{\infty} \widetilde{r}_{-1-j}(x, \xi, \lambda, \mu).$$

In Sect. 4 of [12], the $\widetilde{\theta}_{1-j}(x, \xi, \lambda)$ are computed recursively. For example,

$$\widetilde{\theta}_1(x, \xi, \lambda) = 2|\xi|, \qquad \widetilde{\theta}_0(x, \xi, \lambda) = \frac{1}{|\xi|}\left(p_1(x, \xi) + i\partial_\xi|\xi| \cdot \partial_x|\xi|\right),$$

where $p_1(x, \xi)$ is defined in (58) below. Moreover, one can see that

$$\widetilde{\theta}_{1-j}(x, -\xi, \lambda) = (-1)^j \widetilde{\theta}_{1-j}(x, \xi, \lambda). \tag{29}$$

We refer to [12] for details. The homogeneous symbols $\widetilde{r}_{-1-j}(x, \xi, \lambda, \mu)$ are computed as follows.

$$\widetilde{r}_{-1}(x, \xi, \lambda, \mu) = (\mu - 2|\xi|)^{-1},$$

$$\widetilde{r}_{-1-j}(x, \xi, \lambda, \mu) = (\mu - 2|\xi|)^{-1} \sum_{k=0}^{j-1} \sum_{|\omega|+l+k=j} \frac{1}{\omega!} \partial_\xi^\omega \widetilde{\theta}_{1-l} \cdot D_x^\omega \widetilde{r}_{-1-k}. \tag{30}$$

*Remark* Here we regard $\lambda$ as a usual constant so that the principal symbols of $R(\lambda)$ and $(\mu - R(\lambda))^{-1}$ are $2|\xi|$ and $(\mu - 2|\xi|)^{-1}$, respectively. However, in Sect. 3 $\lambda$ is a parameter of weight 2 so that $R(\lambda)$ is an elliptic operator with parameter of weight 2 (for the definition, see the Appendix of [2] or [19]), and the principal symbols of $R(\lambda)$ and $(\mu - R(\lambda))^{-1}$ are $2\sqrt{|\xi|^2 + \lambda}$ and $(\mu - 2\sqrt{|\xi|^2 + \lambda})^{-1}$, respectively.

Equation (30) shows that

$$\widetilde{r}_{-1-j}(x, -\xi, \lambda, \mu) = (-1)^j \widetilde{r}_{-1-j}(x, \xi, \lambda, \mu), \tag{31}$$

which shows that for $j$ odd, $\widetilde{r}_{-1-j}(x, \xi, \lambda)$ is an odd function with respect to $\xi$. It is well known [7, 16, 18] that for $t \to 0$,

$$\operatorname{Tr} e^{-tR_g(\lambda)} \sim \sum_{j=0}^{\infty} v_j(\lambda) t^{-(m-1)+j} + \sum_{l=1}^{\infty} w_l(\lambda) t^l \ln t, \tag{32}$$

where $v_j(\lambda)$ for $0 \leq j \leq m-1$ is computed by integration of some local density $v_j(\lambda)(x)$ consisting of the homogeneous symbols of $(\mu - R(\lambda))^{-1}$. It is also known [5, 18] that

$$\zeta_{R_g(\lambda)}(0) = v_{m-1}(0) = \frac{1}{(2\pi)^{m-1}} \int_N \int_{T_x^*N} \frac{1}{2\pi i} \int_\gamma e^{-\mu} \operatorname{Tr} \widetilde{r}_{-1-j}(x, \xi, \lambda, \mu) \, d\mu \, d\xi \, dx. \tag{33}$$

Hence, if $m-1$ is odd, $\zeta_{R_g(\lambda)}(0) = 0$ since $\widetilde{r}_{-1-(m-1)}(x, \xi, \lambda, \mu)$ is an odd function with respect to $\xi$. Summarizing this argument, we have the following result, which is obtained for the case of a warped manifold in [11].

**Theorem 3** *Let $N$ be an odd dimensional closed Riemannian manifold. Then, for $\lambda > 0$,*

$$\zeta_{R_g(\lambda)}(0) = 0.$$

*Let $\ell = \dim \ker R(0) = \dim \ker \Delta_M$. If $\lambda = 0$, then,*

$$\zeta_{R_g(0)}(0) + \ell = 0.$$

In the remaining part of this section, we are going to discuss analogues of Theorems 2 and 3 for the operator used in the Steklov eigenvalue problem on a compact manifold with boundary. Let $(W, Y; g)$ be a compact Riemannian manifold with boundary $Y$. We denote by $\Delta_{W,N}$, $\Delta_{W,D}$ the scalar Laplacian with the Neumann and Dirichlet boundary conditions, respectively. For $0 \leq \lambda \in \mathbb{R}$, we define $Q(\lambda) : C^\infty(Y) \to C^\infty(Y)$, which is also called a Dirichlet-to-Neumann operator, as follows. For $f \in C^\infty(Y)$, choose $\psi \in C^\infty(W)$ such that

$$(\Delta_M + \lambda)\psi = 0, \qquad \psi|_Y = f.$$

We define

$$Q(\lambda) f = \left(\nabla_{\partial_Y} \psi\right)\big|_Y, \tag{34}$$

where $\partial_Y$ is the outward unit normal vector field along $Y$. Then, $Q(\lambda)$ is a self-adjoint elliptic $\Psi$DO with parameter $\lambda$ of order 1. It is proved in Lemma 4.1 of [13] that there exists a real polynomial $S(\lambda) = \sum_{j=0}^{[\frac{m-1}{2}]} b_j \lambda^j$ such that

$$\ln \operatorname{Det}\left(\Delta_{W,N} + \lambda\right) - \operatorname{Det}\left(\Delta_{W,D} + \lambda\right) = \sum_{j=0}^{[\frac{m-1}{2}]} b_j \lambda^j + \ln \operatorname{Det} Q(\lambda). \tag{35}$$

We now consider the metric rescaling from $g$ to $c^2 g$ for $c > 0$. By (18) and (35), we have

$$
\begin{aligned}
&\ln \operatorname{Det}\left(\Delta_{W,N}(c^2 g) + \lambda\right) - \operatorname{Det}\left(\Delta_{W,D}(c^2 g) + \lambda\right) \qquad (36)\\
&= -2 \ln c \left\{\zeta_{(\Delta_{W,N}(g)+c^2\lambda)}(0) - \zeta_{(\Delta_{W,D}(g)+c^2\lambda)}(0)\right\}\\
&\quad + \left\{\ln \operatorname{Det}(\Delta_{W,N}(g) + c^2\lambda) - \ln \operatorname{Det}(\Delta_{W,D}(g) + c^2\lambda)\right\}\\
&= -2 \ln c \left\{\zeta_{(\Delta_{W,N}(g)+c^2\lambda)}(0) - \zeta_{(\Delta_{W,D}(g)+c^2\lambda)}(0)\right\}\\
&\quad + \sum_{j=0}^{[\frac{m-1}{2}]} b_j(g) c^{2j} \lambda^j + \ln \operatorname{Det} Q_g(c^2\lambda)\\
&= \sum_{j=0}^{[\frac{m-1}{2}]} b_j(c^2 g) \lambda^j + \ln \operatorname{Det} Q_{c^2 g}(\lambda).
\end{aligned}
$$

By the same reasoning as for (20) we have

$$
Q_{c^2 g}(\lambda) = \frac{1}{c} Q_g(c^2\lambda), \qquad (37)
$$

which shows that

$$
\ln \operatorname{Det} Q_{c^2 g}(\lambda) = -\ln c \cdot \zeta_{Q_g(c^2\lambda)}(0) + \ln \operatorname{Det} Q_g(c^2\lambda). \qquad (38)
$$

By (36)–(38) we have

$$
\begin{aligned}
&-\ln c \cdot \left\{2\left(\zeta_{(\Delta_{W,N}(g)+c^2\lambda)}(0) - \zeta_{(\Delta_{W,D}(g)+c^2\lambda)}(0)\right) - \zeta_{Q_g(c^2\lambda)}(0)\right\} \qquad (39)\\
&= \sum_{j=0}^{\nu-1} \left(b_j(c^2 g) - b_j(g) c^{2j}\right) \lambda^j.
\end{aligned}
$$

The homogeneous symbol of $Q(\lambda)$ is computed in (2.27) of [12] (cf. [15, 18]), which enables us to compute the symbol of $(\mu - Q(\lambda))^{-1}$. Proceeding as for (27), we have

$$
b_k(c^2 g) = c^{2k} b_k(g), \qquad (40)
$$

which shows that

$$
2\left\{\zeta_{(\Delta_{W,N}(g)+\lambda)}(0) - \zeta_{(\Delta_{W,D}(g)+\lambda)}(0)\right\} = \zeta_{Q_g(\lambda)}(0). \qquad (41)
$$

Since $\ker Q_g(0) = \{\phi|_Y \mid \phi \in \ker \Delta_{W,N}(g)\}$, we have, for $\ell = \dim\ker \Delta_{W,N}$,

$$2\left\{(\zeta_{\Delta_{W,N}(g)}(0) + \ell) - \zeta_{\Delta_{W,D}(g)}(0)\right\} = \zeta_{Q_g(0)}(0) + \ell. \tag{42}$$

Summarizing this argument, we obtain the following result.

**Theorem 4** *Let $(W, Y; g)$ be a compact Riemannian manifold with boundary $Y$ and $Q_g(\lambda)$ be an operator defined in (34). Let $\ell = \dim\ker \Delta_{W,N} = \dim\ker Q(0)$. Then, we have the following equalities.*

$$\begin{aligned}\zeta_{Q_g(\lambda)}(0) &= 2\left\{\zeta_{(\Delta_{W,N}(g)+\lambda)}(0) - \zeta_{(\Delta_{W,D}(g)+\lambda)}(0)\right\},\\ \zeta_{Q_g(0)}(0) + \ell &= 2\left\{(\zeta_{\Delta_{W,N}(g)}(0) + \ell) - \zeta_{\Delta_{W,D}(g)}(0)\right\}.\end{aligned}$$

## 3 Proof of Theorem 1

In this section, we are going to discuss the coefficient $\pi_2(g)$ in (10) when $\dim M = 3$. The density $\pi_2(x; g)$ computing $\pi_2(g)$ satisfies

$$\pi_2(x; c^2 g) = c^{-2}\pi_2(x; g), \tag{43}$$

which shows that $\pi_2(x; g)$ is an endomorphism valued invariant of weight 2. It is shown (Lemma 3.1.11 in [6]) that $\pi_2(x; g)$ is spanned by

$$\{L_{\alpha\alpha}L_{\beta\beta}\,\mathrm{Id},\ L_{\alpha\beta}L_{\alpha\beta}\,\mathrm{Id},\ R_{ijji}\,\mathrm{Id},\ R_{\alpha 33\alpha}\,\mathrm{Id},\ E\}, \tag{44}$$

where $1 \le \alpha, \beta \le 2$ and $1 \le i, j \le 3$. We denote the principal curvatures on $\mathcal{N}$ by $\kappa_1,\ \kappa_2$ and denote the mean curvatures by

$$H_1 = \frac{1}{2}(\kappa_1 + \kappa_2), \qquad H_2 = \kappa_1\kappa_2. \tag{45}$$

Then, we may write

$$L_{\alpha\alpha}L_{\beta\beta} = (\kappa_1 + \kappa_2)^2 = 4H_1^2, \quad L_{\alpha\beta}L_{\alpha\beta} = \kappa_1^2 + \kappa_2^2 = 4H_1^2 - 2H_2, \quad R_{ijji} = \tau_M, \tag{46}$$

where $\tau_M$ is the scalar curvature on $M$. It is also shown in Lemma 3.2 of [12] that

$$R_{\alpha 33\alpha} = \frac{1}{2}(\tau_M - \tau_{\mathcal{N}}) + H_2, \tag{47}$$

where $\tau_{\mathcal{N}}$ is the scalar curvature on $\mathcal{N}$. For some constants $b_1,\ b_2,\ b_3,\ b_4,\ b_5$ depending only on the dimension of $\mathcal{N}$, $\pi_2(x; g)$ is expressed as follows.

$$\begin{aligned}
&\pi_2(x; g) &(48)\\
&= b_1 R_{ijji}(g) + b_2 R_{\alpha 33\alpha}(g) + b_3 L_{\alpha\alpha}(g) L_{\beta\beta}(g)\\
&\quad + b_4 L_{\alpha\beta}(g) L_{\alpha\beta}(g) + b_5 \operatorname{Tr} E(g)\\
&= \left(b_1 + \frac{1}{2} b_2\right) \tau_M(x; g) - \frac{1}{2} b_2 \tau_{\mathcal{N}}(x; g) + (4b_3 + 4b_4) H_1^2(x; g)\\
&\quad + (b_2 - 2b_4) H_2(x; g) + b_5 \operatorname{Tr} E(x; g).
\end{aligned}$$

*Remark* The coefficients in the heat trace asymptotics do not depend on the dimension of the underlying manifold (Lemma 3.1.12 in [6]). In this case, however, simple computation shows that the coefficients $b_j$ may depend on the dimension of the underlying manifold (cf. [18]).

For a 2-dimensional Riemannian manifold $(\mathcal{N}, h)$ and a smooth function $f : [a, b] \to \mathbb{R}^+$ with $f(c) = 1$, let $(M, g)$ be a warped product manifold $[a, b] \times_f \mathcal{N}$, where $g$ is a warped product metric with respect to $h$ and a warping function $f$. We also consider a trivial bundle $\mathcal{E}$ on $[a, b] \times_f \mathcal{N}$ of rank $r_0$ and a Laplacian acting on smooth sections of this bundle. If we consider the BFK-gluing formula with respect to the partitioning $[a, b] \times_f \mathcal{N} = [a, c] \times_f \mathcal{N} \cup_{\{c\} \times \mathcal{N}} [c, b] \times_f \mathcal{N}$, it is shown ((3.26) in [12], Corollary 3.7 in [9]) that

$$\pi_2(x; g) = r_0 \cdot \frac{\tau_{\mathcal{N}}}{24\pi} \ln 2 - \frac{r_0}{4\pi} \left\{ \ln 2 \left( \frac{1}{2} f''(c) + \frac{1}{4} f'(c)^2 \right) - \left( \frac{3}{16} f'(c)^2 + \frac{1}{4} f''(c) \right) \right\}. \tag{49}$$

It is straightforward that

$$f''(c) = \frac{1}{4} \left( -\tau_M + \tau_{\mathcal{N}} - 2 f'(c)^2 \right), \qquad \kappa_1 = \kappa_2 = -f'(c), \tag{50}$$

which shows that

$$\begin{aligned}
\pi_2(x; g) &= r_0 \left( \frac{1}{32\pi} \ln 2 - \frac{1}{64\pi} \right) \tau_M(x; g)\\
&\quad + r_0 \left( \frac{1}{96\pi} \ln 2 + \frac{1}{64\pi} \right) \tau_{\mathcal{N}}(x; g) + \frac{r_0}{64\pi} H_1^2(x; g).
\end{aligned} \tag{51}$$

Comparing (48) and (51) with $H_1^2 = H_2$, we have the following result.

**Lemma 4**

$$b_1 = r_0 \cdot \frac{1}{24\pi} \ln 2, \qquad b_2 = r_0 \left( -\frac{1}{48\pi} \ln 2 - \frac{1}{32\pi} \right),$$

$$4b_3 + 2b_4 = r_0 \left( \frac{1}{48\pi} \ln 2 + \frac{3}{64\pi} \right).$$

To determine $b_3$ and $b_4$, we need one more relation distinguishing $H_1^2$ and $H_2$. For this purpose, we are going to consider a simple example at which $H_1^2$ is not equal to $H_2$.

Let $X$ be a closed 2-dimensional Riemannian manifold and $Y$ be a hypersurface such that $Y$ is isometric to the round circle $S^1$ of length $2\pi$ and the principal curvature $\kappa_1$ on $Y$ is non-zero. In view of (50), $X$ can be constructed by using the warped product method. We consider a Riemannian product $M := X \times S^1$ and denote this product metric by $g$. We let $\mathcal{N} := Y \times S^1$. Then, $\mathcal{N}$ is isometric to a flat torus $S^1 \times S^1$ and principal curvatures on $\mathcal{N}$ are $\kappa_1 \neq 0$, $\kappa_2 = 0$. Hence,

$$H_1 = \frac{1}{2}\kappa_1 \neq 0, \qquad H_2 = \kappa_1 \kappa_2 = 0. \tag{52}$$

We denote by $M_0$ the closure of $M - \mathcal{N} = (X - Y) \times S^1$ and let $\mathcal{E}$, $\mathcal{E}_0$, $\Delta_M$, $\Delta_{M_0}$ and $\Delta_{M_i}$ be as before. We consider the BFK-gluing formula for zeta-determinants and compute $\pi_2(g)$ in this case by using the formula (12). Since the symbol of the Laplacian $\Delta_{\mathcal{N}}$ on $\mathcal{N}$ does not depend on the space variables, the homogeneous symbol of $R_g(\lambda)$ and hence that of $(\mu - R_g(\lambda))^{-1}$ become quite simple.

Let $x = (x_1, x_2)$ be the coordinate system on $\mathcal{N}$ such that $g_{\alpha\beta} = \delta_{\alpha\beta}$. We introduce a boundary normal coordinate system $(x, x_3) = ((x_1, x_2), x_3)$ on a collar neighborhood $U_0$ of $\mathcal{N}$ in $M$ as in Sect. 2 in [12] such that $g_{\alpha\beta,3} := \partial_{x_3} g_{\alpha\beta}$ satisfies the following relation.

$$g_{\alpha\beta,3}(p) = \begin{cases} -2\kappa_\alpha & \text{for} \quad \alpha = \beta \\ 0 & \text{for} \quad \alpha \neq \beta. \end{cases} \tag{53}$$

It is not difficult to see that $g^{\alpha\beta,3}(x) = -g_{\alpha\beta,3}(x)$ for $x \in \mathcal{N}$. On $U_0$ we fix a local frame $\{\psi_1, \cdots, \psi_{r_0}\}$ on $\mathcal{E}|_{U_0}$ and denote by $\omega : C^\infty\left((T^*M \otimes \mathcal{E})|_{U_0}\right) \to C^\infty(\mathcal{E}|_{U_0})$ the connection form of $\nabla$ on $U_0$ with respect to $\{\psi_1, \cdots, \psi_{r_0}\}$, i.e.

$$\nabla_{\partial_{x_j}} \psi_k = \omega_j \psi_k, \qquad \text{where} \qquad \omega_j := \omega(\partial_{x_j}). \tag{54}$$

The Laplacian $\Delta_M + \lambda$ on $U_0$ is described as follows. The given formulas are valid for general dimension $m$, but eventually we will set $m = 3$. The Laplacian can be decomposed as

$$\Delta_M + \lambda = -\partial_{x_m}^2 \operatorname{Id} + (A(x, x_m) - 2\omega_m)\,\partial_{x_m} + D\left(x, x_m, \frac{\partial}{\partial x}, \lambda\right) \tag{55}$$
$$- \left(\partial_{x_m}\omega_m + \omega_m\omega_m - A(x, x_m)\omega_m\right),$$

where Id is an $r_0 \times r_0$ identity matrix and

$$A(x, x_m) = \left\{ -\frac{1}{2} \sum_{\alpha,\beta=1}^{m-1} g^{\alpha\beta}(x, x_m)\, g_{\alpha\beta,m}(x, x_m) \right\} \mathrm{Id}, \tag{56}$$

$$\begin{aligned} D\left(x, x_m, \frac{\partial}{\partial x}, \lambda\right) = & \left\{ \left( - \sum_{\alpha,\beta=1}^{m-1} g^{\alpha\beta}(x, x_m) \partial_{x_\alpha} \partial_{x_\beta} + \lambda \right) \right. \\ & \left. - \sum_{\alpha,\beta=1}^{m-1} \left( \frac{1}{2} g^{\alpha\beta}(x, x_m)(\partial_{x_\alpha} \ln |g|(x, x_m)) + (\partial_{x_\alpha} g^{\alpha\beta}(x, x_m)) \right) \partial_{x_\beta} \right\} \mathrm{Id} \\ & -2 \sum_{\alpha,\beta=1}^{m-1} g^{\alpha\beta}(x, x_m) \omega_\beta \partial_{x_\alpha} \\ & - \sum_{\alpha,\beta=1}^{m-1} g^{\alpha\beta}(x, x_m) \left( \partial_{x_\alpha} \omega_\beta + \omega_\alpha \omega_\beta - \sum_{\gamma=1}^{m-1} \Gamma^{\gamma}_{\alpha\beta} \omega_\gamma \right) - E. \end{aligned}$$

Setting the symbol of $D\left(x, x_3, \frac{\partial}{\partial x}, \lambda\right)$ as

$$\sigma\left( D\left( x, x_m, \frac{\partial}{\partial x}, \lambda \right) \right) = p_2(x, x_m, \xi, \lambda) + p_1(x, x_m, \xi) + p_0(x, x_m, \xi), \tag{57}$$

we have the following.

$$p_2(x, x_m, \xi, \lambda) = \left( \sum_{\alpha,\beta=1}^{m-1} g^{\alpha\beta}(x, x_m) \xi_\alpha \xi_\beta + \lambda \right) \mathrm{Id} \;=\; \left( |\xi|^2 + \lambda \right) \mathrm{Id}, \tag{58}$$

$$\begin{aligned} p_1(x, x_m, \xi) = & -i \sum_{\alpha,\beta=1}^{m-1} \left( \frac{1}{2} g^{\alpha\beta}(x, x_m) \partial_{x_\alpha} \ln |g|(x, x_m) + \partial_{x_\alpha} g^{\alpha\beta}(x, x_m) \right) \xi_\beta \,\mathrm{Id} \\ & -2i \sum_{\alpha,\beta=1}^{m-1} g^{\alpha\beta} \omega_\alpha \xi_\beta, \end{aligned}$$

$$p_0(x, x_m, \xi) = - \sum_{\alpha\beta=1}^{m-1} g^{\alpha\beta} \left( \partial_{x_\alpha} \omega_\beta + \omega_\alpha \omega_\beta - \sum_{\gamma=1}^{m-1} \Gamma^{\gamma}_{\alpha\beta} \omega_\gamma \right) - E,$$

where $g_{\alpha\beta}(x, 0) = \delta_{\alpha\beta}$ on $\mathcal{N}$. Since $\mathcal{N}$ is a flat torus, we have $|\xi|^2 = \xi_1^2 + \xi_2^2$, which does not depend on the space variables. The following equalities are well known [18].

**Lemma 5** *We consider the boundary normal coordinate system on an open neighborhood $U_0$ of $x \in \mathcal{N}$ with metric tensor $g = (g_{ij})$ and $(x, 0) = (x_1, x_2, 0)$. Then, for $m = 3$ we have the following equalities:*

(1) $g^{\alpha\beta,m}(x) = 2\kappa_\alpha \delta_{\alpha\beta} = -g_{\alpha\beta,m}(x)$,

(2) $\sum_{\alpha=1}^{m-1} g^{\alpha\alpha,mm}(x) = 8 \sum_{\alpha=1}^{m-1} \kappa_\alpha^2 - \sum_{\alpha=1}^{m-1} g_{\alpha\alpha,mm}(x)$,

(3) $\sum_{\alpha=1}^{m-1} g_{\alpha\alpha,mm}(x) = - \left\{ \tau_M(x) - \tau_{\mathcal{N}}(x) - 2(m-1)^2 H_1^2(x) + 3(m-1)(m-2) H_2(x) \right\}$,

*where $\tau_M(x)$ and $\tau_{\mathcal{N}}(x)$ are scalar curvatures of $M$ and $\mathcal{N}$ at $x \in \mathcal{N}$, respectively.*

The homogeneous symbol of $R_g(\lambda)$ is computed in [12]. In particular, the first three terms are given as follows ((2.30) in [12]). We refer to [12] for details. We denote the homogeneous symbol of $R_g(\lambda)$ defined on $\mathcal{N}$ by

$$\sigma\left(R_g(\lambda)\right)(x, 0, \xi, \lambda) \sim \sum_{j=0}^{\infty} \theta_{1-j}(x, \xi, \lambda). \tag{59}$$

Then, from [12] we have for the considered setting

$$\theta_1 = 2\sqrt{|\xi|^2 + \lambda}\ \mathrm{Id}, \qquad \theta_0 = \frac{-2i\omega_\alpha \xi_\alpha}{\sqrt{|\xi|^2 + \lambda}}, \tag{60}$$

$$\begin{aligned}
\theta_{-1} = \frac{1}{2\sqrt{|\xi|^2+\lambda}} \Bigg\{ & \frac{2(\partial_{x_\alpha}\omega_\beta)\xi_\alpha\xi_\beta}{|\xi|^2+\lambda} + \frac{2(\omega_\alpha\omega_\beta)\xi_\alpha\xi_\beta}{|\xi|^2+\lambda} \\
& - \frac{\left( A(x, x_3)\sqrt{|\xi|^2+\lambda} - \frac{\partial_{x_3}|\xi|^2}{2\sqrt{|\xi|^2+\lambda}}\ \mathrm{Id} \right)^2}{2(|\xi|^2+\lambda)} \\
& + 2p_0(x, \xi) + A(x, x_3)\left( A(x, x_3) - \frac{\partial_{x_3}|\xi|^2}{2(|\xi|^2+\lambda)}\ \mathrm{Id} \right) \\
& - \partial_{x_3}\left( A(x, x_3) - \frac{\partial_{x_3}|\xi|^2}{2(|\xi|^2+\lambda)}\ \mathrm{Id} \right) \Bigg\}\Bigg|_{x_3=0}.
\end{aligned}$$

From the flatness of $\mathcal{N}$ and the boundary normal coordinate system around $(x, 0)$, we get

$$\tau_{\mathcal{N}}(x, 0) = 0, \qquad \Gamma_{\alpha\alpha}^{\gamma}(x, 0) = 0,$$

which together with Lemma 5 leads to the following result.

$$
\begin{aligned}
p_0(x,0,\xi) &= -\sum_{\alpha=1}^{2}\left(\partial_{x_\alpha}\omega_\alpha + \omega_\alpha\omega_\alpha - \sum_{\gamma=1}^{2}\Gamma_{\alpha\alpha}^{\gamma}\omega_\gamma\right) - E \\
&= -\sum_{\alpha=1}^{2}\left(\partial_{x_\alpha}\omega_\alpha + \omega_\alpha\omega_\alpha + E\right), \\
\partial_{x_3}|\xi|^2 &= g^{11,3}\xi_1^2 + g^{22,3}\xi_2^2 \ = \ 2\kappa_1\xi_1^2, \\
A(x,0) &= -\frac{1}{2}\left(g_{11,3}+g_{22,3}\right) \ \mathrm{Id} \ = \kappa_1 \ \mathrm{Id}\ , \\
\partial_{x_3}A(x,x_3)\big|_{x_3=0} &= -\frac{1}{2}\left(g^{\alpha\alpha,3}g_{\alpha\alpha,3} + g_{\alpha\alpha,33}\right) \\
&= \left\{2\kappa_1^2 + \frac{1}{2}\left(\tau_M - \tau_N - 8H_1^2\right)\right\} \ \mathrm{Id} \ = \ \left\{\frac{1}{2}\tau_M + 4H_1^2\right\} \ \mathrm{Id}\ , \\
\partial_{x_3}\left(\frac{\partial_{x_3}|\xi|^2}{|\xi|^2+\lambda}\right) &= \frac{g^{\alpha\beta,33}\xi_\alpha\xi_\beta}{|\xi|^2+\lambda} - \frac{4\kappa_1^2\xi_1^4}{(|\xi|^2+\lambda)^2} = \frac{g^{\alpha\beta,33}\xi_\alpha\xi_\beta}{|\xi|^2+\lambda} - \frac{16H_1^2\xi_1^4}{(|\xi|^2+\lambda)^2}.
\end{aligned}
$$

We use the above equalities to express $\theta_{-1}$ as follows.

$$
\begin{aligned}
\theta_{-1} = \frac{1}{\sqrt{|\xi|^2+\lambda}}\Bigg\{&\frac{(\partial_{x_\alpha}\omega_\beta)\xi_\alpha\xi_\beta}{|\xi|^2+\lambda} + \frac{(\omega_\alpha\omega_\beta)\xi_\alpha\xi_\beta}{|\xi|^2+\lambda} - (\frac{1}{4}\tau_M + H_1^2)\ \mathrm{Id}\ + \frac{g^{\alpha\beta,33}\xi_\alpha\xi_\beta}{4(|\xi|^2+\lambda)}\ \mathrm{Id} \\
&- 5H_1^2\frac{\xi_1^4}{(|\xi|^2+\lambda)^2}\ \mathrm{Id}\ - \left(\partial_{x_\alpha}\omega_\alpha + \omega_\alpha\omega_\alpha + E\right)\Bigg\}.
\end{aligned} \tag{61}
$$

If we denote the homogeneous symbol of $(\mu - R_g(\lambda))^{-1}$ by

$$
\sigma\left((\mu - R_g(\lambda))^{-1}\right) \sim \sum_{j=0}^{\infty} r_{-1-j}(x,\xi,\lambda,\mu), \tag{62}
$$

it is computed in (2.34) in [12] that

$$
r_{-1} = \left(\mu - 2\sqrt{|\xi|^2+\lambda}\right)^{-1}\ \mathrm{Id}\ , \qquad r_{-2} \ = \ \left(\mu - 2\sqrt{|\xi|^2+\lambda}\right)^{-2}\cdot\theta_0, \tag{63}
$$

$$
r_{-3} = \left(\mu - 2\sqrt{|\xi|^2+\lambda}\right)^{-1}\cdot\left\{\partial_\xi\theta_1\cdot D_x r_{-2} + \theta_0\cdot r_{-2} + \theta_{-1}\cdot r_{-1}\right\}
$$

$$= \left(\mu - 2\sqrt{|\xi|^2+\lambda}\right)^{-1} \cdot \partial_\xi \theta_1 \cdot D_x r_{-2}$$
$$+ \left(\mu - 2\sqrt{|\xi|^2+\lambda}\right)^{-3} \cdot \theta_0^2 + \left(\mu - 2\sqrt{|\xi|^2+\lambda}\right)^{-2} \cdot \theta_{-1}.$$

The following equalities are straightforward.

**Lemma 6**

$$\frac{1}{2\pi i}\int_\gamma \frac{\mu^{-s}}{\mu - z} d\mu = z^{-s}, \quad \frac{1}{2\pi i}\int_\gamma \frac{\mu^{-s}}{(\mu - z)^2} d\mu = -s z^{-s-1},$$
$$\frac{1}{2\pi i}\int_\gamma \frac{\mu^{-s}}{(\mu - z)^3} d\mu = \frac{1}{2} s(s+1) z^{-s-2},$$
$$\frac{1}{4\pi^2}\int_{\mathbb{R}^2} (|\xi|^2+1)^{-\frac{s}{2}-1} d\xi = \frac{1}{2\pi}\frac{1}{s},$$
$$\frac{1}{4\pi^2}\int_{\mathbb{R}^2} \xi_1^2 (|\xi|^2+1)^{-\frac{s}{2}-2} d\xi = \frac{1}{2\pi}\frac{1}{s(s+2)},$$
$$\frac{1}{4\pi^2}\int_{\mathbb{R}^2} \xi_1^4 (|\xi|^2+1)^{-\frac{s}{2}-3} d\xi = \frac{3}{2\pi}\frac{1}{s(s+2)(s+4)}.$$

We are now going to compute $\pi_2(x)$ by using (12). We recall that

$$\pi_2(x) = -\frac{\partial}{\partial s}\bigg|_{s=0} \left( \frac{1}{(2\pi)^2}\int_{T_x^*N} \frac{1}{2\pi i}\int_\gamma \mu^{-s}\, \mathrm{Tr}\, r_{-3}\left(x, \xi, \frac{\lambda}{|\lambda|}, \mu\right) d\mu d\xi \right). \tag{64}$$

Direct computations lead to the following results.

$$(1) = \frac{1}{(2\pi)^2}\int_{T_x^*N} \frac{1}{2\pi i}\int_\gamma \mu^{-s}(\mu - 2\sqrt{|\xi|^2+1})^{-1} \cdot \mathrm{Tr}\left(\partial_\xi\theta_1 \cdot D_x r_{-2}\right) d\mu\, d\xi$$
$$= \frac{1}{(2\pi)^2}\int_{T_x^*N} \frac{1}{2\pi i}\int_\gamma \mu^{-s}(\mu - 2\sqrt{|\xi|^2+1})^{-3} \cdot \frac{(-4)\ \mathrm{Tr}(\partial_{x_\alpha}\omega_\beta)\xi_\alpha\xi_\beta}{|\xi|^2+1} d\mu\, d\xi$$
$$= \frac{1}{(2\pi)^2}\int_{T_x^*N} -\frac{1}{2}\mathrm{Tr}(\partial_{x_\alpha}\omega_\alpha)\cdot s(s+1)\cdot 2^{-s}\cdot \xi_1^2 \cdot (|\xi|^2+1)^{-\frac{s}{2}-2} d\xi$$
$$= -\frac{1}{4\pi}\mathrm{Tr}(\partial_{x_\alpha}\omega_\alpha)\cdot 2^{-s}\cdot \frac{s+1}{s+2}.$$

$$(2) = \frac{1}{(2\pi)^2}\int_{T_x^*N} \frac{1}{2\pi i}\int_\gamma \mu^{-s}(\mu - 2\sqrt{|\xi|^2+1})^{-3}\ \mathrm{Tr}\,\theta_0^2\, d\mu\, d\xi$$

$$= \frac{1}{(2\pi)^2} \int_{T_x^*N} -\frac{1}{2} \operatorname{Tr}(\omega_\alpha \omega_\alpha) \cdot s(s+1) \cdot 2^{-s} \cdot \xi_1^2 (|\xi|^2+1)^{-\frac{s}{2}-2} \, d\xi$$

$$= -\frac{1}{4\pi} \operatorname{Tr}(\omega_\alpha \omega_\alpha) \cdot 2^{-s} \cdot \frac{s+1}{s+2}.$$

$$(3) = \frac{1}{(2\pi)^2} \int_{T_x^*N} \frac{1}{2\pi i} \int_\gamma \mu^{-s} (\mu - 2\sqrt{|\xi|^2+1})^{-2} \frac{\operatorname{Tr}(\partial_{x_\alpha} \omega_\beta) \xi_\alpha \xi_\beta}{(|\xi|^2+1)^{\frac{3}{2}}} \, d\mu \, d\xi$$

$$= \frac{1}{(2\pi)^2} \int_{T_x^*N} -\frac{1}{2} \operatorname{Tr}(\partial_{x_\alpha} \omega_\alpha) \cdot s \cdot 2^{-s} \cdot \xi_1^2 \cdot (|\xi|^2+1)^{-\frac{s}{2}-2} \, d\xi$$

$$= -\frac{1}{4\pi} \operatorname{Tr}(\partial_{x_\alpha} \omega_\alpha) \cdot 2^{-s} \cdot \frac{1}{s+2}.$$

$$(4) = \frac{1}{(2\pi)^2} \int_{T_x^*N} \frac{1}{2\pi i} \int_\gamma \mu^{-s} (\mu - 2\sqrt{|\xi|^2+1})^{-2} \frac{\operatorname{Tr}(\omega_\alpha \omega_\beta) \xi_\alpha \xi_\beta}{(|\xi|^2+1)^{\frac{3}{2}}} \, d\mu \, d\xi$$

$$= \frac{1}{(2\pi)^2} \int_{T_x^*N} -\frac{1}{2} \cdot s \cdot 2^{-s} \cdot \operatorname{Tr}(\omega_\alpha \omega_\alpha) \cdot \xi_1^2 (|\xi|^2+1)^{-\frac{s}{2}-2} \, d\xi$$

$$= -\frac{1}{4\pi} \operatorname{Tr}(\omega_\alpha \omega_\alpha) \cdot 2^{-s} \cdot \frac{1}{s+2}.$$

$$(5) = \frac{1}{(2\pi)^2} \int_{T_x^*N} \frac{1}{2\pi i} \int_\gamma \mu^{-s} (\mu - 2\sqrt{|\xi|^2+1})^{-2}$$

$$\times \left\{ \frac{r_0 \, (-\frac{1}{4}\tau_M - H_1^2) - \operatorname{Tr}(\partial_{x_\alpha} \omega_\alpha + \omega_\alpha \omega_\alpha + E)}{\sqrt{|\xi|^2+1}} \right\} d\mu \, d\xi$$

$$= \frac{1}{(2\pi)^2} \int_{T_x^*N} \left\{ r_0 \left( -\frac{1}{4}\tau_M - H_1^2 \right) - \operatorname{Tr}((\partial_{x_\alpha} \omega_\alpha + \omega_\alpha \omega_\alpha + E) \right\}$$

$$\times \left( -\frac{1}{2} \right) \cdot s \cdot 2^{-s} \cdot (|\xi|^2+1)^{-\frac{s}{2}-1} \, d\xi$$

$$= \left\{ r_0 \left( \frac{1}{16\pi} \tau_M + \frac{1}{4\pi} H_1^2 \right) + \frac{1}{4\pi} \operatorname{Tr}(\partial_{x_\alpha} \omega_\alpha + \omega_\alpha \omega_\alpha + E) \right\} \cdot 2^{-s}.$$

$$(6) = \frac{r_0}{(2\pi)^2} \int_{T_x^*N} \frac{1}{2\pi i} \int_\gamma \mu^{-s} (\mu - 2\sqrt{|\xi|^2+1})^{-2} \frac{g^{\alpha\beta,33} \xi_\alpha \xi_\beta}{4(|\xi|^2+1)^{\frac{3}{2}}} \, d\mu \, d\xi$$

$$= -\frac{1}{8} \cdot g^{\alpha\alpha,33} \cdot s \cdot 2^{-s} \cdot \frac{r_0}{(2\pi)^2} \int_{T_x^*N} \xi_1^2 (|\xi|^2+1)^{-\frac{s}{2}-2} \, d\xi$$

$$= -\frac{1}{8} \cdot (\tau_M + 24 H_1^2) \cdot s \cdot 2^{-s} \cdot \frac{r_0}{(2\pi)^2} \int_{T_x^*N} \xi_1^2 (|\xi|^2+1)^{-\frac{s}{2}-2} \, d\xi$$

$$= -r_0 \left( \frac{\tau_M}{16\pi} + \frac{3H_1^2}{2\pi} \right) \cdot 2^{-s} \cdot \frac{1}{s+2}.$$

$$(7) = \frac{r_0}{(2\pi)^2} \int_{T_x^* N} \frac{1}{2\pi i} \int_\gamma \mu^{-s} (\mu - 2\sqrt{|\xi|^2 + 1})^{-2} \, (-5H_1^2) \frac{\xi_1^4}{(|\xi|^2 + 1)^{\frac{5}{2}}} \, d\mu \, d\xi$$

$$= \frac{r_0}{(2\pi)^2} \int_{T_x^* N} \frac{5}{2} H_1^2 \cdot s \cdot 2^{-s} \cdot \xi_1^4 (|\xi|^2 + 1)^{-\frac{s}{2} - 3} \, d\xi$$

$$= r_0 \, \frac{15}{4\pi} H_1^2 \cdot 2^{-s} \cdot \frac{1}{(s+2)(s+4)}.$$

Adding up these equalities, we obtain the following.

$$\begin{aligned}(1) + (2) + (3) + (4) + (5) + (6) + (7) = {} & r_0 \, \frac{\tau_M}{16\pi} \cdot 2^{-s} \cdot \frac{s+1}{s+2} \\ & + r_0 \, \frac{H_1^2}{4\pi} \cdot 2^{-s} \cdot \left( \frac{s-4}{s+2} + \frac{15}{(s+2)(s+4)} \right) \\ & + \frac{1}{4\pi} (\mathrm{Tr}\, E) \cdot 2^{-s}.\end{aligned}$$

Taking the derivative of the above equality at $s = 0$, we obtain the following result.

**Lemma 7**

$$\pi_2(x; g) = r_0 \left( \frac{1}{32\pi} \ln 2 - \frac{1}{64\pi} \right) \tau_M + r_0 \left( -\frac{1}{32\pi} \ln 2 - \frac{3}{128\pi} \right) H_1^2 + \frac{1}{4\pi} (\mathrm{Tr}\, E) \ln 2.$$

If we compare (48) with Lemma 7, we obtain

$$4b_3 + 4b_4 = r_0 \left( -\frac{1}{32\pi} \ln 2 - \frac{3}{128\pi} \right) \quad \text{and} \quad b_5 = \frac{1}{4\pi} \ln 2,$$

which together with Lemma 4 leads to the following result.

**Theorem 5** *Let $b_i$ $(1 \leq i \leq 5)$ be the coefficients for the span of $\pi_2(x)$ defined on a closed 2-dimensional Riemannian manifold $N$ as (48). Then,*

$$b_1 = r_0 \cdot \frac{1}{24\pi} \ln 2, \qquad b_2 = r_0 \left( -\frac{1}{48\pi} \ln 2 - \frac{1}{32\pi} \right),$$

$$b_3 = r_0\left(\frac{7}{384\pi}\ln 2 + \frac{15}{512\pi}\right), \qquad b_4 = r_0\left(-\frac{5}{192\pi}\ln 2 - \frac{9}{256\pi}\right),$$

$$b_5 = \frac{1}{4\pi}\ln 2.$$

Theorem 5 with (48) recovers Theorem 1.

## 4 Conclusions

In this contribution we have analyzed the geometric information contained in the BFK-gluing formula [2]. Using homogeneity properties under conformal scaling the detailed geometric content is obtained in terms of curvature tensors. Unknown universal multipliers are found from suitable particular cases where answers can be found comparatively easily. This leads to results in full generality. This approach follows the strategy very successfully applied in the context of the small-$t$ asymptotic expansion of the heat-kernel [6, 8].

## References

1. N. Berlin, E. Getzler, M. Vergne, *Heat Kernels and Dirac Operators* (Springer, Berlin, 1992)
2. D. Burghelea, L. Friedlander, T. Kappeler, Mayer-Vietoris type formula for determinants of elliptic differential operators. J. Funct. Anal. **107**, 34–66 (1992)
3. G. Carron, Déterminant relatif et function Xi. Am. J. Math. **124**, 307–352 (2002)
4. G. Fucci, K. Kirsten, The spectral zeta function for Laplace operators on warped product manifolds of type $I \times_f N$. Comm. Math. Phys. **317**, 635–665 (2013)
5. P.B. Gilkey, *Invariance Theory, the Heat Equation, and the Atiyah-Singer Index Theorem*, 2nd edn. (CRC Press Inc., New York, 1994)
6. P.B. Gilkey, *Asymptotic Formulae in Spectral Geometry* (Chapman and Hall/CRC, New York, 2003)
7. G. Grubb, R. Seeley, Weakly parametric pseudodifferential operators and Atiyah-Patodi-Singer boundary problems. Invent. Math. **121**, 481–529 (1995)
8. K. Kirsten, *Spectral Functions in Mathematics and Physics* (Chapman and Hall/CRC, New York, 2002)
9. K. Kirsten, Y. Lee, The Burghelea-Friedlander-Kappeler-gluing formula for zeta-determinants on a warped product manifold and a product manifold. J. Math. Phys. **58**(12), 123501, 19p. (2015)
10. K. Kirsten, Y. Lee, The BFK-gluing formula and relative determinants on manifolds with cusps. J. Geom. Phys. **117**, 197–213 (2017)
11. K. Kirsten, Y. Lee, The polynomial associated with the BFK-gluing formula of the zeta-determinant on a compact warped product manifold. J. Geom. Anal. **28**, 3856–3891 (2018)
12. K. Kirsten, Y. Lee, The BFK-gluing formula and the curvature tensors on a 2−dimensional compact hypersurface. J. Spectr. Theory **10**, 1007–1051 (2020)
13. K. Kirsten, Y. Lee, *The Zeta-determinant of the Dirichlet-to-Neumann Operator for the Steklov Problem on Forms*. in preparation
14. Y. Lee, Mayer-Vietoris formula for the determinant of a Laplace operator on an even-dimensional manifold. Proc. Am. Math. Soc. **123**(6), 1933–1940 (1995)

15. J. Lee, G. Uhlmann, Determining isotropic real-analytic conductivities by boundary measurements. Comm. Pure Appl. Math. **42**, 1097–1112 (1989)
16. G. Liu, Asymptotic expansion of the trace of heat kernel associated to the Dirichlet-to-Neumann opetrator. J. Diff. Equations **259**, 2499–2545 (2015)
17. J. Müller, W. Müller, Regularized determinants of Laplace type operators, analytic surgery and relative determinants. Duke. Math. J. **133**, 259–312 (2006)
18. I. Polterovich, D.A. Sher, Heat invariants of the Stekelov problem. J. Geom. Anal. **25**, 924–950 (2015)
19. M.A. Shubin, *Pseudodifferential Operators and Spectral Theory* (Springer, Berlin, 1987)
20. A. Voros, Spectral functions, special functions and Selberg zeta function. Comm. Math. Phys. **110**, 439–465 (1987)

# New Representations of the Laguerre–Sobolev and Jacobi–Sobolev Orthogonal Polynomials

**Clemens Markett**

**Abstract** In this paper, we explicitly determine the coefficients of the Laguerre–Sobolev and Jacobi–Sobolev polynomials, both in two alternative versions. These polynomial systems are orthogonal with respect to discrete Sobolev inner products built upon the classical Laguerre and Jacobi measures on the intervals $[0, \infty)$ and $[-1, 1]$, respectively. In addition, they are equipped with two point masses at a finite end point of the interval involving functions and derivatives of any finite order. In general, such polynomials are represented as a linear combination of four polynomials reflecting the influence of the point masses. Moreover, it is shown that the Laguerre–Sobolev polynomials can be deduced from the Jacobi–Sobolev polynomials via a confluent limit relation.

**Keywords** Discrete Sobolev inner product · Sobolev orthogonal polynomials · Sobolev–Laguerre polynomials · Sobolev–Jacobi polynomials · Eigenfunctions

**Mathematics Subject Classification** 33C47 · 33C45 · 34L10

## 1 Introduction

For more than three decades, the study of orthogonal polynomial systems with respect to Sobolev inner products has been attracting enormous interest. In particular, numerous authors investigated the so-called discrete Sobolev (or Sobolev type) inner products given, in general, by

$$(f, g)_S = \int_{\mathbb{R}} f(x)g(x)d\mu + \sum_{r=0}^{T} S_r \int_{\mathbb{R}} f^{(r)}(x)g^{(r)}(x)d\mu_r, \tag{1.1}$$

C. Markett (✉)
Lehrstuhl A für Mathematik, RWTH Aachen, Aachen, Germany
e-mail: markett@matha.rwth-aachen.de; clemens.markett@t-online.de



F. Gesztesy, A. Martinez-Finkelshtein (eds.), *From Operator Theory to Orthogonal Polynomials, Combinatorics, and Number Theory*, Operator Theory: Advances and Applications 285, https://doi.org/10.1007/978-3-030-75425-9_16

where $T \in \mathbb{N}$, $S_r \geq 0$, $0 \leq r \leq T-1$, $S_T > 0$, $d_\mu$ is a positive Borel measure with continuous support on the real line, and $d\mu_r$, $0 \leq r \leq T$, are positive Borel measures supported on finite subsets of $\mathbb{R}$. The corresponding orthogonal polynomials turn out to be of a very rich structure. Among their many features discovered so far are algebraic properties, representations of different kind, recurrence relations, asymptotic properties, distributions of zeros, Fourier expansions, or analytic aspects. Moreover, these polynomials often arise as eigenfunctions of (higher-order) ordinary or partial differential equations. For old and new developments in this fascinating field see the excellent survey article by F. Marcellán and Y. Xu [25] or other informative papers as [2, 5, 11, 15, 22, 23, 26, 30–32]. Just recently, another useful overview is given in [21].

In order to investigate or illustrate the various properties of Sobolev-type orthogonal polynomials, several authors have focussed their attention on the Sobolev inner products which combine the classical Laguerre or Jacobi measures

$$\begin{aligned} d\mu_\alpha(x) &:= \frac{1}{\Gamma(\alpha+1)} e^{-x} x^\alpha dx, \ \alpha > -1, \ 0 < x < \infty, \\ d\mu_{\alpha,\beta}(x) &:= h_{\alpha,\beta}^{-1}(1-x)^\alpha(1+x)^\beta dx, \\ h_{\alpha,\beta}^{-1} &= \frac{\Gamma(\alpha+\beta+2)}{2^{\alpha+\beta+1}\Gamma(\alpha+1)\Gamma(\beta+1)}, \ \alpha, \beta > -1, \ -1 < x < 1, \end{aligned} \tag{1.2}$$

with one or two perturbations at a finite point, respectively. In fact, for $R \in \mathbb{N}_0$, $R < T \in \mathbb{N}$, $S_R$, $S_T \geq 0$ and suitable functions $f, g$, the inner products are defined by

$$(f, g)_{\alpha,S_R,S_T} = \int_0^\infty f(x)g(x)d\mu_\alpha(x) + S_R f^{(R)}(0)g^{(R)}(0) + S_T f^{(T)}(0)g^{(T)}(0), \tag{1.3}$$

$$(f, g)_{\alpha,\beta,S_R,S_T} = \int_{-1}^1 f(x)g(x)d\mu_{\alpha,\beta}(x) + S_R f^{(R)}(1)g^{(R)}(1) + S_T f^{(T)}(1)g^{(T)}(1). \tag{1.4}$$

The purpose of the present paper is to establish new explicit representations of the corresponding Laguerre–Sobolev and Jacobi–Sobolev orthogonal polynomials, denoted here by $\{L_n^{\alpha,S_R,S_T}(x)\}_{n=0}^\infty$ and $\{P_n^{\alpha,\beta,S_R,S_T}(x)\}_{n=0}^\infty$, respectively. If both point masses $S_R$, $S_T$ vanish, the two polynomial systems obviously reduce to the classical Laguerre and Jacobi polynomials defined, e.g., by their (confluent) hypergeometric expansions for $n \in \mathbb{N}_0$, see e.g. [14, 10.8, 10.12],

$$L_n^{\alpha,0,0}(x) = L_n^\alpha(x) := \frac{(\alpha+1)_n}{n!} {}_1F_1(-n; \alpha+1; x), \tag{1.5}$$

$$P_n^{\alpha,\beta,0,0}(x) = P_n^{\alpha,\beta}(x) := \frac{(\alpha+1)_n}{n!} {}_2F_1\left(-n, n+\alpha+\beta+1; \alpha+1; \frac{1-x}{2}\right). \tag{1.6}$$

As usual, $(a)_0 = 1$, $(a)_m = a(a+1)\cdots(a+m-1)$, $a \in \mathbb{R}$, $m \in \mathbb{N}$, is Pochhammer's symbol for the shifted factorials.

The Laguerre–Sobolev polynomials were first treated by R. Koekoek [16], who even admitted arbitrary many point masses $S_r$, $0 \le r \le T$, $T \in \mathbb{N}$, in the inner product, cf. (1.1),

$$(f, g)_{\alpha,\{S_r\}} = \int_0^\infty f(x)g(x)d\mu_\alpha(x) + \sum_{r=0}^{T} S_r f^{(r)}(0)g^{(r)}(0). \tag{1.7}$$

Looking for a representation of the orthogonal polynomials for all $n \ge T+1$ in either of the two (slightly modified) forms

$$L_n^{\alpha,\{S_r\}}(x) = \sum_{k=0}^{T+1} A_k L_{n-k}^{\alpha+k}(x) = \sum_{k=0}^{T+1} B_k x^k L_{n-k}^{\alpha+2k}(x),\ 0 \le x < \infty, \tag{1.8}$$

Koekoek showed that, due to the orthogonality of the classical Laguerre polynomials (1.5), each of the two sets of coefficients $A_k$, $B_k$, $0 \le k \le T+1$, has to satisfy a homogeneous system of linear equations for $0 \le r \le T$, namely

$$\sum_{k=r+1}^{T+1} f_{k,n}^r A_k + S_r\, g_n^{\alpha,r} \sum_{k=0}^{T+1,n-r} h_{k,n}^{\alpha,r} A_k = 0, \tag{1.9}$$

$$\sum_{k=r+1}^{T+1} f_{k,n}^r (r+\alpha+1)_k B_k + S_r\, g_n^{\alpha,r} \sum_{k=0}^{r} \hat{h}_{k,n}^{\alpha,r} B_k = 0. \tag{1.10}$$

Here and in the following we use the abbreviations $M, N := min\{M, N\}$ and

$$\begin{aligned} f_{k,n}^r &:= \frac{(n-k+1)_{k-r-1}}{(k-r-1)!}, & g_n^{\alpha,r} &:= \frac{(-1)^r(\alpha+1)_n\, r!}{(\alpha+1)_r^2\,(n-r)!}, \\ h_{k,n}^{\alpha,r} &:= \frac{(-1)^k(r-n)_k}{(r+\alpha+1)_k}, & \hat{h}_{k,n}^{\alpha,r} &:= \frac{(-r)_k(n+\alpha+1)_k}{(r+\alpha+1)_k}. \end{aligned} \tag{1.11}$$

Consequently, a nontrivial solution must exist in both situations. By solving (1.9) in the simplest case $T = 1$, the coefficients $A_0, A_1, A_2$ have been calculated in [17], see also [18]. Later, Bavinck [4] determined the set $\{A_k\}$ in case there is only one point mass $S_R > 0$ in the inner product (1.3), see also [23, Thm. 1].

In his search for higher-order differential equations with Sobolev-type polynomials as their eigenfunctions, Bavinck developed another useful approach to describe the polynomials, see [4, 6–8] and the literature cited there. Quite generally, inner products with two linear perturbations were dealt with in [6], and later, the theory was applied both to the Laguerre–Sobolev case (1.3) [7] and a certain

Jacobi–Sobolev case [8]. Reflecting the influence of the parameters $S_R, S_T$, the Laguerre–Sobolev polynomials were given as a linear combination of the Laguerre polynomials and three related polynomials (cf. [7, Sec. 2]),

$$L_n^{\alpha,S_R,S_T}(x) = L_n^\alpha(x) + S_R T_n^{\alpha,R}(x) + S_T T_n^{\alpha,T}(x) + S_R S_T T_n^{\alpha,R,T}(x),\ n \in \mathbb{N}_0, \tag{1.12}$$

where,

$$T_n^{\alpha,t}(x) = K_{n-1}^{(t,t)}(0,0)\, L_n^\alpha(x) - (-1)^t L_{n-t}^{\alpha+t}(0)\, K_{n-1}^{(0,t)}(x,0),\ t \in \{R,T\},$$

$$T_n^{\alpha,R,T}(x) = \begin{vmatrix} L_n^\alpha(x) & K_{n-1}^{(0,R)}(x,0) & K_{n-1}^{(0,T)}(x,0) \\ (-1)^R \frac{(R+\alpha+1)_{n-R}}{(n-R)!} & K_{n-1}^{(R,R)}(0,0) & K_{n-1}^{(R,T)}(0,0) \\ (-1)^T \frac{(T+\alpha+1)_{n-T}}{(n-T)!} & K_{n-1}^{(R,T)}(0,0) & K_{n-1}^{(T,T)}(0,0) \end{vmatrix},$$

and the kernel functions are given, for $0 \le r \le t$, by

$$K_{n-1}^{(r,t)}(x,y) = \sum_{k=0}^{n-1} \frac{(L_k^\alpha)^{(r)}(x)(L_k^\alpha)^{(t)}(y)}{\binom{k+\alpha}{k}} = \sum_{k=t}^{n-1} \frac{(-1)^{r+t} k!\, L_{k-r}^{\alpha+r}(x) L_{k-t}^{\alpha+t}(y)}{(\alpha+1)_k}. \tag{1.13}$$

In particular, the numbers $K_{n-1}^{(r,t)}(0,0)$, $r,t \in \{R,T\}$, may be written as certain ${}_3F_2-$ hypergeometric sums. A similar representation of the Jacobi–Sobolev polynomials is even more sophisticated. Nevertheless, this approach has been successfully applied to investigate the properties of these polynomials in various directions, see, e.g. [1, 3, 10, 20, 21, 24].

Finally we point out that A. J. Durán and M. D. de la Iglesia used the powerful concept of so-called $\mathcal{D}$-operators introduced by the first-named author, to establish higher-order differential equations for the Laguerre–Sobolev and Jacobi–Sobolev orthogonal polynomials, see [12, 13] and the literature cited there. The authors actually considered the discrete bilinear forms based on the classical weight functions and $m \times m$—matrices, $m \in \mathbb{N}$, which consist of the functions and their derivatives up to the order $m-1$ at one or two end points of the interval of orthogonality. In a first step, the corresponding orthogonal polynomials were expressed as a linear combination of the classical polynomials with certain Casorati determinants as coefficients. This could be used to construct an algebra of differential operators possessing the new polynomials as eigenfunctions.

The first aim of this paper is to present the Laguerre–Sobolev polynomials $\{L_n^{\alpha,S_R,S_T}(x)\}_{n=0}^\infty$ in the two versions (1.8). According to (1.12), we first separate the dependence on $S_R, S_T$ by expanding the latter three components into sums with new coefficients to be determined, i.e.

$$T_n^{\alpha,R}(x) = \sum_{k=0}^{R+1} a_k L_{n-k}^{\alpha+k}(x) \equiv \sum_{k=0}^{R+1} \hat{a}_k x^k L_{n-k}^{\alpha+2k}(x),\ n \geq R+1,$$

$$T_n^{\alpha,T}(x) = \sum_{k=0}^{T+1} b_k L_{n-k}^{\alpha+k}(x) \equiv \sum_{k=0}^{T+1} \hat{b}_k x^k L_{n-k}^{\alpha+2k}(x),\ n \geq T+1, \tag{1.14}$$

$$T_n^{\alpha,R,T}(x) = \sum_{k=0}^{T+1} c_k L_{n-k}^{\alpha+k}(x) \equiv \sum_{k=0}^{T+1} \hat{c}_k x^k L_{n-k}^{\alpha+2k}(x),\ n \geq T+1.$$

The polynomials of lower degree are supposed to vanish identically. The clue is now to relate the coefficients in each of the two sums (1.8) with those in (1.14) by setting

$$A_0 = 1 + a_0 S_R + b_0 S_T + c_0 S_R S_T,\ B_0 = 1 + \hat{a}_0 S_R + \hat{b}_0 S_T + \hat{c}_0 S_R S_T,$$

$$A_k = a_k S_R + b_k S_T + c_k S_R S_T,\ B_k = \hat{a}_k S_R + \hat{b}_k S_T + \hat{c}_k S_R S_T,\ 1 \leq k \leq R+1,$$

$$A_k = b_k S_T + c_k S_R S_T,\ B_k = \hat{b}_k S_T + \hat{c}_k S_R S_T,\ R+2 \leq k \leq T+1. \tag{1.15}$$

Concerning the $A_k's$, we insert their substitutes into the $T + 1$ homogeneous equations (1.9), where $S_r = 0$ for all $r \neq R, T$. Collecting the terms of equal powers of $(S_R)^l (S_T)^m$, $0 \leq l, m \leq 2$, and observing that the quantities $S_R, S_T$ are arbitrary, the resulting expressions have to vanish simultaneously. Thus we arrive at 14 conditions to be satisfied by the coefficients $a_k,\ b_k,\ c_k$ in (1.14). As it turns out, the first 13 conditions successively determine the quantities we are looking for, and fortunately, the last one holds as well.

The coefficients $\hat{a}_k,\ \hat{b}_k,\ \hat{c}_k$ of the second sums in (1.14) can be found in almost the same way. The two final results are stated in Theorems 2.1 and 2.2, respectively. While we prove the first theorem in detail, it suffices to point out some specifics of the second one. Theorem 2.1 is particularly useful to deduce the leading coefficients of the Laguerre–Sobolev polynomials, while their values at the origin can be read off from Theorem 2.2. This aspect can be used to show how the coefficients in the two different representations interact.

In Sect. 3, we represent the Jacobi–Sobolev polynomials $\{P_n^{\alpha,\beta,S_R,S_T}(x)\}_{n=0}^{\infty}$. Of course, this task proves to be more involved because of the additional parameter $\beta$ as a new degree of freedom. So in contrast to the Laguerre–Sobolev polynomials, much less is known. Nevertheless, our new approach applies here as well. Starting off from the Jacobi–Sobolev inner product with again arbitrary many point masses $S_r$, $0 \leq r \leq T$, the appropriate analogues of the expansions (1.8) are given, in terms of the Jacobi polynomials (1.6), by

$$P_n^{\alpha,\beta,\{S_r\}}(x) = \sum_{k=0}^{T+1} A_k P_{n-k}^{\alpha+k,\beta}(x) = \sum_{k=0}^{T+1} B_k\Big(\frac{1-x}{2}\Big)^k P_{n-k}^{\alpha+2k,\beta}(x),\ n \in \mathbb{N}_0. \tag{1.16}$$

The next step is to establish the homogeneous equations corresponding to (1.9–1.10). This is carried out in Proposition 3.1. Similarly to (1.12) and (1.14), we then split up the Jacobi–Sobolev polynomials into the four terms

$$P_n^{\alpha,\beta,S_R,S_T}(x) = P_n^{\alpha,\beta}(x) + S_R Q_n^{\alpha,\beta,R}(x) + S_T Q_n^{\alpha,\beta,T}(x) + S_R S_T Q_n^{\alpha,\beta,R,T}(x) \tag{1.17}$$

and expand each of the latter three components into sums of the form (1.16). Employing the relationships (1.15) once more, we obtain the appropriate conditions to determine the new coefficients which now depend on $k, n, \alpha, \beta, R, T$. The resulting Jacobi–Sobolev polynomials are given in Theorems 3.1 and 3.2, respectively.

Needless to say, our new representations of the Laguerre–Sobolev and Jacobi–Sobolev polynomials can easily be implemented in a symbolic computer program. In fact, we used MAPLE to confirm all four theorems for a large variety of parameter values.

Our results are particularly suited to establish a relationship between the two polynomial systems, which generalizes the confluent limit relation between the Jacobi and Laguerre polynomials,

$$\lim_{\beta\to\infty} P_n^{\alpha,\beta}\Big(1-\frac{2x}{\beta}\Big) = L_n^{\alpha}(x),\ \alpha > -1,\ 0 \le x < \infty. \tag{1.18}$$

Notice that by formally substituting $x \to \xi := 1-2x/\beta$ in the Jacobi–Sobolev inner product (1.4) and extending the two point masses to $S_r(\beta) := (2/\beta)^{2r} S_r$, $r = R, T$, one arrives at the Laguerre–Sobolev inner product (1.3) in the limit $\beta \to \infty$. This clearly suggests that

$$\lim_{\beta\to\infty} P_n^{\alpha,\beta,S_R(\beta),S_T(\beta)}\Big(1-\frac{2x}{\beta}\Big) = L_n^{\alpha,S_R,S_T}(x),\ \alpha > -1,\ 0 \le x < \infty. \tag{1.19}$$

Indeed, the identity (1.19) will be verified in Corollary 3.1.

The results of the present paper are suited to stimulate further research in various respects. First of all it should be worthwhile to verify the equivalence of our new representations of the Sobolev-type polynomials with those obtained by other methods. Furthermore, our results in Theorems 2.2 and 3.2 are particularly useful to find the higher-order differential equations having the polynomials as eigenfunctions. Such equations have recently been discovered by the author [28, 29], see also [27]. Last but not least, the new representations provide a promising tool to further examine the properties of the polynomials.

The new techniques may apply to other Sobolev-type polynomials as well. For instance, inner products like (1.4) have been treated, which combine the Jacobi measure $\mu_{\alpha,\beta}$ with point masses located at both end points of the interval $x = \pm 1$, cf., e.g. [3, 19]. A particular example is the inner product associated with the Gegenbauer–Sobolev polynomials in case $\alpha = \beta$, which belong to the larger class of symmetric Sobolev-type polynomials dealt with in [9]. Very likely, the Gegenbauer–Sobolev polynomials can be traced back to the Jacobi–Sobolev polynomials via a quadratic transformation of the argument, which would immediately allow applying our results. Finally, our methods might be useful to treat discrete orthogonal polynomial systems like the Meixner–Sobolev or Hahn–Sobolev polynomials.

## 2 Two Representations of the Laguerre–Sobolev Polynomials

Here is our first result on the orthogonal polynomials associated with the inner product (1.3). Partly recalling (1.11), we set

$$g_n^{\alpha,r} := \frac{(-1)^r(\alpha+1)_n\, r!}{(\alpha+1)_r^2\,(n-r)!},\quad h_{k,n}^{\alpha,r} := \frac{(-1)^k(r-n)_k}{(r+\alpha+1)_k},\quad q_{k,n}^{r} := \frac{(k-n)_{r+1-k}}{(r+1-k)!}. \tag{2.1}$$

**Theorem 2.1** *Let $\alpha > -1$, $R \in \mathbb{N}_0$, $R < T \in \mathbb{N}$, and $S_R, S_T \geq 0$. For $n \in \mathbb{N}_0$, $0 \leq x < \infty$, the Laguerre–Sobolev polynomials are given by (1.12), i.e.*

$$L_n^{\alpha,S_R,S_T}(x) = L_n^{\alpha}(x) + S_R T_n^{\alpha,R}(x) + S_T T_n^{\alpha,T}(x) + S_R S_T T_n^{\alpha,R,T}(x),$$

*where $T_n^{\alpha,R}(x) \equiv 0$ if $n \leq R$, $T_n^{\alpha,T}(x) \equiv T_n^{\alpha,R,T}(x) \equiv 0$ if $n \leq T$, and, for higher values of $n$,*

$$T_n^{\alpha,R}(x) = \sum_{k=0}^{R+1} a_k L_{n-k}^{\alpha+k}(x),\ a_k = a_{k,n}^{\alpha,R} := \left\{\begin{array}{l} g_n^{\alpha,R} \displaystyle\sum_{j=1}^{R+1,n-R} h_{j,n}^{\alpha,R}\, q_{j,n}^{R},\ k=0 \\ -\, g_n^{\alpha,R}\, q_{k,n}^{R},\ 1 \leq k \leq R+1 \end{array}\right\}, \tag{2.2}$$

$$T_n^{\alpha,T}(x) = \sum_{k=0}^{T+1} b_k L_{n-k}^{\alpha+k}(x),\ b_k = b_{k,n}^{\alpha,T} := a_{k,n}^{\alpha,T},\ 0 \leq k \leq T+1, \tag{2.3}$$

$$T_n^{\alpha,R,T}(x) = \sum_{k=0}^{T+1} c_k L_{n-k}^{\alpha+k}(x),\ c_k = c_{k,n}^{\alpha,R,T},\ \textit{where}$$

$$c_{k,n}^{\alpha,R,T} := \left\{ \begin{array}{l} a_0\, b_0 - \sum_{j=1}^{R+1,n-T} h_{j,n}^{\alpha,T} a_j \sum_{j=1}^{T+1,n-R} h_{j,n}^{\alpha,R} b_j,\ k = 0 \\ b_k \sum_{j=0}^{R+1,n-T} h_{j,n}^{\alpha,T} a_j + a_k \sum_{j=0}^{T+1,n-R} h_{j,n}^{\alpha,R} b_j,\ 1 \le k \le T+1 \\ \textit{with } a_k = 0,\ k \ge R+2 \end{array} \right\}. \tag{2.4}$$

***Proof*** Following the method outlined in Sect. 1, we successively insert the settings (1.15) into the Eqs. (1.9). For $0 \le r \le R-1$, we have $S_r = 0$, so that

$$S_R \sum_{k=r+1}^{R+1} f_{k,n}^r\, a_k + S_T \sum_{k=r+1}^{T+1} f_{k,n}^r\, b_k + S_R S_T \sum_{k=r+1}^{T+1} f_{k,n}^r\, c_k = 0$$

for any $S_R, S_T$. Hence, each of the three sums must vanish, yielding the conditions

$$(C1)-(C3) \quad \sum_{k=r+1}^{R+1} f_{k,n}^r\, a_k = 0, \quad \sum_{k=r+1}^{R+1} f_{k,n}^r\, b_k = 0, \quad \sum_{k=r+1}^{R+1} f_{k,n}^r\, c_k = 0. \tag{2.5}$$

For $r = R$, both sums in (1.9) contribute to the identity

$$\begin{aligned} 0 =& a_{R+1} S_R + \sum_{k=R+1}^{T+1} f_{k,n}^R \left[ b_k S_T + c_k S_R S_T \right] \\ &+ S_R\, g_n^{\alpha,R} \left\{ 1 + \sum_{k=0}^{R+1,n-R} h_{k,n}^{\alpha,R} a_k S_R + \sum_{k=0}^{T+1,n-R} h_{k,n}^{\alpha,R} \left[ b_k S_T + c_k S_R S_T \right] \right\}. \end{aligned} \tag{2.6}$$

Collecting the terms of equal powers $(S_R)^l(S_T)^m$ and equating them to zero, we obtain five more conditions with respect to

$$\begin{aligned}
&S_R: && (C4) && a_{R+1} = -g_n^{\alpha,R} = \frac{(-1)^{R+1}(\alpha+1)_n}{(\alpha+1)_R^2}\frac{R!}{(n-R)!} \\
&S_T: && (C5) && \sum_{k=R+1}^{T+1} f_{k,n}^R\, b_k = 0 \\
&S_R S_T: && (C6) && \sum_{k=R+1}^{T+1} f_{k,n}^R\, c_k + g_n^{\alpha,R} \sum_{k=0}^{T+1,n-R} h_{k,n}^{\alpha,R}\, b_k = 0 \\
&(S_R)^2: && (C7) && \sum_{k=0}^{R+1,n-R} h_{k,n}^{\alpha,R}\, a_k = 0 \\
&(S_R)^2 S_T: && (C8) && \sum_{k=0}^{T+1,n-R} h_{k,n}^{\alpha,R}\, c_k = 0.
\end{aligned} \tag{2.7}$$

If $R+1 \le r \le T-1$, it follows that

$$(C9),\ (C10) \qquad \sum_{k=r+1}^{T+1} f_{k,n}^r\, b_k = 0, \qquad \sum_{k=r+1}^{T+1} f_{k,n}^r\, c_k = 0. \tag{2.8}$$

Finally, with $r = T$, we arrive at an equation of the form (2.6),

$$\begin{aligned}
0 =& b_{T+1} S_T + c_{T+1} S_R S_T \\
&+ S_T\, g_n^{\alpha,T} \Big\{ 1 + \sum_{k=0}^{R+1,n-T} h_{k,n}^{\alpha,T} a_k S_R + \sum_{k=0}^{T+1,n-T} h_{k,n}^{\alpha,T} \big[ b_k S_T + c_k S_R S_T \big] \Big\}.
\end{aligned}$$

Analogously to the identities (2.7), this implies

$$\begin{aligned}
&(C11) && b_{T+1} = -g_n^{\alpha,T} = \frac{(-1)^{T+1}(\alpha+1)_n}{(\alpha+1)_T^2}\frac{T!}{(n-T)!} \\
&(C12) && c_{T+1} + g_n^{\alpha,T} \sum_{k=0}^{R+1,n-T} h_{k,n}^{\alpha,T}\, a_k = 0 \\
&(C13),\ (C14) && \sum_{k=0}^{T+1,n-T} h_{k,n}^{\alpha,T}\, b_k = 0, \qquad \sum_{k=0}^{T+1,n-T} h_{k,n}^{\alpha,T}\, c_k = 0.
\end{aligned} \tag{2.9}$$

Starting off now from the value of $a_{R+1,n}^{\alpha,R}$ in (C4) and employing (C1), we obtain the required representation (2.2) for $1 \le k \le R$, i.e.

$$a_{k,n}^{\alpha,R} = q_{k,n}^{R}\, a_{R+1,n}^{\alpha,R} = -g_n^{\alpha,R}\, q_{k,n}^{R}. \tag{2.10}$$

In fact, with these values, condition (C1) is fulfilled for all $0 \le r \le R-1$ in view of

$$\sum_{k=r+1}^{R+1} f_{r,k}^{r}\, a_{k,n}^{\alpha,R} = -g_n^{\alpha,R} \sum_{k=r+1}^{R+1} \frac{(n-k+1)_{k-r-1}}{(k-r-1)!} \frac{(k-n)_{R+1-k}}{(R+1-k)!}$$

$$= C \sum_{k=0}^{R-r} \frac{(r-R)_k}{k!} = C \frac{(r-R+1)_{R-r}}{(R-r)!} = 0.$$

Throughout, $C$ is a constant which may differ at each occurrence. Via (C7) we get

$$a_{0,n}^{\alpha,R} = - \sum_{k=1}^{R+1,n-R} h_{k,n}^{\alpha,R}\, a_k = g_n^{\alpha,R} \sum_{k=1}^{R+1,n-R} h_{k,n}^{\alpha,R}\, q_{k,n}^{R}.$$

Analogously, we obtain the representation (2.3) of the coefficients $b_{k,n}^{\alpha,T}$ by successively applying (C11) for $k = T+1$, a combination of (C2), (C5), (C9) for $1 \le k \le T$, and (C13) for $k = 0$.

Concerning the coefficients $c_{k,n}^{\alpha,R,T}$, $0 \le k \le T+1$, stated in (2.4), (C12) yields

$$c_{T+1} = -g_n^{\alpha,T} \sum_{j=0}^{R+1,n-T} h_{j,n}^{\alpha,T}\, a_j = b_{T+1} \sum_{j=0}^{R+1,n-T} h_{j,n}^{\alpha,T}\, a_j.$$

Analogously to (2.10) and in view of (C10), we conclude that for $R+2 \le k \le T+1$,

$$c_k = q_{k,n}^{T}\, c_{T+1} = q_{k,n}^{T}\, b_{T+1} \sum_{j=0}^{R+1,n-T} h_{j,n}^{\alpha,T}\, a_j = b_k \sum_{j=0}^{R+1,n-T} h_{j,n}^{\alpha,T}\, a_j.$$

Furthermore, (C6) implies that

$$c_{R+1} = - \sum_{k=R+2}^{T+1} f_{k,n}^{R}\, c_k - g_n^{\alpha,R} \sum_{k=0}^{T+1,n-R} h_{k,n}^{\alpha,R}\, b_k =: G + H, \tag{2.11}$$

where, again in combination with (C4) and (C5),

$$G = -\sum_{k=R+2}^{T+1} f_{k,n}^{R}\, b_k \sum_{j=0}^{R+1,n-T} h_{j,n}^{\alpha,T}\, a_j = b_{R+1} \sum_{j=0}^{R+1,n-T} h_{j,n}^{\alpha,T}\, a_j,$$

$$H = a_{R+1} \sum_{j=0}^{T+1,n-R} h_{j,n}^{\alpha,R}\, b_j.$$

In order to show that the representation (2.11) carries over to $c_k$, $1 \le k \le R$, we first apply the argument used in (2.10) to any set $\{d_k\}_{k=1}^{T+1}$ satisfying

$$\sum_{k=r+1}^{T+1} f_{k,n}^{r}\, d_k = 0,\ \ 0 \le r \le T-1.$$

Then $d_k = q_{k,n}^{T}\, d_{T+1}$, $1 \le k \le T$. Hence, setting $d_k = c_k$, $R+2 \le k \le T+1$, and invoking (2.11), we get the two identities

$$d_{R+1} = -\sum_{k=R+2}^{T+1} f_{k,n}^{R}\, c_k = c_{R+1} - H, \quad d_k = q_{k,n}^{T}\, C_{T+1},\ 1 \le k \le T.$$

On the other hand, it follows by (C3) and our assumption on $\{d_k\}_{k=1}^{T+1}$ that

$$0 = \sum_{k=r+1}^{T+1} f_{k,n}^{r}\, c_k = \sum_{k=r+1}^{R+1} f_{k,n}^{r}\, c_k + \sum_{k=R+2}^{T+1} f_{k,n}^{r}\, d_k = \sum_{k=r+1}^{R+1} f_{k,n}^{r}\, (c_k - d_k)$$

for $0 \le r \le R-1$, whence $c_k - d_k = q_{k,n}^{R}\,(c_{R+1} - d_{R+1}) = q_{k,n}^{R} H$. Consequently,

$$\begin{aligned} c_k &= q_{k,n}^{T}\, c_{T+1} + q_{k,n}^{R} H \\ &= q_{k,n}^{T}\, b_{T+1} \sum_{j=0}^{R+1,n-T} h_{j,n}^{\alpha,T} a_j + q_{k,n}^{R}\, a_{R+1} \sum_{j=0}^{T+1,n-R} h_{j,n}^{\alpha,R}\, b_j \\ &= b_k \sum_{j=0}^{R+1,n-T} h_{j,n}^{\alpha,T} a_j + a_k \sum_{j=0}^{T+1,n-R} h_{j,n}^{\alpha,R}\, b_j. \end{aligned}$$

Using (C8) and then (C7) again, we finally arrive at

$$
\begin{aligned}
c_0 &= -\Big\{ \sum_{k=1}^{R+1,n-R} + \sum_{k=R+2}^{T+1,n-R} \Big\} h_{k,n}^{\alpha,R}\, c_k \\
&= - \sum_{k=1}^{T+1,n-R} h_{k,n}^{\alpha,R}\, b_k \sum_{j=0}^{R+1,n-T} h_{j,n}^{\alpha,T}\, a_j - \sum_{k=1}^{R+1,n-R} h_{k,n}^{\alpha,R}\, a_k \sum_{j=0}^{T+1,n-R} h_{j,n}^{\alpha,R}\, b_j \\
&= - \sum_{k=1}^{T+1,n-R} h_{k,n}^{\alpha,R}\, b_k \Big\{ a_0 + \sum_{j=1}^{R+1,n-T} h_{j,n}^{\alpha,T}\, a_j \Big\} + a_0 \Big\{ b_0 + \sum_{j=1}^{T+1,n-R} h_{j,n}^{\alpha,R}\, b_j \Big\} \\
&= a_0\, b_0 - \sum_{j=1}^{T+1,n-R} h_{j,n}^{\alpha,R}\, b_j \sum_{j=1}^{R+1,n-T} h_{j,n}^{\alpha,T}\, a_j .
\end{aligned}
$$

It remains to verify that (C14) holds as well. In fact, the latter identity and (C13) imply that

$$
\begin{aligned}
&\sum_{k=1}^{T+1,n-T} h_{k,n}^{\alpha,T}\, c_k \\
&= \sum_{k=1}^{T+1,n-R} h_{k,n}^{\alpha,T}\, b_k \sum_{j=0}^{R+1,n-T} h_{j,n}^{\alpha,T}\, a_j + \sum_{k=1}^{R+1,n-R} h_{k,n}^{\alpha,T}\, a_k \sum_{j=0}^{T+1,n-R} h_{j,n}^{\alpha,R}\, b_j \\
&= -b_0 \Big\{ a_0 + \sum_{j=1}^{R+1,n-T} h_{j,n}^{\alpha,T}\, a_j \Big\} + \sum_{j=1}^{R+1,n-R} h_{j,n}^{\alpha,T}\, a_j \Big\{ b_0 + \sum_{j=1}^{T+1,n-R} h_{j,n}^{\alpha,R}\, b_j \Big\} \\
&= -a_0\, b_0 + \sum_{j=1}^{T+1,n-R} h_{j,n}^{\alpha,R}\, b_j \sum_{j=1}^{R+1,n-T} h_{j,n}^{\alpha,T}\, a_j = -c_0 .
\end{aligned}
$$

This concludes the proof of Theorem 2.1. □

An equivalent version of the Laguerre–Sobolev polynomials (1.12) is obtained by determining the coefficients of the second sums in (1.14). As in (1.11), (2.1), set

$$
g_n^{\alpha,r} = \frac{(-1)^r(\alpha+1)_n\, r!}{(\alpha+1)_r^2\,(n-r)!},\quad \hat{h}_{k,n}^{\alpha,r} = \frac{(-r)_k(n+\alpha+1)_k}{(r+\alpha+1)_k},\quad q_{k,n}^{r} = \frac{(k-n)_{r+1-k}}{(r+1-k)!}. \tag{2.12}
$$

**Theorem 2.2** *Let $\alpha, R, T, S_R, S_T$ be as in Theorem 2.1. Then there holds*

$$
L_n^{\alpha,S_R,S_T}(x) = L_n^{\alpha}(x) + S_R\hat{T}_n^{\alpha,R}(x) + S_T\hat{T}_n^{\alpha,T}(x) + S_R S_T\hat{T}_n^{\alpha,R,T}(x),\ n \in \mathbb{N}_0, \tag{2.13}
$$

*where* $\hat{T}_n^{\alpha,R}(x) \equiv 0$ *if* $n \leq R$, $\hat{T}_n^{\alpha,T}(x) \equiv \hat{T}_n^{\alpha,R,T}(x) \equiv 0$ *if* $n \leq T$, *and, for higher values of* $n$,

$$\hat{T}_n^{\alpha,R}(x) = \sum_{k=0}^{R+1} \hat{a}_k x^k L_{n-k}^{\alpha+2k}(x),\ \hat{a}_{k,n}^{\alpha,R} := \left\{ \begin{array}{ll} g_n^{\alpha,R} \sum_{j=1}^{R} \frac{\hat{h}_{j,n}^{\alpha,R} q_{j,n}^R(\alpha+2j)}{(R+\alpha+1)_{j+1}},\ k=0 \\ -g_n^{\alpha,R} \frac{q_{k,n}^R(\alpha+2k)}{(R+\alpha+1)_{k+1}},\ 1 \leq k \leq R+1 \end{array} \right\} \tag{2.14}$$

$$\hat{T}_n^{\alpha,T}(x) = \sum_{k=0}^{T+1} \hat{b}_k x^k L_{n-k}^{\alpha+2k}(x), \quad \hat{b}_k = \hat{b}_{k,n}^{\alpha,T} := \hat{a}_{k,n}^{\alpha,T},\ k = 0, \ldots, T+1, \tag{2.15}$$

$$\hat{T}_n^{\alpha,R,T}(x) = \sum_{k=0}^{T+1} \hat{c}_k x^k L_{n-k}^{\alpha+2k}(x), \quad \hat{c}_k = \hat{c}_{k,n}^{\alpha,R,T},\ \textit{where}$$

$$\hat{c}_{k,n}^{\alpha,R,T} := \left\{ \begin{array}{l} \hat{a}_0 \hat{b}_0 - \sum_{j=1}^{R+1} \hat{h}_{j,n}^{\alpha,T} \hat{a}_j \sum_{j=1}^{R} \hat{h}_{j,n}^{\alpha,R} \hat{b}_j,\ k=0 \\ \hat{b}_k \sum_{j=0}^{R+1} \hat{h}_{j,n}^{\alpha,T} \hat{a}_j + \hat{a}_k \sum_{j=0}^{R} \hat{h}_{j,n}^{\alpha,R} \hat{b}_j,\ 1 \leq k \leq T+1 \\ \textit{with } \hat{a}_k = 0,\ k \geq R+2 \end{array} \right\}. \tag{2.16}$$

***Proof*** Proceeding as in the proof of Theorem 2.1, we substitute the $B_k$'s in the system of Eqs. (1.10) by their settings in (1.15). Again this leads to 14 conditions, now for the coefficients $\hat{a}_k$, $\hat{b}_k$, $\hat{c}_k$. The counterparts of (C1)–(C3), (C5), (C9)–(C10) read

$$(C1a) \qquad \sum_{k=r+1}^{R+1} f_{k,n}^r (r+\alpha+1)_k \hat{a}_k = 0,\ 0 \leq r \leq R-1,$$

$$(C2a), (C5a), (C9a) \quad \sum_{k=r+1}^{T+1} f_{k,n}^r (r+\alpha+1)_k \hat{b}_k = 0,\ 0 \leq r \leq T,$$

$$(C3a), (C10a) \qquad \sum_{k=r+1}^{T+1} f_{k,n}^r (r+\alpha+1)_k \hat{c}_k = 0,\ 0 \leq r \leq T,\ r \neq R.$$

Combining the first two lines with

$$\text{(C4}a\text{), (C11}a\text{)} \qquad \hat{a}_{R+1} = -\frac{g_n^{\alpha,R}}{(R+\alpha+1)_{R+1}}, \qquad \hat{b}_{T+1} = -\frac{g_n^{\alpha,T}}{(T+\alpha+1)_{T+1}},$$

$$\text{(C7}a\text{), (C13}a\text{)} \qquad \sum_{k=0}^{R} \hat{h}_{k,n}^{\alpha,R}\,\hat{a}_k = 0, \quad \sum_{k=0}^{T} \hat{h}_{k,n}^{\alpha,T}\,\hat{b}_k = 0,$$

we end up with the representations (2.14) and (2.15). In particular, the analogue of (2.10) reads

$$\hat{a}_{k,n}^{\alpha,R} = q_{k,n}^{\alpha,R}\,\frac{(R+\alpha+1)_{R+1}\,(\alpha+2k)}{(R+\alpha+1)_{k+1}}\,\hat{a}_{R+1,n}^{\alpha,R} = -g_n^{\alpha,R}\,\frac{q_{k,n}^{\alpha,R}(\alpha+2k)}{(R+\alpha+1)_{k+1}}. \tag{2.17}$$

Inserting the values (2.17) into the left-hand side of (C1a) we essentially get a terminating ${}_3F_2$—hypergeometric sum which vanishes for all $0 \le r \le R-1$ in view of Dixon's theorem [14, 4.4.(5)], see also [8, (3.4)]. In fact,

$$\begin{aligned}
&-g_n^{\alpha,R} \sum_{k=r+1}^{R+1} f_{k,n}^{r}\,(r+\alpha+1)_k\,\frac{(k-n)_{R+1-k}\,(\alpha+2k)}{(R+1-k)!\,(r+\alpha+1)_{k+1}} \\
&= C \sum_{k=0}^{R-r} \frac{(r-R)_k\,(2r+\alpha+2)_k\,(2k+2r+\alpha+2)}{k!\,(r+R+\alpha+3)_k} \\
&= C\,\frac{(2r+\alpha+2)_{R-r+1}\,(r-R+1)_{R-r}}{(R-r)!\,(r+R+\alpha+3)_{R-r}} = 0.
\end{aligned} \tag{2.18}$$

The new conditions (C8a), (C14a) are of the form (C7a), (C13a) with the coefficients $\hat{a}_k$ and $\hat{b}_k$ being replaced by $\hat{c}_k$. Finally, there are the two mixed-type conditions

$$\text{(C6}a\text{)} \qquad \sum_{k=R+1}^{T+1} f_{k,n}^{R}\,(R+\alpha+1)_k\,\hat{c}_k + g_n^{\alpha,R} \sum_{k=0}^{R} \hat{h}_{k,n}^{\alpha,R}\,\hat{b}_k = 0,$$

$$\text{(C12}a\text{)} \qquad \hat{c}_{T+1} + \frac{g_n^{\alpha,T}}{(T+\alpha+1)_{T+1}} \sum_{k=0}^{R+1} \hat{h}_{k,n}^{\alpha,T}\,\hat{a}_k = 0.$$

Following the arguments used in Theorem 2.1 we arrive at the representation (2.16) of $\hat{c}_{k,n}^{\alpha,R,T}$. Moreover, the conditions (C8a) and (C14a) turn out to be equivalent. □

# 3 New Representations of the Jacobi–Sobolev Polynomials

First of all, we need appropriate analogues of the linear equations (1.9)–(1.10), now associated with the Jacobi–Sobolev case.

**Proposition 3.1** *For $\alpha, \beta > -1$ let the Jacobi–Sobolev inner product (1.4) be extended to arbitrary many point masses $S_m$, $0 \le m \le T \in \mathbb{N}$, and suppose that for all $n > T$, the corresponding orthogonal polynomials are given in either of the two forms (1.16), i.e.*

$$P_n^{\alpha,\beta,\{S_m\}}(x) = \sum_{k=0}^{T+1} A_k P_{n-k}^{\alpha+k,\beta}(x) = \sum_{k=0}^{T+1} B_k \left(\frac{1-x}{2}\right)^k P_{n-k}^{\alpha+2k,\beta}(x). \tag{3.1}$$

*Then the coefficients $A_k$, $B_k$, $0 \le k \le T+1$, satisfy the systems of linear equations*

$$\sum_{k=r+1}^{T+1} f_{k,n}^r \phi_{k,n}^{\alpha,\beta,r} A_k + S_r \frac{g_n^{\alpha,r}}{2^{2r}} \sum_{k=0}^{T+1,n-r} h_{k,n}^{\alpha,r} A_k = 0, \tag{3.2}$$

$$\sum_{k=r+1}^{T+1} f_{k,n}^r (r+\alpha+1)_k \chi_{k,n}^{\alpha,\beta,r} B_k + S_r \frac{g_n^{\alpha,r}}{2^{2r}} \sum_{k=0}^{r} \hat{h}_{k,n}^{\alpha,r} \psi_{k,n}^{\alpha,\beta,r} B_k = 0, \tag{3.3}$$

*both for $0 \le r \le T$. Here $f_{k,n}^r$, $g_n^{\alpha,r}$, $h_{k,n}^{\alpha,r}$, $\hat{h}_{k,n}^{\alpha,r}$ are defined in (1.11) and*

$$\begin{aligned}\phi_k &:= \frac{(\beta+1)_{n-k}}{(\alpha+\beta+2)_{n+r-k}(n+\alpha+\beta+1)_r}, \qquad \chi_k := \frac{(\beta+1)_{n-k}}{(\alpha+\beta+2)_{n+r}}, \\ \psi_k &:= \frac{(n+\alpha+\beta+1)_r}{(n+\alpha+\beta+1)_k}.\end{aligned} \tag{3.4}$$

***Proof*** By assumption, the polynomials $P_n^{\alpha,\beta,\{S_m\}}(x)$ with $0 \le m \le T < n$, are orthogonal with respect to the (generalized) Jacobi–Sobolev inner product to any polynomial of lower degree. So if $p_r(x) = ((1-x)/2)^r$, $0 \le r \le n-1$, we have

$$\begin{aligned}&I_{n,r} + \Sigma_{n,r}^T \\ &:= \int_{-1}^{1} P_n^{\alpha,\beta,\{S_m\}}(x)\, p_r(x) d\mu_{\alpha,\beta}(x) + \sum_{m=0}^{T} S_m (P_n^{\alpha,\beta,\{S_m\}})^{(m)}(1)\, p_r^{(m)}(1) = 0.\end{aligned} \tag{3.5}$$

In order to evaluate the two expressions on the left-hand side of (3.5), we use some well-known properties of the Jacobi polynomials, see [14, 10.8]: For $\gamma, \delta > -1$,

$$P_n^{\gamma,\delta}(1) = \frac{(\gamma+1)_n}{n!}, \ n \in \mathbb{N}_0,$$

$$\frac{\mathrm{d}^m}{\mathrm{d}x^m} P_n^{\gamma,\delta}(x) = 2^{-m}(n+\gamma+\delta+1)_m \, P_{n-m}^{\gamma+m,\delta+m}(x), \ 1 \le m \le n,$$

$$(2n+\gamma+\delta+1)\, P_n^{\gamma,\delta}(x) = (n+\gamma+\delta+1)\, P_n^{\gamma+1,\delta}(x) - (n+\delta)\, P_{n-1}^{\gamma+1,\delta}(x), \ n \in \mathbb{N},$$

$$P_n^{\gamma+1,\delta}(x) = \sum_{j=0}^{n} \frac{(j+\delta+1)_{n-j}\,(2j+\gamma+\delta+1)}{(j+\gamma+\delta+1)_{n+1-j}} P_j^{\gamma,\delta}(x), \ n \in \mathbb{N}_0.$$

In particular, the latter two (contiguous) relations can be iterated to

$$P_n^{\gamma,\delta}(x) = \sum_{j=0}^{s} c_{j,n,s}^{\gamma,\delta} P_{n-j}^{\gamma+s,\delta}(x), \ 1 \le s \le n, \ \text{where}$$
$$c_{j,n,s}^{\gamma,\delta} = \frac{(-s)_j\,(n+\gamma+\delta+1)_{s-j}\,(n+\delta+1-j)_j}{j!\,(2n+\gamma+\delta+1-j)_{s-j}\,(2n+s+\gamma+\delta+2-2j)_j}, \tag{3.6}$$

$$P_n^{\gamma+s,\delta}(x) = \sum_{j=0}^{n} d_{j,n,s}^{\gamma,\delta} P_j^{\gamma,\delta}(x), \ s \in \mathbb{N}, \ \text{where}$$
$$d_{j,n,s}^{\gamma,\delta} = \frac{(n-j+1)_{s-1}\,(j+n+\gamma+\delta+2)_{s-1}\,(j+\delta+1)_{n-j}\,(2j+\gamma+\delta+1)}{(s-1)!\,(j+\gamma+\delta+1)_{n+s-j}}. \tag{3.7}$$

Inserting the first representation of (3.1) into (3.5), we immediately get

$$\Sigma_{n,r}^{T} = \sum_{m=0}^{T} S_m\, p_r^{(m)}(1) \sum_{k=0}^{T+1} A_k\,(P_{n-k}^{\alpha+k,\beta})^{(m)}(1) = 0, \ \text{if } r \ge T+1,$$

while for $0 \le r \le T$,

$$\begin{aligned}\Sigma_{n,r}^{T} &= S_r\, r! \left(-\frac{1}{2}\right)^r \sum_{k=0}^{T+1} A_k\,(P_{n-k}^{\alpha+k,\beta})^{(r)}(1) \\ &= S_r\, r! \left(-\frac{1}{2}\right)^r \sum_{k=0}^{T+1,n-r} A_k \frac{(n+\alpha+\beta+1)_r\,(k+r+\alpha+1)_{n-r-k}}{2^r\,(n-r-k)!} \\ &= \frac{(-1)^r\,(n+\alpha+\beta+1)_r\,(\alpha+1)_n\, r!\, S_r}{2^{2r}\,(\alpha+1)_r\,(n-r)!} \sum_{k=0}^{T+1,n-r} \frac{(-1)^k (r-n)_k}{(r+\alpha+1)_k} A_k.\end{aligned} \tag{3.8}$$

Concerning the integral term $I_{n,r}$ we employ (3.6) and (3.7) to adjust the parameters of the Jacobi polynomials to those of the weight function. Hence, by the orthogonality of the Jacobi polynomials,

$$\begin{aligned} I_{n,r} &= \sum_{k=0}^{T+1} \frac{A_k}{h_{\alpha,\beta} 2^r} \int_{-1}^{1} P_{n-k}^{\alpha+k,\beta}(x)(1-x)^{\alpha+r}(1+x)^{\beta} dx \\ &= \sum_{k=0}^{T+1,r} \frac{A_k}{h_{\alpha,\beta} 2^r} \sum_{j=0}^{r-k} c_{j,n-k,r-k}^{\alpha+k,\beta} \int_{-1}^{1} P_{n-k-j}^{\alpha+r,\beta}(x)(1-x)^{\alpha+r}(1+x)^{\beta} dx \\ &\quad + \sum_{k=r+1}^{T+1} \frac{A_k}{h_{\alpha,\beta} 2^r} \sum_{j=0}^{n-k} d_{j,n-k,k-r}^{\alpha+r,\beta} \int_{-1}^{1} P_{j}^{\alpha+r,\beta}(x)(1-x)^{\alpha+r}(1+x)^{\beta} dx \\ &= \sum_{k=r+1}^{T+1} \frac{A_k \, h_{\alpha+r,\beta}}{h_{\alpha,\beta} 2^r} d_{0,n-k,k-r}^{\alpha+r,\beta} = \sum_{k=r+1}^{T+1} \frac{f_{k,n}^{r} (\alpha+1)_r (\beta+1)_{n-k}}{(\alpha+\beta+2)_{n+r-k}} A_k . \end{aligned} \tag{3.9}$$

Notice that $I_{n,r} = 0$ for $T+1 \le r \le n-1$. Combining (3.8) and (3.9) and dividing by $(\alpha+1)_r$ and $(n+\alpha+\beta+1)_r$, we get the Eqs. (3.2) for all $0 \le r \le T$.

The Eqs. (3.3) are obtained in a similar way. Here we insert the second representation of (3.1) into the two terms of (3.5) to achieve

$$\begin{aligned} \Sigma_{n,r}^{T} &= S_r \, r! \left(-\frac{1}{2}\right)^r \sum_{k=0}^{T+1} B_k \left( \left(\frac{1-x}{2}\right)^k P_{n-k}^{\alpha+2k,\beta}(x) \right)^{(r)} \Bigg|_{x=1} \\ &= \frac{(-1)^r (\alpha+1)_n \, r! \, S_r}{2^{2r} (\alpha+1)_r \, (n-r)!} \sum_{k=0}^{T+1,n-r} \frac{(-r)_k \, (n+\alpha+1)_k}{(r+\alpha+1)_k} \frac{(n+\alpha+\beta+1)_r}{(n+\alpha+\beta+1)_k} B_k , \end{aligned}$$

$$\begin{aligned} I_{n,r} &= \sum_{k=0}^{T+1} \frac{B_k}{h_{\alpha,\beta} \, 2^{k+r}} \int_{-1}^{1} P_{n-k}^{\alpha+2k,\beta}(x)(1-x)^{\alpha+k+r}(1+x)^{\beta} dx \\ &= \sum_{k=r+1}^{T+1} \frac{B_k \, h_{\alpha+k+r,\beta}}{h_{\alpha,\beta} \, 2^{k+r}} d_{0,n-k,k-r}^{\alpha+k+r,\beta} = \sum_{k=r+1}^{T+1} \frac{f_{k,n}^{r} (\alpha+1)_{k+r} (\beta+1)_{n-k}}{(\alpha+\beta+2)_{n+r}} B_k . \end{aligned}$$

This settles the second part of Proposition 3.1. □

**Theorem 3.1** *Let $\alpha, \beta > -1$, $R \in \mathbb{N}_0$, $R < T \in \mathbb{N}$, and let $S_R, S_T \ge 0$ be some point masses in the inner product (1.4). Furthermore, let $g_n^{\alpha,r}$, $h_{k,n}^{\alpha,r}$, $q_{k,n}^{r}$ be defined in (2.1) and*

$$\Omega_{k,n}^{\alpha,\beta,r} := \frac{(n+\alpha+\beta)_{r+1-k} \, (\alpha+\beta+2)_{n+r-1}}{2^{2r} (\beta+1)_{n-k}} . \tag{3.10}$$

*For $-1 \leq x \leq 1$, $n \in \mathbb{N}_0$, the Jacobi–Sobolev polynomials are given by (1.17), i.e.*

$$P_n^{\alpha,\beta,S_R,S_T}(x) = P_n^{\alpha,\beta}(x) + S_R\, Q_n^{\alpha,\beta,R}(x) + S_T\, Q_n^{\alpha,\beta,T}(x) + S_R S_T\, Q_n^{\alpha,\beta,R,T}(x),$$

*where $Q_n^{\alpha,\beta,R}(x) \equiv 0$ if $n \leq R$, $Q_n^{\alpha,\beta,T}(x) \equiv Q_n^{\alpha,\beta,R,T}(x) \equiv 0$ if $n \leq T$, and, for higher values of $n$,*

$$Q_n^{\alpha,\beta,R}(x) = \sum_{k=0}^{R+1} a_k P_{n-k}^{\alpha+k,\beta}(x),\ a_{k,n}^{\alpha,\beta,R} := \left\{ \begin{array}{l} g_n^{\alpha,R} \displaystyle\sum_{j=1}^{R+1,n-R} h_{j,n}^{\alpha,R} q_{j,n}^R \Omega_{j,n}^{\alpha,\beta,R},\ k=0 \\ -\, g_n^{\alpha,R} q_{k,n}^R \Omega_{k,n}^{\alpha,\beta,R},\ 1 \leq k \leq R+1 \end{array} \right\}, \tag{3.11}$$

$$Q_n^{\alpha,\beta,T}(x) = \sum_{k=0}^{T+1} b_k P_{n-k}^{\alpha+k,\beta}(x), \quad b_k = b_{k,n}^{\alpha,\beta,T} := a_{k,n}^{\alpha,\beta,T},\ 0 \leq k \leq T+1, \tag{3.12}$$

$$Q_n^{\alpha,\beta,R,T}(x) = \sum_{k=0}^{T+1} c_k P_{n-k}^{\alpha+k,\beta}(x), \quad c_k = c_{k,n}^{\alpha,\beta,R,T},\ \textit{where}$$

$$c_{k,n}^{\alpha,\beta,R,T} := \left\{ \begin{array}{l} a_0 b_0 - \displaystyle\sum_{j=1}^{R+1,n-T} h_{j,n}^{\alpha,T} a_j \sum_{j=1}^{T+1,n-R} h_{j,n}^{\alpha,R} b_j,\ k=0 \\ b_k \displaystyle\sum_{j=0}^{R+1,n-T} h_{j,n}^{\alpha,T} a_j + a_k \sum_{j=0}^{T+1,n-R} h_{j,n}^{\alpha,R} b_j,\ 1 \leq k \leq T+1 \\ \textit{with } a_k = 0,\ k \geq R+2 \end{array} \right\}. \tag{3.13}$$

***Proof*** Observe that the system of Eqs. (3.2) differs from the system (1.9) in the Laguerre–Sobolev case by the quantities $\phi_{k,n}^{\alpha,\beta,r}$ and the factors $2^{-2r}$. So we can essentially follow the lines of proof of Theorem 2.1. Using again the substitutions (1.15), the conditions (C1)–(C3), (C5), (C9)–(C10) are modified by multiplying each of the coefficients $a_k, b_k, c_k$ by $\phi_{k,n}^{\alpha,\beta,r}$, while (C7)–(C8), (C13)–(C14) remain formally unchanged. The conditions (C4), (C11), (C6), (C12) turn into

$$(C4b),\ (C11b)\ a_{R+1,n}^{\alpha,\beta,R} = -\frac{g_n^{\alpha,R}}{2^{2R}\phi_{R+1,n}^{\alpha,\beta,R}} = -g_n^{\alpha,R}\,\Omega_{R+1,n}^{\alpha,\beta,R}, \quad b_{T+1,n}^{\alpha,\beta,T} = -g_n^{\alpha,T}\,\Omega_{T+1,n}^{\alpha,\beta,T},$$

$$(C6b) \qquad \sum_{k=R+1}^{T+1} f_{k,n}^R\,\phi_{k,n}^{\alpha,\beta,R}\,c_k + \frac{g_n^{\alpha,R}}{2^{2R}} \sum_{k=0}^{T+1,n-R} h_{k,n}^{\alpha,R}\,b_k = 0,$$

$$(C12b) \qquad \phi_{T+1,n}^{\alpha,\beta,T}\,c_{T+1} + \frac{g_n^{\alpha,T}}{2^{2T}} \sum_{k=0}^{R+1,n-T} h_{k,n}^{\alpha,T}\,a_k = 0.$$

All these conditions are satisfied by the values of $a_{k,n}^{\alpha,\beta,R}$, $b_{k,n}^{\alpha,\beta,T}$, $c_{k,n}^{\alpha,\beta,R,T}$ stated in (3.11)–(3.13). For instance, it follows by (C1b) and (C4b) that for $1 \le k \le R+1$,

$$a_{k,n}^{\alpha,\beta,R} = q_{k,n}^R\,\frac{(n+\alpha+\beta)_{R+1-k}}{(n-R+\beta)_{R+1-k}}\,a_{R+1,n}^{\alpha,\beta,R} = -q_{k,n}^R\,g_n^{\alpha,R}\,\Omega_{k,n}^{\alpha,\beta,R}.$$

In fact, condition (C1b) is valid in view of the Chu-Vandermonde summation formula yielding

$$\sum_{k=r+1}^{R+1} f_{n,k}^r\,\phi_{k,n}^{\alpha,\beta,r}\,a_{k,n}^{\alpha,\beta,R} = C \sum_{k=r+1}^{R+1} \frac{(-1)^k}{(k-r-1)!\,(R+1-k)!}\,\frac{(n+\alpha+\beta)_{R+1-k}}{(\alpha+\beta+2)_{n+r-k}}$$

$$= C\,{}_2F_1\left(\begin{matrix} r-R,\ n+\alpha+\beta \\ n+\alpha+\beta+r-R+1 \end{matrix};1\right) = C\,\frac{(r-R+1)_{R-r}}{(n+\alpha+\beta+r-R+1)_{R-r}} = 0.$$

The value of $a_{0,n}^{\alpha,\beta,R}$ in (3.11) is obtained by means of (C7b). As in the proof of Theorem 2.1, (2.3)–(2.4), we finally arrive at the representations of $Q_n^{\alpha,\beta,T}(x)$ and $Q_n^{\alpha,\beta,R,T}(x)$ as asserted. □

**Corollary 3.1** *Replacing $S_r$ by $S_r(2/\beta)^{2r}S_r$ for $r \in \{R,T\}$, $\beta > 0$, and substituting $\xi = 1 - 2x/\beta$, the Jacobi–Sobolev polynomials (1.17) reduce to the Laguerre–Sobolev polynomials (1.12) via the limit relation (1.19), i.e.*

$$P_n^{\alpha,\beta,S_R,S_T}(\xi) = \Big\{P_n^{\alpha,\beta}(\xi) + S_R(2\beta)^{2R}\,Q_n^{\alpha,\beta,R}(\xi)$$

$$+ S_T(2\beta)^{2T}\,Q_n^{\alpha,\beta,T}(\xi) + S_R S_T(2\beta)^{2R+2T}\,Q_n^{\alpha,\beta,R,T}(\xi)\Big\}$$

$$\to_{\beta\to\infty} \left\{L_n^\alpha(x) + S_R T_n^{\alpha,R}(x) + S_T T_n^{\alpha,T}(x) + S_R S_T T_n^{\alpha,R,T}(x)\right\} = L_n^{\alpha,S_R,S_T}(x).$$

***Proof*** By definition (3.10) we have

$$\lim_{\beta\to\infty}(2\beta)^{2r}\,\Omega_{k,n}^{\alpha,\beta,r}=\lim_{\beta\to\infty}\frac{(n+\alpha+\beta)_{r+1-k}\,(\alpha+\beta+2)_{n+r-1}}{\beta^{2r}\,(\beta+1)_{n-k}}=1.$$

Hence, the coefficients in Theorem 3.1 reduce to those in Theorem 2.1 via

$$\lim_{\beta\to\infty}(2\beta)^{2R}a_{k,n}^{\alpha,\beta,R}=a_{k,n}^{\alpha,R},\qquad \lim_{\beta\to\infty}(2\beta)^{2T}b_{k,n}^{\alpha,\beta,T}=b_{k,n}^{\alpha,T},$$
$$\lim_{\beta\to\infty}(2\beta)^{2R+2T}c_{k,n}^{\alpha,\beta,R,T}=c_{k,n}^{\alpha,R,T}.$$

Applying the relation (1.18) to each term then yields the required formula. □

Finally, we establish the counterpart of Theorem 2.2 in the Jacobi–Sobolev case.

**Theorem 3.2** *Let* $\alpha,\beta,R,T,S_R,S_T$ *be given as in Theorem 3.2. Furthermore,* $g_n^{\alpha,r}$, $\hat{h}_{k,n}^{\alpha,r}$, $q_{k,n}^{r}$ *are defined in (2.12), and*

$$\hat{\Omega}_{k,n}^{\alpha,\beta,r}:=\frac{(n+\alpha+\beta+1)_r\,(\alpha+\beta+2)_{n+r}}{2^{2r}(\beta+1)_{n-k}}. \tag{3.14}$$

*For* $-1\le x\le 1$, $n\in\mathbb{N}_0$, *the Jacobi–Sobolev polynomials are given by*

$$P_n^{\alpha,\beta,S_R,S_T}(x)=P_n^{\alpha,\beta}(x)+S_R\,\hat{Q}_n^{\alpha,\beta,R}(x)+S_T\,\hat{Q}_n^{\alpha,\beta,T}(x)+S_RS_T\,\hat{Q}_n^{\alpha,\beta,R,T}(x), \tag{3.15}$$

*where* $\hat{Q}_n^{\alpha,\beta,R}(x)\equiv 0$ *if* $n\le R$, $\hat{Q}_n^{\alpha,\beta,T}(x)\equiv\hat{Q}_n^{\alpha,\beta,R,T}(x)\equiv 0$ *if* $n\le T$, *and, for higher values of* $n$,

$$\hat{Q}_n^{\alpha,\beta,R}(x)=\sum_{k=0}^{R+1}\hat{a}_k\left(\frac{1-x}{2}\right)^k P_{n-k}^{\alpha+2k,\beta}(x),\quad \hat{a}_k=\hat{a}_{k,n}^{\alpha,\beta,R},\ \text{where}$$

$$\hat{a}_{k,n}^{\alpha,\beta,R}:=\left\{\begin{array}{ll} g_n^{\alpha,R}\displaystyle\sum_{j=1}^{R}\frac{\hat{h}_{j,n}^{\alpha,R}\,q_{j,n}^{R}\,(\alpha+2j)\,\hat{\Omega}_{j,n}^{\alpha,\beta,R}}{(R+\alpha+1)_{j+1}\,(n+\alpha+\beta+1)_j}, & k=0\\ -g_n^{\alpha,R}q_{k,n}^{R}\dfrac{(\alpha+2k)}{(R+\alpha+1)_{k+1}}\hat{\Omega}_{k,n}^{\alpha,\beta,R}, & 1\le k\le R+1\end{array}\right\}, \tag{3.16}$$

$$\hat{Q}_n^{\alpha,\beta,T}(x)=\sum_{k=0}^{T+1}\hat{b}_k\left(\frac{1-x}{2}\right)^k P_{n-k}^{\alpha+2k,\beta}(x),\ \hat{b}_k=\hat{b}_{k,n}^{\alpha,\beta,T}:=\hat{a}_{k,n}^{\alpha,\beta,T},\ 0\le k\le T+1, \tag{3.17}$$

$$\hat{Q}_n^{\alpha,\beta,R,T}(x) = \sum_{k=0}^{T+1} \hat{c}_k \left(\frac{1-x}{2}\right)^k P_{n-k}^{\alpha+2k,\beta}(x), \quad \hat{c}_k = \hat{c}_{k,n}^{\alpha,\beta,R,T}, \text{ where}$$

$$\hat{c}_{k,n}^{\alpha,\beta,R,T} := \left\{ \begin{array}{l} \hat{a}_0\,\hat{b}_0 - \sum_{j=1}^{R+1} \frac{\hat{h}_{j,n}^{\alpha,T}\,\hat{a}_j}{(n+\alpha+\beta+1)_j} \sum_{j=1}^{R} \frac{\hat{h}_{j,n}^{\alpha,R}\,\hat{b}_j}{(n+\alpha+\beta+1)_j},\ k=0 \\ \hat{b}_k \sum_{j=0}^{R+1} \frac{\hat{h}_{j,n}^{\alpha,T}\,\hat{a}_j}{(n+\alpha+\beta+1)_j} + \hat{a}_k \sum_{j=0}^{R} \frac{\hat{h}_{j,n}^{\alpha,R}\,\hat{b}_j}{(n+\alpha+\beta+1)_j},\ 1 \le k \le T+1 \\ \text{with } \hat{a}_k = 0,\ k \ge R+2 \end{array} \right\}. \tag{3.18}$$

***Proof*** Proceeding as in the proof of Theorem 2.2, we have to take account of the two specific quantities $\chi_{k,n}^{\alpha,\beta,r}$ and $2^{-2r}\psi_{k,n}^{\alpha,\beta,r}$ arising in (3.3). To this end, we modify the conditions (C1a)–(C3a), (C5a), (C9a)–(C10a) by multiplying each of the coefficients $\hat{a}_k, \hat{b}_k, \hat{c}_k$ by $\chi_{k,n}^{\alpha,\beta,r}$, while the conditions (C7a)–(C8a), (C13a)–(C14a) are altered by multiplication with $\psi_{k,n}^{\alpha,\beta,r}$. Furthermore, the conditions (C4a), (C6a), (C11a), (C12a) turn into

$$(C4c) \qquad \hat{a}_{R+1} = -\frac{g_n^{\alpha,R}\,\psi_{0,n}^{\alpha,\beta,R}}{2^{2R}\,(R+\alpha+1)_{R+1}\,\chi_{R+1,n}^{\alpha,\beta,R}} = -\frac{g_n^{\alpha,R}\,\hat{\Omega}_{R+1,n}^{\alpha,\beta,R}}{(R+\alpha+1)_{R+1}},$$

$$(C6c) \qquad \sum_{k=R+1}^{T+1} f_{k,n}^{R}(R+\alpha+1)_k\,\chi_{k,n}^{\alpha,\beta,R}\,\hat{c}_k + \frac{g_n^{\alpha,R}}{2^{2R}} \sum_{k=0}^{R} \hat{h}_{k,n}^{\alpha,R}\,\psi_{k,n}^{\alpha,\beta,R}\,\hat{b}_k = 0,$$

$$(C11c) \qquad \hat{b}_{T+1} = -\frac{g_n^{\alpha,T}\,\hat{\Omega}_{T+1,n}^{\alpha,\beta,T}}{(T+\alpha+1)_{T+1}},$$

$$(C12c) \qquad \hat{c}_{T+1} + \frac{g_n^{\alpha,T}}{2^{2T}\,\chi_{T+1,n}^{\alpha,\beta,T}\,(T+\alpha+1)_{T+1}} \sum_{k=0}^{R+1} \hat{h}_{k,n}^{\alpha,T}\,\psi_{k,n}^{\alpha,\beta,T}\,\hat{a}_k = 0.$$

Analogously to (2.17)–(2.18), a combination of (C1c) and (C4c) yields, for $1 \le k \le R+1$,

$$\hat{a}_k = \frac{q_{k,n}^{\alpha,R}\,(\alpha+2k)\,(R+\alpha+1)_{R+1}}{(R+\alpha+1)_{k+1}\,(n-R+\beta)_{R+1-k}}\,\hat{a}_{R+1} = -g_n^{\alpha,R}\,\frac{q_{k,n}^{\alpha,R}\,(\alpha+2k)\,\hat{\Omega}_{k,n}^{\alpha,\beta,R}}{(R+\alpha+1)_{k+1}}$$

and thus the representation of $\hat{a}_0$ via (C7c). In the same way we obtain (3.17). Concerning (3.18) we adopt again the arguments used in the proof of Theorem 2.1. □

## References

1. M. Alfaro, F. Marcellán, M.L. Rezola, A. Ronveaux, On orthogonal polynomials of Sobolev-type: algebraic properties and zeros. SIAM J. Math. Anal. **23**, 737–757 (1992)
2. M. Alfaro, F. Marcellán, M.L. Rezola, Orthogonal polynomials on Sobolev spaces: old and new directions. J. Comput. Appl. Math. **48**, 113–131 (1993)
3. J. Arvesú, R. Álvarez-Nodarse, F. Marcellán, K. Pan, Jacobi–Sobolev-type orthogonal polynomials: Second-order differential equation and zeros. J. Comput. Appl. Math. **90**, 135–156 (1998)
4. H. Bavinck, Differential operators having Sobolev type Laguerre polynomials as eigenfunctions. Proc. Am. Math. Soc. **125**, 3561–3567 (1997)
5. H. Bavinck, Differential operators having Laguerre type and Sobolev type Laguerre polynomials as eigenfunctions: a survey, in *Special Functions and Differential Equations*, ed. by K. Srinivasa Rao, et al. (Allied Publishers, New Dehli, 1998), pp. 102–118
6. H. Bavinck, Differential and difference operators having orthogonal polynomials with two linear perturbations as eigenfunctions. J. Comput. Appl. Math. **92**, 85–95 (1998)
7. H. Bavinck, Differential operators having Sobolev–Laguerre polynomials as eigenfunctions: new developments. J. Comput. Appl. Math. **133**, 183–193 (2001)
8. H. Bavinck, Differential operators having Sobolev-type Jacobi polynomials as eigenfunctions. J. Comput. Appl. Math. **151**, 271–295 (2003)
9. H. Bavinck, J. Koekoek, Differential operators having symmetric orthogonal polynomials as eigenfunctions. J. Comput. Appl. Math. **106**, 369–393 (1999)
10. H. Dueñas, F. Marcellán, The Laguerre–Sobolev-type orthogonal polynomials. J. Approx. Theory **162**, 421–440 (2010)
11. H. Dueñas, E.J. Huertas, F. Marcellán, Asymptotic properties of Laguerre–Sobolev-type orthogonal polynomials. Numer. Algor. **60**, 51–73 (2012)
12. A.J. Durán, M.D. De la Iglesia, Differential equations for discrete Laguerre–Sobolev orthogonal polynomials. J. Approx. Theory **195**, 70–88 (2015)
13. A.J. Durán, M.D. De la Iglesia, Differential equations for discrete Jacobi–Sobolev orthogonal polynomials. J. Spectr. Theory. **8**, 191–234 (2018)
14. A. Erdélyi, W. Magnus, F. Oberhettinger, F.G. Tricomi, *Higher Transcendental Functions*, vols. I, II (McGraw-Hill, New York, 1953)
15. W.N. Everitt, K.H. Kwon, L.L. Littlejohn, R. Wellman, Orthogonal polynomial solutions of linear ordinary differential equations. J. Comput. Appl. Math. **133**, 85–109 (2001)
16. R. Koekoek, Generalizations of Laguerre polynomials. J. Math. Anal. Appl. **153**, 576–590 (1990)
17. R. Koekoek, H.G. Meijer, A generalization of Laguerre polynomials. SIAM J. Math. Anal. **24**, 768–782 (1993)
18. J. Koekoek, R. Koekoek, H. Bavinck, On differential equations for Sobolev-type Laguerre polynomials. Trans. Am. Math. Soc. **350**, 347–393 (1998)
19. L.L. Littlejohn, J.F. Mañas-Mañas, J.J. Moreno-Balcázar, R. Wellman, Differential operator for discrete Gegenbauer–Sobolev orthogonal polynomials: eigenvalues and asymptotics. J. Approx. Theory **230**, 32–49 (2018)
20. J.F. Mañas-Mañas, F. Marcellán, J.J. Moreno-Balcázar, Aymptotic behavior of varying discrete Jacobi–Sobolev orthogonal polynomials. J. Comput. Appl. Math. **300**, 341–353 (2016)
21. J.F. Mañas-Mañas, J.J. Moreno-Balcázar, R. Wellman, Eigenvalue problem for discrete Jacobi–Sobolev orthogonal polynomials. Mathematics **8**, 182 (2020)
22. F. Marcellán, J.J. Moreno-Balcázar, Asymptotics and zeros of Sobolev orthogonal polynomials on unbounded supports. Acta Appl. Math. **94**, 163–192 (2006)
23. F. Marcellán, F.R. Rafaeli, Monotonicity and asymptotics of zeros of Laguerre–Sobolev-type orthogonal polynomials of higher order derivatives. Proc. Am. Math. Soc. **139**, 3929–3936 (2011)

24. F. Marcellán, A. Ronveaux, On a class of polynomials orthogonal with respect to a discrete Sobolev inner product. Indag. Math. N.S. **1**, 451–464 (1990)
25. F. Marcellán, Y. Xu, On Sobolev orthogonal polynomials. Expo. Math. **33**, 308–352 (2015)
26. F. Marcellán, R. Zejnullahu, B. Fejzullahu, E. Huertas, On orthogonal polynomials with respect to certain discrete Sobolev inner products. Pacific J. Math. **257**, 167–188 (2012)
27. C. Markett, The higher-order differential operator for the generalized Jacobi polynomials—new representation and symmetry. Indag. Math. **30**, 81–93 (2019)
28. C. Markett, On the differential equation for the Laguerre-Sobolev polynomials. J. Approx. Theory **247**, 48–67 (2019)
29. C. Markett, Symmetric differential operators for Sobolev orthogonal polynomials of Laguerre- and Jacobi-type, in *Proceedings of the 15th International Symposium on Orthogonal Polynomials, Special Functions and Applications OPSFA15, Hagenberg, Austria 2019* Integral Transforms and Special Functions (2021)
30. A. Martínez-Finkelshtein, Asymptotic properties of Sobolev orthogonal polynomials. J. Comput. Appl. Math. **99**, 491–510 (1998)
31. A. Martínez-Finkelshtein, Analytic aspects of Sobolev orthogonal polynomials revisited. J. Comput. Appl. Math. **127**, 255–266 (2001)
32. A. Ronveaux, Sobolev inner products and orthogonal polynomials of Sobolev type. Numer. Algorithms **3**, 393–400 (1992)

# Compactness, or Lack Thereof, for the Harmonic Double Layer

**Dorina Mitrea, Irina Mitrea, and Marius Mitrea**

*The authors are delighted to dedicate this article to Lance Littlejohn, whose warm friendship they much value and cherish, on the occasion of his 70-th birthday.*

**Abstract** Ahlfors regular domains which are infinitesimally flat, in a scale-invariant fashion, constitute the most general geometric context in which the harmonic double layer potential as well as other similar singular integral operators are compact in the framework of Lebesgue spaces. We review recent progress in this direction and, working with a drop-shaped domain, we give a direct, self-contained proof of the fact that the harmonic double layer potential fails to be compact on Lebesgue spaces in the presence of a single corner (or conical) singularity.

**Keywords** Laplacian · Harmonic double layer · Compact operator · Fredholm theory · Helmholtz operator · Acoustic double layer · Infinitesimally flat Ahlfors regular domain · Spectrum of double layer

## 1 Compactness of the Harmonic Double Layer Operator on Lebesgue Spaces

As part of his Ph.D. Thesis, Erik Ivar Fredholm has pioneered in 1898 the use of what nowadays is generally referred to as Fredholm theory as a tool to solve boundary value problems using singular integral operators of boundary layer type. A

D. Mitrea (✉) · M. Mitrea
Department of Mathematics, Baylor University, Waco, TX, USA
e-mail: Dorina_Mitrea@Baylor.edu; Marius_Mitrea@Baylor.edu

I. Mitrea
Department of Mathematics, Temple University, Philadelphia, PA, USA
e-mail: imitrea@temple.edu



F. Gesztesy, A. Martinez-Finkelshtein (eds.), *From Operator Theory to Orthogonal Polynomials, Combinatorics, and Number Theory*, Operator Theory: Advances and Applications 285, https://doi.org/10.1007/978-3-030-75425-9_17

key aspect in the implementation of this approach is establishing the compactness of various double layer potential operators in the functional analytic setting considered. In turn, the quality of being compact turns out to be subtly connected to the geometry (or rather, regularity) of the domain in question. To elaborate on this aspect, we need some basic definitions.

Throughout, $\mathcal{L}^n$ and $\mathcal{H}^{n-1}$ stand, respectively, for the $n$-dimensional Lebesgue measure and the $(n-1)$-dimensional Hausdorff measure in $\mathbb{R}^n$.

**Definition 1** Let $\Sigma \subseteq \mathbb{R}^n$ be an arbitrary closed set.

(i) Call $\Sigma$ `lower Ahlfors regular` provided there exists a constant $c \in (0, \infty)$ such that

$$c\,r^{n-1} \leq \mathcal{H}^{n-1}\big(B(x,r) \cap \Sigma\big) \text{ for each } x \in \Sigma \text{ and } r \in \big(0, 2\,\mathrm{diam}\,(\Sigma)\big). \tag{1}$$

(ii) Call $\Sigma$ `upper Ahlfors regular` if there exists $C \in (0, \infty)$ with the property that

$$\mathcal{H}^{n-1}\big(B(x,r) \cap \Sigma\big) \leq C\,r^{n-1} \text{ for each } x \in \Sigma \text{ and } r > 0. \tag{2}$$

(iii) Finally, call $\Sigma$ simply `Ahlfors regular` if it is both lower and upper Ahlfors regular.

The definition in (iii) has been first introduced by L. Ahlfors for planar curves, then subsequently considered by G. David for subsets of Euclidean spaces of arbitrary dimension. It should be pointed out that Ahlfors regularity is not a regularity property per se, but rather a scale-invariant way of expressing the fact that the set in question is $(n-1)$-dimensional in a uniform, scale-invariant fashion, involving the Hausdorff measure.

We continue by introducing the notion of uniform rectifiability of G. David and S. Semmes. The following is a slight variant of the original definition in [10, p. 13].

**Definition 2** Call $\Sigma \subset \mathbb{R}^n$ a `uniformly rectifiable` (UR) `set` provided $\Sigma$ is closed, upper Ahlfors regular, and has Big Pieces of Lipschitz Images (BPLI). The latter property signifies the existence of $\varepsilon > 0$ and $M \in (0, \infty)$ (collectively referred to as the UR constants of $\Sigma$) such that, for each location $x \in \Sigma$ and each scale $r \in (0, 2\,\mathrm{diam}\,\Sigma)$, one can find a Lipschitz map $\Phi : B_{n-1}(0', r) \to \mathbb{R}^n$ (where $B_{n-1}(0', r)$ is the $(n-1)$-dimensional ball of radius $r$ centered at the origin $0'$ in $\mathbb{R}^{n-1}$), having Lipschitz constant $\leq M$, with the property that

$$\mathcal{H}^{n-1}\Big(\Sigma \cap B(x,r) \cap \Phi\big(B_{n-1}(0', r)\big)\Big) \geq \varepsilon r^{n-1}. \tag{3}$$

Let us now introduce the class of uniformly rectifiable domains (UR domains, for short). To set the stage, recall that $\partial_* \Omega$ denotes the measure theoretic boundary of a Lebesgue measurable set $\Omega \subseteq \mathbb{R}^n$, i.e., (see, e.g., [13, Definition p. 208])

$$\partial_* \Omega := \Big\{ x \in \mathbb{R}^n : \limsup_{r \to 0^+} \frac{\mathcal{L}^n(B(x,r) \cap \Omega)}{r^n} > 0 \text{ and } \limsup_{r \to 0^+} \frac{\mathcal{L}^n(B(x,r) \setminus \Omega)}{r^n} > 0 \Big\}. \tag{4}$$

Thus, near points in $\partial_* \Omega$ there is enough mass both in $\Omega$ and in $\mathbb{R}^n \setminus \Omega$ (relative to the scale).

Next, as in [22, § 2.2] we make the following definitions.

**Definition 3** Call a nonempty open subset $\Omega$ of $\mathbb{R}^n$ an `Ahlfors regular domain` if $\partial \Omega$ is an Ahlfors regular set and the geometric measure theoretic boundary of $\Omega$ has full $\mathcal{H}^{n-1}$-measure into the topological boundary of $\Omega$, i.e., $\mathcal{H}^{n-1}(\partial \Omega \setminus \partial_* \Omega) = 0$.

Hence, a nonempty open set $\Omega \subseteq \mathbb{R}^n$ is an Ahlfors regular domain provided $\partial \Omega$ is an Ahlfors regular set (which, in particular, renders $\Omega$ a set of locally finite perimeter) and the geometric measure theoretic outward unit normal $\nu$ to $\Omega$ is well defined at $\mathcal{H}^{n-1}$-a.e. point on $\partial \Omega$.

Following [16, Definition 3.7, p. 2631], we now make the following definition:

**Definition 4** Call a nonempty open subset $\Omega$ of $\mathbb{R}^n$ a UR `domain` provided $\partial \Omega$ is a UR set (cf. Definition 2) and $\partial_* \Omega$ has full measure (relative to the $(n-1)$-dimensional Hausdorff measure) in the topological boundary $\partial \Omega$, i.e.,

$$\mathcal{H}^{n-1}(\partial \Omega \setminus \partial_* \Omega) = 0. \tag{5}$$

We emphasize that, by definition, the topological boundary of any UR domain is an Ahlfors regular set. As such, any UR domain is an Ahlfors regular domain. Also, any open set whose boundary may be locally described in terms of functions with first-order derivatives in $\mathrm{BMO}(\mathbb{R}^{n-1})$ happens to be a UR domain. In particular, any Lipschitz domain is a UR domain. More generally, work in [9] guarantees that

> any open set $\Omega \subseteq \mathbb{R}^n$ with an upper Ahlfors regular boundary and satisfying a two-sided corkscrew condition (as defined in [17]) is a UR domain. (6)

In the two-dimensional setting, any bounded chord-arc domain in $\mathbb{C} \equiv \mathbb{R}^2$ is a UR domain. In this regard, recall that a set $\Sigma \subseteq \mathbb{R}^2$ is a `chord-arc curve` provided $\Sigma$ is a simple locally rectifiable closed curve satisfying

$$\sup_{\substack{z_1 \neq z_2 \\ z_1, z_2 \in \Sigma}} \frac{\ell(z_1, z_2)}{|z_1 - z_2|} < +\infty, \tag{7}$$

where $\ell(z_1, z_2)$ is the length of the shorter subarc of $\Sigma$ joining $z_1$ with $z_2$. Also, an open set $\Omega \subseteq \mathbb{R}^2$ is called a `chord-arc domain` provided $\Omega$ is either the inner domain or the outer domain of a chord-arc curve.

Further examples of two-dimensional UR domains are offered by the following result from [23].

**Proposition 1** *Assume that $\Omega \subseteq \mathbb{C}$ is a connected, bounded, open set, whose boundary is a finite union of mutually disjoint, upper Ahlfors regular, Jordan curves, each of which is the boundary of a connected component of $\mathbb{C} \setminus \Omega$. Then $\Omega$ is a UR domain.*

Returning to the mainstream discussion, let us fix some $p \in (1, \infty)$ and consider the $L^p$-Dirichlet problem for the Laplacian $\Delta := \partial_1^2 + \cdots + \partial_n^2$ in a UR domain $\Omega$ of $\mathbb{R}^n$ with arbitrary data $f \in L^p(\partial\Omega, \mathcal{H}^{n-1})$, namely:

$$\begin{cases} u \in \mathscr{C}^\infty(\Omega), \\ \Delta u = 0 \ \text{ in } \ \Omega, \\ \mathcal{N}_\kappa u \in L^p(\partial\Omega, \mathcal{H}^{n-1}), \\ u\Big|_{\partial\Omega}^{\kappa-\text{n.t.}} = f \ \text{ at } \mathcal{H}^{n-1}\text{-a.e. point on } \ \partial\Omega. \end{cases} \tag{8}$$

Above, $\kappa \in (0, \infty)$ is a fixed aperture parameter, used to define the nontangential maximal operator

$$\big(\mathcal{N}_\kappa u\big)(x) := \|u\|_{L^\infty(\Gamma_\kappa(x), \mathcal{L}^n)} \ \text{ for each } \ x \in \partial\Omega, \tag{9}$$

where $\Gamma_\kappa(x)$ is the nontangential approach region with vertex at $x$, given by

$$\Gamma_\kappa(x) := \big\{y \in \Omega : \ |y - x| < (1 + \kappa)\,\text{dist}\,(y, \partial\Omega)\big\} \ \text{ for each } \ x \in \partial\Omega. \tag{10}$$

Also, for each point $x \in \partial\Omega$ such that $x \in \overline{\Gamma_\kappa(x)}$ (i.e., $x$ is an accumulation point for the nontangential approach region $\Gamma_\kappa(x)$) we shall say that the nontangential limit of $u$ at $x$ from within $\Gamma_\kappa(x)$ exists, and its value is $\big(u\big|_{\partial\Omega}^{\kappa-\text{n.t.}}\big)(x) \in \mathbb{C}$, provided

$$\begin{gathered} \text{for each } \varepsilon > 0 \text{ there exists } r > 0 \text{ so that } \Big|u(y) - \big(u\big|_{\partial\Omega}^{\kappa-\text{n.t.}}\big)(x)\Big| < \varepsilon \\ \text{for } \mathcal{L}^n\text{-a.e. point } y \in \Gamma_\kappa(x) \cap B(x, r). \end{gathered} \tag{11}$$

The essence of the boundary integral method is to seek a solution for (8) in the form

$$u(x) := \frac{1}{\omega_{n-1}} \int_{\partial\Omega} \frac{\langle \nu(y), y - x\rangle}{|x - y|^n} g(y)\, d\mathcal{H}^{n-1}(y), \qquad x \in \Omega, \tag{12}$$

where $\omega_{n-1}$ is the area of the unit sphere in $\mathbb{R}^n$, $\nu$ is the outward unit normal vector to $\Omega$ (considered in a geometric measure theoretic sense; see the discussion in, e.g., [13]), and the function $g \in L^p(\partial\Omega, \mathcal{H}^{n-1})$ is yet to be determined. The benefit of this choice is then apparent from the realization that, irrespective of the choice of $g$, the function $u$ in (12) satisfies all but the last condition in (8). In addition, as far as this final property is concerned, it turns out that (see [24] for a proof)

$$u\Big|_{\partial\Omega}^{\kappa-\mathrm{n.t.}} = \left(\tfrac{1}{2}I + K\right)g \ \text{ at } \mathcal{H}^{n-1}\text{-a.e. point on } \partial\Omega, \tag{13}$$

where $I$ denotes the identity operator and $K$, the harmonic double layer potential operator, is the principal-value singular integral operator given by

$$Kg(x) := \lim_{\varepsilon\to 0^+} \frac{1}{\omega_{n-1}} \int_{\partial\Omega\setminus\overline{B(x,\varepsilon)}} \frac{\langle \nu(y), y-x\rangle}{|x-y|^n} g(y)\, d\mathcal{H}^{n-1}(y), \quad x \in \partial\Omega. \tag{14}$$

In view of (13), matters are then reduced to solving the boundary integral equation

$$\left(\tfrac{1}{2}I + K\right)g = f \ \text{ in } \ L^p(\partial\Omega, \mathcal{H}^{n-1}). \tag{15}$$

If $K$ happens to be compact on $L^p(\partial\Omega, \mathcal{H}^{n-1})$ then $\frac{1}{2}I + K$ is a Fredholm operator with index zero on $L^p(\partial\Omega, \mathcal{H}^{n-1})$, hence its invertibility on this space hinges on whether said operator is injective, or has dense range. Since the latter properties are more tractable, this provides a feasible plan of attack for establishing the solvability of (8), in the form (12), for the choice $g := \left(\frac{1}{2}I + K\right)^{-1} f$.

When $\Omega$ is a bounded domain of class $\mathscr{C}^{1,\alpha}$ for some $\alpha \in (0,1)$ (aka Lyapunov domain), its topological boundary $\partial\Omega$ may be locally described as the graph of a real-valued function $\phi \in \mathscr{C}^{1,\alpha}$. Bearing in mind that, under such an identification, we have $\nu(y) = \frac{(\nabla\phi(y'),-1)}{\sqrt{1+|\nabla\phi(y')|^2}}$ for each $y = (y', \phi(y')) \in \partial\Omega \cap \operatorname{graph}\phi$, whenever $x, y \in \partial\Omega \cap \operatorname{graph}\phi$ we may estimate

$$\begin{aligned}
|\langle \nu(y), y-x\rangle| &\le \left|\left\langle \nu\big(y', \phi(y')\big), \big(y'-x', \phi(y')-\phi(x')\big)\right\rangle\right| \\
&\le \left|\big(\nabla\phi(y'), -1\big)\cdot\big(y'-x', \phi(y')-\phi(x')\big)\right| \\
&= \left|\phi(x') - \phi(y') - \nabla\phi(y')\cdot(x'-y')\right| \\
&= \left|\int_0^1 \frac{d}{dt}\big[\phi(y'+t(x'-y'))\big]\,dt - \nabla\phi(y')\cdot(x'-y')\right|
\end{aligned}$$

$$\leq |x'-y'|\left(\int_0^1 \left|(\nabla\phi)(y'+t(x'-y'))-\nabla\phi(y')\right| dt\right)$$

$$\leq \|\nabla\phi\|_{\dot{\mathscr{C}}^\alpha(\mathbb{R}^{n-1})}|x-y|^{1+\alpha}, \tag{16}$$

where $\|\nabla\phi\|_{\dot{\mathscr{C}}^\alpha(\mathbb{R}^{n-1})}$ is the homogeneous Hölder semi-norm of $\nabla\phi$ in $\mathbb{R}^{n-1}$, i.e.,

$$\|\nabla\phi\|_{\dot{\mathscr{C}}^\alpha(\mathbb{R}^{n-1})} := \sup_{\substack{x',y'\in\mathbb{R}^{n-1}\\ x'\neq y'}} \frac{|(\nabla\phi)(x')-(\nabla\phi)(y')|}{|x'-y'|^\alpha} < +\infty. \tag{17}$$

Note that (16) is better than what the Cauchy–Schwarz estimate for the dot product of two vectors would ordinarily give, namely $|\langle\nu(y), y-x\rangle| \leq |x-y|$. Employing the Lebesgue Number Theorem (for the finite open cover of the compact set $\partial\Omega$ constituting of graphs of various functions of class $\mathscr{C}^{1,\alpha}$), we see that a similar inequality as the one derived in (16) is also valid when $x, y \in \partial\Omega$ are "far" apart (simply by adjusting constants). As a result, the integral kernel of $K$ is $O(|x-y|^{n-1-\alpha})$ as $|x-y| \to 0^+$, hence weakly singular. Granted this, elementary considerations (see [16, Lemma 2.20, p. 2608]) then show that $K$ is compact on $L^p(\partial\Omega, \mathcal{H}^{n-1})$ in this scenario. That $K$ has a weakly singular kernel remains true when the boundary has defining functions whose first derivatives have a Dini modulus of continuity, but fails for boundaries that are merely of class $\mathscr{C}^1$ (see also [3, 18, 20] and the references therein for some related, early work). Nonetheless, using more sophisticated tools, Fabes, Jodeit, and Rivière have succeeded to show in [14] that, in spite of not being weakly singular, $K$ does remain compact on $L^p(\partial\Omega, \mathcal{H}^{n-1})$ for each $p \in (1,\infty)$ whenever $\Omega$ is a bounded $\mathscr{C}^1$ domain. The main idea in [14] was to construct a sequence of bounded $\mathscr{C}^\infty$ domains $\{\Omega_j\}_{j\in\mathbb{N}}$ such that $\Omega_j \nearrow \Omega$ as $j \nearrow \infty$, with the property that if $K_j$ is the harmonic double layer associated with $\Omega_j$ suitably transported (via changes of variables) to $\partial\Omega$, then $\lim_{j\to\infty} \|K_j - K\|_{L^p(\partial\Omega,\mathcal{H}^{n-1})\to L^p(\partial\Omega,\mathcal{H}^{n-1})} = 0$. Given that, as noted earlier, each $K_j$ is compact, and given that the space of compact operators on $L^p(\partial\Omega, \mathcal{H}^{n-1})$ is closed (in the operator norm), this ultimately implies that $K$ is compact on $L^p(\partial\Omega, \mathcal{H}^{n-1})$.

For rougher sets, the very boundedness of $K$ on the space $L^p(\partial\Omega, \mathcal{H}^{n-1})$ is in question. In this direction, Calderón initiated in [4] a breakthrough, proving $L^p$-bounds for Cauchy integral operators on Lipschitz curves with small Lipschitz constant (in this vein, it is worth recalling that the harmonic double layer is the real part of the Cauchy operator acting on real-valued functions). Subsequently, Coifman, McIntosh, and Meyer have extended in [5] Calderón's result on Cauchy integrals to general Lipschitz curves, and applied this to such results as the boundedness of $K$ from (14) on $L^p(\partial\Omega, \mathcal{H}^{n-1})$ for each $p \in (1,\infty)$, whenever $\Omega$ is an arbitrary Lipschitz domain in $\mathbb{R}^n$. Since then there have been further developments regarding surfaces more general than the boundaries of Lipschitz domains.

A fundamental question emerging from this body of work is that of identifying optimal geometric demands on the domain $\Omega \subseteq \mathbb{R}^n$ guaranteeing the $L^p$-boundedness of such singular integral operators as $K$. From the work of David [7, 8], of David-Jerison [9], of David-Semmes [10, 11], of Semmes [28], and of Nazarov, Tolsa, Volberg [25], we now know that the most general class of "surfaces" on which said operators are bounded on Lebesgue spaces is the category of uniformly rectifiable sets (see Definition 2). This work has interfaced tightly with geometric measure theory, and has been applied to problems in partial differential equations for the first time in [16].

As is apparent from the theorem below (obtained by combining [10, Theorem, pp. 10–14] with the main result in [25]), there is a strong, two-way link, between singular integral operators on Lebesgue spaces and uniform rectifiability.

**Theorem 1** *Given a closed set $\Sigma \subseteq \mathbb{R}^n$ which is Ahlfors regular, the following conditions are equivalent:*

*(i) $\Sigma$ is a UR set;*
*(ii) The maximal singular integral operator*

$$\big(T_{k,\max} f\big)(x) := \sup_{\varepsilon>0}\Big|\int_{y\in\Sigma\setminus\overline{B(x,\varepsilon)}} k(x-y)f(y)\,\mathcal{H}^{n-1}(y)\Big| \quad \textit{for } x\in\Sigma, \tag{18}$$

*is bounded on $L^2\big(\Sigma,\mathcal{H}^{n-1}\big)$ for each integral kernel*

$$\begin{aligned} &k\in\mathscr{C}^\infty(\mathbb{R}^n\setminus\{0\}) \ \textit{ which is odd and satisfies}\\ &\sup_{x\in\mathbb{R}^n\setminus\{0\}}|x|^{n-1+\ell}|(\nabla^\ell k)(x)|<+\infty \ \textit{ for all } \ \ell\in\mathbb{N}_0. \end{aligned} \tag{19}$$

*(iii) The maximal singular integral operator $T_{k,\max}$ from (18) is bounded on the space $L^2\big(\Sigma,\mathcal{H}^{n-1}\big)$ for each integral kernel*

$$\begin{aligned} &k\in\mathscr{C}^\infty(\mathbb{R}^n\setminus\{0\}) \ \textit{ which is odd and}\\ &\textit{positive homogeneous of degree } 1-n. \end{aligned} \tag{20}$$

*(iv) The maximal singular integral operator $T_{k,\max}$ from (18) is bounded on the space $L^2\big(\Sigma,\mathcal{H}^{n-1}\big)$ for each integral kernel of the form $k(x) := x_j/|x|^n$ for $x\in\mathbb{R}^n\setminus\{0\}$, where $j\in\{1,\dots,n\}$.*

*Furthermore, if $\Sigma$ is a UR set then the principal-value versions of the maximal operators considered above are all given by well-defined limits at $\mathcal{H}^{n-1}$-a.e. point on $\Sigma$.*

This settles the issue of $L^p$-boundedness. Dealing with the stronger property of compactness on $L^p$ requires that we consider singular integral operators on boundaries of UR domains in $\mathbb{R}^n$ whose kernels depend linearly on the geometric measure theoretic outward unit normal of the domain in question (this is a harmless assumption; see (22) below). Within this class of singular integral operators, it turns out that those who actually become *compact* when considered on bounded smooth domains are precisely the ones which have a similar algebraic format to the harmonic double layer operator (14). Specifically, the following result has been proved in [24]:

**Proposition 2** *Pick some $n \in \mathbb{N}$ with $n \geq 2$ along with $N \in \mathbb{N}$, and consider a vector-valued function*

$$\mathbf{k} \in \left[\mathscr{C}^N(\mathbb{R}^n \setminus \{0\})\right]^n, \quad \textit{odd, positive homogeneous of degree } 1-n. \tag{21}$$

*Also, for each UR domain $\Omega \subseteq \mathbb{R}^n$ consider the principal-value singular integral operator*

$$\begin{gathered} T_\Omega : L^2(\partial\Omega, \mathcal{H}^{n-1}) \longrightarrow L^2(\partial\Omega, \mathcal{H}^{n-1}) \ \textit{defined} \\ \textit{for each } \ f \in L^2(\partial\Omega, \mathcal{H}^{n-1}) \ \textit{and } \mathcal{H}^{n-1}\textit{-a.e. } \ x \in \partial\Omega \ \textit{ as} \\ (T_\Omega f)(x) := \lim_{\varepsilon \to 0^+} \int\limits_{\partial\Omega \setminus \overline{B(x,\varepsilon)}} \langle \nu_\Omega(y), \mathbf{k}(x-y) \rangle f(y)\, d\mathcal{H}^{n-1}(y) \end{gathered} \tag{22}$$

*where $\nu_\Omega$ is the geometric measure theoretic outward unit normal vector to $\Omega$.*
*Then*

$$\begin{gathered} \textit{the operator } T_\Omega \textit{ is compact whenever} \\ \Omega \subseteq \mathbb{R}^n \textit{ is a bounded domain of class } \mathscr{C}^\infty \end{gathered} \tag{23}$$

*if and only if there exists*

$$\begin{gathered} k \in \mathscr{C}^N(\mathbb{R}^n \setminus \{0\}) \textit{ scalar-valued function, even,} \\ \textit{positive homogeneous of degree } -n, \textit{ and satisfying} \\ \mathbf{k}(x) = x\, k(x) \ \textit{ at each point } \ x \in \mathbb{R}^n \setminus \{0\}. \end{gathered} \tag{24}$$

The structural property in the last line of (24) implies that the principal-value singular integral operator $T_\Omega$ from the last line of (22) acquires the form

$$(T_\Omega f)(x) = \lim_{\varepsilon \to 0^+} \int\limits_{\partial\Omega \setminus \overline{B(x,\varepsilon)}} \frac{\langle \nu_\Omega(y), x-y \rangle}{|x-y|^n} k\Big(\tfrac{x-y}{|x-y|}\Big) f(y)\, d\mathcal{H}^{n-1}(y) \tag{25}$$

for each $f \in L^2(\partial\Omega, \mathcal{H}^{n-1})$ and $\mathcal{H}^{n-1}$-a.e. $x \in \partial\Omega$, which should be thought of as a "generalized" harmonic double layer (this reduces to (14) when $k(z) := \omega_{n-1}^{-1}|z|^{-n}$ for each $z \in \mathbb{R}^n \setminus \{0\}$).

Proposition 2 brings to prominence the class of singular integral operators whose integral kernel contains, as a factor, the dot product between the unit normal $\nu(y)$ and the chord $x - y$. These have been called "chord-dot-normal" singular integral operators in [24].

To proceed, we briefly discuss the John-Nirenberg space of functions of bounded mean oscillations. Consider a closed set $\Sigma \subseteq \mathbb{R}^n$ which is Ahlfors regular and, with each function $f \in L^1_{\rm loc}(\Sigma, \mathcal{H}^{n-1})$ associate the quantity (which may happen to be $+\infty$)

$$\|f\|_{\dot{\rm BMO}(\Sigma,\mathcal{H}^{n-1})} := \sup_{x\in\Sigma,\, r>0} ⨍_{\Sigma\cap B(x,r)} \Big| f - ⨍_{\Sigma\cap B(x,r)} f\, d\mathcal{H}^{n-1} \Big|\, d\mathcal{H}^{n-1}, \tag{26}$$

where a barred integral indicates integral average. Going further, for each function $f \in L^1_{\rm loc}(\Sigma, \mathcal{H}^{n-1})$ let us set

$$\|f\|_{{\rm BMO}(\Sigma,\mathcal{H}^{n-1})} := \begin{cases} \|f\|_{\dot{\rm BMO}(\Sigma,\mathcal{H}^{n-1})} & \text{if } \Sigma \text{ is unbounded,} \\ \|f\|_{\dot{\rm BMO}(\Sigma,\mathcal{H}^{n-1})} + \Big| \int_\Sigma f\, d\mathcal{H}^{n-1} \Big| & \text{if } \Sigma \text{ is bounded.} \end{cases} \tag{27}$$

We are now in a position to introduce ${\rm BMO}(\Sigma, \mathcal{H}^{n-1})$, the space of functions of bounded mean oscillations on $\Sigma$ as

$${\rm BMO}(\Sigma, \mathcal{H}^{n-1}) := \Big\{ f \in L^1_{\rm loc}(\Sigma, \mathcal{H}^{n-1}) : \|f\|_{{\rm BMO}(\Sigma,\mathcal{H}^{n-1})} < +\infty \Big\}. \tag{28}$$

Finally, define the Sarason space ${\rm VMO}(\Sigma, \mathcal{H}^{n-1})$ of functions of vanishing mean oscillations on $\Sigma$ (cf. [27]) as the closure of

$$\big\{ f \in {\rm BMO}(\Sigma, \mathcal{H}^{n-1}) : f \text{ uniformly continuous} \big\} \tag{29}$$

in the space ${\rm BMO}(\Sigma, \mathcal{H}^{n-1})$.

Our next theorem accomplishes the task of estimating the essential norm of chord-dot-normal singular integral operators acting on Lebesgue spaces in terms of the distance (measured in the John-Nirenberg space of functions with bounded mean oscillations) from the geometric measure theoretic outward unit normal to the Sarason space of functions with vanishing mean oscillations. In essence, this offers a quantified version of the heuristic principle stating that: *the flatter (at an infinitesimal level) the boundary, the closer the chord-dot-normal singular integral operator to being compact*. To state it, we first introduce some notation and make some remarks. Specifically, with $e$ denoting the base of natural logarithms, for each $m \in \mathbb{N}_0$ and $t \in [0, \infty)$ let us define

$$t^{\langle 0 \rangle} := 1 \tag{30}$$

and, if $m \geq 1$,

$$t^{\langle m \rangle} := \begin{cases} 0 & \text{if } t = 0, \\ t \cdot \underbrace{\ln\Big(\cdots \ln\big(\ln(}_{m \text{ natural logarithms}} 1/t)\big) \cdots\Big) & \text{if } 0 < t \leq ({}^m e)^{-1}, \\ ({}^m e)^{-1} & \text{if } t > ({}^m e)^{-1}, \end{cases} \tag{31}$$

where ${}^m e$ is the $m$-th tetration of $e$ (involving $m$ copies of $e$, combined by exponentiation), i.e.,

$${}^m e := \underbrace{e^{e^{\cdot^{\cdot^{\cdot^{e}}}}}}_{m \text{ copies of } e}, \quad \text{the } m\text{-th fold exponentiation of } e. \tag{32}$$

We also agree to set ${}^0 e := 1$. Ergo, inductively, for each $m \in \mathbb{N}_0$ and each $t \in [0, \infty)$ we have

$$t^{\langle m+1 \rangle} = \begin{cases} 0 & \text{if } t = 0, \\ t \cdot \ln\big(t^{\langle m \rangle}/t\big) & \text{if } 0 < t \leq ({}^{m+1} e)^{-1}, \\ ({}^{m+1} e)^{-1} & \text{if } t > ({}^{m+1} e)^{-1}. \end{cases} \tag{33}$$

Elementary calculus gives that this function enjoys the following properties:

$$\begin{gathered} [0, \infty) \ni t \longmapsto t^{\langle m \rangle} \in [0, \infty) \ \text{ is continuous, non-decreasing,} \\ \text{and vanishes continuously at the origin,} \end{gathered} \tag{34}$$

$$\begin{gathered} t^{\langle m \rangle} \leq t^{\langle m-1 \rangle} \leq \cdots \leq t^{\langle 1 \rangle} \leq \big(e^{\varepsilon - 1}/\varepsilon\big) \cdot t^{1-\varepsilon} \\ \text{for each } \ t \in [0, \infty), \ m \in \mathbb{N}, \ \varepsilon \in (0, 1), \end{gathered} \tag{35}$$

$$t \leq \max\{1, ({}^m e) t\} \cdot t^{\langle m \rangle} \quad \text{for all } \ t \in [0, \infty) \ \text{ and } \ m \in \mathbb{N}_0, \tag{36}$$

$$(\lambda t)^{\langle m \rangle} \leq \lambda t^{\langle m \rangle} \quad \text{for all } \ t \in [0, \infty), \ m \in \mathbb{N}_0, \ \text{ and } \ \lambda \in [1, \infty), \tag{37}$$

$$\begin{gathered} (t^{\alpha})^{\langle m \rangle} \leq t^{\alpha} \cdot \underbrace{\ln\Big(\cdots \ln\big(\ln(}_{m \text{ natural logarithms}} 1/\min\{t, ({}^m e)^{-1}\})\big) \cdots\Big) \\ \text{for all numbers } \ t \in [0, \infty), \ m \in \mathbb{N}, \ \text{ and } \ \alpha \in (0, 1). \end{gathered} \tag{38}$$

We are now ready to state the major result alluded to above (a proof of which appears in [24]):

**Theorem 2** *Let $\Omega \subseteq \mathbb{R}^n$ be a UR domain with compact boundary and denote by $\nu$ the geometric measure theoretic outward unit normal to $\Omega$. Fix an integrability exponent $p \in (1, \infty)$, and consider a sufficiently large integer $N = N(n) \in \mathbb{N}$. Given a complex-valued function $k \in \mathscr{C}^N(\mathbb{R}^n \setminus \{0\})$ which is even and positive homogeneous of degree $-n$, consider the singular integral operators $T$, $T^{\#}$ whose actions on each given function $f \in L^p(\partial\Omega, \mathcal{H}^{n-1})$ are, respectively, defined for $\mathcal{H}^{n-1}$-a.e. $x \in \partial\Omega$ as*

$$Tf(x) := \lim_{\varepsilon\to 0^+} \int_{\substack{y\in\partial\Omega \\ |x-y|>\varepsilon}} \langle x - y, \nu(y)\rangle k(x - y) f(y)\, d\mathcal{H}^{n-1}(y) \tag{39}$$

*and*

$$T^{\#} f(x) := \lim_{\varepsilon\to 0^+} \int_{\substack{y\in\partial\Omega \\ |x-y|>\varepsilon}} \langle y - x, \nu(x)\rangle k(y - x) f(y)\, d\mathcal{H}^{n-1}(y). \tag{40}$$

*Then for each $m \in \mathbb{N}$ there exists some $C_m \in (0, \infty)$ which depends only on $m$, $n$, $p$, the UR constants of $\partial\Omega$, and the relative flatness ratio $\wp(\Omega)/\mathrm{diam}\,\partial\Omega$ (cf. [24] for a definition of the geometric entity $\wp(\Omega)$) such that, with the piece of notation introduced in* (31)*, one has*

$$\begin{aligned} &\mathrm{dist}\big(T\,,\ \mathrm{Cp}(L^p(\partial\Omega, \mathcal{H}^{n-1}))\big) \\ &\qquad \le C_m\Big(\sum_{|\alpha|\le N} \sup_{S^{n-1}} |\partial^\alpha k|\Big)\Big[\mathrm{dist}\big(\nu\,,\ \mathrm{VMO}(\partial\Omega, \mathcal{H}^{n-1})\big)\Big]^{\langle m\rangle} \end{aligned} \tag{41}$$

*and*

$$\begin{aligned} &\mathrm{dist}\big(T^{\#}\,,\ \mathrm{Cp}(L^p(\partial\Omega, \mathcal{H}^{n-1}))\big) \\ &\qquad \le C_m\Big(\sum_{|\alpha|\le N} \sup_{S^{n-1}} |\partial^\alpha k|\Big)\Big[\mathrm{dist}\big(\nu\,,\ \mathrm{VMO}(\partial\Omega, \mathcal{H}^{n-1})\big)\Big]^{\langle m\rangle}. \end{aligned} \tag{42}$$

*Above,* $\mathrm{Cp}(L^p(\partial\Omega, \mathcal{H}^{n-1}))$ *is the space of compact operators on $L^p(\partial\Omega, \mathcal{H}^{n-1})$, the distances in the left-hand side are measured in the operator norm in the space of linear and bounded operators on $L^p(\partial\Omega, \mathcal{H}^{n-1})$, and the distances in the right-hand side are measured in* $\mathrm{BMO}(\partial\Omega, \mathcal{H}^{n-1})$.

The above result identifies a general class of sets providing a natural environment where the size of singular integral operators (of "chord-dot-normal" type) may be directly linked to geometry. In this regard, we recall the following definition, which appears in [24].

**Definition 5** Fix $n \in \mathbb{N}$ with $n \geq 2$ and consider an Ahlfors regular domain $\Omega \subseteq \mathbb{R}^n$ with compact boundary and denote by $\nu$ the geometric measure theoretic outward unit normal to $\Omega$.

(1) Given $\delta \in (0,1)$, call $\Omega$ a $\delta$-`oscillating AR domain` (or $\delta$-`osc AR`, for short) provided

$$\operatorname{dist}\big(\nu\,,\ \mathrm{VMO}(\partial\Omega, \mathcal{H}^{n-1})\big) < \delta, \tag{43}$$

where the distance is measured in the John-Nirenberg space $\mathrm{BMO}(\partial\Omega, \sigma)$.

(2) Call $\Omega$ an `infinitesimally flat AR domain` (or `inf-flat AR`, for short) provided

$$\nu \in \mathrm{VMO}(\partial\Omega, \mathcal{H}^{n-1}). \tag{44}$$

We emphasize that, by definition, any $\delta$-oscillating AR domain is an open set with compact boundary. Clearly, an infinitesimally flat AR domain is a $\delta$-oscillating AR domain for any $\delta \in (0,1)$.

To offer some significant examples of $\delta$-oscillating AR domains, and infinitesimally flat AR domains, stay in the two-dimensional setting. Let $\Omega \subseteq \mathbb{R}^2$ be a chord-arc domain with compact boundary, and denote by $\ell(\cdot,\cdot)$ the shortest arc-length between points on $\partial\Omega$. Since $\Omega$ is known to be an Ahlfors regular domain, it makes sense to consider the geometric measure theoretic outward unit normal $\nu$ to $\Omega$. Finally, introduce

$$\varkappa := \lim_{R\to 0^+}\Bigg\{\sup_{\substack{z_1,z_2\in\partial\Omega\\ |z_1-z_2|<R}}\Big(\frac{\ell(z_1,z_2)}{|z_1-z_2|}-1\Big)\Bigg\} \in [0,+\infty). \tag{45}$$

In this context, work in [24] implies that, with the distance measured in the John-Nirenberg space $\mathrm{BMO}(\partial\Omega, \mathcal{H}^1)$, we have

$$\operatorname{dist}\Big(\nu\,,\ \mathrm{VMO}(\partial\Omega, \mathcal{H}^1)\Big) = O(\varkappa^{1/2}) \ \text{ as } \ \varkappa \to 0^+, \tag{46}$$

where the implicit constant depends exclusively on the upper Ahlfors regularity constant of $\partial\Omega$. In particular,

$$\Omega \text{ is a } \delta\text{-oscillating AR domain with } \ \delta = O(\varkappa^{1/2}) \ \text{ as } \ \varkappa \to 0^+. \tag{47}$$

Recall that a chord-arc domain $\Omega \subseteq \mathbb{C}$ is said to have `vanishing constant` provided $\varkappa$ in (45) is zero (hence, the manner in which the length of the chord may be used as a substitute for the length of the arc improves indefinitely as the distance between the two points on the curve tends to zero; note that this is not the case in the presence of a corner). Then, as a corollary of the above results,

any chord-arc domain with vanishing constant and compact boundary is an infinitesimally flat AR domain in the plane. (48)

In the converse direction, suppose $\Omega \subseteq \mathbb{R}^2$ is a $\delta$-oscillating AR domain, assumed to be simply connected when $\Omega$ is bounded and, respectively, connected when $\Omega$ is unbounded. Denote by $\nu$ the geometric measure theoretic outward unit normal to $\Omega$. Then work in [24] guarantees that $\Omega$ is a chord-arc domain with

$$\lim_{R\to 0^+}\Bigg\{\sup_{\substack{z_1,z_2\in\partial\Omega\\ |z_1-z_2|<R}}\Big(\frac{\ell(z_1,z_2)}{|z_1-z_2|}-1\Big)\Bigg\}=O\big(\delta\ln(1/\delta)\big)\ \text{ as }\ \delta\to 0^+. \tag{49}$$

In particular,

any bounded, simply connected, infinitesimally flat AR domain in the plane is a chord-arc domain with vanishing constant. (50)

Finally, from (48) and (50) we conclude that

the class of bounded, simply connected, infinitesimally flat AR domains in the plane coincides with the class of bounded chord-arc domains with vanishing constant. (51)

In view of Definition 5 and the fact that any infinitesimally flat AR domain is also a UR domain (cf. [24]), Theorem 2 readily implies the following compactness result on Lebesgue spaces for chord-dot-normal singular integral operators on the boundaries of infinitesimally flat AR domains.

**Corollary 1** *Let $\Omega \subseteq \mathbb{R}^n$ be an infinitesimally flat AR domain (that is, $\Omega$ is an Ahlfors regular domain with compact boundary and such that the geometric measure theoretic outward unit normal $\nu$ to $\Omega$ belongs to the Sarason space $\mathrm{VMO}(\partial\Omega,\mathcal{H}^{n-1})$). Fix a sufficiently large integer $N = N(n) \in \mathbb{N}$ and consider a complex-valued function $k \in \mathscr{C}^N(\mathbb{R}^n\setminus\{0\})$ which is even and positive homogeneous of degree $-n$. Bring in the operators $T$, $T^\#$ associated with $k$ and $\Omega$ as in* (39), (40). *Then for each integrability exponent $p \in (1,\infty)$ one has*

$$T,T^\#\in \mathrm{Cp}\big(L^p(\partial\Omega,\mathcal{H}^{n-1})\big), \tag{52}$$

*where, as before, $\mathrm{Cp}(L^p(\partial\Omega,\mathcal{H}^{n-1}))$ denotes the space of compact operators on $L^p(\partial\Omega,\mathcal{H}^{n-1})$.*

In the two-dimensional setting, we see from (52) and (48) that

$$\text{if } \Omega \subseteq \mathbb{R}^2 \text{ is a chord-arc domain with vanishing constant and compact boundary then } T, T^{\#} \in \mathrm{Cp}\big(L^p(\partial\Omega, \mathcal{H}^1)\big) \text{ for each integrability exponent } p \in (1, \infty). \tag{53}$$

Similar compactness results are valid for a multitude of other function spaces, such as Muckenhoupt weighted Lebesgue spaces, Lorentz spaces, Morrey spaces, Sobolev spaces, and Hölder spaces.

As a special case, Corollary 1 yields the following result, regarding the compactness of the harmonic double layer potential operator and of its transpose.

**Corollary 2** *Let $\Omega \subseteq \mathbb{R}^n$ be an infinitesimally flat AR domain. Recall the harmonic double layer potential operator $K$ associated with $\Omega$ as in* (14)*, and define the transpose harmonic double layer potential operator $K^{\#}$ acting on real-valued functions $f$ defined on $\partial\Omega$ according to*

$$K^{\#} f(x) := \lim_{\varepsilon \to 0^+} \frac{1}{\omega_{n-1}} \int_{\partial\Omega \setminus \overline{B(x,\varepsilon)}} \frac{\langle \nu(x), x - y \rangle}{|x - y|^n} f(y)\, d\mathcal{H}^{n-1}(y), \quad x \in \partial\Omega, \tag{54}$$

*where, as before, $\nu$ is the geometric measure theoretic outward unit normal to the set $\Omega$. Then for each integrability exponent $p \in (1, \infty)$ one has*

$$K, K^{\#} \in \mathrm{Cp}\big(L^p(\partial\Omega, \mathcal{H}^{n-1})\big). \tag{55}$$

*In particular, corresponding to the two-dimensional setting,*

$$\text{if } \Omega \subseteq \mathbb{R}^2 \text{ is a chord-arc domain with vanishing constant and compact boundary then } K, K^{\#} \in \mathrm{Cp}\big(L^p(\partial\Omega, \mathcal{H}^1)\big) \text{ for each integrability exponent } p \in (1, \infty). \tag{56}$$

A salient feature of infinitesimally flat AR domains, which is a key ingredient in the proof of Theorem 2, is the fact that the dot product of the unit normal with the normalized chord tends to zero, as the scale goes to zero. More specifically, the following result is a consequence of work in [24]:

**Proposition 3** *Let $\Omega \subset \mathbb{R}^n$ be an infinitesimally flat AR domain and denote by $\nu$ the geometric measure theoretic outward unit normal to $\Omega$. Then*

$$\lim_{R \to 0^+} \Big\{ \sup_{z \in \partial\Omega} \sup_{x, y \in \Delta(z,R)} R^{-1} \big\langle x - y, \nu_{\Delta(z,R)} \big\rangle \Big\} = 0, \tag{57}$$

*where $\nu_{\Delta(z,R)} := \fint_{\partial\Omega \cap B(z,R)} \nu \, d\mathcal{H}^{n-1}$ for each $z \in \partial\Omega$ and $R > 0$.*

In particular, (57) is violated if $\Omega$ is, say, an Ahlfors regular domain with a corner. We shall work in such a scenario and show, in Theorems 3 and 4 stated in the next section, that $K$ fails to be compact when acting on Lebesgue spaces in this geometric setting.

Moving on, fix a wave number $k \in (0, \infty)$ and consider the Helmholtz operator $\Delta + k^2$ in $\mathbb{R}^n$. It is known (see the discussion in [19]) that the only fundamental solution of the Helmholtz operator $\Delta + k^2$ in $\mathbb{R}^n$ which satisfies Sommerfeld's radiation condition is the distribution of function type given at each $x \in \mathbb{R}^n \setminus \{0\}$ by

$$\Phi_k(x) := c_n k^{(n-2)/2} \frac{H^{(1)}_{(n-2)/2}(k|x|)}{|x|^{(n-2)/2}}, \quad \text{with } c_n := \frac{1}{4i(2\pi)^{(n-2)/2}}, \tag{58}$$

where, generally speaking, $H^{(1)}_\lambda(\cdot)$ denotes the Hankel function of the first kind with index $\lambda \in \mathbb{R}$ (cf. [1, § 9.1]). Corresponding to the limiting case $k = 0$ (when the Helmholtz operator $\Delta + k^2$ is simply the Laplacian $\Delta$), let us define

$$\Phi_0(x) := \begin{cases} \dfrac{1}{(2-n)\omega_{n-1}} |x|^{2-n}, & x \in \mathbb{R}^n \setminus \{0\}, \ \text{if } n \geq 3, \\ \dfrac{1}{2\pi} \ln |x|, & x \in \mathbb{R}^2 \setminus \{0\}, \ \text{if } n = 2. \end{cases} \tag{59}$$

As is well known, $\Phi_0$ is a fundamental solution for the Laplacian in $\mathbb{R}^n$. The following result, elucidating the behavior of the gradient of $\Phi_k(x)$ as $|x| \to 0$, has been proved in [15].

**Lemma 1** *Fix $k \in (0, \infty)$ and $n \in \mathbb{N}$, $n \geq 2$. Also, pick some $R > 0$. Then the function $\Phi_k$ satisfies the following estimates for each $x \in \mathbb{R}^n$ with $0 < |x| < R$:*

$$|(\partial_j \Phi_k)(x) - (\partial_j \Phi_0)(x)| \leq \begin{cases} C, & n = 2, 3, \\ C\,|x|^{3-n}, & n \geq 4, \end{cases} \tag{60}$$

*where $C = C(R, n, k) > 0$ is a finite constant, and $j \in \{1, \ldots, n\}$ is arbitrary.*

In turn, Lemma 1 is a basic ingredient in the proof of the next proposition, concerning the compactness of the difference between the double layers associated with the Laplacian and the Helmholtz operator.

**Proposition 4** *Let $\Omega \subseteq \mathbb{R}^n$ be a UR domain with compact boundary and denote by $\nu$ the geometric measure theoretic outward unit normal to $\Omega$. Recall the harmonic double layer potential operator $K$ associated with $\Omega$ as in* (14). *Also, having fixed a wave number $k \in (0, \infty)$, introduce the principal-value singular integral operator acting on real-valued functions $f$ defined on $\partial\Omega$ as*

$$K_k f(x) := \lim_{\varepsilon \to 0^+} \frac{1}{\omega_{n-1}} \int_{\partial\Omega \setminus \overline{B(x,\varepsilon)}} \partial_{\nu(y)} \big[\Phi_k(x - y)\big] f(y)\, d\mathcal{H}^{n-1}(y) \tag{61}$$

*at $\mathcal{H}^{n-1}$-a.e. point $x \in \partial\Omega$, where $\partial_\nu$ is the directional derivative along $\nu$, and $\Phi_k$ is as in* (58).

*Then for each integrability exponent $p \in (1, \infty)$ the difference $K - K_k$ is a compact operator on $L^p(\partial\Omega, \mathcal{H}^{n-1})$.*

***Proof*** From (14) and (61) we see that the integral kernel of $K - K_k$ is the function

$$\Theta(x, y) := -\sum_{j=1}^{n} \nu_j(y)\big[(\partial_j \Phi_0)(x-y) - (\partial_j \Phi_k)(x-y)\big], \qquad x, y \in \partial\Omega. \tag{62}$$

In particular, Lemma 1 applies and gives that

$$|\Theta(x, y)| \le \begin{cases} C, & n = 2, 3, \\ C\,|x|^{3-n}, & n \ge 4, \end{cases} \qquad x, y \in \partial\Omega. \tag{63}$$

Hence, $K - K_k$ is a weakly singular integral operator. As such, [16, Lemma 2.20, p. 2608] applies and gives that $K - K_k$ is compact on $L^p(\partial\Omega, \mathcal{H}^{n-1})$ for each $p \in (1, \infty)$. □

From Proposition 4, Corollary 2, and the fact that any infinitesimally flat AR domain is a UR domain (cf. [24]) we conclude the following:

**Corollary 3** *Let $\Omega \subseteq \mathbb{R}^n$ be an infinitesimally flat AR domain and fix a wave number $k \in (0, \infty)$. Recall the acoustic double layer potential operator $K_k$ associated with $\Omega$ as in* (61)*. Then for each integrability exponent $p \in (1, \infty)$ one has*

$$K_k \in \mathrm{Cp}\big(L^p(\partial\Omega, \mathcal{H}^{n-1})\big). \tag{64}$$

*As a consequence, corresponding to the two-dimensional setting,*

> *if $\Omega \subseteq \mathbb{R}^2$ is a chord-arc domain with vanishing constant and compact boundary then $K_k \in \mathrm{Cp}\big(L^p(\partial\Omega, \mathcal{H}^1)\big)$ for each integrability exponent $p \in (1, \infty)$ and each wave number $k \in (0, \infty)$.* (65)

This result is particularly relevant in the context of boundary value problems in acoustic scattering.

## 2 Failure of Compactness for the Harmonic Double Layer Operator

In the introduction of his influential 1911 monograph [26] J. Plemelj writes that "*one of the most important tasks of modern Potential Theory is to explore the invertibility of boundary singular integral equations arising from single and double layer potentials.*"

The spectrum of the classical harmonic double layer potential operator on Lebesgue spaces considered on boundaries of curvilinear polygons in the plane has been explicitly identified by Shelepov in [29]. To state such a result, for each $\theta \in (0, 2\pi)$ and $p \in (1, \infty)$ define the following ("bow" shaped) closed contour given by the parametric representation

$$\mathcal{B}_\theta(p) := \left\{ \pm \frac{\sin\Big((\pi - \theta)z\Big)}{\sin(\pi z)} : z \in \tfrac{1}{p} + i\,\mathbb{R} \right\}. \tag{66}$$

Also, the reader is reminded that a curvilinear polygon is a piecewise smooth domain in the plane, without cusps.

**Proposition 5** *Let $\Omega \subseteq \mathbb{R}^2$ be a bounded, simply connected, curvilinear polygon with internal angles $\theta_j \in (0, 2\pi)$, labeled by $j \in \{1, \dots, N\}$, and fix an integrability exponent $p \in (1, \infty)$. For each $j \in \{1, \dots, N\}$, consider the bow-shaped curve $\mathcal{B}_{\theta_j}(p)$ associated with the given $p$ and the angle $\theta_j$ as in* (66)*. Also, denote by $\widehat{\mathcal{B}_{\theta_j}(p)}$ the (closed) two-dimensional region encompassed by $\mathcal{B}_{\theta_j}(p)$. Let $K$ stand for the harmonic double layer potential operator on $\partial\Omega$. Then the spectrum of $K$ on $L^p(\partial\Omega, \mathcal{H}^1)$ has the following structure:*

$$\operatorname{Spec}\Big(K; L^p(\partial\Omega, \mathcal{H}^1)\Big) = \left( \bigcup_{1 \le j \le N} \widehat{\mathcal{B}_\theta(p)} \right) \bigcup \{\lambda_k\}_k, \tag{67}$$

*where $\{\lambda_k\}_k$, the collection of eigenvalues of $K$ on $L^p(\partial\Omega, \mathcal{H}^1)$, is a finite subset of $(-1/2, 1/2]$. Moreover, for $z \in \bigcup_{j=1}^N \mathcal{B}_{\theta_j}(p)$ the operator $zI - K$ is not Fredholm on $L^p(\partial\Omega, \mathcal{H}^1)$, whereas for $z \in \mathbb{C} \setminus \Big(\bigcup_{j=1}^N \mathcal{B}_{\theta_j}(p)\Big)$ the operator $zI - K$ is Fredholm on $L^p(\partial\Omega, \mathcal{H}^1)$ and its index is given by*

$$\operatorname{index}\Big(zI - K; L^p(\partial\Omega, \mathcal{H}^1)\Big) = \sum_{j=1}^N W\big(z, \mathcal{B}_{\theta_j}(p)\big), \tag{68}$$

*where $W\big(z, \mathcal{B}_{\theta_j}(p)\big)$ denotes the winding number of $z$ with respect to the curve $\mathcal{B}_{\theta_j}(p)$.*

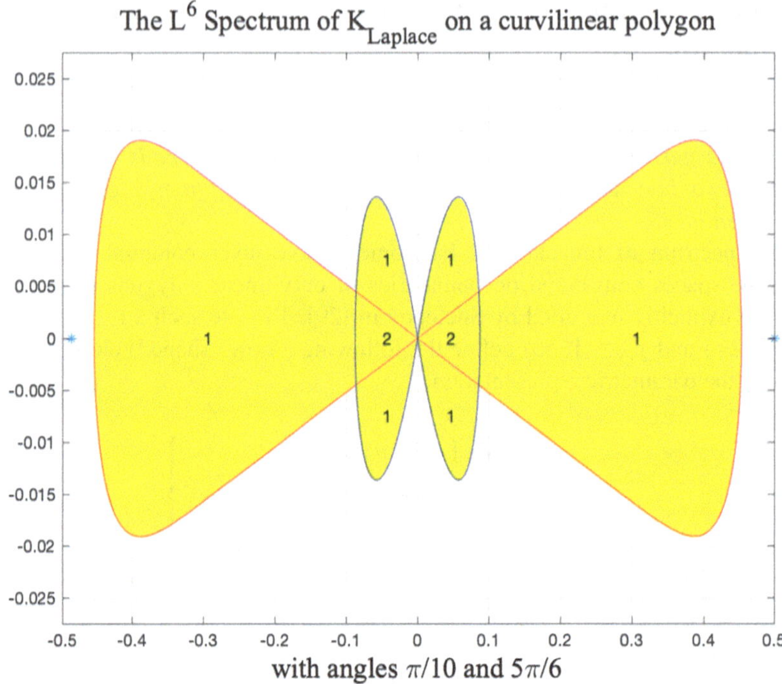


**Fig. 1** $L^6$ spectra of the harmonic layer potential on a curvilinear polygon. The numbers indicate the value of the index of the operator $zI - K$ for $z$ belonging to each region of the spectrum

Figure 1 depicts the spectrum of $K$ in a concrete situation.

Shelepov's result has been subsequently extended to other partial differential operators in [21]. Among other things, Proposition 5 may be used to conclude that that

$$\text{for any nonsmooth, simply connected, curvilinear polygon } \Omega \text{ in } \mathbb{R}^2\text{, the corresponding harmonic double layer potential operator } K \text{ fails to be compact on } L^p(\partial\Omega, \mathcal{H}^1) \text{ for each } p \in (1, \infty). \tag{69}$$

Indeed, if $K$ were to be compact on $L^2(\partial\Omega, \mathcal{H}^1)$, then so would its transpose $K^\#$. As such, the spectrum of $K^\#$ on $L^2(\partial\Omega, \mathcal{H}^1)$ would consist of just eigenvalues, plus possibly the point 0. The fact that the eigenvalues of $K^\#$ on $L^2(\partial\Omega, \mathcal{H}^1)$ are real and belong to the interval $\left[-\frac{1}{2}, \frac{1}{2}\right]$ has been established by J. Plemelj in 1911 and O. Kellogg in 1929, initially in smooth domains but via an approach which works for more inclusive classes of sets. A general spectral result of this type appears in [22, Proposition 5.9, p. 194], where D. Mitrea, I. Mitrea, M. Mitrea, and M. Taylor have dealt with uniformly rectifiable subdomains of Riemannian manifolds. Earlier, this result has been noted in the class of Lipschitz domains by E. Fabes, L. Escauriaza, and G. Verchota in [12]. All together, this shows that

$$\operatorname{Spec}\Big(K;\, L^2(\partial\Omega, \mathcal{H}^1)\Big) = \overline{\operatorname{Spec}\Big(K^{\#};\, L^2(\partial\Omega, \mathcal{H}^1)\Big)} \subseteq \big[-\tfrac{1}{2}, \tfrac{1}{2}\big], \tag{70}$$

contradicting (67). This contradiction then establishes (69) in view of Krasnosel-skiĭ's theorem (see, e.g., [2, Theorem 2.9, p. 203]) which allows us to replace the integrability exponent 2 with any other $p \in (1, \infty)$.

Another indirect proof of (69) is via an estimate which shows that, for each curvilinear polygon $\Omega \subseteq \mathbb{R}^2$ and each $p \in (1, \infty)$, the degree of proximity of $K$ to the space of compact operators on $L^p(\partial\Omega, \mathcal{H}^1)$ dominates the size of the worst singularity on $\partial\Omega$ (see (74) below for a precise formulation). To elaborate on this, on the one hand, we recall from [24] that there exists some $c \in (0, \infty)$ which depends only on $p$ and the Ahlfors regularity constants of $\partial\Omega$ with the property that

$$\operatorname{dist}\big(K\,,\ \operatorname{Cp}(L^p(\partial\Omega, \mathcal{H}^1))\big) \geq c \cdot \operatorname{dist}\Big(\nu\,,\ \operatorname{VMO}(\partial\Omega, \mathcal{H}^1)\Big), \tag{71}$$

where the distance in the left-hand side is measured in the operator norm in the space of linear and bounded operators on $L^p(\partial\Omega, \mathcal{H}^1)$, and the distance in the right-hand side is measured in $\operatorname{BMO}(\partial\Omega, \mathcal{H}^1)$. This should be compared with (41) (in relation to which (71) serves as an opposite inequality of sorts). On the other hand, from [16, Proposition 2.22, p. 2611] we know that

$$\begin{aligned} &\operatorname{dist}\big(\nu,\ \operatorname{VMO}(\partial\Omega, \mathcal{H}^1)\big) \\ &\approx \limsup_{r\to 0^+} \left\{ \sup_{x\in\partial\Omega} ⨍_{\partial\Omega\cap B(x,r)} ⨍_{\partial\Omega\cap B(x,r)} |\nu(y) - \nu(z)|\, d\mathcal{H}^1(y)\, d\mathcal{H}^1(z) \right\} \end{aligned} \tag{72}$$

where the implicit proportionality constants depend only on the Ahlfors regularity constants of $\partial\Omega$. If $\theta_j \in (0, 2\pi)$ and $x_j \in \partial\Omega$, with $j \in \{1, \dots, N\}$, are the internal angles and the vertices, respectively, of the curvilinear polygon $\Omega \subseteq \mathbb{R}^2$, then straightforward calculus yields

$$\lim_{r\to 0^+} ⨍_{\partial\Omega\cap B(x_j,r)} ⨍_{\partial\Omega\cap B(x_j,r)} |\nu(y) - \nu(z)|\, d\mathcal{H}^1(y)\, d\mathcal{H}^1(z) = \big|\cos(\theta_j/2)\big|, \tag{73}$$

for each $j \in \{1, \dots, N\}$. After combining (71) with (72) and (73) we arrive at the conclusion that

$$\operatorname{dist}\big(K\,,\ \operatorname{Cp}(L^p(\partial\Omega, \mathcal{H}^1))\big) \geq c \cdot \max_{1\leq j\leq N} \big|\cos(\theta_j/2)\big|, \tag{74}$$

where the distance in the left-hand side is measured in the operator norm in the space of linear and bounded operators on $L^p(\partial\Omega, \mathcal{H}^1)$, and $c \in (0, \infty)$ depends only on

$p$ and the Ahlfors regularity constants of $\partial\Omega$. Of course, (74) readily implies (69) (in fact, one should think of (74) as a quantitative embodiment of (69)).

One last comment pertaining to (74) is that

$$\begin{array}{l}\text{given any simply connected curvilinear polygon } \Omega \subseteq \mathbb{R}^2 \text{ and}\\ \text{any } p \in (1,\infty)\text{, the compactness of the corresponding harmonic}\\ \text{double layer potential operator } K \text{ on } L^p(\partial\Omega, \mathcal{H}^1) \text{ is equivalent}\\ \text{to } \Omega \text{ actually being a domain of class } \mathscr{C}^1.\end{array} \tag{75}$$

Indeed, if $\Omega$ is a domain of class $\mathscr{C}^1$ then work in [14] shows that $K$ is compact on $L^p(\partial\Omega, \mathcal{H}^1)$. In the opposite direction, if $K$ is compact on $L^p(\partial\Omega, \mathcal{H}^1)$ then (74) forces all internal angles of $\Omega$ to be equal to $\pi$, hence $\Omega$ is a domain of class $\mathscr{C}^1$.

Our main goal in this section is to give a direct, self-contained proof of (69), i.e., the fact that the harmonic double layer potential fails to be compact on Lebesgue spaces in the presence of a single corner singularity. We shall work with a drop-shaped domain $\Omega$ described below (cf. (76)). Note that by Proposition 5 the $L^2$ spectrum of the harmonic double layer potential in a drop-shaped domain with angle $\alpha_0 = \pi/3$ looks as in Fig. 2, where the red curve is precisely the essential spectrum of $K$ on $L^2(\partial\Omega, \mathcal{H}^1)$.

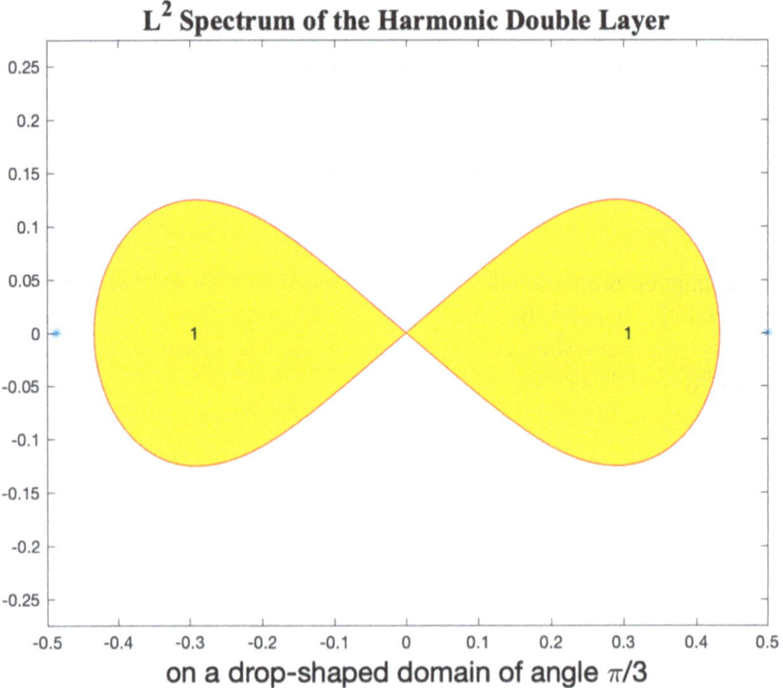


**Fig. 2** The $L^2$ spectrum of the harmonic layer potential on a drop-shaped domain. The numbers indicate the value of the index of the operator $zI - K$ for $z$ belonging to each region of the spectrum

Here is the result just advertised:

**Theorem 3** *Fix* $\alpha_0 \in (0, \pi)$ *and* $r_0 \in (1, \infty)$. *Let* $\Omega \subset \mathbb{R}^2$ *be a bounded open set whose boundary is smooth except at the origin and such that*

$$\Omega \cap B(0, r_0) = \big\{(r\cos t, r\sin t) : 0 < r < r_0,\ t \in (0, \alpha_0)\big\}. \tag{76}$$

*Denote by* $K$ *the harmonic double layer potential operator associated with the set* $\Omega$ *as in* (14).

*Then for each* $p \in (1, \infty)$, *the operator* $K : L^p(\partial\Omega, \mathcal{H}^1) \to L^p(\partial\Omega, \mathcal{H}^1)$ *is not compact.*

***Proof*** We will reason by contradiction. Suppose there exists $p \in (1, \infty)$ for which $K$ is a compact operator when acting from $L^p(\partial\Omega, \mathcal{H}^1)$ into itself. Let $L_1 := \partial\Omega \cap \{(x, 0) : x \in (0, r_0)\}$ and $L_2 := \partial\Omega \cap \{(r\cos\alpha_0, r\sin\alpha_0) : r \in (0, r_0)\}$. For each set $A$, denote by $\mathbf{1}_A$ the characteristic function of $A$. Then, the operator $\widetilde{K} := \mathbf{1}_{L_2} \cdot K$ is also compact on $L^p(\partial\Omega, \mathcal{H}^1)$ (Fig. 3).

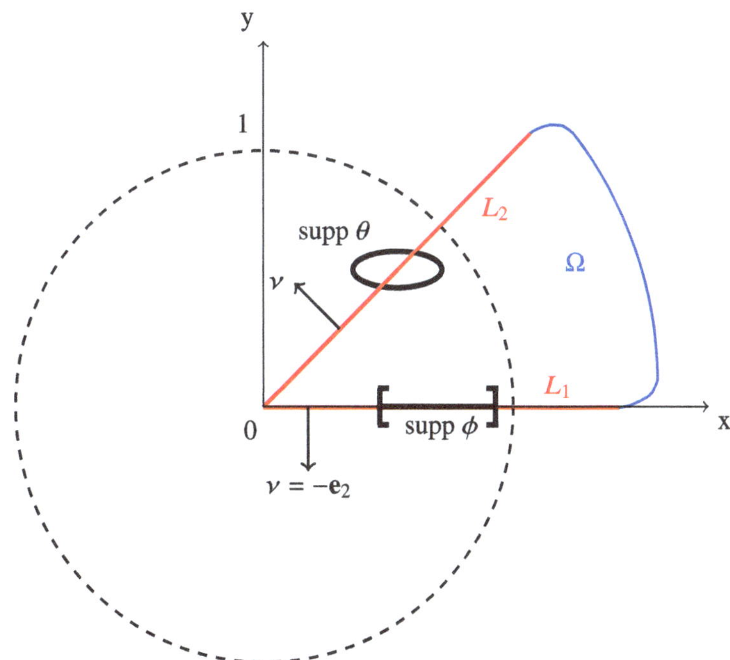


**Fig. 3** The domain $\Omega$

To proceed, fix an arbitrary function

$$\phi \in L^p(\partial\Omega, \mathcal{H}^1) \text{ with } \operatorname{supp}\phi \text{ a compact subset of } L_1 \cap B(0,1), \tag{77}$$

and observe that for each $x \in \partial\Omega$ we have

$$(\widetilde{K}\phi)(x) = \begin{cases} \dfrac{1}{2\pi}\displaystyle\int_{L_1} \dfrac{\langle -\mathbf{e}_2, y-x\rangle}{|x-y|^2}\phi(y)\,d\mathcal{H}^1(y) & \text{if } x \in L_2, \\ 0 \text{ if } x \in \partial\Omega \setminus L_2, \end{cases} \tag{78}$$

where $\mathbf{e}_2 := (0,1) \in \mathbb{R}^2$. We also observe that since $\operatorname{supp}\phi$ is a compact subset of $L_1 \cap B(0,1)$, there exists some constant $c = c(\phi) \in (0,\infty)$ such that $|x-y| \geq c$ for all $x \in L_2$ and all $y \in L_1 \cap \operatorname{supp}\phi$. This further implies that at each $x \in L_2$ we have

$$\begin{aligned} K\phi(x) &= \lim_{\varepsilon\to 0^+} \frac{1}{2\pi}\int_{\partial\Omega\setminus\overline{B(x,\varepsilon)}} \frac{\langle \nu(y), y-x\rangle}{|x-y|^2}\phi(y)\,d\mathcal{H}^1(y) \\ &= \frac{1}{2\pi}\int_{y\in L_1\cap\operatorname{supp}\phi} \frac{\langle \nu(y), y-x\rangle}{|x-y|^2}\phi(y)\,dy. \end{aligned} \tag{79}$$

From (79) we now see that

$$K\phi \text{ is continuous on } L_2. \tag{80}$$

Next, for each $\varepsilon \in (0,1)$ define

$$\phi_{(\varepsilon)}(x) := \begin{cases} \varepsilon^{-1/p}\phi(x/\varepsilon) & \text{if } x \in \partial\Omega \cap B(0,\varepsilon), \\ 0 \text{ otherwise,} \end{cases} \tag{81}$$

at $\mathcal{H}^1$-a.e. point $x \in \partial\Omega$. Note that the non-standard normalization employed in (81) is designed to ensure that

$$\big\|\phi_{(\varepsilon)}\big\|_{L^p(\partial\Omega,\mathcal{H}^1)} = \|\phi\|_{L^p(\partial\Omega,\mathcal{H}^1)} \text{ for each } \varepsilon \in (0,1). \tag{82}$$

In concert, (82) and the fact that $\widetilde{K}$ is assumed to be compact on $L^p(\partial\Omega, \mathcal{H}^1)$, imply the existence of a function $g \in L^p(\partial\Omega, \mathcal{H}^1)$ and of a sequence of numbers $\{\varepsilon_j\}_{j\in\mathbb{N}} \subset (0,1)$ with $\lim\limits_{j\to\infty} \varepsilon_j = 0$ and such that

$$\widetilde{K}\phi_{(\varepsilon_j)} \to g \text{ in } L^p(\partial\Omega, \mathcal{H}^1) \text{ as } j \to \infty. \tag{83}$$

We make the claim that

$$g = 0 \text{ at } \mathcal{H}^1\text{-a.e. point on } \partial\Omega. \tag{84}$$

Let us see how, by assuming this claim, we can finish the proof of the theorem. First, (83)–(84) imply

$$\widetilde{K}\phi_{(\varepsilon_j)} \to 0 \ \text{ in } \ L^p(\partial\Omega, \mathcal{H}^1) \ \text{ as } \ j \to \infty. \tag{85}$$

Second, (78), (77), and the fact that $\operatorname{supp}\phi_{(\varepsilon)} = \varepsilon \cdot \operatorname{supp}\phi$ for every $\varepsilon \in (0, 1)$, allow us to write, for each $j \in \mathbb{N}$,

$$\begin{aligned}
\|\widetilde{K}\phi_{(\varepsilon_j)}\|^p_{L^p(\partial\Omega,\mathcal{H}^1)} &\geq \|\widetilde{K}\phi_{(\varepsilon_j)}\|^p_{L^p(\varepsilon_j L_2,\mathcal{H}^1)} \\
&= \int_{\varepsilon_j L_2} \left| \frac{1}{2\pi} \int_{L_1 \cap B(0,\varepsilon_j)} \frac{\langle -\mathbf{e}_2, y - x\rangle}{|x-y|^2} \phi_{(\varepsilon_j)}(y)\, d\mathcal{H}^1(y) \right|^p d\mathcal{H}^1(x) \\
&= \int_{L_2} \left| \frac{1}{2\pi} \int_{L_1 \cap B(0,1)} \frac{\langle -\mathbf{e}_2, y - x\rangle}{|x-y|^2} \phi(y)\, d\mathcal{H}^1(y) \right|^p d\mathcal{H}^1(x) \\
&= \|\mathbf{1}_{L_2} \cdot K\phi\|^p_{L^p(\partial\Omega,\mathcal{H}^1)},
\end{aligned} \tag{86}$$

where the second equality in (86) follows from a suitable change of variables in both integrals. Combining the estimate in (86) with (85), it follows that $K\phi = 0$ at $\mathcal{H}^1$-a.e. point on $L_2$, and since $K\phi$ is continuous on $L_2$, that $K\phi = 0$ on $L_2$. Hence,

$$\int_{L_1} \frac{\langle \nu(y), y - x\rangle}{|x-y|^2} \phi(y)\, d\mathcal{H}^1(y) = 0 \ \text{ for every } \ x \in L_2. \tag{87}$$

In turn, condition (87) being true for each $\phi$ as in (77) implies

$$\frac{\langle -\mathbf{e}_2, y - x\rangle}{|x-y|^2} = 0 \ \text{ for every } \ x \in L_2 \ \text{ and } \ \mathcal{H}^1\text{-a.e. } \ y \in L_1. \tag{88}$$

By continuity, this ultimately forces

$$\langle \mathbf{e}_2, y - x\rangle = 0 \ \text{ for every } \ x \in L_2 \ \text{ and every } \ y \in L_1. \tag{89}$$

This is clearly false leading to a contradiction and proving that the assumption that $K$ is a compact operator from $L^p(\partial\Omega, \mathcal{H}^1)$ is not true.

To finish the proof of the theorem it remains to prove the claim in (84). To this end, we first observe that the convergence in (83) implies

$$\widetilde{K}\phi_{(\varepsilon_j)} \to g \ \text{ at } \mathcal{H}^1\text{-a.e. point in } \ \partial\Omega \ \text{ as } \ j \to \infty. \tag{90}$$

In light of (78), from (90) we obtain that $g = 0$ at $\mathcal{H}^1$-a.e. point in $\partial\Omega \setminus L_2$. Thus, we are left with showing that $g = 0$ at $\mathcal{H}^1$-a.e. point in $L_2$. To continue, pick

$$\theta \in \mathscr{C}^\infty_c(\mathbb{R}^2) \ \text{ with } \ 0 \notin \operatorname{supp}\theta \ \text{ and } \ \operatorname{supp}\theta \cap \partial\Omega \subseteq L_2. \tag{91}$$

Then, with $K^{\#}$ denoting the transpose of $K$ (cf. (54)), we have

$$\begin{aligned}\int_{\partial\Omega} g\theta\, d\mathcal{H}^1 &= \lim_{j\to\infty}\int_{\partial\Omega}(\widetilde{K}\phi_{(\varepsilon_j)})\theta\, d\mathcal{H}^1 = \lim_{j\to\infty}\int_{\partial\Omega}(K\phi_{(\varepsilon_j)})\theta\, d\mathcal{H}^1\\ &= \lim_{j\to\infty}\int_{\partial\Omega}\phi_{(\varepsilon_j)}K^{\#}\theta\, d\mathcal{H}^1\\ &= \lim_{j\to\infty}\int_{L_1\cap B(0,\varepsilon)}\varepsilon_j^{-1/p}\phi(x/\varepsilon_j)(K^{\#}\theta)(x)\, d\mathcal{H}^1(x)\\ &= \lim_{j\to\infty}\varepsilon_j^{1-1/p}\int_{L_1\cap B(0,1)}\phi(y)(K^{\#}\theta)(\varepsilon_j y)\, d\mathcal{H}^1(y)\\ &= 0\cdot(K^{\#}\theta)(0)\cdot\int_{L_1\cap B(0,1)}\phi(y)\, d\mathcal{H}^1(y) = 0. \end{aligned} \tag{92}$$

The first equality in (92) is a consequence of (83) and the fact that, if $p'$ denotes the Hölder conjugate of $p$, we have $\theta|_{\partial\Omega} \in L^{p'}(\partial\Omega, \mathcal{H}^1)$. Also, the second equality in (92) relies on the support condition $\operatorname{supp}\theta \cap \partial\Omega \subseteq L_2$ and the definition of $\widetilde{K}$. The fourth equality in (92) uses (81), the fifth one is based on the change of variables $y := x/\varepsilon_j$, while the sixth equality relies on Lebesgue's Dominated Convergence Theorem (here we also note that $0 \notin \operatorname{supp}\theta$ and $K^{\#}\theta$ is continuous at 0, as seen from an argument similar to the one that led to (80)). Invoking now the arbitrariness of $\theta$ as in (91), from (92) it follows that, as wanted, $g = 0$ at $\mathcal{H}^1$-a.e. point in $L_2$. The proof of Theorem 3 is therefore complete. □

A higher-dimensional version of Theorem 3 is also true, and this is made precise below.

**Theorem 4** *Fix $\alpha \in (0, \pi/2)$ and let $\Omega \subset \mathbb{R}^n$, where $n \in \mathbb{N}$ with $n \geq 2$, be a bounded open set whose boundary is smooth except at the origin and such that*

$$\partial\Omega \cap B(0,1) = \big\{x = (x', x_n) \in \mathbb{R}^{n-1}\times[0,\infty) : |x| < 1,\ x_n = (\cos\alpha)|x|\big\}. \tag{93}$$

*Let $K$ be the harmonic double layer potential operator associated with the set $\Omega$ as in* (14).

*Then for each $p \in (1,\infty)$, the operator $K : L^p(\partial\Omega, \mathcal{H}^{n-1}) \to L^p(\partial\Omega, \mathcal{H}^{n-1})$ is not compact.*

***Proof*** The argument proceeds along similar lines as in the proof of Theorem 3, with a number of alterations designed to accommodate the more general $n$-dimensional setting considered here. Specifically, reasoning by contradiction, assume there exists some integrability exponent $p \in (1,\infty)$ such that $K$ is a compact operator when acting from the space $L^p(\partial\Omega, \mathcal{H}^{n-1})$ into itself. To proceed, consider the sets $L_1 :=$

$\{x \in \partial\Omega \cap B(0,1) : x_{n-1} > 0\}$ and $L_2 := \{x \in \partial\Omega \cap B(0,1) : x_{n-1} < 0\}$. In particular, $\{L_1, L_2\}$ is a partition of $\partial\Omega \cap B(0,1)$ up to an $\mathcal{H}^{n-1}$-null set, since $\mathcal{H}^{n-1}\big((\partial\Omega \cap B(0,1)) \setminus (L_1 \cup L_2)\big) = 0$. The geometry of $\Omega$ implies that at points $x \in \partial\Omega \cap B(0,1) \setminus \{0\}$ the outward unit normal $\nu(x)$ is in the span of $\{x, \mathbf{e}_n\}$, where $\mathbf{e}_n$ denotes the unit vector in $\mathbb{R}^n$ with 1 on the $n$-th component. The latter, when combined with the fact that $\nu(x)$ is orthogonal to $x$ and has norm 1, yields the following description:

$$\nu(x) = (\cos\alpha)\frac{x}{|x'|} - \frac{|x|}{|x'|}\mathbf{e}_n \quad \text{for all} \quad x = (x', x_n) \in \partial\Omega \cap B(0,1) \setminus \{0\}. \tag{94}$$

As a consequence of (94) we have that

$$\nu(\varepsilon x) = \nu(x) \quad \text{for all} \quad x \in \partial\Omega \cap B(0,1) \setminus \{0\} \quad \text{and every} \quad \varepsilon \in (0,1). \tag{95}$$

As before, define $\widetilde{K} := \mathbf{1}_{L_2} \cdot K$ which by the current assumption is compact on $L^p(\partial\Omega, \mathcal{H}^{n-1})$. Pick $\phi \in L^p(\partial\Omega, \mathcal{H}^{n-1})$ whose support is a compact subset of $L_1 \cap B(0,1)$ and observe that for each $x \in \partial\Omega$ we have

$$(\widetilde{K}\phi)(x) = \begin{cases} \dfrac{1}{\omega_{n-1}} \displaystyle\int_{L_1} \frac{\langle \nu(y), y - x\rangle}{|x - y|^n} \phi(y)\, d\mathcal{H}^{n-1}(y) & \text{if } x \in L_2, \\ 0 & \text{if } x \in \partial\Omega \setminus L_2. \end{cases} \tag{96}$$

In addition, similar to the reasoning in (79) we obtain that $K\phi$ is continuous on $L_2$. Next, for each $\varepsilon \in (0,1)$ define

$$\phi_{(\varepsilon)}(x) := \begin{cases} \varepsilon^{-(n-1)/p}\phi(x/\varepsilon) & \text{if } x \in \partial\Omega \cap B(0,\varepsilon), \\ 0 & \text{otherwise}, \end{cases} \tag{97}$$

at $\mathcal{H}^{n-1}$-a.e. point $x \in \partial\Omega$, so that $\|\phi_{(\varepsilon)}\|_{L^p(\partial\Omega, \mathcal{H}^{n-1})} = \|\phi\|_{L^p(\partial\Omega, \mathcal{H}^{n-1})}$ for each $\varepsilon \in (0,1)$. From this point on, the reasoning in the proof Theorem 3 continues to work virtually verbatim, and we once again reach a contradiction after arriving at the conclusion that

$$\frac{\langle \nu(y), y - x\rangle}{|x - y|^n} = 0 \quad \text{for every} \quad x \in L_2 \quad \text{and every} \quad y \in L_1. \tag{98}$$

Hence, the desired conclusion follows. □

A cursory inspection reveals that the same type of argument as in the proof of Theorem 4 may be adapted to the more general class of singular integral operators $T, T^{\#}$ considered in (39)–(40). In this situation, in place of (98) we now reach the conclusion that

$$\langle \nu(y), y - x\rangle k(x - y) = 0 \quad \text{for every} \quad x \in L_2 \quad \text{and every} \quad y \in L_1. \tag{99}$$

In turn, this yields a contraction if the kernel $k \in \mathscr{C}^N(\mathbb{R}^n \setminus \{0\})$, which is even and positive homogeneous of degree $-n$, is also assumed not vanish identically in the set

$$\{x = (x', x_n) \in \mathbb{R}^{n-1} \times \mathbb{R} : |x_n| < (\cos\alpha)|x|\} \tag{100}$$

(incidentally, the latter property ensures that the operators $T, T^{\#}$ are not identically zero). In addition, a result similar in spirit to the one just mentioned above is true in the case when the kernel $k \in \mathscr{C}^N(\mathbb{R}^n \setminus \{0\})$ is matrix-valued (a scenario in which the singular integral operators $T, T^{\#}$ from (39)–(40) act on vector-valued functions $f$). This contains the special case of the double layer potential operator for the Lamé system in $\mathbb{R}^n$, corresponding to the so-called pseudostress conormal derivative.

We end with an application to scattering. To set the stage, recall from Corollary 3 that the acoustic double layer potential operator $K_k$ is compact on Lebesgue spaces when the underlying "surface" is the boundary of an infinitesimally flat AR domain. In [6, p. 51], Colton and Kress write "*However, the situation changes considerably if the boundary is allowed to have edges and corners since this affects the compactness of the [acoustic] double layer integral operator.*" Below we provide a proof of this heuristic claim, working with the domain from Theorem 4.

**Corollary 4** *Bring back the domain $\Omega$ from Theorem 4, and fix an arbitrary wave number $k \in (0, \infty)$. Denote by $K_k$ the acoustic double layer potential operator associated with $\Omega$ as in* (61)*. Then for each integrability exponent $p \in (1, \infty)$, the operator $K_k : L^p(\partial\Omega, \mathcal{H}^{n-1}) \to L^p(\partial\Omega, \mathcal{H}^{n-1})$ is not compact.*

***Proof*** This is a direct consequence of Theorem 4 and Proposition 4, bearing in mind that $\Omega$ is a bounded Lipschitz domain (hence, a UR domain with compact boundary). □

**Acknowledgments** The authors gratefully acknowledge partial support from the Simons Foundation (through grants # 426669, # 637481), as well as NSF (grant # 1900938).

## References

1. M. Abramowitz, I.A. Stegun, *Handbook of Mathematical Functions* (Dover, New York, 1972)
2. C. Bennett, R. Sharpley, Interpolation of operators, in *Pure and Applied Mathematics*, vol. 129 (Academic Press, New York, 1988)
3. Yu.D. Burago, V.G. Maz'ya, Potential theory and function theory for irregular regions, in *Translated from Russian, Seminars in Mathematics, V.A. Steklov Mathematical Institute, Leningrad*, vol. 3, (Consultants Bureau. New York, 1969)
4. A.P. Calderón, Cauchy integrals on Lipschitz curves and related operators. Proc. Nat. Acad. Sci. USA. **74**(4), 1324–1327 (1977)
5. R. Coifman, A. McIntosh, Y. Meyer, L'intégrale de Cauchy definit un opérateur borné sur $L^2$ pour les courbes lipschitziennes. Ann. Math. **116**, 361–388 (1982)
6. D. Colton, R. Kress, Inverse acoustic and electromagnetic scattering theory, in *Applied Mathematical Sciences*, vol. 93 (Springer, Berlin, 1992)

7. G. David, Opérateurs d'intégrale singulière sur les surfaces régulières. Ann. Scient. Ecole Norm. Sup. **21**, 225–258 (1988)
8. G. David, Wavelets and singular integrals on curves and surfaces, in *Lecture Notes in Mathematics*, vol. 1465 (Springer, Berlin, 1991)
9. G. David, D. Jerison, Lipschitz approximation to hypersurfaces, harmonic measure, and singular integrals. Indiana Univ. Math. J. **39**(3), 831–845 (1990)
10. G. David, S.W. Semmes, Singular integrals and rectifiable sets in $\mathbb{R}^n$: Beyond lipschitz graphs. Astérisque **193**, 13 (1991)
11. G. David, S.W. Semmes, Analysis of and on uniformly rectifiable sets, in *Mathematical Surveys and Monographs* (AMS Series, New York, 1993)
12. L. Escauriaza, E.B. Fabes, G. Verchota, On a regularity theorem for weak solutions to transmission problems with internal Lipschitz boundaries. Proc. Am. Math. Soc. **115**, 1069–1076 (1992)
13. L.C. Evans, R.F. Gariepy, Measure theory and fine properties of functions, in *Studies in Advanced Mathematics* (CRC Press, Boca Raton, 1992)
14. E.B. Fabes, M. Jodeit Jr, N.M. Rivière, Potential techniques for boundary value problems on $C^1$-domains. Acta Math. **141**(3–4), 165–186 (1978)
15. F. Gesztesy, M. Mitrea, Generalized Robin boundary conditions, Robin-to-Dirichlet maps, and Krein-type resolvent formulas for Schrödinger operators on bounded Lipschitz domains, in *Proceedings of Symposia in Pure Mathematics*, vol. 79 (American Mathematical Society, New York, 2008)
16. S. Hofmann, M. Mitrea, M. Taylor, Singular Integrals and Elliptic Boundary Problems on Regular Semmes-Kenig-Toro Domains. Int. Math. Res. Not. IMRN. **14**, 2567–2865 (2010)
17. D.S. Jerison, C.E. Kenig, Boundary behavior of harmonic functions in nontangentially accessible domains. Adv. Math. **46**(1), 80–147 (1982)
18. J. KrálJ., Integral operators in potential theory, in *Lecture Notes in Mathematics*, vol. 823 (Springer, Berlin, 1980)
19. E. Marmolejo-Olea, D. Mitrea, I. Mitrea, M. Mitrea, Radiation conditions and integral representations for Clifford algebra-valued null-solutions of the Helmholtz operator. J. Math. Sci. **231**(3), 367–472 (2018)
20. V.G. Maz'ya, Boundary integral equations, in *Analysis, IV, Encyclopaedia Mathematical Science*, vol. 27, ed. by V.G. Maz'ya, S.M. Nikol'skii (Springer, Berlin, 1991), pp. 127–222
21. I. Mitrea, On the spectra of elastostatic and hydrostatic layer potentials on curvilinear polygons. J. Fourier Anal. Appl. **8**(5), 443–487 (2002)
22. D. Mitrea, I. Mitrea, M. Mitrea, M. Taylor, The Hodge-Laplacian: boundary value problems on Riemannian manifolds. Studies Math. **64**, De Gruyter (2016)
23. I. Mitrea, M. Mitrea, M. Taylor, *Riemann-Hilbert Problems, Cauchy Integrals, and Toeplitz Operators on Uniformly Rectifiable Domains*. Book manuscript, in preparation (2017)
24. D. Mitrea, I. Mitrea, M. Mitrea, *A Sharp Divergence Theorem with Non-Tangential Pointwise Traces and Applications to Singular Integrals and Boundary Problems*. Book manuscript (2020)
25. F. Nazarov, X. Tolsa, A. Volberg, On the uniform rectifiability of AD-regular measures with bounded Riesz transform operator: the case of codimension 1. Acta Math. **213**, 237–321 (2014)
26. J. Plemelj, Potentialtheoretische Untersuchungen, in *Preisschriften der Fürstlich Jablonowskischen Gesellschaft zu Leipzig*. (Teubner, Leipzig, 1911)
27. D. Sarason, Functions of vanishing mean oscillation. Trans. Am. Math. Soc. **207**, 391–405 (1975)
28. S.W. Semmes, A criterion for the boundedness of singular integrals on hypersurfaces. Trans. Am. Math. Soc. **311**(2), 501–513 (1989)
29. V.Yu. Shelepov, On the index of an integral operator of the potential type in the space $L_p$. Dokl. Akad. Nauk SSSR **186**(6), 1266–1268, (1969). (English translation: Sov. Math. Dokl. **10/A**, 754–757)

# Weighted Chebyshev Polynomials on Compact Subsets of the Complex Plane

**Galen Novello, Klaus Schiefermayr, and Maxim Zinchenko**

**Abstract** We study weighted Chebyshev polynomials on compact subsets of the complex plane with respect to a bounded weight function. We establish existence and uniqueness of weighted Chebyshev polynomials and derive weighted analogs of Kolmogorov's criterion, the alternation theorem, and a characterization due to Rivlin and Shapiro. We derive invariance of the Widom factors of weighted Chebyshev polynomials under polynomial pre-images and a comparison result for the norms of Chebyshev polynomials corresponding to different weights. Finally, we obtain a lower bound for the Widom factors in terms of the Szegő integral of the weight function and discuss its sharpness.

**Keywords** Weighted Chebyshev polynomials · Bernstein–Walsh inequality · Szegő lower bound

**Mathematics Subject Classification (2020)** 41A50, 30C10, 30E10

## 1 Introduction

Let $K \subset \mathbb{C}$ be a compact set in the complex plane. In this paper, we consider *weighted Chebyshev polynomials* with respect to a weight function $w : K \to [0, \infty)$. For this purpose, we denote by $T_{n,w}^{(K)}(z)$ the weighted Chebyshev polynomial of degree $n$ on $K$ with respect to $w$, that is, the minimizer of $\|P_n\|_{K,w} := \|wP_n\|_K$ among all monic polynomials $P_n$ of degree $n$, where $\|\cdot\|_K$ is the supremum norm

G. Novello · M. Zinchenko (✉)
Department of Mathematics and Statistics, University of New Mexico, Albuquerque, NM, USA
e-mail: gnovello@unm.edu; maxim@math.unm.edu

K. Schiefermayr
University of Applied Sciences Upper Austria, Wels, Austria
e-mail: klaus.schiefermayr@fh-wels.at



F. Gesztesy, A. Martinez-Finkelshtein (eds.), *From Operator Theory to Orthogonal Polynomials, Combinatorics, and Number Theory*, Operator Theory: Advances and Applications 285, https://doi.org/10.1007/978-3-030-75425-9_18

on $K$. Moreover, we are interested in the norm of $T_{n,w}^{(K)}$, denoted by

$$t_n(K, w) := \|T_{n,w}^{(K)}\|_{K,w}. \tag{1}$$

To avoid the trivial case, we assume that $w$ is nonzero on at least $n + 1$ points of $K$ in which case $\|\cdot\|_{K,w}$ defines a norm on the space of degree $n$ polynomials and hence $t_n(K, w) > 0$. When $K$ has positive logarithmic capacity cap$(K)$ we are also interested in the so-called *Widom factors*, defined by

$$\mathcal{W}_n(K, w) := \frac{t_n(K, w)}{\text{cap}(K)^n}. \tag{2}$$

Our contribution in this paper is twofold. First, in Sect. 2, we discuss existence, uniqueness, and characterization of weighted Chebyshev polynomials for rather general weight functions $w$ on a compact set $K \subset \mathbb{C}$. In particular, we derive weighted analogs of Kolmogorov's criterion, the Rivlin–Shapiro characterization, and the alternation theorem in the case of $K \subset \mathbb{R}$. Our discussion parallels the classical case of the constant weight $w_0 = 1$. While the topic of weighted Chebyshev polynomials is certainly well known to experts we were not able to find a suitable reference that addresses the basic questions in full generality. As an application of Kolmogorov's criterion we also derive an invariance result for the weighted Widom factors under polynomial pre-images.

Our second contribution consists of several inequalities for norms and Widom factors of weighted Chebyshev polynomials. In Sect. 3 we derive a comparison result for Chebyshev norms corresponding to two different weights. Applied to the case of a constant weight the comparison result provides a way of estimating weighted Chebyshev norms in terms of the unweighted ones. Then we discuss a class of Szegő weights and derive a weighted analog of the Bernstein–Walsh inequality. We finish the paper with a detailed discussion of a Szegő-type lower bound for the weighted Widom factors,

$$\mathcal{W}_n(K, w) \geq \exp\left[\int \log w(\zeta)\, d\rho_K(\zeta)\right], \tag{3}$$

where $\rho_K$ denotes the equilibrium measure of $K$. We derive a characterization for equality in the above lower bound for a given $n$ and as a consequence show that the inequality is always strict when $K \subset \mathbb{R}$. In the real setting we also show that unlike the unweighted case the above weighted lower bound is sharp in the class of polynomial weights $w$.

## 2 Existence, Uniqueness, and Characterization of Weighted Chebyshev Polynomials

In this section we discuss existence, uniqueness, and characterization of weighted Chebyshev polynomials. Throughout the section we assume that $K \subset \mathbb{C}$ is a compact set and $w : K \to [0, \infty)$ is a bounded weight function nonzero on at least $n + 1$ points of $K$. We note that $w$ is allowed to have discontinuities and zeros on $K$. The above assumptions on $w$ are the minimal ones that guarantee that $\|\cdot\|_{K,w}$ is a norm on the space of polynomials of degree $n$. However, for our discussion of weighted Chebyshev polynomials it will be convenient to assume that $w$ is upper semi-continuous (i.e., $w(z) \geq \limsup_{K \ni \zeta \to z} w(\zeta)$ for each accumulation point $z$ of $K$). The following lemma shows that this can be done without loss of generality by replacing $w$ with $\hat{w} : K \to [0, \infty)$ defined by

$$\hat{w}(z) = \lim_{r \to 0^+} \sup_{\zeta \in D(z,r) \cap K} w(\zeta), \quad z \in K, \tag{4}$$

where $D(z, r)$ denotes the open disk of radius $r$ centered at $z$.

**Lemma 1** *The weight function $\hat{w}$ is upper semi-continuous on $K$, $\hat{w} \geq w$ on $K$, and $\|wf\|_K = \|\hat{w}f\|_K$ for any continuous function $f$ on $K$.*

***Proof*** Let $z_0 \in K$ and $\varepsilon > 0$. Then, by (4), there exists $r > 0$ such that $w(z) \leq \hat{w}(z_0) + \varepsilon$ for all $z \in D(z_0, r) \cap K$. This together with (4) implies that $\hat{w}(z) \leq \hat{w}(z_0) + \varepsilon$ for all $z \in D(z_0, r) \cap K$. Thus, $\hat{w}$ is upper semi-continuous on $K$. We also get the inequality $w(z_0) \leq \hat{w}(z_0)$.

Let $f$ be a continuous function on $K$. Since $\hat{w}$ is upper semi-continuous and $K$ is compact, there exists $z_0 \in K$ such that $\hat{w}(z_0)|f(z_0)| = \|\hat{w}f\|_K$. Using (4), pick a sequence $\{z_n\} \subset K$ such that $z_n \to z_0$ and $w(z_n) \to \hat{w}(z_0)$. Then $\|wf\|_K \geq \hat{w}(z_0)|f(z_0)| = \|\hat{w}f\|_K$. The opposite inequality $\|wf\|_K \leq \|\hat{w}f\|_K$ follows from $w \leq \hat{w}$. □

Let $\mathbb{P}_n$ denote the space of polynomials of degree at most $n$. By assumption $w(z) \neq 0$ for at least $n+1$ points $z \in K$, hence $\|\cdot\|_{K,w}$ is a norm on $\mathbb{P}_n$. Since every finite dimensional normed linear space over $\mathbb{C}$ is complete, there exists $p_{n-1} \in \mathbb{P}_{n-1}$ that minimizes $\|z^n - p_{n-1}\|_{K,w}$. Thus, there always exists a weighted Chebyshev polynomial on $K$ given by $T_{n,w}^{(K)} = z^n - p_{n-1}$.

Next, we show that Kolmogorov's criterion holds for weighted Chebyshev polynomials. We call a point $z \in K$ a *$w$-extremal point* of a polynomial $P$ on $K$ if $w(z)|P(z)| = \|wP\|_K = \|P\|_{K,w}$. In general a polynomial $P$ might have no $w$-extremal points on $K$ due to discontinuities of $w$, however the next lemma shows that such pathological cases do not occur when $w$ is upper semi-continuous.

**Lemma 2** *If $w$ is an upper semi-continuous weight on $K$, then for any polynomial $P$ the set of all its $w$-extremal points $K_0$ is a nonempty compact set. If $\|P\|_{K,w} > 0$ the restriction of $w$ to $K_0$ is a positive continuous function.*

***Proof*** Since $w$ is upper semi-continuous, so is $w|P|$ and hence $w|P|$ attains a maximum on the compact set $K$. Thus, the set $K_0 \subset K$ of $w$-extremal points is nonempty. Since $w|P| \leq \|wP\|_K$ with equality attained on $K_0$, it follows from the upper semi-continuity of $w|P|$ that any limit point of $K_0$ is a $w$-extremal point so $K_0$ is closed and hence compact. If $\|wP\|_K > 0$ we have $|P| > 0$ on $K_0$ hence $1/|P|$ is continuous on $K_0$ and so the restriction $w|_{K_0} = \|wP\|_K/|P|$ is a positive continuous function. □

**Theorem 3** *Suppose $w$ is an upper semi-continuous weight on $K$ and $P_n$ is a monic polynomial of degree $n$. Then $P_n$ is a weighted Chebyshev polynomial on $K$ if and only if for each $q \in \mathbb{P}_{n-1}$,*

$$\max_{z_0 \in K_0} \operatorname{Re}\left[P_n(z_0)\overline{q(z_0)}\right] \geq 0, \tag{5}$$

*where $K_0$ is the compact set of $w$-extremal points of $P_n$ on $K$.*

*Remark 4* In (5) one can replace $P_n(z_0)$ with $\operatorname{sgn}(P_n(z_0))$, where the sign function is defined by $\operatorname{sgn}(z) = z/|z|$.

***Proof*** Suppose (5) holds and let $z_0 \in K_0$ denote a point for which the maximum is attained. Then for any monic polynomial $Q$ of degree $n$ we have $Q = P_n + q$ for some $q \in \mathbb{P}_{n-1}$ and so, by (5),

$$|Q(z_0)|^2 = |P_n(z_0)|^2 + |q(z_0)|^2 + 2\operatorname{Re}\left[P_n(z_0)\overline{q(z_0)}\right] \geq |P_n(z_0)|^2. \tag{6}$$

It follows that

$$\|Q\|_{K,w} \geq |w(z_0)Q(z_0)| \geq |w(z_0)P_n(z_0)| = \|P_n\|_{K,w}, \tag{7}$$

so $P_n$ is a weighted Chebyshev polynomial on $K$.

Conversely, suppose (5) does not hold, that is, there exists $q \in \mathbb{P}_{n-1}$ such that $\operatorname{Re}[P_n(z_0)\overline{q(z_0)}] < 0$ for all $z_0 \in K_0$. Then since $K_0$ is compact, there exists an open set $U \supset K_0$ and $M > 0$ such that $\operatorname{Re}[P_n(z)\overline{q(z)}] \leq -M$ for all $z \in \overline{U}$. Hence for all sufficiently small $\varepsilon > 0$ the $n$-th degree monic polynomial $Q_\varepsilon = P_n + \varepsilon q$ satisfies

$$\begin{aligned} |Q_\varepsilon(z)|^2 &= |P_n(z)|^2 + \varepsilon^2|q(z)|^2 + 2\varepsilon \operatorname{Re}\left[P_n(z)\overline{q(z)}\right] \\ &\leq |P_n(z)|^2 + \varepsilon^2\|q\|_K^2 - 2\varepsilon M < |P_n(z)|^2, \quad z \in \overline{U} \cap K. \end{aligned} \tag{8}$$

Since $w$ is upper semi-continuous, so is $w|Q_\varepsilon|$ and hence $w|Q_\varepsilon|$ attains a maximum on the compact set $\overline{U} \cap K$, that is, $\|Q_\varepsilon\|_{\overline{U}\cap K,w} = w(z_1)|Q_\varepsilon(z_1)|$ for some $z_1 \in \overline{U} \cap K$. Then, by (8), $\|Q_\varepsilon\|_{\overline{U}\cap K,w} < w(z_1)|P_n(z_1)| \leq \|P_n\|_{K,w}$ if $w(z_1) \neq 0$ and otherwise $\|Q_\varepsilon\|_{\overline{U}\cap K,w} = 0 < \|P_n\|_{K,w}$. In either case, $\|Q_\varepsilon\|_{\overline{U}\cap K,w} < \|P_n\|_{K,w}$.

Now we estimate $\|Q_\varepsilon\|_{K\setminus U,w}$. Since $K\setminus U$ is compact and $w|P_n|$ is upper semi-continuous, there exists $z_2 \in K\setminus U$ such that $\|P_n\|_{K\setminus U,w} = w(z_2)|P_n(z_2)|$. Since

$K_0 \subset U$, $z_2 \notin K_0$ and hence $\|P_n\|_{K\setminus U,w} = w(z_2)|P_n(z_2)| < \|P_n\|_{K,w}$. Then $\|Q_\varepsilon\|_{K\setminus U,w} \le \|P_n\|_{K\setminus U,w} + \varepsilon\|q\|_{K\setminus U,w} < \|P_n\|_{K,w}$ for sufficiently small $\varepsilon > 0$. Combining with the earlier inequality $\|Q_\varepsilon\|_{\overline{U}\cap K,w} < \|P_n\|_{K,w}$ we get $\|Q_\varepsilon\|_{K,w} < \|P_n\|_{K,w}$, so $P_n$ is not a weighted Chebyshev polynomial on $K$. □

**Corollary 5** *If $w$ is upper semi-continuous on $K$, then a weighted Chebyshev polynomial of degree $n$ on $K$ has at least $n+1$ $w$-extremal points.*

***Proof*** Suppose by contradiction that a weighted Chebyshev polynomial $T_{n,w}^{(K)}$ has $m \le n$ $w$-extremal points $z_1, \ldots, z_m$ on $K$. By Lagrange interpolation there exists $q \in \mathbb{P}_{n-1}$ such that $q(z_j) = -T_{n,w}^{(K)}(z_j)$ for all $j = 1, \ldots, m$. Since $\|T_{n,w}^{(K)}\|_{K,w} > 0$ we have $T_{n,w}^{(K)}(z_j) \neq 0$, and hence,

$$\mathrm{Re}\left[T_{n,w}^{(K)}(z_j)\overline{q(z_j)}\right] = -|T_{n,w}^{(K)}(z_j)|^2 < 0, \quad j = 1, \ldots, m, \tag{9}$$

which is a contradiction to Kolmogorov's criterion. □

**Corollary 6** *The weighted Chebyshev polynomial $T_{n,w}^{(K)}$ is unique.*

***Proof*** By Lemma 1 we may assume without loss of generality that $w$ is upper semi-continuous on $K$. Let $P_n$ be a weighted Chebyshev polynomials of degree $n$ on $K$, then $t_n(K, w) = \|T_{n,w}^{(K)}\|_{K,w} = \|P_n\|_{K,w}$.

Let $Q = \frac{1}{2}(P_n + T_{n,w}^{(K)})$, then $Q$ is a monic polynomial of degree $n$ and, by triangle inequality, $\|Q\|_{K,w} \le t_n(K, w)$. Thus, $Q$ is also a weighted Chebyshev polynomial on $K$ and so, by the previous corollary, $Q$ has at least $n+1$ $w$-extremal points $z_1, \ldots, z_{n+1}$. Then, by triangle inequality, we get for each $j = 1, \ldots, n+1$,

$$\begin{aligned}\frac{t_n(K, w)}{w(z_j)} &= |Q(z_j)| = \frac{1}{2}\left|P_n(z_j) + T_{n,w}^{(K)}(z_j)\right| \\ &\le \frac{1}{2}\Big(\left|P_n(z_j)\right| + \left|T_{n,w}^{(K)}(z_j)\right|\Big) = \frac{t_n(K, w)}{w(z_j)},\end{aligned} \tag{10}$$

which implies $P_n(z_j) = T_{n,w}^{(K)}(z_j)$, $j = 1, \ldots, n+1$, and hence, $P_n \equiv T_{n,w}^{(K)}$. □

In the special case $K \subset \mathbb{R}$ the weighted Kolmogorov criterion yields a weighted version of the well known Alternation Theorem:

**Theorem 7** *Suppose $K \subset \mathbb{R}$ and let $w$ be an upper semi-continuous weight on $K$. Then:*

(i) *The $w$-Chebyshev polynomial $T_{n,w}^{(K)}$ is real.*
(ii) *Let $P_n$ be a real monic polynomial of degree $n$. Then $P_n = T_{n,w}^{(K)}$ if and only if $P_n$ has $n$ sign changes on the set of its $w$-extremal points on $K$, that is, there exist points $x_0 < x_1 < \cdots < x_n$ on $K$ such that*

$$w(x_j)P_n(x_j) = (-1)^{n-j}\|P_n\|_{K,w}, \quad j = 0, 1, \ldots, n. \tag{11}$$

***Proof***

(i) Since $|\mathrm{Re}[T_{n,w}^{(K)}(x)]| \le |T_{n,w}^{(K)}(x)|$ and $\mathrm{Re}[T_{n,w}^{(K)}(x)]$ is a monic polynomial of $x \in \mathbb{R}$, it follows from uniqueness of the weighted Chebyshev polynomial, Corollary 6, that $T_{n,w}^{(K)}$ must be real.

(ii) Let $P_n$ be a real monic polynomial of degree $n$ such that $P_n$ alternates sign on some $w$-extremal points $x_0 < x_1 < \cdots < x_n$ and suppose by contradiction that $P_n$ is not the $w$-Chebyshev polynomial. Then by Kolmogorov's criterion, Theorem 3, there exists a polynomial $q \in \mathbb{P}_{n-1}$ such that

$$\mathrm{Re}\left[P_n(x_j)\overline{q(x_j)}\right] < 0, \quad j = 0, \ldots, n. \tag{12}$$

Then, the real polynomial $r(x) = \mathrm{Re}[q(x)]$ of degree at most $n-1$ alternates sign on $x_0 < x_1 < \cdots < x_n$ hence $r(x)$ has at least $n$ zeros and so $r(x)$ must be identically zero, a contradiction.

Now, let $P_n = T_{n,w}^{(K)}$ and suppose by contradiction that $P_n$ has fewer than $n$ sign changes on the set of its $w$-extremal points. Partition the real line by the real zeros of $P_n$ into $m \le n+1$ sets (open intervals). If $P_n$ has the same sign on two consecutive sets, replace the two sets by their union and repeat the process until $P_n$ takes alternating signs on consecutive sets. Then if a set does not contain any $w$-extremal point replace the set and its nearest neighbors by their union. This will result in $m \le n$ sets since by assumption $P_n$ has at most $n-1$ sign changes on its $w$-extremal points. Now let $q$ be the degree $m-1$ polynomial with zeros between consecutive sets and leading coefficient $-1$. Then $q \in \mathbb{P}_{n-1}$ and by construction $q$ has the opposite sign to $P_n$ at each $w$-extremal point of $P_n$. The existence of such a polynomial $q$ contradicts (5). Thus, $P_n$ must have $n$ real simple zeros which partition $\mathbb{R}$ into $n+1$ intervals each of which contains a $w$-extremal point and hence $P_n$ has $n$ sign changes on the set of its $w$-extremal points. □

We also mention a reformulation of Kolmogorov's criterion due to Rivlin and Shapiro. The following weighted version follows from Theorem 3 as in the classical unweighted case [4, Thm. 2.4].

**Theorem 8** *Suppose $w$ is an upper semi-continuous weight on $K$ and $P_n$ is a monic polynomial of degree $n$. Then $P_n$ is a weighted Chebyshev polynomial on $K$ if and only if there exist $m \le 2n+1$ $w$-extremal points $z_1, \ldots, z_m$ and positive number $\alpha_1, \ldots, \alpha_m$ such that $\sum_{j=1}^m \alpha_j = 1$ and for each $q \in \mathbb{P}_{n-1}$,*

$$\sum_{j=1}^{m} \alpha_j P_n(z_j)\overline{q(z_j)} = 0. \tag{13}$$

**Corollary 9** *Suppose $w$ is an upper semi-continuous weight on $K$ and $P_n$ is a monic polynomial of degree $n$. Then $P_n$ is the weighted Chebyshev polynomial on $K$ if and only if there exist $n+1 \le m \le 2n+1$ $w$-extremal points $z_1, \ldots, z_m$ and* *positive numbers* $\lambda_1, \ldots, \lambda_m$ *such that*

$$\sum_{j=1}^{m} \lambda_j z_j^k \operatorname{sgn} \overline{P_n(z_j)} = 0, \quad k = 0, \ldots, n-1, \tag{14}$$

*where* $\operatorname{sgn}(z) = z/|z|$.

***Proof*** This is a straightforward restatement of the previous theorem. First, note that (13) may hold only if $m > n$ since otherwise, by Lagrange interpolation, there exists $q \in \mathbb{P}_{n-1}$ such that $q(z_j) = P_n(z_j)$ for all $j = 1, \ldots, m$ making the left hand-side of (13) strictly positive.

By linearity, it suffices to consider in (13) only the monomials $q(z) = z^k$, $k = 0, \ldots, n-1$. In addition, since (13) is unaffected by an overall factor, the normalization $\sum_{j=1}^m \alpha_j = 1$ can be dropped. Then equivalence of (13) and (14) follows from setting $\lambda_j = \alpha_j |P_n(z_j)|$ and taking complex conjugation. □

We finish this section with a weighted analog of the invariance of Chebyshev norms and Widom factors under polynomial pre-images [5], [3, Sect. 6].

**Theorem 10** *Let* $L \subset \mathbb{C}$ *be a compact set containing at least* $n+1$ *points,* $w_L$ *an upper semi-continuous weight on* $L$*, and* $p(z) = a_m z^m + \cdots + a_0$ *a polynomial of degree* $m$*. Define* $K = p^{-1}(L)$ *and* $w_K(z) = w_L(p(z))$*. Then*

$$T_{nm,w_K}^{(K)}(z) = T_{n,w_L}^{(L)}(p(z))/a_m^n, \tag{15}$$

$$t_{nm}(K, w_K) = t_n(L, w_L)/|a_m^n|, \tag{16}$$

*and if, in addition,* $\operatorname{cap}(L) > 0$,

$$\mathcal{W}_{nm}(K, w_K) = \mathcal{W}_n(L, w_L). \tag{17}$$

***Proof*** Suppose by contradiction that the monic polynomial of degree $nm$,

$$R(z) = T_{n,w_L}^{(L)}(p(z))/a_m^n \tag{18}$$

is not the weighted Chebyshev polynomial on $K$ with respect to the weight $w_K$. Then by Kolmogorov's criterion, Theorem 3, there exists a polynomial $q(z) = c_{nm-1}z^{nm-1} + \cdots + c_0$ such that

$$\max_{z_0 \in K_0} \operatorname{Re}\left[R(z_0)\overline{q(z_0)}\right] < 0, \tag{19}$$

where $K_0$ is the set of $w_K$-extremal points of $R(z)$ on $K$.

For each $\zeta$ let $z_1, \ldots, z_m$ denote the solutions of $p(z) = \zeta$ and define

$$S(\zeta) = \sum_{k=1}^{m} q(z_k) = \sum_{j=0}^{nm-1} c_j S_j(\zeta), \tag{20}$$

where $S_j(\zeta) := z_1^j + \cdots + z_m^j$. We will show that $S(\zeta)$ is a polynomial of degree at most $n-1$. By Newton's identities for the power sums $S_j(\zeta)$ of the roots of the polynomial $p(z) - \zeta$, we have

$$\sum_{j=1}^{k} a_{m-k+j} S_j(\zeta) - k a_{m-k} = 0, \quad k < m, \tag{21}$$

$$\sum_{j=k-m+1}^{k} a_{m-k+j} S_j(\zeta) + (a_0 - \zeta) S_{k-m}(\zeta) = 0, \quad k \geq m. \tag{22}$$

Since $a_m \neq 0$, it follows from (21) that $S_k(\zeta)$ is independent of $\zeta$ for each $k = 1, \ldots, m-1$. Then (22) implies that $S_k(\zeta)$ is a polynomials of degree at most 1 for each $k = m, \ldots, 2m-1$. Proceeding by induction, shows that $S_k(\zeta)$ is a polynomial of degree at most $d$ for each $k = dm, \ldots, (d+1)m - 1$. Thus, by (20), $S(\zeta)$ is a polynomial of degree at most $n-1$.

Let $L_0$ be the set of $w_L$-extremal points of $T_{n,w_L}^{(L)}$ on $L$. Then for each $\zeta \in L_0$ the solutions $z_1, \ldots, z_m$ of $p(z) = \zeta$ are the $w_K$-extremal points of $R(z)$ on $K$ by (18). Then by (19) and (18),

$$\operatorname{Re}\left[T_{n,w_L}^{(L)}(\zeta)\overline{q(z_j)}\right] = \operatorname{Re}\left[R(z_j)\overline{q(z_j)}\right] < 0, \quad j = 1, \ldots, m. \tag{23}$$

Summing the inequalities and using (20) yield $\operatorname{Re}[T_{n,w_L}^{(L)}(\zeta)\overline{S(\zeta)}] < 0, \zeta \in L_0$. Then since $S \in \mathbb{P}_{n-1}$, Kolmogorov's criterion implies that $T_{n,w_L}^{(L)}$ is not the $w_L$-Chebyshev polynomial on $L$ which is a contradiction. Thus, $R(z)$ must be the $w_K$-Chebyshev polynomial on $K$, that is, (15) holds. Equality for the Chebyshev norms (16) then follows from (15). By Ransford [8, Thm. 5.2.5], $\operatorname{cap}(K) = (\operatorname{cap}(L)/|a_m|)^{1/m}$ hence (17) follows from (16) and (2). □

## 3 Bounds for Weighted Chebyshev Polynomials

In this section we derive several inequalities for weighted Chebyshev polynomials. Throughout this section we assume that $K \subset \mathbb{C}$ is an infinite compact set and $w : K \to [0, \infty)$ is a bounded weight function nonzero at infinitely many points of $K$. We start with an elementary comparison theorem.

**Theorem 11** *Let $w_1$, $w_2$ be two weights on a compact set $K \subset \mathbb{C}$ and suppose that $w_2(z) \neq 0$ for all $z \in K$. Then*

$$t_n(K, w_1) \leq \sup_{z \in K} \frac{w_1(z)}{w_2(z)} t_n(K, w_2), \quad n \in \mathbb{N}. \tag{24}$$

*In particular, if $w$ is a weight on $K$ then comparison to the constant weight $w_0 = 1$ gives the bounds*

$$\inf_{z \in K} w(z)\, t_n(K, w_0) \leq t_n(K, w) \leq \sup_{z \in K} w(z)\, t_n(K, w_0), \quad n \in \mathbb{N}. \tag{25}$$

***Proof*** Let $T_{n,w_j}^{(K)}$, $j = 1, 2$, be minimizers for the two weighted problems on $K$. Then

$$\begin{aligned} t_n(K, w_1) &= \|w_1 T_{n,w_1}^{(K)}\|_K \leq \|w_1 T_{n,w_2}^{(K)}\|_K = \|(w_1/w_2) w_2 T_{n,w_2}^{(K)}\|_K \\ &\leq \sup_{z \in K} \frac{w_1(z)}{w_2(z)} \|w_2 T_{n,w_2}^{(K)}\|_K = \sup_{z \in K} \frac{w_1(z)}{w_2(z)} t_n(K, w_2). \end{aligned} \tag{26}$$

Applying (24) to $w_1 = w$, $w_2 = 1$ then yields the upper bound in (25). If $w$ has a zero on $K$ the lower bound in (25) is trivial, otherwise applying (24) to $w_1 = 1$, $w_2 = w$ yield the lower bound in (25). □

The lower bound in (25) can be improved for a class of Szegő weights which includes weights with $\inf_{z \in K} w(z) = 0$. Before we turn to a discussion of such weights we recall several concepts from potential theory that can be found in [6–9, 12].

In the following we assume that $K \subset \mathbb{C}$ has positive logarithmic capacity $\text{cap}(K) > 0$ and denote by $\Omega$ the *outer domain* of $K$, that is, the unbounded component of $\overline{\mathbb{C}} \backslash K$. The *polynomial convex hull* of $K$ is the set $\hat{K} := \mathbb{C} \backslash \Omega$ and $O\partial K := \partial \Omega = \partial \hat{K}$ is the *outer boundary* of $K$. The assumption of positive capacity guarantees that there exists a unique *equilibrium measure* $\rho_K$ of $K$ supported on $O\partial K$ and a *Green function* $g_K(z)$ for the domain $\Omega$, given by (see e.g., [9, Eq. (I.4.8)]),

$$g_K(z) = -\log \text{cap}(K) + \int \log |z - \zeta|\, d\rho_K(\zeta). \tag{27}$$

By the above formula, the Green function $g_K(z)$ extends to the entire complex plane with $g_K(z) \geq 0$ for all $z \in \overline{\mathbb{C}}$ and $g_K(z) = 0$ for all $z \in \mathbb{C} \backslash \overline{\Omega}$ and quasi-every $z \in \partial \Omega$, see [9, Cor. I.4.5]. The points $z \in \partial \Omega$ where $g_K(z)$ is continuous are called *regular*. By Saff and Totik [9, Thms. I.4.4, II.3.5], $z \in \partial \Omega$ is a regular point if and only if $g_K(z) = 0$. We also recall a similar notion of the Green function for $\Omega$ with a logarithmic pole at a finite point $a \in \Omega$ which we denote by $g_K(z, a)$. By Sect. II.4

and, in particular, equation (4.31) in [9], we have

$$g_K(z,a) = \log\frac{1}{|z-a|} + \int \log|z-\zeta|\,d\omega_a(\zeta) + g_K(a), \tag{28}$$

where $\omega_z$ is the *harmonic measure* for $\Omega$. By Saff and Totik [9, Thm. II.4.4], $g_K(\cdot, a)$ given by the above formula extends to the entire complex plane with $g_K(z,a) \geq 0$ for all $z \in \overline{\mathbb{C}}$ and $g_K(z,a) = 0$ for all $z \in \mathbb{C}\backslash\overline{\Omega}$ and all regular points $z \in \partial\Omega$.

For a bounded Borel measurable weight $w$ on $K$ we define the exponential Szegő integral of $w$ by

$$S(w) := \exp\left[\int \log w(\zeta)\,d\rho_K(\zeta)\right]. \tag{29}$$

We say that a bounded weight $w$ is of *Szegő class* if $\log w \in L^1(d\rho_K)$, equivalently, $S(w) > 0$. For explicit computations, it is convenient to note that $S(\cdot)$ is multiplicative, that is, $S(f^\alpha g^\beta) = S(f)^\alpha S(g)^\beta$ and for the special weight $w(z) = |z - z_0|$ we have, by (27),

$$S(|z - z_0|) = e^{g_K(z_0)}\,\mathrm{cap}(K). \tag{30}$$

This, in particular, allows us to evaluate $S(|f|)$ for polynomial and rational functions $f$.

Finally, we recall that $\rho_K = \omega_\infty$ by Ransford [8, Thm. 4.3.14] and that all harmonic measures for $\Omega$ are comparable, $c_1(z)\rho_K \leq \omega_z \leq c_2(z)\rho_K$, by Ransford [8, Cor. 4.3.5]. Then for any $f \in L^1(d\rho_K)$ we have a well defined generalized Poisson integral,

$$\mathcal{PI}_\Omega(f, z) := \int f(\zeta)\,d\omega_z(\zeta), \quad z \in \Omega. \tag{31}$$

For later reference we also note that

$$\mathcal{PI}_\Omega(\log f; \infty) = \log S(f). \tag{32}$$

Our first result for Szegő class weights is the following weighted analog of the Bernstein–Walsh inequality. For a recent review and an extension of the classical unweighted Bernstein–Walsh inequality see [11].

**Theorem 12** *Let $K \subset \mathbb{C}$ be a compact set with* $\mathrm{cap}(K) > 0$ *and $w$ be a Szegő class weight on $K$. Then for any polynomial $P_n$ of degree $n$,*

$$|P_n(z)| \leq \|P_n\|_{K,w} \exp[-\mathcal{PI}_\Omega(\log w, z) + n g_K(z)], \quad z \in \Omega. \tag{33}$$

***Proof*** WLOG assume that $P_n$ is monic and let $P_n(\zeta) = \prod_{j=1}^{n}(\zeta - z_j)$. Then, by (31), (28), and linearity of $\mathcal{PI}_\Omega(\cdot, z)$ we have

$$\mathcal{PI}_\Omega(\log|P_n|, z) = \int \log\left(\prod_{j=1}^{n}|\zeta - z_j|\right) d\omega_z(\zeta) = \sum_{j=1}^{n}\int \log|\zeta - z_j|\, d\omega_z(\zeta)$$

$$= \sum_{j=1}^{n}\left[g_K(z_j, z) - \log\frac{1}{|z - z_j|} - g_K(z)\right]$$

$$= \log|P_n(z)| + \sum_{j=1}^{n} g_K(z_j, z) - n g_K(z), \quad z \in \Omega, \tag{34}$$

and since $\log[w|P_n|] \le \log\|P_n\|_{K,w}$ on $K$ and $\omega_z$ is a probability measure,

$$\log\|P_n\|_{K,w} \ge \mathcal{PI}_\Omega(\log[w|P_n|], z) = \mathcal{PI}_\Omega(\log w, z) + \mathcal{PI}_\Omega(\log|P_n|, z)$$

$$= \mathcal{PI}_\Omega(\log w, z) + \log|P_n(z)| + \sum_{j=1}^{n} g_K(z_j, z) - n g_K(z)$$

$$\ge \mathcal{PI}_\Omega(\log w, z) + \log|P_n(z)| - n g_K(z), \quad z \in \Omega. \tag{35}$$

Exponentiating and rearranging then yields (33). □

Next we derive a Szegő-type lower bound for the weighted Widom factors and show that it is sharp even if $K \subset \mathbb{R}$.

**Theorem 13** *Let $K \subset \mathbb{C}$ be a compact set with $\mathrm{cap}(K) > 0$ and $w$ be a Szegő class weight on $K$. Then for all $n \in \mathbb{N}$,*

$$\mathcal{W}_n(K, w) \ge S(w). \tag{36}$$

*Equality in (36) is attained for some $n$ if and only if there exists a monic polynomial $P_n$ of degree $n$ with all zeros on the set $\{z \in \mathbb{C} : g_K(z) = 0\}$ and such that $w|P_n| = \|P_n\|_{K,w}$ $\rho_K$-a.e, in which case $P_n = T_{n,w}^{(K)}$.*

*If $K \subset \mathbb{R}$ or more generally if $\hat{K}$ has empty interior, then the inequality (36) is strict for all $n \in \mathbb{N}$. Nevertheless, if $K \subset \mathbb{R}$ then for each $n \in \mathbb{N}$ the lower bound (36) is sharp in the class of polynomial weights on $K$, that is,*

$$\inf_{w} \frac{\mathcal{W}_n(K, w)}{S(w)} = 1, \tag{37}$$

*where the infimum is taken over polynomials $w(z)$ positive on $K$.*

***Proof*** To derive (36) we apply Theorem 12 to the polynomial $P_n = T_{n,w}^{(K)}$. Taking $z \to \infty$ in (35), recalling (32), and noting that $\log|z| - g_K(z) \to \log\mathrm{cap}(K)$ and therefore $\log|P_n(z)| - ng_K(z) \to n \log \operatorname{cap}(K)$ we obtain

$$\log \|P_n\|_{K,w} \geq \log S(w|P_n|) = \log S(w) + \sum_{j=1}^{n} g_K(z_j) + n \log \operatorname{cap}(K)$$
$$\geq \log S(w) + n \log \operatorname{cap}(K), \tag{38}$$

where $z_1, \ldots, z_n$ are the zeros of $P_n$. Exponentiation then gives (36).

The equality $\|P_n\|_{K,w} = S(w)\operatorname{cap}(K)^n$ holds if and only if equality is attained in both inequalities of (38), that is, $g_K(z_j) = 0$ for all zeros $z_j$ of $P_n$ and

$$\log S(w|P_n|) = \int \log[w|P_n|]\, d\rho_K = \log \|P_n\|_{K,w}. \tag{39}$$

Since $w|P_n| \leq \|P_n\|_{K,w}$ on $K$, (39) is equivalent to the maximality of the integrand $\rho_K$-a.e., that is, $w|P_n| = \|P_n\|_{K,w}$ $\rho_K$-a.e. Finally, if a monic polynomial $P_n$ satisfies $\|P_n\|_{K,w} = S(w)\operatorname{cap}(K)^n$, then by (36) we have $\|T_{n,w}^{(K)}\|_{K,w} \geq S(w)\operatorname{cap}(K)^n = \|P_n\|_{K,w}$ which, by uniqueness of the Chebyshev polynomial, Corollary 6, implies $P_n = T_{n,w}^{(K)}$.

If $\hat{K}$ has empty interior, then $\overline{\Omega} = \overline{\mathbb{C}}$. Since $\operatorname{supp}(\rho_K) \subset \partial\Omega$ it follows that the complement of the $\operatorname{supp}(\rho_K)$ is connected and hence $g_K > 0$ outside of the support of $\rho_K$. Then we have $\{z \in \mathbb{C} : g_K(z) = 0\} \subset \operatorname{supp}(\rho_K)$ and hence, by the previous part, $T_{n,w}^{(K)}$ has a zero $z_0$ in the support of $\rho_K$. Then every neighborhood of $z_0$ has positive $\rho_K$ measure and since $w$ is bounded $w|T_{n,w}^{(K)}| < \|T_{n,w}^{(K)}\|_{K,w}$ on a sufficiently small neighborhood of $z_0$. Thus, by the previous part, equality in (36) does not occur in this case.

Next we show (37). Fix $n \in \mathbb{N}$ and, by shifting $K$ if necessary, assume that $0 \in K$ and that 0 is a regular point of $K$, that is, $g_K(z) \to 0$ as $z \to 0$. Consider the weights $w_\varepsilon(z) = (z^2 + \varepsilon^2)^{-n/2}$. Then using (30) we obtain

$$S(w_\varepsilon) = S\big(|z + i\varepsilon|^{-n/2}|z - i\varepsilon|^{-n/2}\big) = S\big(|z + i\varepsilon|\big)^{-n/2} S\big(|z - i\varepsilon|\big)^{-n/2}$$
$$= e^{-\frac{n}{2}[g_K(i\varepsilon) + g_K(-i\varepsilon)]} \operatorname{cap}(K)^{-n} \to \operatorname{cap}(K)^{-n} \quad \text{as } \varepsilon \to 0. \tag{40}$$

Since

$$t_n(K, w_\varepsilon) \leq \|x^n\|_{K,w_\varepsilon} = \sup_{x \in K} |w_\varepsilon x^n| = \sup_{x \in K} \left| \frac{x^2}{x^2 + \varepsilon^2} \right|^{n/2} \leq 1, \tag{41}$$

we have $\mathcal{W}_n(K, w_\varepsilon) \leq \operatorname{cap}(K)^{-n}$ and hence

$$\limsup_{\varepsilon \to 0} \mathcal{W}_n(K, w_\varepsilon)/S(w_\varepsilon) \leq 1. \tag{42}$$

Next, fix $\varepsilon \in (0, 1)$ and approximate $w_\varepsilon$ by polynomials $w_j$ in the uniform norm on $K$. Since $w_\varepsilon \geq c > 0$ on $K$ we may assume $w_j > 0$ on $K$ for each $j$ and hence we have

$$\left\| 1 - \frac{w_j}{w_\varepsilon} \right\|_K \leq \frac{1}{c} \|w_\varepsilon - w_j\|_K \to 0 \text{ as } j \to \infty. \tag{43}$$

Then, $\log(w_j/w_\varepsilon) \to 0$ uniformly on $K$ hence

$$\frac{S(w_j)}{S(w_\varepsilon)} = \exp\left[\int \log \frac{w_j(x)}{w_\varepsilon(x)} d\rho_K\right] \to 1 \text{ as } j \to \infty. \tag{44}$$

By (24), we have

$$\limsup_{j\to\infty} t_n(K, w_j) \leq \limsup_{j\to\infty} \|w_j/w_\varepsilon\|_K t_n(K, w_\varepsilon) = t_n(K, w_\varepsilon), \tag{45}$$

so $\limsup_{j\to\infty} \mathcal{W}_n(K, w_j) \leq \mathcal{W}_n(K, w_\varepsilon)$. Combined with (44) this yields

$$\limsup_{j\to\infty} \frac{\mathcal{W}_n(K, w_j)}{S(w_j)} \leq \frac{\mathcal{W}_n(K, w_\varepsilon)}{S(w_\varepsilon)}. \tag{46}$$

Finally, (37) follows from (36), (42), and (46). □

*Remark 14*

(i) A similar lower bound for $L^p$-extremal polynomials, $p < \infty$, has been recently obtained in [2, Thm. 2.1]. In fact, (36) can be alternatively derived from that result.

(ii) For the constant weight $w_0 = 1$ we have $S(w_0) = 1$. In this case it is shown in [3, Thm. 1.2] that the lower bound (36) is saturated, that is, $\mathcal{W}_n(K, w_0) = 1$ if and only if the outer boundary of $K$ is a degree $n$ lemniscate (a polynomial pre-image of the unit circle under a degree $n$ polynomial). A similar saturation result for $L^p(d\rho_K)$-extremal polynomials, $p < \infty$, is derived in [1, Thm. 4.2].

(iii) In the weighted case, the lower bound (36) can be saturated on any set $K$ whose polynomial convex hull has nonempty interior. Indeed, if $\hat{K}$ has nonempty interior, let $P_n$ be a degree $n$ monic polynomial with all zeros in the interior of $\hat{K}$ and set $w(z) = 1/|P_n(z)|$ for all $z \in \partial\hat{K}$ and $w(z) = 0$ everywhere else on $K$. Then since $\text{supp}(\rho_K) \subset \partial\hat{K}$, it follows from Theorem 13 that equality is attained in (36) in this case. In addition, Theorem 13 shows that nonempty interior of $\hat{K}$ is a necessary condition for the equality in (36).

(iv) For the constant weight $w_0 = 1$, it is known [10] that the lower bound doubles for subsets of the real line, that is, $\mathcal{W}_n(K, w_0) \geq 2$ if $K \subset \mathbb{R}$. Theorem 13 shows that no doubling of the lower bound occurs in the class of weighted Chebyshev polynomials on subsets of the real line.

(v) However, there are nonconstant weights $w$ on $K \subset \mathbb{R}$ for which the lower bound (36) doubles. For example, if $K \subset \mathbb{R}$ is a regular compact set and $w(z) = |P_d(z)|$ for some monic polynomial $P_d(z)$ of degree $d$ with all zeros on $K$, then

$$t_n(K, w) = \|P_d T_{n,w}^{(K)}\|_K \geq \|T_{n+d,w_0}^{(K)}\|_K = t_{n+d}(K, w_0), \tag{47}$$

and hence, by the doubled lower bound for the constant weight $w_0 = 1$ and (30), we get

$$\mathcal{W}_n(K, w) \geq \mathcal{W}_{n+d}(K, w_0)\,\mathrm{cap}(K)^d \geq 2\,\mathrm{cap}(K)^d = 2S(w). \tag{48}$$

We pose it as an open problem to find a characterization of weights $w$ on $K = [-1, 1]$ (and more generally on $K \subset \mathbb{R}$) for which the doubled lower bound $\mathcal{W}_n(K, w) \geq 2S(w)$ holds.

**Acknowledgments** M.Z. was supported in part by Simons Foundation grant CGM–581256.

## References

1. G. Alpan, M. Zinchenko, On the Widom factors for $L_p$ extremal polynomials. J. Approx. Theory **259**, 105480 (2020)
2. G. Alpan, M. Zinchenko, Sharp lower bounds for the Widom factors on the real line. J. Math. Anal. Appl. **484**(1), 123729 (2020)
3. J.S. Christiansen, B. Simon, M. Zinchenko, Asymptotics of Chebyshev polynomials, III. Sets saturating Szegő, Schiefermayr, and Totik–Widom bounds. Oper. Theory Adv. Appl. **276**, 231–246 (2020)
4. R.A. DeVore, G.G. Lorentz, Constructive approximation, in *Grundlehren der Mathematischen Wissenschaften [Fundamental Principles of Mathematical Sciences]*, vol. 303 (Springer, Berlin, 1993)
5. S.O. Kamo, P.A. Borodin, Chebyshev polynomials for Julia sets. Moscow Univ. Math. Bull. **49**, 44–45 (1994)
6. N.S. Landkof, *Foundations of Modern Potential Theory* (Springer, Berlin, 1972)
7. A. Martínez-Finkelshtein, Equilibrium problems of potential theory in the complex plane, in *Orthogonal Polynomials and Special Functions*. Lecture Notes in Mathematical, vol. 1883. (Springer, Berlin, 2006), pp. 79–117
8. T. Ransford, *Potential Theory in the Complex Plane* (Cambridge University, Cambridge, 1995)
9. E.B. Saff, V. Totik, Logarithmic potentials with external fields, in *Grundlehren der Mathematischen Wissenschaften [Fundamental Principles of Mathematical Sciences]*, vol. 316 (Springer, Berlin, 1997)
10. K. Schiefermayr, A lower bound for the minimum deviation of the Chebyshev polynomial on a compact real set. East J. Approx. **14**, 223–233 (2008)
11. K. Schiefermayr, The growth of polynomials outside of a compact set—the Bernstein-Walsh inequality revisited. J. Approx. Theory **223**, 9–18 (2017)
12. M. Tsuji, *Potential Theory in Modern Function Theory* (Chelsea Publishing Co., New York, 1975). Reprinting of the 1959 original

# The Eichler Integral of $E_2$ and $q$-brackets of $t$-hook Functions

**Ken Ono**

*For my friend Lance Littlejohn in celebration of his distinguished career*

**Abstract** For functions $f : \mathcal{P} \mapsto \mathbb{C}$ on partitions, Bloch and Okounkov defined a power series $\langle f \rangle_q$ that is the "weighted average" of $f$. As Fourier series in $q = e^{2\pi iz}$, such $q$-brackets generate the ring of quasimodular forms, and the modular forms that are powers of Dedekind's eta-function. Using work of Berndt and Han, we build modular objects from

$$f_t(\lambda) := t \sum_{h \in \mathcal{H}_t(\lambda)} \frac{1}{h^2},$$

weighted sums over partition hook numbers that are multiples of $t$. We find that $\langle f_t \rangle_q$ is the Eichler integral of $(1 - E_2(tz))/24$, which we modify to construct a function $M_t(z)$ that enjoys weight 0 modularity properties. As a consequence, the non-modular Fourier series

$$H_t^*(z) := \sum_{\lambda \in \mathcal{P}} f_t(\lambda) q^{|\lambda| - \frac{1}{24}}$$

inherits weight $-1/2$ modularity properties. These are sufficient to imply a Chowla-Selberg type result, generalizing the fact that weight $k$ algebraic modular forms evaluated at discriminant $D < 0$ points $\tau$ are algebraic multiples of $\Omega_D^k$, the $k$th

The author thanks the support of the Thomas Jefferson Fund and the NSF (DMS-1601306)

K. Ono (✉)
Department of Mathematics, University of Virginia, Charlottesville, VA, USA
e-mail: ko5wk@virginia.edu



F. Gesztesy, A. Martinez-Finkelshtein (eds.), *From Operator Theory to Orthogonal Polynomials, Combinatorics, and Number Theory*, Operator Theory: Advances and Applications 285, https://doi.org/10.1007/978-3-030-75425-9_19

power of the canonical period. If we let $\Psi(\tau) := -\pi i \left(\frac{\tau^2-3\tau+1}{12\tau}\right) - \frac{\log(\tau)}{2}$, then for $t = 1$ we prove that

$$H_1^*(-1/\tau) - \frac{1}{\sqrt{-i\tau}} \cdot H_1^*(\tau) \in \overline{\mathbb{Q}} \cdot \frac{\Psi(\tau)}{\sqrt{\Omega_D}}.$$

**Keywords** $t$-hooks · Partitions · $q$-brackets · Eichler integrals

## 1 Introduction and Statement of Results

A *partition* of a non-negative integer $n$ is any nonincreasing sequence of positive integers which sum to $n$. As usual, if $\lambda = \lambda_1 + \lambda_2 + \cdots + \lambda_s$ is a partition of size $|\lambda| = n$, then we associate the *Ferrers-Young* diagram

$$\begin{array}{llll} \bullet & \bullet & \bullet \cdots \bullet & \leftarrow \lambda_1 \text{ many nodes} \\ \bullet & \bullet & \ldots \quad \bullet & \leftarrow \lambda_2 \text{ many nodes} \\ \vdots & \vdots & \vdots & \\ \bullet & \ldots & \bullet & \leftarrow \lambda_s \text{many nodes.} \end{array}$$

Each node has a *hook number*. For a node in row $i$ and column $j$, it is the positive integer $h(i, j) := (\lambda_i - i) + (\lambda'_j - j) + 1$, where $\lambda'_j$ denotes the number of nodes in column $j$.

Ferrers-Young diagrams and their hook numbers play significant roles in representation theory (for example, see [13]). Indeed, partitions of size $n$ are used to define *Young tableaux*, and their combinatorial properties encode the representation theory of the symmetric group $S_n$. For example, the *t-core partitions* of size $n$ play an important role in number theory (for example, see [9, 10]) and the modular representation theory of $S_n$ and $A_n$ (for example, see Chapter 2 of [13], and [8, 10]). Recall that a partition is a $t$-core if none of its hook numbers are multiples of $t$. If $p$ is prime, then the existence of a $p$-core of size $n$ is equivalent to the existence of a defect 0 $p$-block for both $S_n$ and $A_n$.

In this note we make use of partitions that are not $t$-core, namely those partitions that have some hook numbers that are multiples of $t$. We shall use these partitions to define partition functions $f_t : \mathcal{P} \mapsto \mathbb{Q}$, which in turn we use to define new modular objects.

To make this precise, we recall the framework of $q$-brackets. For functions $f : \mathcal{P} \mapsto \mathbb{C}$ on the integer partitions, Bloch and Okounkov [3] defined the formal power series

$$\langle f \rangle_q := \frac{\sum_{\lambda \in \mathcal{P}} f(\lambda) q^{|\lambda|}}{\sum_{\lambda \in \mathcal{P}} q^{|\lambda|}} \in \mathbb{C}[[q]], \tag{1}$$

which can conceptually be thought of as the "weighted average" of $f$. Schneider [16] developed a "multiplicative theory of partitions" based on these $q$-brackets, which includes partition theoretic analogs of many constructions in classical multiplicative number theory, such as Möbius inversion, Dirichlet convolution, as well as incarnations of sieve methods.

Viewed as Fourier expansions, Bloch and Okounkov [3] showed that the ring of quasimodular forms is generated by the $q$-brackets of distinguished functions $f$ associated to shifted symmetric polynomials, work which was later expanded and refined by Zagier [19]. Recently, Griffin, Jameson, and Trebat-Leder [11] developed a theory of $p$-adic modular forms in the context of these specific $q$-brackets for these shifted symmetric polynomials.

Nekrasov and Okounkov [14] and Westbury [18][1] later defined further functions, say $D_s$, formulated in terms of partition hook numbers. They used these functions to give a partition theoretic description of every power (including complex) of Dedekind's eta-function $\eta(z) := q^{\frac{1}{24}} \prod_{n=1}^{\infty}(1-q^n)$. For $s \in \mathbb{C}$, define the function $D_s : \mathcal{P} \mapsto \mathbb{C}$ by

$$D_s(\lambda) := \prod_{h \in \mathcal{H}(\lambda)} \left(1 - \frac{s}{h^2}\right),$$

where $\mathcal{H}(\lambda)$ denotes the multiset of hook numbers of the partition $\lambda$. A slight reformulation of their formula (see (6.12) of [14]), using Euler's partition generating function

$$\sum_{n=0}^{\infty} p(n) q^n = \sum_{\lambda \in \mathcal{P}} q^{|\lambda|} = \prod_{n=1}^{\infty} \frac{1}{(1-q^n)}, \tag{2}$$

asserts that $q^{\frac{s}{24}} \cdot \langle D_s \rangle_q = \eta(z)^s$. If $s \in \mathbb{Z}$, then we have a weight $s/2$ modular form (see Chapter 1 of [15]), thereby encoding a particularly important family of modular forms.

In view of these constructions of quasimodular forms and powers of Dedekind's eta-function, it is natural to ask whether further modular objects arise from $q$-brackets. To this end, we recall one of the simplest $q$-brackets, the formal power series associated to $f(\lambda) := |\lambda|$, the "size" function. Thanks to (2), a straightforward logarithmic differentiation calculation reveals that

$$\langle |\cdot| \rangle_q = \frac{\sum_{\lambda \in \mathcal{P}} |\lambda| q^{|\lambda|}}{\sum_{\lambda \in \mathcal{P}} q^{|\lambda|}} = \sum_{n=1}^{\infty} \sigma_1(n) q^n = \frac{1 - E_2(z)}{24}, \tag{3}$$

[1] Westbury discovered (see Prop. 6.1 and 6.2 of [18]) the Nekrasov-Okounkov formula concurrently.

where $\sigma_v(n) := \sum_{1\le d|n} d^v$, and $E_2(z) = 1 - 24\sum_{n=1}^{\infty}\sigma_1(n)q^n$ is the weight 2 quasimodular Eisenstein series. Although $E_2(z)$ is not a modular form, it is well known that

$$E_2^*(z) := -\frac{3}{\pi \cdot \mathrm{Im}(z)} + E_2(z)$$

is a non-holomorphic weight 2 modular form (for example, see Chapter 6 of [4]). Therefore, its *modified Eichler integral*

$$\mathcal{E}(z) := \sum_{n=1}^{\infty} \frac{\sigma_1(n)}{n} \cdot q^n = \sum_{n=1}^{\infty} \sigma_{-1}(n) q^n \tag{4}$$

enjoys certain weight 0 modularity properties. These were determined by Berndt in the 1970s [1]. Therefore, in view of (3), it is natural to ask whether $\mathcal{E}(z)$ arises as a $q$-bracket.

We show that this is indeed the case, and the construction makes use of $t$-hooks, the hook numbers which are multiples of $t$. To this end, for each $t \in \mathbb{Z}^+$ we define $f_t : \mathcal{P} \mapsto \mathbb{Q}$ by

$$f_t(\lambda) := t \sum_{h \in \mathcal{H}_t(\lambda)} \frac{1}{h^2}, \tag{5}$$

where $\mathcal{H}_t(\lambda)$ is the multiset of hook numbers which are multiples of $t$.

*Example 1* We consider the partition $\lambda = 4 + 3 + 1$, which has Ferrers-Young diagram

$$\begin{array}{llll} \bullet_6 & \bullet_4 & \bullet_3 & \bullet_1 \\ \bullet_4 & \bullet_2 & \bullet_1 & \\ \bullet_1 & & & \end{array}$$

(the subscripts denote the hook numbers). We find that $\mathcal{H}(\lambda)=\{1, 1, 1, 2, 3, 4, 4, 6\}$, $\mathcal{H}_2(\lambda) = \{2, 4, 4, 6\}$, and $\mathcal{H}_3(\lambda) = \{3, 6\}$. Therefore, we find that

$$f_1(\lambda) = 1 + 1 + 1 + \frac{1}{4} + \frac{1}{9} + \frac{1}{16} + \frac{1}{16} + \frac{1}{36} = \frac{253}{72},$$

$$f_2(\lambda) = 2\left(\frac{1}{4} + \frac{1}{16} + \frac{1}{16} + \frac{1}{36}\right) = \frac{29}{36},$$

$$f_3(\lambda) = 3\left(\frac{1}{9} + \frac{1}{36}\right) = \frac{5}{12}.$$

For convenience, we define

$$H_t(z) := \sum_{\lambda \in \mathcal{P}} f_t(\lambda) q^{|\lambda|}. \tag{6}$$

Work of Han [12] allows us to describe the $q$-brackets of the $f_t$ in terms of $\mathcal{E}(z)$.

**Theorem 1** *If $t$ is a positive integer, then we have*

$$\langle f_t \rangle_q = \prod_{n=1}^{\infty}(1-q^n) \cdot H_t(z) = \mathcal{E}(tz).$$

*Remark 1* In a follow-up to the present note, the author, Bringmann, and Wagner [5] will describe a general framework for Eichler integrals of arbitrary weight Eisenstein series as $q$-brackets of "weighted" $t$-hook functions on partitions. These results will include results where these $q$-brackets are completed to obtain harmonic Maass forms, sesquiharmonic Maass forms (in the case of $E_2$), and holomorphic quantum modular forms.

We now turn to the problem of determining the modularity properties of these $q$-brackets. Using the crucial functions $P_t(z)$ and $L_t(z)$ defined by

$$P_t(z) := -t\left(t + \frac{\pi i}{12}\right) z + \frac{1}{z} \quad \text{and} \quad L_t(z) := -\frac{1}{4} \cdot \log(tz), \tag{7}$$

we define

$$M_t(z) := P_t(z) + L_t(z) + \langle f_t \rangle_q. \tag{8}$$

These functions enjoy weight 0 modularity properties with respect to translation $z \to z+1$ and inversion $z \to -1/t^2 z$.

**Theorem 2** *If $t$ is a positive integer, then the following are true for all $z \in \mathbb{H}$.*

1. *We have that*

$$M_t(z+1) - M_t(z) = -t\left(t + \frac{\pi i}{12}\right) - \frac{1}{z(z+1)} + \frac{1}{4} \log\left(\frac{z}{z+1}\right).$$

2. *We have that*

$$M_t(z) = M_t\left(-\frac{1}{t^2 z}\right).$$

***Two Remarks***

*(1) We use the branch of* $\log z$ *with* $-\pi \leq \arg(z) < \pi$ *(resp.* $\sqrt{z}$ *that is positive on* $\mathbb{R}^+$*).*

*(2) Obviously,* $\langle f_t \rangle_q = \mathcal{E}(tz)$ *is invariant under* $z \to z+1$*. The definition of* $M_t(z)$ *is* motivated by the desire to obtain the more difficult invariance under the inversion $z \to -1/t^2 z$.

Theorem 2 implies modularity properties for the individual series $H_t(z)$ defined in (6). We modify these generating functions to define

$$H_t^*(z) := q^{-\frac{1}{24}} \cdot H_t(z) = \sum_{\lambda \in \mathcal{P}} f_t(\lambda) q^{|\lambda| - \frac{1}{24}}, \tag{9}$$

so that $\langle f_t \rangle_q = \eta(z) \cdot H_t^*(z)$. This allows us to make use of the modularity of Dedekind's eta-function. To ease notation, we define the auxiliary function

$$\Psi(z) := -\pi i \cdot \left( \frac{z^2 - 3z + 1}{12z} \right) - \frac{1}{2} \log(z). \tag{10}$$

For $H_1^*(z)$, these transformation laws are described in terms of $\Psi(z)$ and Dedekind's $\eta(-1/z)$.

**Corollary 1** *If* $z \in \mathbb{H}$*, then the following are true.*

*1. We have that*

$$H_1^*(z+1) = e^{-\frac{\pi i}{12}} \cdot H_1^*(z).$$

*2. We have that*

$$H_1^*(-1/z) - \frac{1}{\sqrt{-iz}} \cdot H_1^*(z) = \frac{\Psi(z)}{\eta(-1/z)}.$$

These results imply a Chowla-Selberg type result, generalizing the classical fact [6, 17] that weight $k$ algebraic modular forms evaluated at discriminant $D < 0$ points $\tau$ are algebraic multiples of $\Omega_D^k$, the $k$th power of a canonical period $\Omega_D$. To make this precise, we let $\overline{\mathbb{Q}}$ denote the algebraic closure of the field of rational numbers. Suppose that $D < 0$ is the fundamental discriminant of the imaginary quadratic field $\mathbb{Q}(\sqrt{D})$. Let $h(D)$ denote the class number of $\mathbb{Q}(\sqrt{D})$, and define $h'(D) := 1/3$ (resp. $1/2$) when $D = -3$ (resp. $-4$), and $h'(D) := h(D)$ when $D < -4$. We then have the canonical period

$$\Omega_D := \frac{1}{\sqrt{2\pi |D|}} \left( \prod_{j=1}^{|D|-1} \Gamma\left( \frac{j}{|D|} \right)^{\chi_D(j)} \right)^{\frac{1}{2h'(D)}}, \tag{11}$$

where $\chi_D(\bullet) := \left(\frac{D}{\bullet}\right)$.

The Chowla-Selberg phenomenon concerns modular forms $f(z)$ with algebraic Fourier coefficients. If $f(z)$ has weight $k \in \mathbb{Z}$ and $\tau \in \mathbb{H} \cap \mathbb{Q}(\sqrt{D})$, then

$$f(\tau) \in \overline{\mathbb{Q}} \cdot \Omega_D^k. \tag{12}$$

We obtain the following generalization of this phenomenon for $H_1^*(-1/z) - \frac{1}{\sqrt{-iz}} \cdot H_1^*(z)$, which is somewhat analogous to similar results obtained by Dawsey and the author [7] in a different partition theoretic setting.

**Corollary 2** *If $\tau \in \mathbb{Q}(\sqrt{D}) \cap \mathbb{H}$, where $D < 0$ is a fundamental discriminant, then*

$$H_1^*(-1/\tau) - \frac{1}{\sqrt{-i\tau}} \cdot H_1^*(\tau) \in \overline{\mathbb{Q}} \cdot \frac{\Psi(\tau)}{\sqrt{\Omega_D}}.$$

*Remark 2* Corollary 2 is a special case of the general fact that

$$\alpha_t(\tau) H_t^*(-1/t^2\tau) - \frac{\beta_t(\tau)}{\sqrt{-i\tau}} \cdot H_t^*(\tau) \in \overline{\mathbb{Q}} \cdot \frac{\Psi(t\tau)}{\sqrt{\Omega_D}},$$

where $\alpha_t(\tau) := \eta(-1/t^2\tau)/\sqrt{\Omega_D}$ and $\beta_t(\tau) := \eta(-1/\tau)/\sqrt{\Omega_D}$ are both algebraic.

To obtain these results, we make use of recent work of Han [12] on extensions of the Nekrasov-Okounkov formula, and work of Berndt [1] on modular transformation properties for generalized Lambert series. These results are recalled in Sect. 2, and in Sect. 3 we prove the theorems and corollaries described above.

## 2 Nuts and Bolts

Here we recall important work by Berndt and Han which are the critical ingredients in the proofs of the results obtained in this note.

### 2.1 A Formula of Han

Han recently expanded and refined the Nekrasov-Okounkov product formula. Theorem 1.3 of [12] offers the striking refinement we require.

**Theorem 3 (Han, 2008)** *We have that*

$$\sum_{\lambda \in \mathcal{P}} q^{|\lambda|} \prod_{h \in \mathcal{H}_t(\lambda)} \left( y - \frac{tyw}{h^2} \right) = \prod_{n=1}^{\infty} \frac{(1-q^{tn})^t}{(1-(yq^t)^n)^{t-w}(1-q^n)}.$$

*Remark 3* Han's formula can be reformulated in terms of the $t$-hook function

$$F_{t,y,w}(\lambda) := \prod_{h\in\mathcal{H}_t(\lambda)} \left(y - \frac{tyw}{h^2}\right).$$

Theorem 3 is equivalent to the identity

$$\langle F_{t,y,w}\rangle_q = \prod_{n=1}^{\infty} \frac{(1-q^{tn})^t}{(1-(yq^t)^n)^{t-w}}.$$

## 2.2 A Formula of Berndt

Using the modularity of the weight 2 nonholomorphic Eisenstein series $E_2^*(z)$, one knows that $\mathcal{E}(z)$ possesses weight 0 type modular transformation laws which can be computed using the method of Eichler integrals (for example, see Sect. 1.4 of [4]). It turns out that Berndt [1] has previously determined the modular transformation properties we require. Here we offer a slight reformulation of the $m = 0$ case of Theorem 2.2 of [1], the key transformation property of $\mathcal{E}(z)$.

**Theorem 4 (Berndt, 1977)** *If $z \in \mathbb{H}$, then*

$$\mathcal{E}(z) - \mathcal{E}(-1/z) = -\Psi(z).$$

# 3 Proofs of Results

Here we employ the work of Berndt and Han to prove our results.

***Proof*** *(Proof of Theorem 1)* Letting $y = 1$ in Theorem 3, we find that

$$\sum_{\lambda\in\mathcal{P}} q^{|\lambda|} \prod_{h\in\mathcal{H}_t(\lambda)} \left(1 - \frac{tw}{h^2}\right) = \prod_{n=1}^{\infty} \frac{1}{(1-q^{tn})^{-w}(1-q^n)}. \tag{13}$$

We recall the classical identity for Euler's partition generating function

$$\sum_{n=0}^{\infty} p(n)q^n = \prod_{n=1}^{\infty} \frac{1}{1-q^n} = \exp\left(\sum_{n=1}^{\infty} \frac{q^n}{n(1-q^n)}\right).$$

Applying this identity to the factor in (13) with exponent $-w$, we find that

$$\sum_{\lambda\in\mathcal{P}} q^{|\lambda|} \prod_{h\in\mathcal{H}_t(\lambda)} \left(1-\frac{tw}{h^2}\right) = \prod_{n=1}^{\infty} \frac{1}{1-q^n} \cdot \exp\left(-w\sum_{n=1}^{\infty} \frac{q^{tn}}{n(1-q^{tn})}\right).$$

By comparing the coefficients in $w$, we find from (6) that

$$H_t(z) = \prod_{n=1}^{\infty} \frac{1}{1-q^n} \cdot \sum_{n=1}^{\infty} \frac{q^{tn}}{n(1-q^{tn})}.$$

A straightforward calculation shows that this is equivalent to the claim that $\langle f_t\rangle_q = \mathcal{E}(tz)$. □

***Proof*** *(Proof of Theorem 2)* By Theorem 1, we have that

$$M_t(z) = P_t(z) + L_t(z) + \mathcal{E}(tz).$$

By letting $z \to tz$ in Theorem 4, we have that

$$\mathcal{E}(tz) - \mathcal{E}(-1/tz) = -\Psi(tz).$$

Therefore, by direct calculation we find that

$$\begin{aligned} M_t(z) - M_t(-1/t^2z) &= (P_t(z) - P_t(-1/t^2z)) \\ &\quad + (L_t(z) - L_t(-1/t^2z)) + (\mathcal{E}(tz) - \mathcal{E}(-1/tz)) \\ &= -\pi i\left(\frac{t^2z^2+1}{12tz}\right) + \frac{\pi i}{4} - \frac{1}{2}\log(tz) - \Psi(tz) = 0. \end{aligned}$$

This confirms the second claim.

The series $\mathcal{E}(tz)$ is invariant in $z \to z+1$, as it is a power series in $q = e^{2\pi i z}$ with integer exponents. Therefore, we find that

$$M_t(z+1) - M_t(z) = (P_t(z+1) + L_t(z+1)) - (P_t(z) + L_t(z)).$$

The first claim follows by direct calculation. □

***Proof*** *(Proof of Corollary 1)* By definition, we have that

$$H_1^*(z) = \sum_{\lambda\in\mathcal{P}} f_t(\lambda) q^{|\lambda|-\frac{1}{24}}.$$

Therefore, the first claim is a triviality.

To establish the second claim, we note that Theorem 2 implies that

$$M_1(z) = M_1(-1/z).$$

Using the fact that $\eta(-1/z) = \sqrt{-iz} \cdot \eta(z)$, we obtain

$$P_1(z) + L_1(z) + \frac{\eta(-1/z)}{\sqrt{-iz}} \cdot H_1^*(z) = P_1(-1/z) + L_1(-1/z) + \eta(-1/z)H_1^*(-1/z).$$

Since the $\eta$-function is nonvanishing on $\mathbb{H}$, this is equivalent to the desired conclusion

$$\begin{aligned} H_1^*(-1/z) - \frac{1}{\sqrt{-iz}} \cdot H_1^*(z) &= \frac{1}{\eta(-1/z)} \cdot (P_1(z) + L_1(z) - P_1(-1/z) - L_1(-1/z)) \\ &= \frac{\Psi(z)}{\eta(-1/z)}. \end{aligned}$$

□

***Proof*** *(Proof of Corollary 2)* By the classical Chowla-Selberg theorem described by (12), since the Dedekind eta-function has weight 1/2, we have that

$$\eta(-1/\tau) \in \overline{\mathbb{Q}} \cdot \sqrt{\Omega_D}.$$

The claimed conclusion is now an immediate consequence of Corollary 1 (2). □

## 4 Some Examples

Here we offer examples of the results of this paper.

*Example 2* We illustrate the $t = 1$ and 2 cases of Theorem 1. By direct calculation, we have

$$\begin{aligned} H_1(z) &= q + \frac{5}{2}q^2 + \frac{29}{6}q^3 + \frac{109}{12}q^4 + \ldots, \\ H_2(z) &= q^2 + q^3 + \frac{7}{2}q^4 + \frac{9}{2}q^5 + \ldots. \end{aligned}$$

Therefore, one finds that

$$\langle f_1 \rangle_q = \prod_{n=1}^{\infty}(1-q^n)\cdot H_1(z) = q + \frac{3}{2}q^2 + \frac{4}{3}q^3 + \frac{7}{4}q^4 + \cdots = \mathcal{E}(z) = \sum_{n=1}^{\infty}\sigma_{-1}(n)q^n,$$

$$\langle f_2 \rangle_q = \prod_{n=1}^{\infty}(1-q^n)\cdot H_2(z) = q^2 + \frac{3}{2}q^4 + \frac{4}{3}q^6 + \frac{7}{4}q^8 + \cdots = \mathcal{E}(2z) = \sum_{n=1}^{\infty}\sigma_{-1}(n)q^{2n}.$$

*Example 3* We illustrate Theorem 2 (2) with $t = 2$ and $z = i$. Ramanujan proved (see p. 326 of [2]) that

$$\eta(i) = \frac{\sqrt{2}\pi^{\frac{1}{4}}}{2\cdot\Gamma(3/4)} \approx 0.7682$$

$$\eta(i/4) = \frac{2^{\frac{1}{4}}\cdot\sqrt{\pi}}{2\cdot\Gamma(3/4)^2} \approx 0.7018.$$

Moreover, we find that

$$H_2^*(i) \approx 4.5395\cdot 10^{-6} \quad \text{and} \quad H_2^*(i/4) \approx 0.06572.$$

Therefore, we find that

$$M_2(i) = P_2(i) + L_2(i) + \eta(i)H_2^*(i) \approx 0.3503 - 5.3926i,$$

$$M_2(i/4) = P_2(i/4) + L_2(i/4) + \eta(i/4)H_2^*(i/4) \approx 0.3503 - 5.3926i.$$

This illustrates the fact that $M_2(i) = M_2(i/4)$.

*Example 4* We now illustrate Corollary 1 (2) and Corollary 2 using $z = \tau = 2i$. By direct calculation, we find that $H_1^*(i/2) \approx 0.05506$ and $H_2^*(2i) \approx 5.8870\cdot 10^{-6}$. Therefore, we have

$$H_1^*(i/2) - \frac{\sqrt{2}}{2}\cdot H_1^*(2i) \approx 0.05506. \tag{14}$$

Ramanujan proved (see p. 326 of [2]) that

$$\eta(i/2) = 2^{\frac{1}{8}}\cdot\sqrt{\Omega_{-4}} = \frac{\pi^{\frac{1}{4}}}{2^{\frac{3}{8}}\cdot\Gamma(3/4)} \approx 0.8377\ldots.$$

By direct calculation, we find that $\Psi(2i) = \frac{\pi}{8} - \frac{\log(2)}{2} \approx 0.04612\ldots$, and so

$$\frac{\Psi(2i)}{\eta(i/2)} \approx 0.05506\ldots.$$

Combined with (14), we find that

$$H_1^*(i/2) - \frac{\sqrt{2}}{2} \cdot H_1^*(2i) = \frac{\Psi(2i)}{\eta(i/2)} = \frac{1}{2^{\frac{1}{8}}} \cdot \frac{\Psi(2i)}{\sqrt{\Omega_{-4}}}.$$

This illustrates Corollary 2, where the algebraic factor is $1/2^{\frac{1}{8}}$.

**Acknowledgments** The author thanks the referee, Madeline Locus Dawsey, Wei-Lun Tsai, Ian Wagner, and Ole Warnaar for their comments on an earlier draft of this note.

## References

1. B.C. Berndt, Modular transformations and generalizations of several formulae of Ramanujan. Rocky Mountain J. Math. **7**, 147–189 (1977)
2. B. C. Berndt, *Ramanujan's Notebooks, Part V* (Springer, New York, 1998)
3. S. Bloch, A. Okounkov, The character of the infinite wedge representation. Adv. Math. **149**, 1–60 (2000)
4. K. Bringmann, A. Folsom, K. Ono, L. Rolen, Harmonic Maass forms and mock modular forms: theory and applications. Am. Math. Soc. Colloquium Series, **64**, 391 (2017). Providence, RI
5. K. Bringmann, K. Ono, I. Wagner, *Eichler Integrals of Arbitrary Weight Eisenstein Series as $q$-brackets of Weighted $t$-hook Functions on Partitions, Ramanujan J.*, accepted for publication.
6. S. Chowla, A. Selberg, On Epstein's zeta-function. J. Reine Angew. Math. **227**, 86–110 (1967)
7. M.L. Dawsey, K. Ono, CM evaluations of the Goswami-Sun series, in *Elliptic Integrals, Elliptic Functions and Modular Forms in Quantum Field Theory* (Springer, New York, 2019), pp. 183–193
8. P. Fong, B. Srinivasan, The blocks of finite general linear groups and unitary groups. Invent. Math. **69**, 109–153 (1982)
9. F. Garvan, D. Kim, D. Stanton, Cranks and t-cores. Invent. Math. **101**, 1–17 (1990)
10. A. Granville, K. Ono, Defect zero p-blocks for finite simple groups. Trans. Am. Math. Soc. **348**, 331–347 (1996)
11. M. Griffin, M. Jameson, S. Trebat-Leder, $p$-adic modular forms and the Bloch-Okounkov theorem. Res. Math. Sci. **3**, Art. 11 (2016)
12. G.-N. Han, The Nekrasov-Okounkov hook length formula: refinement, elementary proof, extension and applications. Ann. Inst. Fourier (Grenoble) **60**, 1–29 (2010)
13. G. James, A. Kerber, *The Representation Theory of the Symmetric Group* (Addison-Wesley, Reading, 1979)
14. N.A. Nekrasov, A. Okounkov, Seiberg-Witten theory and random partitions, in *The Unity of Mathematics*, vol 244. Program of Mathematical (Birkhauser, Boston, 2006), pp. 525–596
15. K. Ono, *The Web of Modularity: Arithmetic of the Coefficients of Modular Forms and $q$-series.* (American Mathematical Society, Providence, 2004)
16. R. Schneider, Arithmetic of partitions and the $q$-bracket operator. Proc. Am. Math. Soc. **145**, 1953–1968 (2017)
17. A. van der Poorten, K. Williams, Values of Dedekind's eta-function at quadratic irrationalities. Canadian J. Math. **51**, 176–224 (1999)
18. B.W. Westbury, Universal characters from the Macdonald identities. Adv. Math. **202**, 50–63 (2006)
19. D. Zagier, Partitions, quasimodular forms, and the Bloch-Okounkov theorem. Ramanujan J. **41**, 345–368 (2016)